Sustainable Aviation

The Sustainable Aviation book series focuses on sustainability in aviation, considering all aspects of the field. The books are developed in partnership with the International Sustainable Aviation Research Society (SARES). They include contributed volumes comprising select contributions to international symposiums and conferences, monographs, and professional books focused on all aspects of sustainable aviation. The series aims at publishing state-of-the-art research and development in areas including, but not limited to:

- Green and renewable energy resources and aviation technologies
- Aircraft engine, control systems, production, storage, efficiency, and planning
- Exploring the potential of integrating renewables within airports
- Sustainable infrastructure development under a changing climate
- Training and awareness facilities with aviation sector and social levels
- Teaching and professional development in renewable energy technologies and sustainability

Mohammad Aslam • Sanjeev Mishra
Jorge Arturo Aburto Anell
Editors

Sustainable Aviation Fuels

Recent Advances and Future Challenges

Springer

Editors
Mohammad Aslam
Department of Chemistry
National Institute of Technology Srinagar
Srinagar (J&K), India

Sanjeev Mishra
Sardar Swaran Singh National Institute
Kapurthala, Punjab, India

Jorge Arturo Aburto Anell
Biomass Conversion Division
Mexican Institute of Petroleum
Mexico City, Distrito Federal, Mexico

ISSN 2730-7778 ISSN 2730-7786 (electronic)
Sustainable Aviation
ISBN 978-3-031-83720-3 ISBN 978-3-031-83721-0 (eBook)
https://doi.org/10.1007/978-3-031-83721-0

This Springer imprint is published by the registered company Springer Nature Switzerland AG
The registered company address is: Gewerbestrasse 11, 6330 Cham, Switzerland

Contents

Chapter 1
Introduction to Green Aviation Fuel: A Sustainable Driver for the Next Generation

Francisco Izaias da Silva Aires, José Roberto de Matos Filho, Letícia da Costa Silva, Maurício Quintas Salamba, Jacob Tchiyeke António Kandjila, Rodrigo da Silva Vieira, and José Cleiton Sousa dos Santos

1 Introduction

Climate change and air pollution are persistent challenges that capture the attention of the global scientific community, mobilizing multidisciplinary efforts for mitigation and adaptation due to their critical importance to environmental and public health [1, 2]. The aviation industry is on the brink of a significant transformation, driven by the urgent need to implement strategies that meet sustainability goals [3, 4]. Traditional aviation fuels, derived from fossil resources, are the primary contributors to greenhouse gas emissions, air pollution, and global climate change, which threaten the health of living beings [5–7]. Aircraft release gases into the atmosphere, such as carbon dioxide (CO_2), sulfate aerosols, nitrogen oxides, particles, and water vapor, contributing to the formation of contrails and negatively impacting the environment [8–10]. As global air traffic continues to grow, the environmental impact of aviation is becoming increasingly severe, intensifying climate change, one of the major concerns of modern society [11, 12]. In response to these challenges, developing and adopting green aviation fuels that are sustainable and renewable alternatives to conventional jet fuel emerged as a pivotal strategy [13–15].

Green aviation fuels, also known as sustainable aviation fuels (SAFs), are derived from various renewable sources, including biomass, waste materials, and synthetic processes that capture atmospheric carbon [16, 17]. These fuels have the potential to significantly reduce the carbon footprint of aviation, contributing to the broader goals of reducing greenhouse gas emissions and mitigating climate change [18, 19].

F. I. da Silva Aires · J. R. de Matos Filho · L. da Costa Silva · M. Q. Salamba · J. T. A. Kandjila · R. da Silva Vieira · J. C. S. dos Santos (✉)
Instituto de Engenharia e Desenvolvimento Sustentável, Universidade da Integração Internacional da Lusofonia Afro-Brasileira, Redenção, Brasil
e-mail: jcs@unilab.edu.br

M. Aslam et al. (eds.), *Sustainable Aviation Fuels*, Sustainable Aviation,
https://doi.org/10.1007/978-3-031-83721-0_1

Moreover, the transition to green aviation fuels aligns with the international aviation industry's commitment to achieving net-zero carbon emissions by 2050, as outlined by the International Air Transport Association (IATA) [20–23].

Adopting green fuels in aviation has been promoted in the aeronautical sector as a crucial strategy to reduce CO_2 emissions [13, 24]. This multifaceted approach foresees benefits in the sustainable energy sector and economic opportunities, such as job creation in the green energy sector and the development of new markets for sustainable resources [25–27]. Bibliometrics is a powerful tool that assesses scientific production and identifies trends, allowing us to understand the evolution of knowledge in this field, identify the main actors and institutions involved, and highlight keywords, emerging areas, and research gaps [28–31]. With this approach, we gain valuable insights into how knowledge is being disseminated and applied and the main challenges and opportunities in developing and implementing fuels for aviation [32, 33].

This study investigates the complexities of green aviation fuels, exploring their types, production processes, benefits, and the challenges associated with their widespread adoption. By understanding the current landscape and future potential of sustainable aviation fuels, stakeholders can better appreciate their role as sustainable drivers of the next generation of aviation. Additionally, this work conducts a quantitative bibliometric analysis of the available data, mapping the current research landscape on green aviation fuels. This allows for identifying trends, knowledge gaps, and emerging research areas, offering a comprehensive view that can guide future investigations and policy formulation in the aviation sector.

2 Methodology

2.1 Data Source

This study used data from the Web of Science database following the methodology described in previous studies [34–41]. Specific classification criteria were used to select this information aligned with the most modern bibliometric analysis methods. The Coordination for the Improvement of Higher Education Personnel (CAPES) facilitated access to the platform through the access provided by the CAPES Periodicals portal.

The search began with the term "Fuel." Subsequently, an additional filter was implemented based on the publication time frame, which spanned from 2004 to 2024. The definition of this period was motivated by the need to delimit the scope of the study and the accessibility of relevant data for the proposed analysis. 2004 was established as the starting point, while 2024 allows for examining the main trends and developments in green aviation fuels over the past two decades. In the Web of Science (WoS) database, the principal terms used were "Fuel," with a total of 393,127 articles; "biofuels," with 8802 articles; "Transport," with 913 articles;

and "Production," with 545 articles. The search terms were carefully chosen to optimize the accuracy of the results and intensify the focus on the studied theme. English was selected due to its prevalence as the lingua franca in international scientific publications. Moreover, the investigation was restricted to original and review articles, aiming for a robust and comprehensive analysis of the available literature.

2.2 Data Analysis

This study employed two cutting-edge data visualization tools for bibliometric analysis: CiteSpace (version 5.8.R6) and VOSviewer (version 1.6.20). Used sequentially, these software programs facilitated a deeper understanding of the collected data. CiteSpace, developed in Philadelphia, PA, USA, and VOSviewer, originating from Leiden, South Holland, Netherlands, allowed the construction of visual maps and networks. These graphical representations are crucial for highlighting emerging patterns, dominant trends, and significant connections within the data corpus. The result is a comprehensive and detailed perspective on the current state of research in green aviation fuels, shedding light on future advancements in the field.

3 Results

3.1 Bibliometric Analysis of Global Publications and Citations over the Past 20 Years

During the research, we analyzed publications and citations of 20 years, from 2004 to 2024, to better understand the data and trends. We observed significant fluctuations over this period. In the early years, there was a relatively low number of publications, possibly due to the gradual development of the research field. From 2017 to 2022, we identified notable peaks in publications and citations. These peaks suggest areas of great academic interest and possibly significant advancements during this interval. The causes of these increases may be related to the emergence of new technologies, increased research funding, or the growth of interdisciplinary collaborations. These results are significant because they can guide future research and collaborations. By identifying areas of high academic activity, researchers can focus on emerging topics and seize opportunities for cooperation with other specialists. Additionally, trend analysis can help predict future research directions, facilitating strategic decision-making in academic and funding institutions. These insights provide a comprehensive view of the publication and citation landscape, highlighting the development and evolution of the field over two decades, as Fig. 1.1 illustrates.

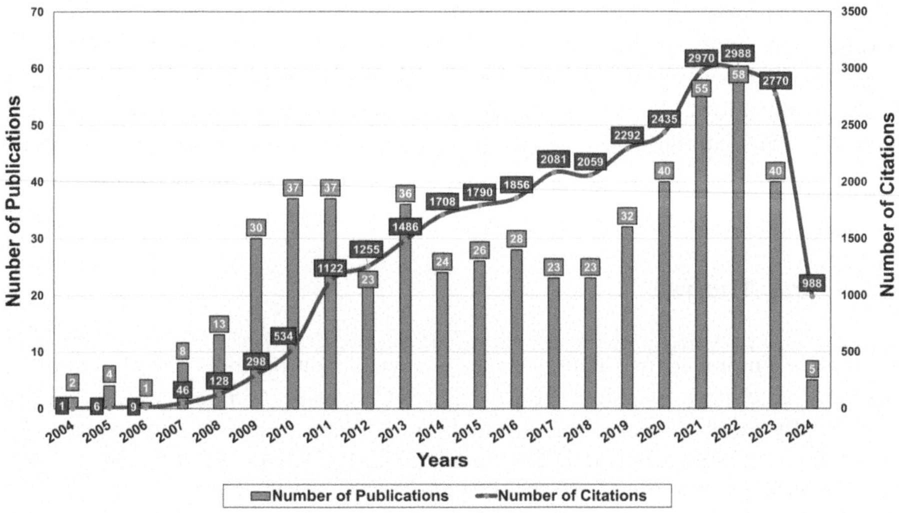

Fig. 1.1 Annual trend of global publications and citations related to aviation fuels research from 2004 to 2024

3.2 Contributions of Countries/Regions and Institutions

Data analysis on green aviation fuel research highlights significant contributions from countries and institutions worldwide. Figure 1.2 shows that of the 72 countries studied, 70 met the criteria of producing at least one document with a minimum of two citations, indicating widespread global engagement in this critical area of research. Leading the ranking, the United States stands out with 81 documents and 4731 citations, demonstrating its extensive research output and influence in advancing green aviation fuel technologies. Close behind, England contributed 64 documents and obtained 3345 citations, highlighting substantial research activity and significant global impact, primarily through institutions like Imperial College London. Australia also plays a notable role, with 25 documents and 2982 citations, emphasizing its research efforts in developing sustainable aviation fuel solutions driven by institutions like the University of Queensland. These findings underscore the crucial role that countries and regions play in shaping the global landscape of green aviation fuel research. Their collective efforts enhance scientific understanding and drive policy initiatives and technological innovations to address environmental challenges in the aviation sector.

As shown in Fig. 1.3, the VOSviewer software allows identifying the leading institutions that have significantly contributed to research on green aviation fuels. Using a minimum threshold of two documents per institution and five citations per institution, out of the 719 institutions analyzed, 172 met these criteria. The United States Department of Energy (DOE) leads with 19 publications. The DOE has been a critical player in the development of sustainable technologies and in promoting research aimed at reducing the aviation carbon footprint. Their efforts include

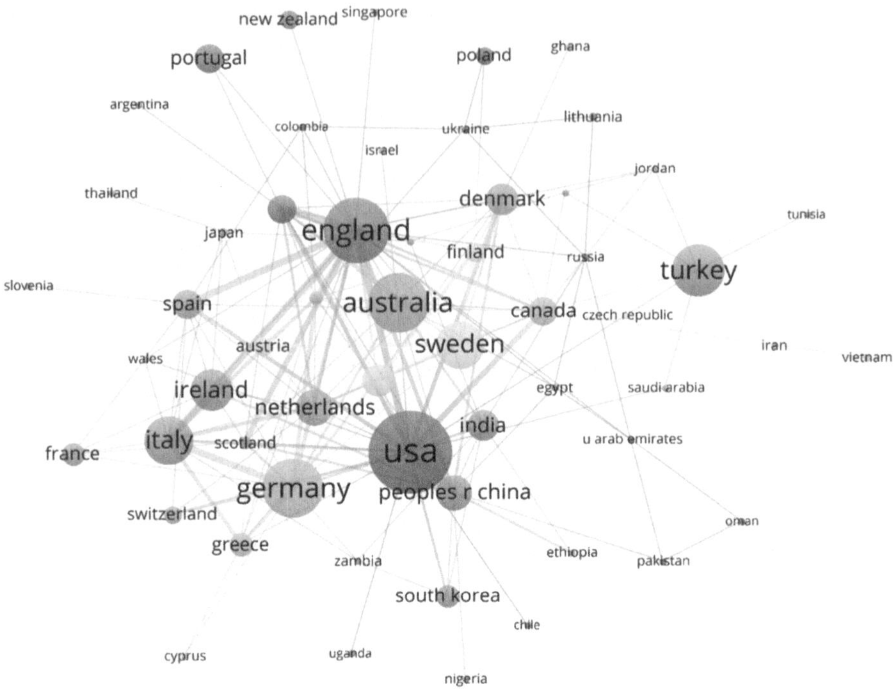

Fig. 1.2 Connections between different countries. The central node, labeled as "England," shows a sharing of connections, indicating a network of knowledge exchange, communication, or other forms of international relations. The larger node representing the United States ("USA") also stands out in the network, reflecting its significant influence and extensive global collaboration in this context

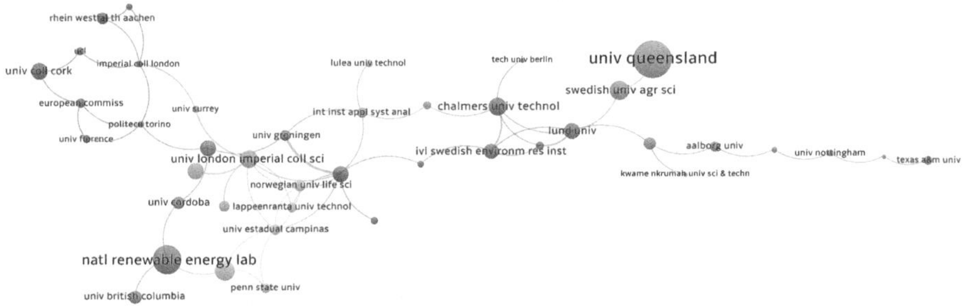

Fig. 1.3 Leading universities that have excelled in information sharing. The larger clusters represent the extent of a university's contribution in terms of publications. The lines indicate the connections and data exchanges between the universities

research and development projects in biofuels, synthetic fuels, and ecological alternatives to fossil fuels. Imperial College London ranks second with 17 publications. This institution is known for pioneering chemical and environmental engineering

contributions, focusing on innovative solutions for sustainable aviation. Their research spans from developing new biofuels to analyzing the life cycle of green fuels.

The Helmholtz Association, with 15 publications, stands out as one of the leading contributors to research on green aviation fuels in Europe. This German association integrates several research institutions collaborating on large-scale projects, focusing on advanced and sustainable aviation technologies. Institutions such as the United States Department of Energy, Imperial College London, and the Helmholtz Association are at the forefront of research on green aviation fuels. Their publications and high number of citations reflect the significant impact of their research, which advances scientific knowledge and helps shape industrial policies and practices. The contribution of these institutions is crucial for the aviation transition to more sustainable fuel alternatives, promoting a greener and environmentally responsible future.

Research on green aviation fuels has gained significant attention over the past two decades. Table 1.1 highlights the top ten institutions that have worked on this research topic. The bibliometric analysis was limited to institutions with more than two publications and at least five citations each in this field from 2004 to 2024. This analysis will focus on the top three institutions: the University of Queensland, the University of Bielefeld, and the University of Karlsruhe.

The University of Queensland leads the list with six publications and a remarkable 2304 citations. This high number of citations indicates their research's significant impact and influence on green aviation fuels. The university's contributions likely cover various aspects of green aviation fuels, such as sustainable fuel sources, advancements in fuel technology, and the environmental impact of aviation fuels. Research from this institution is fundamental to driving innovations and policies in the green aviation sector. Despite having only two publications, the University of Bielefeld boasts an impressive 1650 citations. This indicates that their work is highly valued and frequently referenced by other researchers in the field. The high number of citations per publication suggests that their studies are foundational or

Table 1.1 Top ten institutions with more than two publications and at least five citations each in green aviation fuel research from 2004 to 2024

Rank	Institutions	Countries/ Regions	Publications	Citations
1	Univ Queensland	Australia	6	2304
2	Univ Bielefeld	Germany	2	1650
3	Univ Karlsruhe	Germany	2	1650
4	Natl Renewable Energy Lab	USA	5	1436
5	Sila Sci	France	4	1161
6	Univ Illinois	USA	3	682
7	Swedish Univ Agr Sci	Sweden	3	638
8	Univ Aquila	Italy	4	581
9	Univ London Imperial Coll Sci Technol & Med	United Kingdom	9	547
10	Chalmers Univ Technol	Sweden	8	546

groundbreaking, possibly addressing critical challenges in green aviation fuels or introducing novel technologies or methodologies. The concentration of their research impact reflects the depth and quality of their contributions. Like the University of Bielefeld, the University of Karlsruhe has two publications with 1650 citations. This parallel performance highlights a robust research output with significant influence. The contributions from the University of Karlsruhe might include advancements in biofuels, synthetic fuels, or life cycle assessments of green aviation fuels. The consistency in citation counts with Bielefeld suggests that Germany is a notable contributor to high-impact research in this domain, possibly fostering collaborations and a robust research environment in green aviation technologies.

Based on the number of publications and citations, the top three institutions in green aviation fuel research are making substantial contributions to the field. The University of Queensland, the University of Bielefeld, and the University of Karlsruhe each exhibit unique strengths, with the former leading in overall publication count and the latter excelling in citation impact per publication. These institutions are at the forefront of developing and promoting sustainable aviation fuels, addressing the urgent need for environmentally friendly solutions in the aviation industry. Their research advances scientific knowledge and informs industry practices and public policy, driving the transition toward greener aviation.

3.3 Contributions of Authors

The bibliometric analysis of the authors revealed six large clusters, as shown in Fig. 1.4, with 142 authors meeting the criteria of at least two publications and five citations. The largest cluster in the blue cluster is led by Hansson, Julia, with four publications and 433 citations, highlighting her significant contribution and influence in the field. Following her, in the green cluster, is Faaij, Andre, with five documents and 380 citations, also demonstrating considerable impact. The third cluster, purple, is led by Wicke, Birca, with three publications and 170 citations. Lensink, Sander stands out in the yellow cluster with two publications and 120 citations. The blue and red clusters comprise authors with only one publication each, yet they still form a significant collaboration network. The larger the cluster, the greater the contribution network among the authors, indicating the importance of these collaborations for advancing research in green aviation fuels.

3.4 Keyword Analysis

Researchers can utilize a keyword analysis to better understand the global scientific community's engagements in research development, inclinations, and imminent challenges. Figure 1.5 illustrates the relationship of keywords cited over the past 4 years, reflecting highly relevant themes in aviation fuel research.

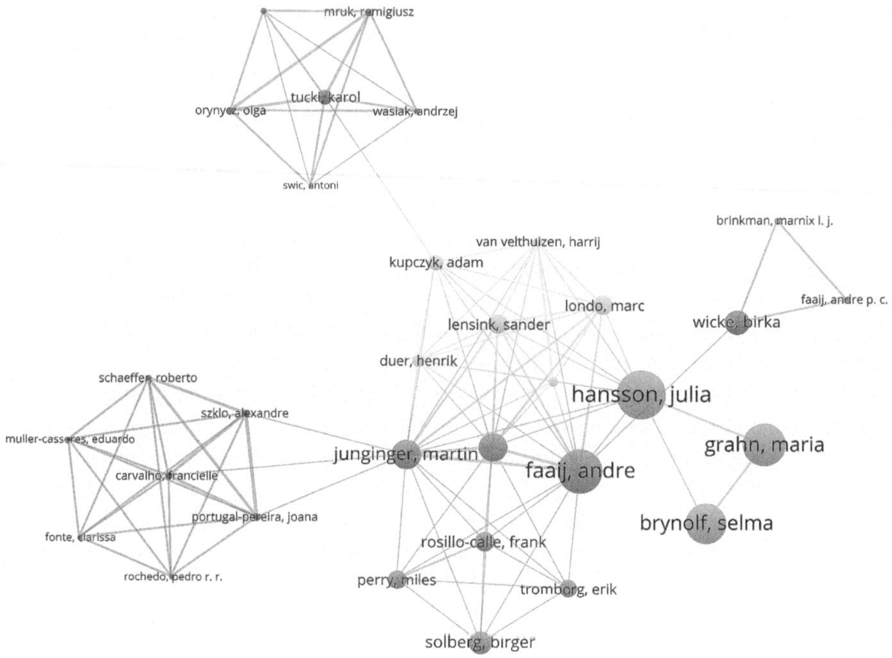

Fig. 1.4 Collaboration network between authors grouped into clusters. Each node represents an author, and the lines indicate collaborations. Clusters suggest areas of close cooperation or related fields of research

The most frequent keywords highlighted in this field include "Biofuels" (232 citations), "Biomass" (124 citations), "Energy" (89 citations), "Emissions" (81 citations), and "Biodiesel" (89 citations). These keywords indicate the main focuses of interest and study in the area, suggesting where most research efforts are concentrated. In addition to these, other keywords present in the figure refer to complementary but equally essential themes involved in aviation fuel research, such as "ethanol," "transesterification," and "fermentation." In Fig. 1.5a, keywords such as "biofuels," "energy," "emissions," and "biodiesel" are highlighted, representing the most commonly used terms in the context of aviation fuel research. The distribution of these keywords and their connections with smaller keywords show how these themes are interconnected, reflecting the interdisciplinary nature of the study. Figure 1.5b presents the frequency with which the collected keywords were mentioned in articles from 2004 to 2024, indicating a strong trend or evolution of these keywords over time, using different colors on the overlay map. This analysis allows researchers to identify the central themes of aviation fuel research, emerging trends, and potential areas of difficulty that may require further investigation. Additionally, understanding these keywords and their interrelationships can help guide future research efforts and interdisciplinary collaboration, promoting a more coordinated and practical advancement in the field.

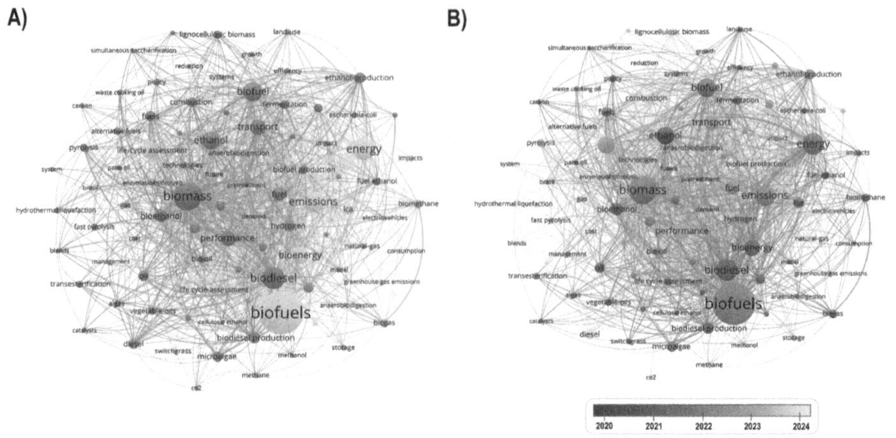

Fig. 1.5 Co-occurrence network generated by VOSviewer. (**a**) Co-occurrence analysis of keywords in VOSviewer for aviation fuel research. The image highlights a central cluster of closely related keywords, indicating strong thematic connections, while more distant keywords suggest broader and less direct cooperation. (**b**) Overlay visualization of the co-occurrence analysis. Clusters with lighter colors represent more recent keywords, whereas darker colors indicate older ones

Figure 1.6 shows a temporal look at the keywords from the studied articles between 2004 and 2024, highlighting those associated with the most important keywords, which means they are associated with the lines to be investigated. Cluster 0 has "climate change" as the main keyword. This cluster encompasses research focused on understanding and mitigating the effects of global climate change. Research in this cluster is of utmost importance for developing strategies to reduce greenhouse gas emissions and promote environmental sustainability. Clusters 1 and 2 are represented by "advanced biofuel," analyzing the production and utilization of second- and third-generation biofuels. These biofuels are produced from nonfood materials, such as agricultural waste, lignocellulose, and algae. Research in this cluster aims at production efficiency, cost reduction, and minimization of the environmental impacts caused by biofuels as much as possible.

Cluster 3, "biomethane," focuses on biogas production through the anaerobic digestion of organic matter. This gas can later be used in engines as fuel, for electricity generation, and for heating. Research in this cluster discusses improvements in anaerobic digestion processes, the integration of biomethane into energy infrastructures, and biogas purification techniques. Cluster 4, "lignocellulose," includes research on converting lignocellulosic biomass, such as wood residues, into biofuel and chemical products. This cluster is significant for developing sustainable biofuels that do not compete with food production. Cluster 5, "greenhouse gas emissions," investigates emission sources and ways to mitigate greenhouse gas emissions. This includes assessing emissions in different sectors. This cluster is crucial for

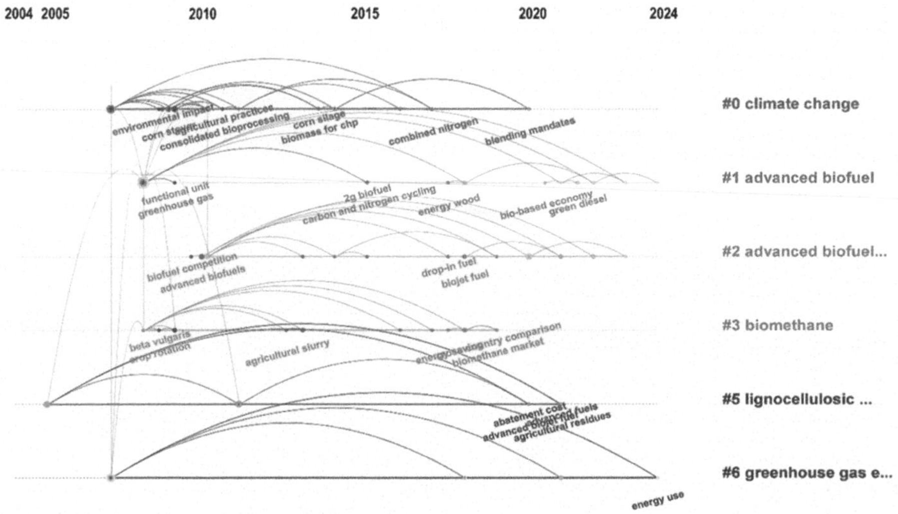

Fig. 1.6 Timeline from 2004 to 2024. The clusters are organized vertically in descending order of size. The temporal evolution is represented by lines of different colors, where the nodes along the lines indicate co-cited keywords. The variable density of nodes in various time intervals reflects dynamic changes within the corresponding clusters along the time axis

Table 1.2 Keyword related to bioenergy and sustainability

Clusters	Items	Main keywords
0	17	Climate change; greenhouse gas emission; transport biofuel; food prices; environmental impact; corn Stover; system expansion; economic frameworks; greenhouse gases energy; crops; dioxide; agricultural; potential heat; tariff; biomass; feed-in; electricity
1	18	Life cycle assessment; advanced biofuel; aqueous phase; hydrothermal liquefaction; energy–food nexus; carbon footprint; land footprint; bio-based economy; water footprint; energy–food nexus footprint; economy; water; nexus; supply vegetable; pyrolysis; blends; lower; EcoDiesel
2	18	Advanced biofuels; biofuel competition; experience curves; technological learning; alternative fuels; analytic hierarchy process; transport sector; drop-in fuel; economic analysis tower; pinch; Fischer–Tropsch; upgrading; Aspen biofuel; blending; prices; sector; transport
3	10	Anaerobic digestion; agricultural slurry; energy crops; biomethane market; *Triticum aestivum*; crop rotation; *Hordeum vulgare*; energy balance

understanding how to limit global warming and reduce its effects as much as possible.

Scientific research in bioenergy and sustainability has been a relevant topic in recent decades. Using keywords in scientific articles provides insights into this field's main themes and trends. Table 1.2 presents four significant clusters identified by CiteSpace, each associated with a specific set of keywords.

Cluster 0 addresses issues related to environmental impact and biofuels. The main keywords include "climate change," "greenhouse gas emissions," "transport biofuels," "food prices," and "environmental impact." Additionally, terms such as "corn residue," "system expansion," and "economic frameworks" are also relevant for research. The analysis of these keywords can guide public policies, business strategies, and academic research, contributing to more conscious and balanced development. Cluster 1 focuses on life cycle assessment and the intersection between energy and food. Important keywords include "carbon footprint," "land footprint," "bioenergy-based economy," "water footprint," and "energy-food nexus." Additionally, terms such as "pyrolysis," "blends," and "EcoDiesel" are also present. Cluster 2 explores advanced biofuels and the associated technological learning. Relevant keywords include "alternative fuels," "analytical hierarchy process," "transportation sector," and "substitute fuels." Furthermore, terms like "economic analysis," "Fischer-Tropsch," and "upgrading" are also discussed. Cluster 3 deals with anaerobic digestion. Important keywords include "*Triticum aestivum*" (wheat), "crop rotation," "*Hordeum vulgare*" (barley), and "energy balance." Additionally, these clusters reflect the main research areas in bioenergy and sustainability, providing a comprehensive overview of the trends and challenges researchers face in this field.

3.5 Rank of the Top Ten Most Cited Articles

Table 1.3 presents the classification of the most cited articles and provides a comprehensive view of the importance and development of research related to green aviation fuel. These articles amount to 7,430 citations over the past 20 years, revealing their extreme importance in the global research landscape.

The article by Schenk, Peer M. et al., published by *BioEnergy Research* in 2008, stands out as the most cited, accumulating a total of 1,504 citations, corresponding to 20.24% of the citations in Table 1.3. This article highlights the advantages of second-generation microalgae in the production of raw materials for biodiesel, bioethanol, and biohydrogen, emphasizing the benefits of these carbon-neutral fuels as a source to meet the transportation fuel demand [42]. In third place, the article from the journal *Journal of Industrial Microbiology & Biotechnology* also addresses the importance of "Microalgae as a raw material for biofuel production." With 1,192 citations (11.92%), Gouveia, Luisa et al. discuss the screening of microalgae and emphasize that freshwater microalgae (*Neochloris oleabundans*) and marine microalgae (*Nannocloropsis* sp.) have proven to be suitable as raw materials for biofuel production (29.0% and 28.7%, respectively) [43]. It is essential to highlight the impact of microalgae in biofuel production processes [44, 45], as seen in the titles of the first and third most cited articles in Table 1.3. The second most cited article, titled "Progress and recent trends in biofuels" by Demirbas, Ayhan, published in the journal *Progress in Energy and Combustion Science*, has accumulated 1,192 citations (16.04%) over the past 20 years. This article highlights the hydrolysis process

for biodiesel production and emphasizes Fischer–Tropsch synthesis using biomass as a raw material, an alternative for producing sustainable fuels [46].

In fourth place, with 760 citations (10.22%), the article on the genetic engineering of algae for enhanced biofuel production highlights the genetic engineering procedures that can be used to enhance microalgae biomass as a biofuel platform for the production of starch-derived alcohols, biohydrogen, and other fossil fuel

Table 1.3 Most cited articles in the past 20 years on green aviation fuel

Rank	Title	Authors	Journal	Publication year	Total citation	Reference
1	Second generation biofuels: high-efficiency Microalgae for biodiesel production	Schenk, Peer M.; Thomas-Hall, Skye R.; Stephens, Evan; Marx, Ute C.; Mussgnug, Jan H.; Posten, Clemens; Kruse, Olaf; Hankamer, Ben	*BioEnergy research*	2008	1504	[42]
2	Progress and recent trends in biofuels	Demirbas, Ayhan	*Progress in energy and Combustion science*	2007	1192	[52]
3	Microalgae as a raw material for biofuels production	Gouveia, Luisa; Oliveira, Ana Cristina	*Journal of Industrial Microbiology & Biotechnology*	2009	886	[43]
4	Genetic engineering of algae for enhanced biofuel production	Radakovits, Randor; Jinkerson, Robert E.; Darzins, Al; Posewitz, Matthew C.	*Eukaryotic cell*	2010	760	[47]
5	Political, economic and environmental impacts of biofuels: a review	Demirbas, Ayhan	*Applied energy*	2009	693	[53]

(continued)

Table 1.3 (continued)

Rank	Title	Authors	Journal	Publication year	Total citation	Reference
6	Lignin valorization through integrated biological funneling and chemical catalysis	Linger, Jeffrey G.; Vardon, Derek R.; Guarnieri, Michael T.; Karp, Eric M.; Hunsinger, Glendon B.; Franden, Mary Ann; Johnson, Christopher W.; Chupka, Gina; Strathmann, Timothy J.; Pienkos, Philip T.; Beckham, Gregg T.	*Proceedings of the National Academy of Sciences of the United States of America*	2014	567	[48]
7	Constraints to commercialization of algal fuels	Chisti, Yusuf	*Journal of biotechnology*	2013	499	[54]
8	Catalysis for conversion of biomass to fuels via pyrolysis and gasification: a review	Bulushev, Dmitri A.; Ross, Julian R. H.	*Catalysis today*	2011	451	[55]
9	Use of algae as biofuel sources	Demirbas, Ayhan	*Energy conversion and management*	2010	451	[46]
10	Legumes for mitigation of climate change and the provision of feedstock for biofuels and biorefineries. A review	Jensen, Erik Steen; Peoples, Mark B.; Boddey, Robert M.; Gresshoff, Peter M.; Hauggaard-Nielsen, Henrik; Alves, Bruno J. R.; Morrison, Malcolm J.	*Agronomy for sustainable development*	2012	427	[56]

substitutes [47]. Like the first and third most cited articles in the ranking, this article also identifies microalgae as a potential raw material for biofuel production, making research on these processes more relevant.

Although it appears in the sixth place, the article by Linger, Jeffrey G., published in 2014 in the journal *Proceedings of the National Academy of Sciences of the United States of America*, addressing the valorization of lignin through integrated biological systems, funneling, and chemical catalysis, stands out as the most recent article in this ranking, accumulating a total of 567 citations (7.63%) [48]. With the keywords "lignocellulosic biofuels," "biorefinery," and "aromatic degradation," this article demonstrates that the use of aromatic catabolic pathways provides a way to valorize lignin, overcoming its natural heterogeneity for the production of fuels, chemicals, and materials.

By analyzing the data in Table 1.3, it is possible to verify the importance of microalgae in biofuel production as a substitute for conventional fuels [49–51]. The word "microalgae" appears in the titles of the first and third most cited articles in the ranking and is presented as a keyword or highlighted in the abstracts of the articles in the fourth, seventh, and ninth positions. Words like "Algae," "Biodiesel," "Biomass," "Hydrogen," "Bio-Oil," "Hydrolysis," "Catalysis," "Pyrolysis," and "Enzyme Engineering" are also frequently found in the ranking of the most cited articles, offering a comprehensive view of the development of research on green fuels. Furthermore, it can be observed that the most cited articles are relatively old, with the most recent one published 10 years ago, revealing that research on green aviation fuels is relatively recent.

4 History and Evolution of Aviation Fuels

4.1 Brief History: Development of Fossil Fuels in Aviation

Since the emergence of the first aircraft, the aviation sector has always been characterized by the use of fossil fuels [57]. In the initial phase of using fossil fuels in aviation, aviation gasoline (avgas) was used [58]. Based on this, the provided refueling infrastructure and the potential of refineries, other fractions of aviation fuels emerged [59]. However, to understand the urgency of using green aviation fuel, it is necessary to understand the history and development of aviation fuels [60]. The emergence of the first "turbojet" engine marked an impactful change in the aviation sector [61]. From this new engine, the economic viability, performance, safety, and availability were analyzed to create an ideal fuel for the turbojet engine [62–65]. According to Maurice et al., gas turbine engines were more compatible with aviation fuel properties than piston engines [66]. However, there was still much to discuss and improve for avgas to be considered a perfect fuel for gas turbine engines [67, 68]. The lighter fractions of gasoline increased octane rating and had low lubrication, which caused various problems for the engine [69, 70]. On the other hand, the heavier fractions with efficient combustion in kerosene showed more excellent compatibility, making the engine more efficient and increasing its lifespan [71, 72].

Thus, the military sector was so engaged in developing aviation fuels that this would bring significant advancements even for commercial aircraft [73].

Over the years, in 1944, JP-1 was established, the first aviation fuel specifically for gas turbines [74]. JP-1 received great acceptance in the US Army Air Corps in the kerosene distillation range. However, it had a boiling point of −60 °C (−76 °F), which hindered its availability [75, 76]. From this initial specification for the ideal turbojet engine fuel, the US Army Air Corps established JP-2 and JP-3, which had the same problems as their predecessor JP-1 [77]. Still aiming to improve the performance of this engine, research continued toward the ideal aviation fuel. Thus, JP-4 emerged, a fraction of gasoline and kerosene, with a low freezing point and greater availability, and monitored by strict pressure restrictions to minimize evaporation losses [78]. JP-4 was considered the most promising aviation fuel at the time.

Despite its great acceptance, JP-4 was highly flammable, which posed significant safety issues for the aviation sector [79]. To minimize safety risks, the US Navy separated only the kerosene fractions, which were more suitable for the Navy's turbojet engines [80]. JP-4 and JP-5 were established for a period of time, and it was believed that they would be suitable fuels for aircraft [81, 82]. However, gas turbine engine aircraft were frequently exposed to extreme conditions, which resulted in disadvantages in engine cooling through deposit formation (coking) [83, 84]. After discovering this phenomenon in turbojet engines, several tests were conducted until a minimum temperature of approximately 163 °C (325 °F) was established, where free hydrocarbons could operate without problems [85]. A new specification was also created, in which military additives were added to reduce impurities (JP-8) [86–89].

Research continued, and the commercial aviation world felt persuaded to join the race toward the ideal aviation fuel, and thus, Jet A specifications emerged, a kerosene-based fuel designated for commercial flights [90, 91]. Subsequently, specifications like JP-8+100 emerged, an improved version of JP-8 intended to make the engine more efficient and reduce costs, and the commercial Jet A-1 became ideal for high-altitude and high-speed travel, emerged [92–95]. The aviation fuel market grew increasingly competitive until it established itself in the global economy. Studies indicate that the transportation sector has become indispensable for humans, and aviation plays a crucial role in national and international mobility [96].

However, the use of fossil fuels has caused environmental damage. According to Chiaramonti et al. (2021), the aviation sector was responsible for emitting 3.8% of carbon dioxide (CO_2) and for total greenhouse gas emissions corresponding to 13.9% in Europe [97]. Furthermore, Wang et al. (2019) stated that the use of fossil energy significantly favors a gradual increase in CO_2 concentration in the atmosphere [98]. This increase has caused global temperature rise, climate change, glacier melting, biodiversity loss, respiratory diseases, and other problems [99–104]. According to Qasem et al. (2024), as the aviation sector has gradually grown, fuel consumption is expected to be three times more by 2050 [105].

4.2 *Regulatory Milestones: Policies and Regulations Driving the Pursuit of More Sustainable Fuels*

Organizations have mobilized to create policies to mitigate CO_2 emissions, with the aviation sector being a key target in this effort. For example, in December 2015, during the 21st Conference of the Parties (COP21) of the United Nations Framework Convention on Climate Change (UNFCCC), the Paris Agreement was signed [106, 107]. One hundred and ninety-six countries initially endorsed this agreement to limit greenhouse gas (GHG) emissions to reduce global warming to 1.5 °C above preindustrial levels [108].

Moreover, the European Union (EU), ambitious with the European Green Deal, views climate change as a global threat and aims to achieve carbon neutrality by 2050 [109]. The deal encompasses various policy strategies such as promoting energy efficiency, encouraging the use of renewable energies, protecting biodiversity, and promoting human health [109, 110]. According to Braun-Unkhoff et al. (2015), the International Air Transport Association (IATA), committed to more sustainable aviation, has pledged to achieve carbon-neutral growth from 2020 and reduce emissions by half by 2050 [111]. Other agreements, such as the Kyoto Protocol, the Kigali Amendment to the Montreal Protocol, and others, also encourage the pursuit of cleaner fuels and sustainable practices in the aviation sector [112–119].

4.3 *Pioneering Initiatives and Projects: Examples of Green Fuel Projects and Initiatives*

Policies and regulations driving the pursuit of more sustainable fuels have been disseminated globally, resulting in new green fuel projects and initiatives [120]. Sustainable aviation fuels (SAFs), such as biofuels and synthetic fuels, have emerged as robust candidates for the energy transition and for reducing dependency on fossil fuels [121, 122]. In addition, hydrogen has been highlighted as a promising solution for green aviation [123]. Oteman et al. (2017) report a significant increase in sustainable energy initiatives in the Netherlands, from 40 to 360 initiatives since 2009 [124]. In 2013, the journal "Biofuels, Bioproducts and Biorefining" reviewed US and EU initiatives for developing, demonstrating, and commercializing lignocellulosic biofuels [125]. Balan et al. state that to achieve its biofuel goals, the EU should have 1.3 billion tons of lignocellulosic biomass available annually. However, due to shortages, it may need to import raw materials from various parts of the world.

Green fuel production projects are already being implemented on a large scale globally. Studies indicate that Norway is leading the way in renewable energies, adopting zero-carbon vehicles for freight transport [126–128]. Additionally, Han, Chen, and Lu Yang (2024), in a strategic study, suggest that China Energy Engineering Corporation is planning a significant investment of billions of dollars

in green energies to reduce CO_2 emissions by 20% [129]. While green fuel projects are a reality, they still face challenges such as needing more government investments, shortages of raw materials for biodiesel, and inadequate infrastructure [130, 131]. However, biofuels, synthetic fuels, and hydrogen can offer the necessary diversification in the energy matrix, promoting environmental sustainability, economic diversification, and complementarity between different initiatives [132].

4.4 Types of Green Aviation Fuels

4.4.1 Biofuels: Origins, Types, and Production Processes

The classification of biofuels is based on the origin of their raw materials, animal or plant, and is divided into several distinct generations. First-generation biofuels, such as ethanol and biodiesel, derive from the fermentation of sugars and animal or vegetable fats, respectively [133]. Although they are produced from renewable resources with the potential to reduce greenhouse gas (GHG) emissions, they face significant challenges, such as competition with food production [134]. Second-generation biofuels use lignocellulosic raw materials to overcome this dilemma, which offers greater availability and lower costs [135]. These biofuels, such as plant biomass and cellulosic ethanol, require technological advances and biorefinery infrastructure for large-scale production despite the potential benefits in terms of sustainability [136, 137].

Third-generation biofuels come primarily from algae and microalgae, which present significant advantages but face technical and economic challenges for large-scale commercialization [138, 139]. Improvements in biomass harvesting methods and the development of more efficient technologies are crucial to increasing the competitiveness of these biofuels [140]. Fourth-generation biofuels, characterized by genetic modifications in microalgae, represent a promising advance [141, 142]. The application of genetic engineering and advanced data analysis in microalgae offers valuable insights to increase biofuel production and meet the growing demand for green fuels more efficiently and sustainably [143, 144]. These innovations are essential to transform the biofuels industry and promote a broader and more effective energy transition.

Hydroprocessed esters and fatty acids (HEFA) and Fischer–Tropsch synthetic paraffinic kerosene (FT-SPK) are specific biofuels widely considered the potential substitutes for fossil fuels in aviation [145, 146]. HEFA can be produced from various raw materials, including vegetable or animal oils and recycled cooking oils [147]. However, the availability of these raw materials can present challenges when it comes to large-scale production [148]. Its production process involves several steps, such as pretreatment of raw materials, catalytic hydrogenation, refining, and separation, followed by its adaptation as aviation fuel [149]. A study focused on optimizing the HEFA production process highlighted that one of the main

disadvantages is the need to improve low-temperature properties, which can result in yield losses due to additives used to adjust the freezing temperature [150].

On the other hand, FT-SPK is an aviation biofuel derived from the conversion of synthesis gas, which can be obtained from biomass, making the process more sustainable [151]. This biofuel is produced through the Fischer–Tropsch process, and its great advantage lies in its ability to mimic the characteristics of fossil fuels more sustainably [152, 153]. Studies like those by Meurer, A., and Kern, J. (2021) highlight that the use of the Fischer–Tropsch reactor in the Power-to-Liquid (PtL) process, which uses renewable energy to generate hydrogen through water electrolysis and nonfossil carbon dioxide as the primary raw material, offers significant flexibility in the production of synthetic liquid hydrocarbon fuels, especially kerosene [154, 155]. The Power-to-Liquid (PtL) route also stands out because these synthetic fuels can be used directly without changing conventional aeronautical systems [156, 157]. In addition to the Fischer–Tropsch route, there is also the methanol route, which uses methanol as an intermediate for producing synthetic fuels, thus showing the diversity of approaches in the search for more sustainable and efficient aviation fuels [158].

Various biofuels offer significant advantages in advancing the energy transition and reducing aviation dependence on fossil fuels. However, hydrogen has increasingly been considered the ideal, clean, and green fuel for aviation as its combustion results only in water vapor, not in the emission of greenhouse gases [159–162]. Despite these advantages, there are several significant barriers to overcome before green hydrogen can be widely adopted as a fuel for aviation on a large scale [163, 164]. Studies indicate that hydrogen's entire production, storage, transport, distribution, and refueling cycle still presents considerable challenges [165, 166]. Comparatively, fossil fuels are still more economically viable, besides using fossil raw materials for large-scale production, as in the case of currently predominant blue and gray hydrogen [167–172].

However, significant advances are being made in new hydrogen production technologies, such as cyanobacterial hydrogen production, steam methane reforming with a proton membrane reformer, water electrolysis facilitated by nickel-molybdenum nitride nanoplates, low-temperature water electrolysis with ionic state solid materials, and photoelectrochemical electrolysis with seawater concentrated by a membrane process [173–178]. These advances promise to reduce the costs associated with green hydrogen production and increase its viability as an alternative to fossil fuels in aviation. Thus, while green hydrogen represents a promising future for sustainable aviation, overcoming infrastructure, cost, and production efficiency challenges is crucial for its large-scale adoption [179–181]. Hydrogen can be stored in various ways, with the aviation sector favoring liquid storage due to its specific advantages [182]. Liquid hydrogen offers higher energy content, high density, and low pressure, making space use more efficient and facilitating large-scale transport [182, 183]. However, it is essential to note that part of the energy contained in hydrogen is consumed in the liquefaction process, which requires additional infrastructure and costs for large-scale storage [184, 185].

Studies highlight that concentrating hydrogen production in specific plants can significantly reduce distribution costs [186–188]. Regarding transport means,

hydrogen can be transported by trucks, ships, pipelines, or trains, depending on the distance and logistical conditions [189]. For longer distances, pipelines are often considered economical, sustainable, and safe solutions for hydrogen transport [190]. These considerations are fundamental to understanding the challenges and opportunities associated with hydrogen fueling aviation and other industrial sectors, highlighting the importance of infrastructure and technologies to enable its large-scale adoption.

5 Production Technologies

In recent years, biofuel production technologies have gained significant attention, mainly to reduce carbon emissions and increase the sustainability of aviation operations [17]. Before addressing the technologies, it is essential to highlight some of the biomasses used in the processing of sustainable aviation fuels (SAFs). Among the main biomasses are agricultural and forestry residues, including woody biomass [191] and forest residues [192]; animal fats, such as chicken fat [193], fish oil [194], and beef tallow [193]; and vegetable oils, such as microalgae oil [195], sunflower oil [196], karanja oil [197], cottonseed oil [196], and palm oil [198], among others. Highlighting these biomass sources is essential to understand the variety of materials available for SAF production and how they contribute to a more sustainable future in aviation.

To better understand the technological processes used, we can talk about gasification and Fischer–Tropsch (FT) [199, 200] process. Gasification converts biomass into synthesis gas (syngas), composed of carbon monoxide (CO) and hydrogen (H_2), resulting in a balanced carbon cycle. In the Fischer–Tropsch process, syngas is transformed into liquid hydrocarbons. Catalysts convert these gases into organic compounds refined into liquid fuels, such as synthetic aviation kerosene, with reduced emissions. Hydroprocessed Esters and Fatty Acids (HEFA) [201]is a process that converts vegetable oils and animal fats into sustainable aviation kerosene. First, these oils and fats are hydrogenated to remove oxygen and transform triglycerides and fatty acids into saturated paraffins. Then, hydrotreating and isomerization adjust the hydrocarbon properties to meet aviation fuel specifications, resulting in a fuel with reduced emissions and compatibility with existing engines. Catalytic Hydrothermolysis (CH) [202] is a process that converts biomass or organic waste into green aviation fuels. Under high temperature and pressure, the biomass is broken down and converted into liquid hydrocarbons in the presence of water and catalysts. These are then refined to meet aviation kerosene specifications.

Alternative fuel production techniques, such as gasification followed by Fischer–Tropsch (FT) synthesis, hydroprocessed esters and fatty acids (HEFA) process, and catalytic hydrothermolysis (CH), present different energy efficiency profiles. Gasification and FT are highly efficient in energy conversion [200]. Hydroprocessed esters and fatty acids (HEFA) are generally more energy efficient and have a lower carbon footprint than traditional fossil fuels. Still, it is worth remembering that sustainability depends on the origin of raw materials and land use for oilseed

cultivation [200, 203]. Catalytic hydrothermolysis (CH) has the potential for high energy efficiency and lower environmental impact. This method can utilize various feedstocks and significantly reduce greenhouse gas emissions compared to fossil fuels [202, 204]. Aviation biofuels have significant growth potential shortly, and several emerging technologies are likely to dramatically impact the production of aviation biofuels in the coming years [17].

6 Challenges and Barriers

Green aviation fuels face significant challenges, particularly regarding energy density. Energy density is crucial because it directly influences aircraft energy consumption, linked to their mass. Currently, several alternatives are being explored: bio-jet fuel, which has an energy density almost identical to fossil kerosene [123], and hydrogen, which, despite having a much lower volumetric energy density, is also considered a viable option [205]. Electrochemical batteries have significantly lower volumetric and mass energy densities [123, 200].

Gasification and the Fischer–Tropsch (FT) process are highly efficient in energy conversion. Still, they can be intensive in energy and water use and may generate CO_2 emissions depending on the carbon source [200]. Bio-kerosene is another alternative under consideration, but its large-scale production requires significant investments, mainly due to the high prices of vegetable oils [206]. Although organic raw materials are the most used currently, sustainability represents a substantial barrier to large-scale application. Power-to-liquids (PtLs or e-fuels) offer a promising solution but are still expensive and under development [207].

Sustainable aviation fuels (SAFs) have the potential to reduce CO_2 emissions in the aviation industry significantly, as well as minimize other direct climate impacts, such as cirrus cloud formation and contrails. However, they compete with other CO_2 mitigation options, such as hydrogen and carbon offset schemes. Additionally, different types of SAFs compete in market prices due to various production processes and raw materials [21]. Despite some of the challenges mentioned above, it is worth noting that future studies may mitigate these challenges and further improve the quality of SAFs. The aviation industry is moving toward increasing the use of renewable fuels to reduce its environmental impact and find alternative supply chains for aviation fuel and its barriers.

7 Case Studies and Current Implementations

The aviation industry faces a significant challenge: reducing greenhouse gas emissions and the potential to contribute to mitigating climate change. Globally, approximately 80 billion gallons of traditional fuels are produced annually, reflecting the impact of increasing air traffic, a challenge that needs to be considered in the search for more sustainable solutions for the future of aviation [15].

Sustainable alternative fuels, known as SAFs, stand out for their reduced carbon footprint and lower greenhouse gas emissions than traditional fossil fuels [208]. To ensure safety and performance in commercial aircraft, the American Society for Testing and Materials International (ASTM International) must certify biofuels from various technological routes before use. Currently, only six biofuel production pathways have certification: Fischer–Tropsch (FT) process, hydroprocessed esters and fatty acids (HEFA), direct sugar to hydrocarbon (DSHC), alcohol to jet (ATJ; using isobutanol or ethanol intermediates), and catalytic hydrothermolysis jet (CHJ) [209].

Sustainable aviation fuels (SAFs) provided by Neste are available at several prominent airports, such as San Francisco International Airport (SFO), Heathrow Airport (LHR), and Frankfurt Airport (FRA). Currently, they are used by several major commercial airlines, including KLM, Lufthansa, Delta, and American Airlines. In addition, strategic partnerships with crucial fuel distributors ensure the delivery of Neste's SAF to the market [15]. Currently, HEFA technology is the most mature and commercially available, capable of being used in blends of up to 50% with fossil fuels in aviation [210]. Similarly, alcohol-based fuels and Fischer–Tropsch fuels can also be used in blends of up to 50% but are still under research [15].

There is extensive research on the processes and pathways that can be used to synthesize biofuels [211], highlighting technological and scientific advancements in the production of alternative fuels for aviation and focusing on hydroprocessing, which is considered the most promising route for alternative fuel production, thermochemical processing, and alcohol routes for jets. The authors comprehensively analyzed various jet biofuel production pathways, both existing and under development. This analysis considered each technology's economic and environmental aspects but did not delve into a detailed review of related policies [212]. Wang and Tao L [213] conducted a comprehensive review of four main pathways for biomass conversion into jet biofuels: alcohol to jet, oil to jet, gas to jet, and sugar to jet. Drawing from multiple studies, the review provided valuable insights into process design, technical–economic performance, and greenhouse gas emissions associated with each technology.

Some companies stand out with promising initiatives regarding the synthesis and use of biofuels. One notable example is SkyNRG, based in Amsterdam, which has one of its main projects, the construction of DSL-01, the first sustainable aviation fuel (SAF) production unit in the Netherlands, supported by Shell. Neste, a Finnish company, also stands out with plans to expand its aviation fuel production. In addition to these biomass-based initiatives, there are plans to build a pilot facility at Rotterdam The Hague Airport to produce synthetic kerosene using direct air capture. More recently, Synkero, a newly launched startup, announced the creation of a synthetic kerosene plant at the port of Amsterdam, utilizing renewable hydrogen and CO_2 from point sources or directly from the air [214]. Sustainable aviation biofuel is a relatively new field and, as such, faces several challenges in its development, certification, and distribution. Enhancing aircraft biofuel production technologies is crucial to reducing greenhouse gas emissions in the aviation sector.

This achievement hinges on synergistic integration among biology, chemical engineering, and the development of suitable energy crops [98].

8 Future Perspectives

As research and development in aviation fuels continue to advance, there is an expected significant increase in investment in promising new technologies that facilitate their market application [215, 216]. This includes the development of more efficient and sustainable biomass conversion methods, such as utilizing microalgae and agricultural residues [217–219]. Interdisciplinary collaboration across biology, chemistry, and engineering will be pivotal in overcoming technical and economic challenges [220, 221].

Market growth is projected to be significant, driven by increasingly stringent regulations and growing environmental awareness [222]. Investments in infrastructure and public–private partnerships are crucial to expanding production capacity and reducing costs, facilitating more significant adoption of biofuels in aviation [223, 224]. Environmental impact is expected to be positively affected by widespread biofuel adoption, promising substantial reductions in greenhouse gas emissions [225]. This aligns with global sustainability goals and enhances the aviation sector's resilience to fluctuations in fossil fuel prices through diversified energy sources. Policy support and incentives are essential drivers expected to continue fostering innovation and expanding the biofuel market in aviation [226, 227]. Government policies and financial incentives for research and development will accelerate technological advancements and market growth [228]. These perspectives outline a promising path toward a more sustainable and resilient aviation sector. Emphasizing technological innovation and a steadfast commitment to environmentally responsible practices are essential for shaping the future of aviation biofuels.

9 Conclusion

The transition to green aviation fuels is crucial for a more sustainable future. This article highlighted the main advantages of biofuels and other sustainable fuels, such as the significant reduction in greenhouse gas emissions and the decreased dependence on fossil fuels. Additionally, we addressed the technical, economic, and regulatory challenges that need to be overcome for the widespread adoption of these fuels. Bibliometric analysis was instrumental in identifying periods with the highest number of publications, revealing areas of great academic interest and significant advancements. This increase in publications can be attributed to the emergence of new technologies, increased research funding, and the growth of interdisciplinary collaborations. These factors have been crucial in driving the development and

implementation of green aviation fuels. Continuous technological advances, favorable government policies, and infrastructure investments will be essential to accelerate this transition. Collaboration between governments, industry, and research institutions is vital to overcoming the remaining obstacles. The United States has been significantly focusing on this area, demonstrating leadership and influence in advancing green aviation fuel technologies. Ultimately, the global commitment to sustainability in aviation is not only an environmental necessity but also an opportunity for innovation and leadership in the sector. As green fuel technologies continue to evolve, they will pave the way for cleaner, more efficient, and responsible aviation, ensuring that future generations can enjoy the benefits of air mobility without compromising the health of our planet.

Acknowledgments The authors would like to express their gratitude for the support that was provided through the Instituto de Engenharias e Desenvolvimento Sustentável (IEDS) at the Universidade da Integração Internacional da Lusofonia Afro-Brasileira (UNILAB). The authors gratefully acknowledge the following Brazilian agencies for scientific and technological development: Fundação Cearense de Apoio ao Desenvolvimento Científico e Tecnológico (FUNCAP) (PS1-0186-00216.01.00/21; 31052.000417/2023; UNI-0210-00537.01.00/23), Conselho Nacional de Desenvolvimento Científico e Tecnológico (CNPq) (311062/2019-9; 440891/2020-5; 307454/2022-3), and Coordenação de Aperfeiçoamento de Ensino Superior (CAPES) (finance code 001).

Declaration of Competing Interests The authors declare that they have no known competing financial interests or personal relationships that could have appeared to influence the work reported in this paper.

Data Availability Data will be made available on request.

References

1. Genovese, M., Schlüter, A., Scionti, E., et al. (2023). Power-to-hydrogen and hydrogen-to-X energy systems for the industry of the future in Europe. *International Journal of Hydrogen Energy, 48*, 16545–16568. https://doi.org/10.1016/j.ijhydene.2023.01.194
2. Amin, M., Shah, H. H., Fareed, A. G., et al. (2022). Hydrogen production through renewable and non-renewable energy processes and their impact on climate change. *International Journal of Hydrogen Energy, 47*, 33112–33134. https://doi.org/10.1016/j.ijhydene.2022.07.172
3. García-Hernández, A. E., Segovia-Hernández, J. G., Sánchez-Ramírez, E., et al. (2024). Sustainable aviation fuel from butanol: A study in optimizing economic and environmental impact through process intensification. *Chemical Engineering and Processing: Process Intensification, 200*, 109769. https://doi.org/10.1016/j.cep.2024.109769
4. Keiser, D., Pupkes, B., & Freitag, M. (2023). Passenger expectations towards a sustainable aviation industry. *Transportation Research Procedia, 75*, 189–197. https://doi.org/10.1016/j.trpro.2023.12.022
5. Gao, Y. (2023). Sustainable aviation fuel as a pathway to mitigate global warming in the aviation industry. *Theoretical and Natural Science, 26*, 60–67. https://doi.org/10.54254/2753-8818/26/20241015
6. Perera, F. (2018). Pollution from fossil-fuel combustion is the leading environmental threat to global pediatric health and equity: Solutions exist. *International Journal of Environmental Research and Public Health, 15*, 1–17. https://doi.org/10.3390/ijerph15010016

7. Soeder, D. J. (2021). Fossil fuels and climate change. *Fracking and the Environment: A scientific assessment of the environmental risks from hydraulic fracturing and fossil fuels, 155–185*. https://doi.org/10.1007/978-3-030-59121-2_9
8. Grewe, V., Gangoli Rao, A., Grönstedt, T., et al. (2021). Evaluating the climate impact of aviation emission scenarios towards the Paris agreement including COVID-19 effects. *Nature Communications, 12*, 1–10. https://doi.org/10.1038/s41467-021-24091-y
9. Lee, D. S., Fahey, D. W., Skowron, A., et al. (2021). The contribution of global aviation to anthropogenic climate forcing for 2000 to 2018. *Atmospheric Environment, 244*, 1–29. https://doi.org/10.1016/j.atmosenv.2020.117834
10. Lai, Y. Y., Christley, E., Kulanovic, A., et al. (2022). Analysing the opportunities and challenges for mitigating the climate impact of aviation: A narrative review. *Renewable and Sustainable Energy Reviews, 156*, 111972. https://doi.org/10.1016/j.rser.2021.111972
11. Zhang, L., Xu, M., Chen, H., et al. (2022). Globalization, green economy and environmental challenges: State of the art review for practical implications. *Frontiers in Environmental Science, 10*, 1–9. https://doi.org/10.3389/fenvs.2022.870271
12. Afonso, F., Sohst, M., Diogo, C. M. A., et al. (2023). Strategies towards a more sustainable aviation: A systematic review. *Progress in Aerospace Sciences, 137*, 100878. https://doi.org/10.1016/j.paerosci.2022.100878
13. Abrantes, I., Ferreira, A. F., Silva, A., & Costa, M. (2021). Sustainable aviation fuels and imminent technologies – CO2 emissions evolution towards 2050. *Journal of Cleaner Production, 313*, 127937. https://doi.org/10.1016/j.jclepro.2021.127937
14. Zhang, L., Butler, T. L., & Yang, B. (2020). Recent trends, opportunities and challenges of sustainable aviation fuel. *Green Energy to Sustainability*, 85–110. https://doi.org/10.1002/9781119152057.ch5
15. Detsios, N., Theodoraki, S., Maragoudaki, L., et al. (2023). Recent advances on alternative aviation fuels/pathways: A critical review. *Energies (Basel), 16*, 1–25. https://doi.org/10.3390/en16041904
16. Grimme, W. (2023). The introduction of sustainable aviation fuels—A discussion of challenges, options and alternatives. *Aerospace, 10*, 1–15. https://doi.org/10.3390/aerospace10030218
17. Peters, M. A., Alves, C. T., & Onwudili, J. A. (2023). A review of current and emerging production technologies for biomass-derived sustainable aviation fuels. *Energies (Basel), 16*, 1–40. https://doi.org/10.3390/en16166100
18. Becken, S., Mackey, B., & Lee, D. S. (2023). Implications of preferential access to land and clean energy for sustainable aviation fuels. *Science of the Total Environment, 886*, 1–11. https://doi.org/10.1016/j.scitotenv.2023.163883
19. Mussatto, S. I., Motta, I. L., Filho, R. M., et al. (2022). 5.16 – Sustainable aviation fuels: Production, use and impact on decarbonization. *Comprehensive Renewable Energy (Second Edition), 1*, 348–371. https://doi.org/10.1016/B978-0-12-819727-1.00057-1
20. Santos, K., & Delina, L. (2021). Soaring sustainably: Promoting the uptake of sustainable aviation fuels during and post-pandemic. *Energy Research and Social Science, 77*, 102074. https://doi.org/10.1016/j.erss.2021.102074
21. Braun, M., Grimme, W., & Oesingmann, K. (2024). Pathway to net zero: Reviewing sustainable aviation fuels, environmental impacts and pricing. *Journal of Air Transport Management, 117*, 102580. https://doi.org/10.1016/j.jairtraman.2024.102580
22. Almena, A., Siu, R., Chong, K., et al. (2024). Reducing the environmental impact of international aviation through sustainable aviation fuel with integrated carbon capture and storage. *Energy Conversion and Management, 303*, 118186. https://doi.org/10.1016/j.enconman.2024.118186
23. Korba, P., Sekelová, I., Koščáková, M., & Behúnová, A. (2023). Passengers' knowledge and attitudes toward green initiatives in aviation. *Sustainability (Switzerland), 15*, 1–25. https://doi.org/10.3390/su15076187

24. Yin, C., Wang, C., Wang, Q., & Ge, Y. (2024). Effects of regional freight structure and energy intensity on CO2 emission of transport – A case study in Yangtze River Delta. *International Journal of Sustainable Transportation, 18*, 379–392. https://doi.org/10.1080/15568318.2023.2299918
25. Ree, K. (2019). Promoting green jobs: Decent work in the transition to low-carbon, green economies. *International Development Policy | Revue internationale de politique de développement, 11*, 248–271. https://doi.org/10.4000/poldev.3107
26. Hanna, R., Heptonstall, P., & Gross, R. (2024). Job creation in a low carbon transition to renewables and energy efficiency: A review of international evidence. *Sustainability Science, 19*, 125–150. https://doi.org/10.1007/s11625-023-01440-y
27. Kumar, J. C. R., & Majid, M. A. (2020). Renewable energy for sustainable development in India: Current status, future prospects, challenges, employment, and investment opportunities. *Energy, Sustainability and Society, 10*, 1–36. https://doi.org/10.1186/s13705-019-0232-1
28. Donthu, N., Kumar, S., Mukherjee, D., et al. (2021). How to conduct a bibliometric analysis: An overview and guidelines. *Journal of Business Research, 133*, 285–296. https://doi.org/10.1016/j.jbusres.2021.04.070
29. Ullah, R., Asghar, I., & Griffiths, M. G. (2023). An integrated methodology for bibliometric analysis: A case study of internet of things in healthcare applications. *Sensors, 23*, 1–28. https://doi.org/10.3390/s23010067
30. McBurney, M. K., & Novak, P. L. (2002). What is bibliometrics and why should you care? *Proceedings IEEE International Professional Communication Conference, 108–114*. https://doi.org/10.1109/IPCC.2002.1049094
31. Bran, R., Tiru, L., Grosseck, G., et al. (2021). Learning from each other – A bibliometric review of research on information disorders. *Sustainability (Switzerland), 13*, 1–38. https://doi.org/10.3390/su131810094
32. López Gómez, M., Posada, J., Silva, V., et al. (2023). Diagnosis of challenges and uncertainties for implementation of sustainable aviation fuel (SAF) in Colombia, and recommendations to move forward. *Energies (Basel), 16*, 1–25. https://doi.org/10.3390/en16155667
33. Koščáková, M., Korba, P., Sekelová, I., & Koščák, P. (2023). Possibilities and challenges for the sustainable development of air transport. *9th International Conference on Mobility, IoT and Smart Cities, 1*, 133–147. https://doi.org/10.1007/978-3-031-28225-6_9
34. Dari, D. N., Freitas, I. S., da Aires, F. I. S., et al. (2024). An updated review of recent applications and perspectives of hydrogen production from biomass by fermentation: A comprehensive analysis. *Biomass (Switzerland), 4*, 132–163. https://doi.org/10.3390/biomass4010007
35. da Silva Aires, F. I., Dari, D. N., Freitas, I. S., et al. (2024). Advanced and prospects in phenol wastewater treatment technologies: Unveiling opportunities and trends. *Discover Water, 4*, 20. https://doi.org/10.1007/s43832-024-00076-y
36. Cavalcante, I. O., Simão Neto, F., da Sousa, S. P., et al. (2024). Evolving sustainable energy technologies and assessments through global research networks: Advancing the role of blue hydrogen for a cleaner future. *RSC Sustainability, 2*, 348–368. https://doi.org/10.1039/D3SU00309D
37. Neto, F. S., de Melo, F., Neta, M. M., Sales, M. B., et al. (2023). Research progress and trends on utilization of lignocellulosic residues as supports for enzyme immobilization via advanced bibliometric analysis. *Polymers (Basel), 15*, 1–31. https://doi.org/10.3390/polym15092057
38. Simão Neto, F., de Sousa Junior, P. G., da Silva Filho, C. J. A., et al. (2024). Green enzymatic synthesis of geranyl butyrate: Process optimization and mechanistic insights. *ACS Omega, 9*, 16992–17001. https://doi.org/10.1021/acsomega.3c08483
39. Nogueira, R. C., Neto, F. S., de Sousa Junior, P. G., et al. (2023). Research trends and perspectives on hydrothermal gasification in producing biofuels. *Energy Nexus, 10*, 100199. https://doi.org/10.1016/j.nexus.2023.100199
40. Guimarães, C. E. C., Neto, F. S., de Castro, B. V., et al. (2023). Sustainable bioethanol production from first- and second-generation sugar-based feedstocks: Advanced biblio-

metric analysis. *Bioresource Technology Reports, 23*, 101543. https://doi.org/10.1016/j.biteb.2023.101543
41. Melo, R. L. F., Sales, M. B., de Castro, B. V., et al. (2023). Recent applications and future prospects of magnetic biocatalysts. *International Journal of Biological Macromolecules, 253*, 126709. https://doi.org/10.1016/j.ijbiomac.2023.126709
42. Schenk, P. M., Thomas-Hall, S. R., Stephens, E., et al. (2008). Second generation biofuels: High-efficiency microalgae for biodiesel production. *Bioenergy Research, 1*, 20–43. https://doi.org/10.1007/s12155-008-9008-8
43. Gouveia, L., & Oliveira, A. C. (2009). Microalgae as a raw material for biofuels production. *Journal of Industrial Microbiology & Biotechnology, 36*, 269. https://doi.org/10.1007/s10295-008-0495-6
44. Neeti, K., Gaurav, K., & Singh, R. (2023). The potential of algae biofuel as a renewable and sustainable bioresource †. *Engineering Proceedings, 37*. https://doi.org/10.3390/ECP2023-14716
45. Chandel, H., Kumar, P., Chandel, A. K., & Verma, M. L. (2024). Biotechnological advances in biomass pretreatment for bio-renewable production through nanotechnological intervention. *Biomass Conversion and Biorefinery, 14*, 2959–2981. https://doi.org/10.1007/s13399-022-02746-0
46. Demirbas, A. (2010). Use of algae as biofuel sources. *Energy Conversion and Management, 51*, 2738–2749. https://doi.org/10.1016/j.enconman.2010.06.010
47. Radakovits, R., Jinkerson, R. E., Darzins, A., & Posewitz, M. C. (2010). Genetic engineering of algae for enhanced biofuel production. *Eukaryotic Cell, 9*, 486–501. https://doi.org/10.1128/ec.00364-09
48. Linger, J. G., Vardon, D. R., Guarnieri, M. T., et al. (2014). Lignin valorization through integrated biological funneling and chemical catalysis. *Proceedings of the National Academy of Sciences of the United States of America, 111*, 12013. https://doi.org/10.1073/pnas.1410657111
49. Sarath, P., Kunchala, V., Shankar, A. J. B., & Sameera, V. S. (2023). Algae – An alternative source of energy (green cell). *Brazilian Journal of Development, 9*, 31756. https://doi.org/10.34117/bjdv9n12-080
50. Hossain, N., Mahlia, T. M. I., & Saidur, R. (2019). Latest development in microalgae-biofuel production with nano-additives. *Biotechnology for Biofuels, 12*, 125. https://doi.org/10.1186/s13068-019-1465-0
51. Milano, J., Ong, H. C., Masjuki, H. H., et al. (2016). Microalgae biofuels as an alternative to fossil fuel for power generation. *Renewable and Sustainable Energy Reviews, 58*, 180–197. https://doi.org/10.1016/j.rser.2015.12.150
52. Demirbas, A. (2007). Progress and recent trends in biofuels. *Progress in Energy and Combustion Science, 33*, 1–18. https://doi.org/10.1016/j.pecs.2006.06.001
53. Demirbas, A. (2009). Political, economic and environmental impacts of biofuels: A review. *Applied Energy, 86*, S108–S117. https://doi.org/10.1016/j.apenergy.2009.04.036
54. Chisti, Y. (2013). Constraints to commercialization of algal fuels. *Journal of Biotechnology, 167*, 201–214. https://doi.org/10.1016/j.jbiotec.2013.07.020
55. Bulushev, D. A., & Ross, J. R. H. (2011). Catalysis for conversion of biomass to fuels via pyrolysis and gasification: A review. *Catalysis Today, 171*, 1–13. https://doi.org/10.1016/j.cattod.2011.02.005
56. Jensen, E. S., Peoples, M. B., Boddey, R. M., et al. (2012). Legumes for mitigation of climate change and the provision of feedstock for biofuels and biorefineries. A review. *Agronomy for Sustainable Development, 32*, 329–364. https://doi.org/10.1007/s13593-011-0056-7
57. Pierrat, E., Rupcic, L., Hauschild, M. Z., & Laurent, A. (2021). Global environmental mapping of the aeronautics manufacturing sector. *Journal of Cleaner Production, 297*, 126603. https://doi.org/10.1016/j.jclepro.2021.126603

58. Kumar, T., Mohsin, R., Ghafir, M. F. A., et al. (2018). Review of alternative fuel initiatives for leaded aviation gasoline (AVGAS) replacement. *Chemical Engineering Transactions, 63*, 175–180. https://doi.org/10.3303/CET1863030
59. Das, D., Sharma, S. K., Parti, R., & Singh, J. (2016). Analyzing the effect of aviation infrastructure over aviation fuel consumption reduction. *Journal of Air Transport Management, 57*, 89. https://doi.org/10.1016/j.jairtraman.2016.07.013
60. Richter, S., Braun-Unkhoff, M., Naumann, C., & Riedel, U. (2018). Paths to alternative fuels for aviation. *CEAS Aeronautical Journal, 9*, 389. https://doi.org/10.1007/s13272-018-0296-1
61. Sato, T., Taguchi, H., Kobayashi, H., et al. (2010). Development study of a precooled turbojet engine. *Acta Astronautica, 66*, 1169. https://doi.org/10.1016/j.actaastro.2009.10.006
62. Akdeniz, H. Y., Balli, O., & Caliskan, H. (2022). Energy, exergy, economic, environmental, energy based economic, exergoeconomic and enviroeconomic (7E) analyses of a jet fueled turbofan type of aircraft engine. *Fuel, 322*, 124165. https://doi.org/10.1016/j.fuel.2022.124165
63. Alrashed, M., Nikolaidis, T., Pilidis, P., et al. (2020). Economic and environmental viability assessment of NASA's turboelectric distribution propulsion. *Energy Reports, 6*, 1685. https://doi.org/10.1016/j.egyr.2020.06.019
64. Balli, O., Caliskan, N., & Caliskan, H. (2023). Aviation, energy, exergy, sustainability, exergoenvironmental and thermoeconomic analyses of a turbojet engine fueled with jet fuel and biofuel used on a pilot trainer aircraft. *Energy, 263*, 126022. https://doi.org/10.1016/j.energy.2022.126022
65. Meher-Homji, C. B. (1998). The development of the whittle turbojet. *Journal of Engineering for Gas Turbines and Power, 120*, 249. https://doi.org/10.1115/1.2818112
66. Maurice, L. Q., Lander, H., Edwards, T., & Harrison, W. E. (2001). Advanced aviation fuels: A look ahead via a historical perspective. *Fuel, 80*, 747. https://doi.org/10.1016/S0016-2361(00)00142-3
67. Dietrich, M., & Filippelli, G. M. (2023). Positive outcomes from U.S. lead regulations, continued challenges, and lessons learned for regulating emerging contaminants. *Environmental Science and Pollution Research, 30*, 57178. https://doi.org/10.1007/s11356-023-26319-4
68. Mills, A., & Peckham, S. (2022). Lead exposure from general aviation emissions in the UK: A review and call for action. *Public Health Challenges, 1*. https://doi.org/10.1002/puh2.27
69. Abdellatief, T. M. M., Ershov, M. A., Savelenko, V. D., et al. (2023). A novelty detection diagnostic methodology for merit function calculations for various individual gasoline components. *Fuel, 344*, 128050. https://doi.org/10.1016/j.fuel.2023.128050
70. Demirbas, A., Balubaid, M. A., Basahel, A. M., et al. (2015). Octane rating of gasoline and octane booster additives. *Petroleum Science and Technology, 33*, 1190. https://doi.org/10.1080/10916466.2015.1050506
71. Akbiyik, T., Kahraman, N., & Taner, T. (2023). Energy and exergy analysis with emissions evaluation of a gasoline engine using different fuels. *Fuel, 345*, 128189. https://doi.org/10.1016/J.FUEL.2023.128189
72. Balli, O. (2022). Thermodynamic, thermoenvironmental and thermoeconomic analyses of piston-prop engines (PPEs) for landing and take-off (LTO) flight phases. *Energy, 250*, 123725. https://doi.org/10.1016/J.ENERGY.2022.123725
73. Smith, P. M., Gaffney, M. J., Shi, W., et al. (2017). Drivers and barriers to the adoption and diffusion of sustainable jet fuel (SJF) in the U.S. Pacific Northwest. *Journal of Air Transport Management, 58*, 113. https://doi.org/10.1016/j.jairtraman.2016.10.004
74. Hamilton, J., Sadat, Y., Dwyer, M., et al. (2021). Thermal stability and impact of alternative fuels. *Aviation Fuels*, 149–218. https://doi.org/10.1016/B978-0-12-818314-4.00003-0
75. Fortin, T. J., & Bruno, T. J. (2022). Heat capacity measurements of conventional aviation fuels. *Fuel Processing Technology, 235*, 107341. https://doi.org/10.1016/j.fuproc.2022.107341
76. Gao, Y., & Casey, J. F. (2012). Lithium isotope composition of ultramafic geological reference materials JP-1 and DTS-2. *Geostandards and Geoanalytical Research, 36*, 75. https://doi.org/10.1111/j.1751-908X.2011.00117.x

77. Zhang, X., & Jia, T. (2020). Development history and basics of aerospace fuels. *High-Energy-Density Fuels for Advanced Propulsion, 1*, 5–38. https://doi.org/10.1002/9783527823789.ch2
78. Vandegrift, S. A., & Kampbell, D. H. (1988). Gas chromatographic determination of aviation gasoline and jp-4 jet fuel in subsurface core samples. *Journal of Chromatographic Science, 26*, 566. https://doi.org/10.1093/chromsci/26.11.566
79. Bendtsen, K. M., Bengtsen, E., Saber, A. T., & Vogel, U. (2021). A review of health effects associated with exposure to jet engine emissions in and around airports. *Environmental Health, 20*, 10. https://doi.org/10.1186/s12940-020-00690-y
80. Cowart, J., Foley, M. P., & Luning Prak, D. (2019). The development and testing of Navy jet fuel (JP-5) surrogates. *Fuel, 249*, 80. https://doi.org/10.1016/j.fuel.2019.03.096
81. Abi Nurazaq, W., Wang, W. C., & Lin, J. K. (2024). The properties of sustainable aviation fuel II: Laminar flame speed. *Energy, 294*, 130946. https://doi.org/10.1016/j.energy.2024.130946
82. Korres, D. M., Karonis, D., Lois, E., et al. (2008). Aviation fuel JP-5 and biodiesel on a diesel engine. *Fuel, 87*, 70. https://doi.org/10.1016/j.fuel.2007.04.004
83. Liu, Z., Pan, H., Feng, S., & Bi, Q. (2015). Dynamic behaviors of coking process during pyrolysis of China aviation kerosene RP-3. *Applied Thermal Engineering, 91*, 408. https://doi.org/10.1016/j.applthermaleng.2015.08.033
84. Wang, R., Ye, T., Bao, Z., & Zhu, Q. (2023). Numerical analysis of pyrolysis and coking of aviation kerosene in a rectangular U-bend tube under one-side heating. *International Journal of Heat and Mass Transfer, 214*, 124379. https://doi.org/10.1016/j.ijheatmasstransfer.2023.124379
85. Kinsey, J. S., Giannelli, R., Howard, R., et al. (2021). Assessment of a regulatory measurement system for the determination of the non-volatile particulate matter emissions from commercial aircraft engines. *Journal of Aerosol Science, 154*, 105734. https://doi.org/10.1016/j.jaerosci.2020.105734
86. Akhtar, M. N., Akhter, F., Ahmed, J., & Pinjaro, M. A. (2023). Enhancing combustion characteristics using JP-8 and JP-10 fuel mixture with different equivalence ratios: Numerical and simulative study. *Chemical Papers, 77*, 5455. https://doi.org/10.1007/s11696-023-02877-2
87. Harris, D., Sakiestewa, D., Titone, D., et al. (2008). JP-8 jet fuel exposure suppresses the immune response to viral infections. *Toxicology and Industrial Health, 24*, 209. https://doi.org/10.1177/0748233708093781
88. Kim, S., Sasmaz, E., Pogaku, R., & Lauterbach, J. (2020). Effects of reaction conditions and organic sulfur compounds on coke formation and HZSM-5 catalyst performance during jet propellant fuel (JP-8) cracking. *Fuel, 259*, 116240. https://doi.org/10.1016/j.fuel.2019.116240
89. McDougal, J. N., & Garrett, C. M. (2007). Gene expression and target tissue dose in the rat epidermis after brief JP-8 and JP-8 aromatic and aliphatic component exposures. *Toxicological Sciences, 97*, 569. https://doi.org/10.1093/toxsci/kfm037
90. Collins, J. M., & McLarty, D. (2020). All-electric commercial aviation with solid oxide fuel cell-gas turbine-battery hybrids. *Applied Energy, 265*, 114787. https://doi.org/10.1016/j.apenergy.2020.114787
91. Jensen, L. L., Bonnefoy, P. A., Hileman, J. I., & Fitzgerald, J. T. (2023). The carbon dioxide challenge facing U.S. aviation and paths to achieve net zero emissions by 2050. *Progress in Aerospace Sciences, 141*, 100921. https://doi.org/10.1016/j.paerosci.2023.100921
92. Heneghan, S. P., Zabarnick, S., Ballal, D. R., & Harrison, W. E. (1996). JP-8+100: The development of high-thermal-stability jet fuel. *Journal of Energy Resources Technology, Transactions of the ASME, 118*, 170. https://doi.org/10.1115/1.2793859
93. Kanikkannan, N., Burton, S., Patel, R., et al. (2001). Percutaneous permeation and skin irritation of JP-8 + 100 jet fuel in a porcine model. *Toxicology Letters, 119*, 133. https://doi.org/10.1016/S0378-4274(00)00311-8
94. Muhammad, F., Brooks, J. D., & Riviere, J. E. (2004). Comparative mixture effects of JP-8(100) additives on the dermal absorption and disposition of jet fuel hydrocarbons in

different membrane model systems. *Toxicology Letters, 150*, 351. https://doi.org/10.1016/j.toxlet.2004.02.012
95. Tsai, S. Y., Su, J. W., Tong, J. T., & Lin, C. P. (2021). Evaluation of suitable mixture proportion compatibility of aviation fuel JET A-1 blended with butanol isomers. *Journal of Thermal Analysis and Calorimetry, 144*, 403. https://doi.org/10.1007/s10973-021-10660-2
96. Hassan, T. H., & Salem, A. E. (2021). The importance of safety and security measures at sharm el sheikh airport and their impact on travel decisions after restarting aviation during the covid-19 outbreak. *Sustainability (Switzerland), 13*. https://doi.org/10.3390/su13095216
97. D C, G T, N S, M P. (2021). The challenge of forecasting the role of biofuel in EU transport decarbonization at 2050: A meta-analysis review of published scenarios. *Renewable & Sustainable Energy Reviews, 139*, 110715. https://doi.org/10.1016/j.rser.2021.110715
98. Wang, M., Raf, D., Maniatis, K., et al. (2019). Biomass-derived aviation fuels: Challenges and perspective. *Progress in Energy and Combustion Science, 74*, 31–49. https://doi.org/10.1016/j.pecs.2019.04.004
99. Ahmadbeyki, A., Ghahderijani, M., Borghaee, A., & Bakhoda, H. (2023). Energy use and environmental impacts analysis of greenhouse crops production using life cycle assessment approach: A case study of cucumber and tomato from Tehran province, Iran. *Energy Reports, 9*, 988. https://doi.org/10.1016/j.egyr.2022.11.205
100. Dong, K., Dong, X., Jiang, Q., & Zhao, J. (2021). Assessing energy resilience and its greenhouse effect: A global perspective. *Energy Economics, 104*, 105659. https://doi.org/10.1016/j.eneco.2021.105659
101. Ghoulem, M., El Moueddeb, K., Nehdi, E., et al. (2019). Greenhouse design and cooling technologies for sustainable food cultivation in hot climates: Review of current practice and future status. *Biosystems Engineering, 183*, 121. https://doi.org/10.1016/j.biosystemseng.2019.04.016
102. Koutsoyiannis, D., & Vournas, C. (2024). Revisiting the greenhouse effect – A hydrological perspective. *Hydrological Sciences Journal, 69*, 151. https://doi.org/10.1080/02626667.2023.2287047
103. Semenov, S. M. (2022). Greenhouse effect and modern climate. *Russian Meteorology and Hydrology, 47*, 725. https://doi.org/10.3103/S1068373922100016
104. Zhou, J., Zheng, Y., Hou, L., et al. (2023). Effects of acidification on nitrification and associated nitrous oxide emission in estuarine and coastal waters. *Nature Communications, 14*, 1380. https://doi.org/10.1038/s41467-023-37104-9
105. Qasem, N. A. A., Mourad, A., Abderrahmane, A., et al. (2024). A recent review of aviation fuels and sustainable aviation fuels. *Journal of Thermal Analysis and Calorimetry, 149*, 4287. https://doi.org/10.1007/s10973-024-13027-5
106. Milkoreit, M. (2019). The Paris agreement on climate change – Made in USA? *Perspectives on Politics, 17*, 1019. https://doi.org/10.1017/S1537592719000951
107. Stagner, J. (2022). Carbon captured: How business and labor control climate politics. *International Journal of Environmental Studies, 79*, 761. https://doi.org/10.1080/00207233.2021.1941676
108. Falkner, R. (2016). The Paris agreement and the new logic of international climate politics. *International Affairs, 92*, 1107. https://doi.org/10.1111/1468-2346.12708
109. Kougias, I., Taylor, N., Kakoulaki, G., & Jäger-Waldau, A. (2021). The role of photovoltaics for the European Green Deal and the recovery plan. *Renewable and Sustainable Energy Reviews, 144*, 111017. https://doi.org/10.1016/j.rser.2021.111017
110. Ryan, M., Giesbers, E., Heffernan, R., et al. (2023). Developing normative criteria for meaningful citizen participation and deliberation in environmental policy. *Innovation: The European Journal of Social Science Research, 37*, 794. https://doi.org/10.1080/13511610.2023.2217520
111. Braun-Unkhoff, M., & Riedel, U. (2015). Alternative fuels in aviation. *CEAS Aeronautical Journal, 6*, 83. https://doi.org/10.1007/s13272-014-0131-2

112. Böhringer, C. (2003). The Kyoto protocol: A review and perspectives. *Oxford Review of Economic Policy, 19*, 451. https://doi.org/10.1093/oxrep/19.3.451
113. Heath, E. A. (2017). Amendment to the Montreal Protocol on substances that deplete the ozone layer (Kigali Amendment). *International Legal Materials, 56*, 193. https://doi.org/10.1017/ilm.2016.2
114. Hillbrand, A., Madan, P., Singh, M., et al. (2022). Scenarios for future Indian HFC demand compared to the Kigali Amendment. *Environmental Research Letters, 17*. https://doi.org/10.1088/1748-9326/ac7538
115. Morgenstern, O., Braesicke, P., Hurwitz, M. M., et al. (2008). The world avoided by the Montreal Protocol. *Geophysical Research Letters, 35*. https://doi.org/10.1029/2008GL034590
116. Tessnow-von Wysocki, I., & Le Billon, P. (2019). Plastics at sea: Treaty design for a global solution to marine plastic pollution. *Environmental Science & Policy, 100*, 94. https://doi.org/10.1016/j.envsci.2019.06.005
117. Velders, G. J. M., Andersen, S. O., Daniel, J. S., et al. (2007). The importance of the Montreal Protocol in protecting climate. *Proceedings of the National Academy of Sciences of the United States of America, 104*, 4814. https://doi.org/10.1073/pnas.0610328104
118. Wu, Y., Polvani, L. M., & Seager, R. (2013). The importance of the Montreal Protocol in protecting Earth's hydroclimate. *Journal of Climate, 26*, 4049. https://doi.org/10.1175/JCLI-D-12-00675.1
119. Smith, H. A., & Sharp, K. (2012). Indigenous climate knowledges. *WIREs Climate Change, 3*, 467–476. https://doi.org/10.1002/wcc.185
120. Srivastava, A. K., Dharwal, M., & Sharma, A. (2022). Green financial initiatives for sustainable economic growth: A literature review. *Materials Today Proceedings, 49*, 3615–3618. https://doi.org/10.1016/j.matpr.2021.08.158
121. Chiaramonti, D. (2019). Sustainable aviation fuels: The challenge of decarbonization. *Energy Procedia, 158*, 1202–1207. https://doi.org/10.1016/j.egypro.2019.01.308
122. Ng, K. S., Farooq, D., & Yang, A. (2021). Global biorenewable development strategies for sustainable aviation fuel production. *Renewable and Sustainable Energy Reviews, 150*, 111502. https://doi.org/10.1016/j.rser.2021.111502
123. Bauen, A., Bitossi, N., German, L., et al. (2020). Sustainable aviation fuels status, challenges and prospects of drop-in liquid fuels, hydrogen and electrification in aviation. *Johnson Matthey Technology Review, 64*, 263. https://doi.org/10.1595/205651320x15816756012040
124. Oteman, M., Kooij, H. J., & Wiering, M. A. (2017). Pioneering renewable energy in an economic energy policy system: The history and development of dutch grassroots initiatives. *Sustainability (Switzerland), 9*. https://doi.org/10.3390/su9040550
125. Balan, V., Chiaramonti, D., & Kumar, S. (2013). Review of US and EU initiatives toward development, demonstration, and commercialization of lignocellulosic biofuels. *Biofuels, Bioproducts and Biorefining, 7*, 732–759. https://doi.org/10.1002/bbb.1436
126. Fridstrøm, L. (2021). The norwegian vehicle electrification policy and its implicit price of carbon. *Sustainability (Switzerland), 13*. https://doi.org/10.3390/su13031346
127. Lieutenant, K., Borissova, A. V., Mustafa, M., et al. (2022). Comparison of "zero emission" vehicles with petrol and hybrid cars in terms of Total CO2 release – A case study for Romania, Poland, Norway and Germany. *Energies (Basel), 15*. https://doi.org/10.3390/en15217988
128. Pfaffenbichler, P., Fearnley, N., Figenbaum, E., & Emberger, G. (2024). Simulating the effects of tax exemptions for plug-in electric vehicles in Norway. *European Transport Research Review, 16*. https://doi.org/10.1186/s12544-024-00648-y
129. Han, C., & Yang, L. (2024). Financing and management strategies for expanding green development projects: A case study of energy corporation in China's renewable energy sector using machine learning (ML) modeling. *Sustainability (Switzerland), 16*. https://doi.org/10.3390/su16114338
130. Azam, Q., Alhaj, A. M., Abidin, M. S. Z., et al. (2023). An outline of alternative aviation fuels from sustainable resources. *J Teknol, 85*, 11. https://doi.org/10.11113/jurnalteknologi.v85.14563

131. Wang, J., Wang, H., & Fan, Y. (2018). Techno-economic challenges of fuel cell commercialization. *Engineering, 4*, 352–360. https://doi.org/10.1016/j.eng.2018.05.007
132. Yilmaz, N., & Atmanli, A. (2017). Sustainable alternative fuels in aviation. *Energy, 140*, 1378. https://doi.org/10.1016/j.energy.2017.07.077
133. Gasparatos, A., Stromberg, P., & Takeuchi, K. (2013). Sustainability impacts of first-generation biofuels. *Animal Frontiers, 3*, 12. https://doi.org/10.2527/af.2013-0011
134. Mohr, A., & Raman, S. (2015). Lessons from first generation biofuels and implications for the sustainability appraisal of second generation biofuels. *Efficiency and Sustainability in Biofuel Production: Environmental and Land-Use Research, 63*, 114. https://doi.org/10.1016/j.enpol.2013.08.033
135. Naik, S. N., Goud, V. V., Rout, P. K., & Dalai, A. K. (2010). Production of first and second generation biofuels: A comprehensive review. *Renewable and Sustainable Energy Reviews, 14*, 578–597. https://doi.org/10.1016/j.rser.2009.10.003
136. Pabbathi, N. P. P., Velidandi, A., Tavarna, T., et al. (2023). Role of metagenomics in prospecting novel endoglucanases, accentuating functional metagenomics approach in second-generation biofuel production: A review. *Biomass Conversion and Biorefinery, 13*, 1371–1398. https://doi.org/10.1007/s13399-020-01186-y
137. Antizar-Ladislao, B., & Turrion-Gomez, J. L. (2008). Second-generation biofuels and local bioenergy systems. *Biofuels, Bioproducts and Biorefining, 2*, 455–469. https://doi.org/10.1002/bbb.97
138. Afzaal, M., Hameed, S., Rasheed, R., & Khan, W. U. D. (2022). Chapter 10: Microalgal biofuels: A sustainable pathway for renewable energy. *Algal Biotechnology, 1*, 187–222. https://doi.org/10.1016/B978-0-323-90476-6.00004-2
139. Maliha, A., & Abu-Hijleh, B. (2023). A review on the current status and post-pandemic prospects of third-generation biofuels. *Energy Systems, 14*, 1185–1216. https://doi.org/10.1007/s12667-022-00514-7
140. Singh, A., Olsen, S. I., & Nigam, P. S. (2011). A viable technology to generate third-generation biofuel. *Journal of Chemical Technology and Biotechnology, 86*, 1349. https://doi.org/10.1002/jctb.2666
141. Shokravi, H., Shokravi, Z., Heidarrezaei, M., et al. (2021). Fourth generation biofuel from genetically modified algal biomass: Challenges and future directions. *Chemosphere, 285*, 131535. https://doi.org/10.1016/j.chemosphere.2021.131535
142. Abdullah, B., Syed Muhammad, S. A. F., Shokravi, Z., et al. (2019). Fourth generation biofuel: A review on risks and mitigation strategies. *Renewable and Sustainable Energy Reviews, 107*, 37–50. https://doi.org/10.1016/j.rser.2019.02.018
143. Li, S., Li, F., Zhu, X., et al. (2022). Biohydrogen production from microalgae for environmental sustainability. *Chemosphere, 291*, 132717. https://doi.org/10.1016/j.chemosphere.2021.132717
144. Sproles, A. E., Fields, F. J., Smalley, T. N., et al. (2021). Recent advancements in the genetic engineering of microalgae. *Algal Research, 53*, 102158. https://doi.org/10.1016/j.algal.2020.102158
145. de Souza, L. M., Mendes, P. A. S., & Aranda, D. A. G. (2020). Oleaginous feedstocks for hydro-processed esters and fatty acids (HEFA) biojet production in southeastern Brazil: A multi-criteria decision analysis. *Renewable Energy, 149*, 1339–1351. https://doi.org/10.1016/J.RENENE.2019.10.125
146. Jürgens, S., Oßwald, P., Selinsek, M., et al. (2019). Assessment of combustion properties of non-hydroprocessed Fischer-Tropsch fuels for aviation. *Fuel Processing Technology, 193*, 232–243. https://doi.org/10.1016/J.FUPROC.2019.05.015
147. Suzihaque, M. U. H., Alwi, H., Kalthum Ibrahim, U., et al. (2022). Biodiesel production from waste cooking oil: A brief review. *Materials Today Proceedings, 63*, S490–S495. https://doi.org/10.1016/j.matpr.2022.04.527

148. Tomás-Pejó, E., Morales-Palomo, S., & González-Fernández, C. (2021). Microbial lipids from organic wastes: Outlook and challenges. *Bioresource Technology, 323*, 124612. https://doi.org/10.1016/j.biortech.2020.124612
149. Pearlson, M., Wollersheim, C., & Hileman, J. (2013). A techno-economic review of hydroprocessed renewable esters and fatty acids for jet fuel production. *Biofuels, Bioproducts and Biorefining, 7*, 89. https://doi.org/10.1002/bbb.1378
150. Starck, L., Pidol, L., Jeuland, N., et al. (2016). Production of Hydroprocessed esters and fatty acids (HEFA) – Optimisation of process yield. *Oil & Gas Science and Technology—Revue d'IFP Energies nouvelles, 71*, 10. https://doi.org/10.2516/ogst/2014007
151. Dry, M. E. (2002). The Fischer–Tropsch process: 1950–2000. *Catalysis Today, 71*, 227–241. https://doi.org/10.1016/S0920-5861(01)00453-9
152. Abatzoglou, N. (2018). Application of plasma technology in Fischer-Tropsch catalysis for the production of synthetic fuels. *Recent Advances in Petrochemical Science, 5*. https://doi.org/10.19080/rapsci.2018.05.555657
153. Guilera, J., Díaz-López, J. A., Berenguer, A., et al. (2022). Fischer-Tropsch synthesis: Towards a highly-selective catalyst by lanthanide promotion under relevant CO2 syngas mixtures. *Applied Catalysis A: General, 629*, 118423. https://doi.org/10.1016/j.apcata.2021.118423
154. Meurer, A., & Kern, J. (2021). Fischer–tropsch synthesis as the key for decentralized sustainable kerosene production. *Energies (Basel), 14*. https://doi.org/10.3390/en14071836
155. Rojas-Michaga, M. F., Michailos, S., Cardozo, E., et al. (2023). Sustainable aviation fuel (SAF) production through power-to-liquid (PtL): A combined techno-economic and life cycle assessment. *Energy Conversion and Management, 292*, 117427. https://doi.org/10.1016/j.enconman.2023.117427
156. Varone, A., & Ferrari, M. (2015). Power to liquid and power to gas: An option for the German Energiewende. *Renewable and Sustainable Energy Reviews, 45*, 207–218. https://doi.org/10.1016/j.rser.2015.01.049
157. White, R. J. (2022). Power-to-liquids – Conversion of CO 2 and renewable H 2 to methanol. *Advances in Energy Storage, 489–520*. https://doi.org/10.1002/9781119239390.ch22
158. Dieterich, V., Buttler, A., Hanel, A., et al. (2020). Power-to-liquid via synthesis of methanol, DME or Fischer–Tropsch-fuels: A review. *Energy & Environmental Science, 13*, 3207–3252. https://doi.org/10.1039/D0EE01187H
159. Gronau, S., Hoelzen, J., Mueller, T., & Hanke-Rauschenbach, R. (2023). Hydrogen-powered aviation in Germany: A macroeconomic perspective and methodological approach of fuel supply chain integration into an economy-wide dataset. *International Journal of Hydrogen Energy, 48*, 5347. https://doi.org/10.1016/j.ijhydene.2022.10.168
160. Huete, J., & Pilidis, P. (2021). Parametric study on tank integration for hydrogen civil aviation propulsion. *International Journal of Hydrogen Energy, 46*, 37049. https://doi.org/10.1016/j.ijhydene.2021.08.194
161. Manigandan, S., Praveenkumar, T. R., Ir Ryu, J., et al. (2023). Role of hydrogen on aviation sector: A review on hydrogen storage, fuel flexibility, flame stability, and emissions reduction on gas turbines engines. *Fuel, 352*, 129064. https://doi.org/10.1016/j.fuel.2023.129064
162. Hoelzen, J., Silberhorn, D., Zill, T., et al. (2022). Hydrogen-powered aviation and its reliance on green hydrogen infrastructure – Review and research gaps. *International Journal of Hydrogen Energy, 47*, 3108–3130. https://doi.org/10.1016/j.ijhydene.2021.10.239
163. Nam, G. D., Vuong, L. D., Sung, H. J., et al. (2021). Conceptual design of an aviation propulsion system using hydrogen fuel cell and superconducting motor. *IEEE Transactions on Applied Superconductivity, 31*, 1. https://doi.org/10.1109/TASC.2021.3064526
164. Degirmenci, H., Uludag, A., Ekici, S., & Karakoc, T. H. (2023). Challenges, prospects and potential future orientation of hydrogen aviation and the airport hydrogen supply network: A state-of-art review. *Progress in Aerospace Sciences, 141*, 100923. https://doi.org/10.1016/j.paerosci.2023.100923

165. Pu, L., Yu, H., Dai, M., et al. (2022). Research progress and application of high-pressure hydrogen and liquid hydrogen in storage and transportation. *Kexue Tongbao/Chinese Science Bulletin, 67*, 2172. https://doi.org/10.1360/TB-2022-0063
166. Nazir, H., Muthuswamy, N., Louis, C., et al. (2020). Is the H2 economy realizable in the foreseeable future? Part II: H2 storage, transportation, and distribution. *International Journal of Hydrogen Energy, 45*, 20693–20708. https://doi.org/10.1016/j.ijhydene.2020.05.241
167. Dorn, F. M. (2024). Towards a multi-color hydrogen production network? Competing imaginaries of development in northern Patagonia, Argentina. *Energy Research & Social Science, 110*, 103457. https://doi.org/10.1016/j.erss.2024.103457
168. Durakovic, G., del Granado, P. C., & Tomasgard, A. (2023). Are green and blue hydrogen competitive or complementary? Insights from a decarbonized European power system analysis. *Energy, 282*, 128282. https://doi.org/10.1016/j.energy.2023.128282
169. Howarth, R. W., & Jacobson, M. Z. (2021). How green is blue hydrogen? *Energy Science & Engineering, 9*, 1676. https://doi.org/10.1002/ese3.956
170. Patel, G. H., Havukainen, J., Horttanainen, M., et al. (2024). Climate change performance of hydrogen production based on life cycle assessment. *Green Chemistry, 26*, 992. https://doi.org/10.1039/d3gc02410e
171. Romano, M. C., Antonini, C., Bardow, A., et al. (2022). Comment on "How green is blue hydrogen?". *Energy Science & Engineering, 10*, 1944. https://doi.org/10.1002/ese3.1126
172. Scovell, M., & Walton, A. (2024). Blue or green? Exploring Australian acceptance and beliefs about hydrogen production methods. *Journal of Cleaner Production, 444*, 141151. https://doi.org/10.1016/j.jclepro.2024.141151
173. El-Shafie, M., Kambara, S., & Hayakawa, Y. (2019). Hydrogen production technologies overview. *Journal of Power and Energy Engineering, 07*, 107. https://doi.org/10.4236/jpee.2019.71007
174. Holladay, J. D., Hu, J., King, D. L., & Wang, Y. (2009). An overview of hydrogen production technologies. *Catalysis Today, 139*, 244–260. https://doi.org/10.1016/j.cattod.2008.08.039
175. Li, Y., Lin, R., O'Shea, R., et al. (2024). A perspective on three sustainable hydrogen production technologies with a focus on technology readiness level, cost of production and life cycle environmental impacts. *Heliyon, 10*, e26637. https://doi.org/10.1016/j.heliyon.2024.e26637
176. Rahim Malik, F., Yuan, H.-B., Moran, J. C., & Tippayawong, N. (2023). Overview of hydrogen production technologies for fuel cell utilization. *Engineering Science and Technology, an International Journal, 43*, 101452. https://doi.org/10.1016/j.jestch.2023.101452
177. Shiva Kumar, S., & Lim, H. (2022). An overview of water electrolysis technologies for green hydrogen production. *Energy Reports, 8*, 13793–13813. https://doi.org/10.1016/j.egyr.2022.10.127
178. Zainal, B. S., Ker, P. J., Mohamed, H., et al. (2024). Recent advancement and assessment of green hydrogen production technologies. *Renewable and Sustainable Energy Reviews, 189*, 113941. https://doi.org/10.1016/j.rser.2023.113941
179. Kumar, S., Arzaghi, E., Baalisampang, T., et al. (2023). Insights into decision-making for offshore green hydrogen infrastructure developments. *Process Safety and Environmental Protection, 174*, 805. https://doi.org/10.1016/j.psep.2023.04.042
180. Smit, R., Weeda, M., & de Groot, A. (2007). Hydrogen infrastructure development in The Netherlands. *International Journal of Hydrogen Energy, 32*, 1387. https://doi.org/10.1016/j.ijhydene.2006.10.044
181. Wang, Y., Pang, Y., Xu, H., et al. (2022). PEM fuel cell and electrolysis cell technologies and hydrogen infrastructure development – A review. *Energy & Environmental Science, 15*, 2288–2328. https://doi.org/10.1039/D2EE00790H
182. Ulucan, T. H., Akhade, S. A., Ambalakatte, A., et al. (2023). Hydrogen storage in liquid hydrogen carriers: Recent activities and new trends. *Progress in Energy, 5*, 12004. https://doi.org/10.1088/2516-1083/acac5c
183. Yang, X., Bulushev, D. A., Yang, J., & Zhang, Q. (2022). New liquid chemical hydrogen storage technology. *Energies (Basel), 15*, 1–18. https://doi.org/10.3390/en15176360

184. Cardella, U., Decker, L., Sundberg, J., & Klein, H. (2017). Process optimization for large-scale hydrogen liquefaction. *International Journal of Hydrogen Energy, 42*, 12339. https://doi.org/10.1016/j.ijhydene.2017.03.167
185. Sadaghiani, M. S., & Mehrpooya, M. (2017). Introducing and energy analysis of a novel cryogenic hydrogen liquefaction process configuration. *International Journal of Hydrogen Energy, 42*, 6033. https://doi.org/10.1016/j.ijhydene.2017.01.136
186. Bonacina, C. N., Gaskare, N. B., & Valenti, G. (2022). Assessment of offshore liquid hydrogen production from wind power for ship refueling. *International Journal of Hydrogen Energy, 47*, 1279. https://doi.org/10.1016/j.ijhydene.2021.10.043
187. Uchman, W., & Kotowicz, J. (2021). Varying load distribution impacts on the operation of a hydrogen generator plant. *International Journal of Hydrogen Energy, 46*, 39095. https://doi.org/10.1016/j.ijhydene.2021.09.166
188. Elsherif, M., Manan, Z. A., & Kamsah, M. Z. (2015). State-of-the-art of hydrogen management in refinery and industrial process plants. *Journal of Natural Gas Science and Engineering, 24*, 346–356. https://doi.org/10.1016/j.jngse.2015.03.046
189. Borsboom-Hanson, T., Patlolla, S. R., Herrera, O. E., & Mérida, W. (2022). Point-to-point transportation: The economics of hydrogen export. *International Journal of Hydrogen Energy, 47*, 31541. https://doi.org/10.1016/j.ijhydene.2022.07.093
190. Zhang, H., Zhao, J., Li, J., et al. (2023). Research progress on corrosion and hydrogen embrittlement in hydrogen–natural gas pipeline transportation. *Natural Gas Industry B, 10*, 570. https://doi.org/10.1016/j.ngib.2023.11.001
191. Lan, K., Ou, L., Park, S., et al. (2021). Dynamic life-cycle carbon analysis for fast pyrolysis biofuel produced from pine residues: Implications of carbon temporal effects. *Biotechnology for Biofuels, 14*, 191. https://doi.org/10.1186/s13068-021-02027-4
192. Nie, Y., & Bi, X. (2018). Life-cycle assessment of transportation biofuels from hydrothermal liquefaction of forest residues in British Columbia. *Biotechnology for Biofuels, 11*, 23. https://doi.org/10.1186/s13068-018-1019-x
193. Alptekin, E., Canakci, M., & Sanli, H. (2014). Biodiesel production from vegetable oil and waste animal fats in a pilot plant. *Waste Management, 34*, 2146–2154. https://doi.org/10.1016/j.wasman.2014.07.019
194. Mrad, N., Varuvel, E. G., Tazerout, M., & Aloui, F. (2012). Effects of biofuel from fish oil industrial residue – Diesel blends in diesel engine. *Energy, 44*, 955–963. https://doi.org/10.1016/j.energy.2012.04.056
195. Çakırca, E. E., Tekin, N., Gilgen, O., & Akın, N. A. (2019). Catalytic activity of CaO-based catalyst in transesterification of microalgae oil with methanol. *Energy & Environment, 30*, 176–187. https://doi.org/10.1177/0958305X18787317
196. Liu, Y., Sotelo-Boyás, R., Murata, K., et al. (2011). Hydrotreatment of vegetable oils to produce bio-hydrogenated diesel and liquefied petroleum gas fuel over catalysts containing sulfided Ni–Mo and solid acids. *Energy & Fuels, 25*, 4675–4685. https://doi.org/10.1021/ef200889e
197. Yenumala, S. R., Maity, S. K., & Shee, D. (2016). Hydrodeoxygenation of karanja oil over supported nickel catalysts: Influence of support and nickel loading. *Catalysis Science & Technology, 6*, 3156–3165. https://doi.org/10.1039/C5CY01470K
198. Ali, A., Li, B., Lu, Y., & Zhao, C. (2019). Highly selective and low-temperature hydrothermal conversion of natural oils to fatty alcohols. *Green Chemistry, 21*, 3059–3064. https://doi.org/10.1039/C9GC01260E
199. Michaga, M. F. R., Michailos, S., Hughes, K. J., et al. (2021). 10 – Techno-economic and life cycle assessment review of sustainable aviation fuel produced via biomass gasification. *Sustainable Biofuels, 1*, 269–303. https://doi.org/10.1016/B978-0-12-820297-5.00012-8
200. Geleynse, S., Brandt, K., Garcia-Perez, M., et al. (2018). The alcohol-to-jet conversion pathway for drop-in biofuels: Techno-economic evaluation. *ChemSusChem, 11*, 3728–3741. https://doi.org/10.1002/cssc.201801690

201. Kurzawska, P. (2021). Overview of sustainable aviation fuels including emission of particulate matter and harmful gaseous exhaust gas compounds. *Transportation Research Procedia, 59*, 38–45. https://doi.org/10.1016/j.trpro.2021.11.095
202. Eswaran, S., Subramaniam, S., Geleynse, S., et al. (2021). Techno-economic analysis of catalytic hydrothermolysis pathway for jet fuel production. *Renewable and Sustainable Energy Reviews, 151*, 111516. https://doi.org/10.1016/j.rser.2021.111516
203. Tao, L., Milbrandt, A., Zhang, Y., & Wang, W.-C. (2017). Techno-economic and resource analysis of hydroprocessed renewable jet fuel. *Biotechnology for Biofuels, 10*, 261. https://doi.org/10.1186/s13068-017-0945-3
204. Chen, P. H., Lee, U., Liu, X., et al. (2024). Life-cycle analysis of sustainable aviation fuel production through catalytic hydrothermolysis. *Biofuels, Bioproducts and Biorefining, 18*, 42–54. https://doi.org/10.1002/bbb.2574
205. Tashie-Lewis, B. C., & Nnabuife, S. G. (2021). Hydrogen production, distribution, storage and power conversion in a hydrogen economy – A technology review. *Chemical Engineering Journal Advances, 8*, 100172. https://doi.org/10.1016/j.ceja.2021.100172
206. Neuling, U., & Kaltschmitt, M. (2015). Conversion routes for production of biokerosene – Status and assessment. *Biomass Conversion and Biorefinery, 5*, 367–385. https://doi.org/10.1007/s13399-014-0154-2
207. Colelli, L., Segneri, V., Bassano, C., & Vilardi, G. (2023). E-fuels, technical and economic analysis of the production of synthetic kerosene precursor as sustainable aviation fuel. *Energy Conversion and Management, 288*, 117165. https://doi.org/10.1016/j.enconman.2023.117165
208. Undavalli, V., Gbadamosi Olatunde, O. B., Boylu, R., et al. (2023). Recent advancements in sustainable aviation fuels. *Progress in Aerospace Sciences, 136*, 100876. https://doi.org/10.1016/J.PAEROSCI.2022.100876
209. Su-ungkavatin, P., Tiruta-Barna, L., & Hamelin, L. (2023). Biofuels, electrofuels, electric or hydrogen?: A review of current and emerging sustainable aviation systems. *Progress in Energy and Combustion Science, 96*, 101073. https://doi.org/10.1016/J.PECS.2023.101073
210. Mawhood, R., Gazis, E., de Jong, S., et al. (2016). Production pathways for renewable jet fuel: A review of commercialization status and future prospects. *Biofuels, Bioproducts and Biorefining, 10*, 462–484. https://doi.org/10.1002/bbb.1644
211. Gutiérrez-Antonio, C., Gómez-Castro, F. I., de Lira-Flores, J. A., & Hernández, S. (2017). A review on the production processes of renewable jet fuel. *Renewable and Sustainable Energy Reviews, 79*, 709–729. https://doi.org/10.1016/J.RSER.2017.05.108
212. Wei, H., Liu, W., Chen, X., et al. (2019). Renewable bio-jet fuel production for aviation: A review. *Fuel, 254*, 115599. https://doi.org/10.1016/J.FUEL.2019.06.007
213. Wang, W. C., & Tao, L. (2016). Bio-jet fuel conversion technologies. *Renewable and Sustainable Energy Reviews, 53*, 801–822. https://doi.org/10.1016/J.RSER.2015.09.016
214. Voor, K., & Programma, B. (2021). *Royal NLR-Netherlands aerospace Centre Feedstocks for sustainable aviation fuels in The Netherlands: A review of feedstock sustainability and availability and identification of knowledge gaps for policy making.*
215. Sahoo, S., Zhao, X., & Kyprianidis, K. (2020). A review of concepts, benefits, and challenges for future electrical propulsion-based aircraft. *Aerospace, 7*, 1–69. https://doi.org/10.3390/aerospace7040044
216. Gegg, P., Budd, L., & Ison, S. (2014). The market development of aviation biofuel: Drivers and constraints. *Journal of Air Transport Management, 39*, 34–40. https://doi.org/10.1016/j.jairtraman.2014.03.003
217. Alao, K. T., Gilani, S. I.-H., Sopian, K., et al. (2024). Biomass and organic waste conversion for sustainable bioenergy: A comprehensive bibliometric analysis of current research trends and future directions. *International Journal of Renewable Energy Development, 13*, 750–782. https://doi.org/10.61435/ijred.2024.60149
218. Okeke, E. S., Ejeromedoghene, O., Okoye, C. O., et al. (2022). Microalgae biorefinery: An integrated route for the sustainable production of high-value-added products. *Energy Conversion and Management: X, 16*, 100323. https://doi.org/10.1016/j.ecmx.2022.100323

219. Sivaramakrishnan, R., Suresh, S., Kanwal, S., et al. (2022). Microalgal biorefinery concepts' developments for biofuel and bioproducts: Current perspective and bottlenecks. *International Journal of Molecular Sciences, 23*, 1–25. https://doi.org/10.3390/ijms23052623
220. Foley, G. (2016). Reflections on interdisciplinarity and teaching chemical engineering on an interdisciplinary degree programme in biotechnology. *Education for Chemical Engineers, 14*, 35–42. https://doi.org/10.1016/j.ece.2015.11.002
221. Iles, A., & Mulvihill, M. J. (2012). Collaboration across disciplines for sustainability: Green chemistry as an emerging multistakeholder community. *Environmental Science & Technology, 46*, 5643–5649. https://doi.org/10.1021/es300803t
222. Oloruntobi, O., Mokhtar, K., Gohari, A., et al. (2023). Sustainable transition towards greener and cleaner seaborne shipping industry: Challenges and opportunities. *Cleaner Engineering and Technology, 13*, 100628. https://doi.org/10.1016/j.clet.2023.100628
223. Panoutsou, C., Germer, S., Karka, P., et al. (2021). Advanced biofuels to decarbonise European transport by 2030: Markets, challenges, and policies that impact their successful market uptake. *Energy Strategy Reviews, 34*, 100633. https://doi.org/10.1016/j.esr.2021.100633
224. Kumar, T., Basakran, G. K., Marsuki, M. Z., et al. (2023). Exploring socioeconomic and political feasibility of aviation biofuel production and usage in Malaysia: A thematic analysis approach using expert opinion from aviation industry. *Green Energy and Intelligent Transportation, 2*, 100111. https://doi.org/10.1016/j.geits.2023.100111
225. Malode, S. J., Prabhu, K. K., Mascarenhas, R. J., et al. (2021). Recent advances and viability in biofuel production. *Energy Conversion and Management: X, 10*, 100070. https://doi.org/10.1016/j.ecmx.2020.100070
226. Ebadian, M., van Dyk, S., McMillan, J. D., & Saddler, J. (2020). Biofuels policies that have encouraged their production and use: An international perspective. *Energy Policy, 147*, 111906. https://doi.org/10.1016/j.enpol.2020.111906
227. Khujamberdiev, R., & Cho, H. M. (2024). Biofuels in aviation: Exploring the impact of sustainable aviation fuels in aircraft engines. *Energies, 17*, 1–17. https://doi.org/10.3390/en17112650
228. Song, L., & Wen, Y. (2023). Financial subsidies, tax incentives and technological innovation in China's integrated circuit industry. *Journal of Innovation and Knowledge, 8*, 1–14. https://doi.org/10.1016/j.jik.2023.100406

Chapter 2
Lignocellulosic Biomass to Sustainable Aviation Fuel

Ling Tao, Calvin Mukarakate, Thomas D. Foust, and Zia Abdullah

1 Introduction

Transportation is one of the largest contributors to greenhouse gas (GHG) emissions, accounting for 28% of total emissions in 2021 in the United States (USA) and 14% globally [1]. Electrification shows considerable promise for reducing GHG emissions in many sectors of transportation, especially the light- and mid-duty sectors. In the transportation sectors that are difficult to electrify, GHG emissions associated with commercial air travel are significant, at 10% of the domestic transportation emissions and 3% of the global GHG emissions [2]. The US Energy Information Administration (EIA) projections show that even with modest annual growth, air transportation is expected to almost double by 2050 [3, 4]. Some aggressive predictions show electrification and hydrogen (H_2) being viable for regional flights by 2030. However, the bulk of carbon dioxide (CO_2) emissions are from larger passenger and freight jet aircraft [5, 6]. These aircraft will be difficult to electrify or power by H_2 because of the longer distance and thus energy requirements for these flights.

Technology projections do not show electrification and H_2 power being commercially viable for these flights until mid-century. Hence, sustainable aviation fuel (SAF) with a much lower GHG emission profile compared to conventional fossil jet fuel is the only option for existing airplanes and is expected to play a key role in the near- and mid-term decarbonization of commercial air travel [7]. Because of the long service life of commercial jet airplanes (20–30 years on average) and hence the long turnover time of the commercial fleet, SAF needs to be "drop-in," meaning it must meet existing standards for commercial jet fuel. By definition, SAF must demonstrate a minimum of 50% reduction in life-cycle GHG emissions compared to conventional jet fuel.

L. Tao (✉) · C. Mukarakate · T. D. Foust · Z. Abdullah
National Renewable Energy Laboratory (NREL), Golden, CO, USA
e-mail: Ling.Tao@nrel.gov

M. Aslam et al. (eds.), *Sustainable Aviation Fuels*, Sustainable Aviation,
https://doi.org/10.1007/978-3-031-83721-0_2

The SAF production industry is a very fledgling industry that is just getting off the ground, and production was 15.8 million gallons/year in 2022 [8]. Although this is considerable growth from a volume of 1.9 million gallons/year in 2019, it is still less than 0.1% of commercial jet fuel use, and to put commercial air travel on a path toward decarbonization, the growth rate of SAF needs to be greatly accelerated. To address this concern, the US government issued an "SAF Grand Challenge." The SAF Grand Challenge is a US government-wide approach to working with the industry to reduce cost, enhance sustainability, and expand production to achieve 3 billion gallons per year of domestic SAF production that achieves a minimum of 50% reduction in life-cycle GHG emissions compared to conventional fuel by 2030 and 100% of projected aviation jet fuel use, or 35 billion gallons of annual production, by 2050.

Figure 2.1 illustrates the scale-up rates required to achieve the SAF Grand Challenge rates of SAF production. To achieve the 2030 target of 3 billion gallons per year (BGPY) from the current level of 23 million gallons (in 2023) per year, production must be scaled up 130 times by 2030, and then, from 2030 to 2050, the scale-up factor would need to be 12 times over 20 years. To put this in perspective of investment capital required, if all this 30-BGPY SAF capacity was produced from greenfield biorefineries and assuming a conservative $10/gallon capital cost, $30 billion of capital would be required, which presents a significant challenge toward achieving the goal.

Given the significant and complex challenges associated with achieving the SAF Grand Challenge, the US Department of Energy (DOE), US Department of Agriculture (USDA), US Environmental Protection Agency (EPA), US Department of Transportation, and Federal Aviation Administration (FAA) collaborated to develop a comprehensive strategy for achieving the objectives and outlined this in the *SAF Grand Challenge Road*map [6]. The roadmap outlines a "whole-of-government" approach with coordinated policies and specific activities that should be undertaken by federal agencies to support the achievement of both the 2030 and 2050 goals of the SAF Grand Challenge. The roadmap lays out six action areas spanning all activities with the potential to impact the SAF Grand Challenge

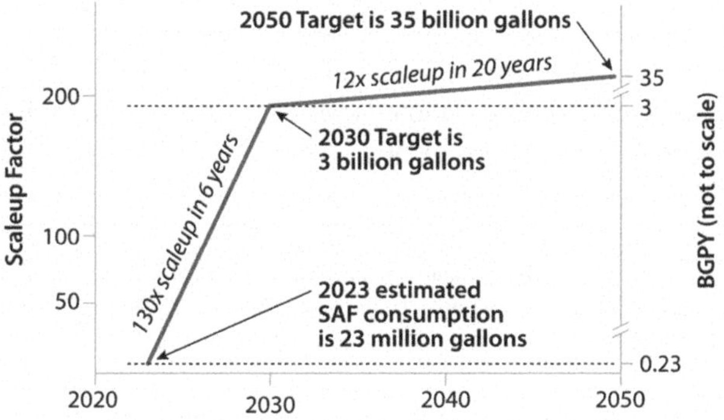

Fig. 2.1 The SAF grand challenge requires 130 times scale-up in production in the next 7 years and 12 times scale-up in the following 20 years

objectives of expanding SAF supply and end use, reducing the cost of SAF, and enhancing the sustainability of SAF:

1. Feedstock innovation
2. Conversion technology innovation
3. Building supply chains
4. Policy and valuation analysis
5. Enabling end use
6. Communicating progress and building support

Within each of the six action areas are work streams that define critical topics to be addressed. Given the limited time to meet the 2030 goal of 3 BGPY, the roadmap calls for an immediate focus on commercially ready or nearly commercially ready conversion technologies and feedstocks. Lipid-based pathways (fats, oils, and greases [FOG]) are expected to be the primary fuel pathway, with a smaller contribution from waste, forest and agricultural residue, and alcohol pathways leading up to 2030.

In summary, producing SAF to meet the Grand Challenge goals is a great opportunity for economic development in the United States. SAF will decarbonize flights within the United States and enable carriers to fly globally with a lower carbon footprint. There are some challenges, such as the need to produce renewable diesel (RD) as well, which will put pressure on the feedstock supply, but there are multiple pathways approved by ASTM that provide considerable feedstock flexibility. SAF biorefineries will have the potential benefit of debottlenecking the United States's capacity-constrained fuel distribution systems if they can be built closer to airports. There are opportunities to make policies (at the state or federal level) to encourage more stable investments for producing SAF. The Bioenergy Technologies Office, Loan Programs Office, and other federal programs such as the FAA's Fueling Aviation's Sustainable Transition are critically important at this nascent stage of the industry as conventional lenders tend to be risk-averse.

2 What Is SAF?

2.1 Definitions of SAF

Because SAF is being developed for commercial aviation and because commercial aviation is global in nature, SAF needs to meet both the US and international definitions for SAF. For the United States, the Commercial Aviation Alternative Fuels Initiative (CAAFI) [9] defines SAF as a Jet A fuel blendstock that:

1. Reduces net life-cycle CO_2 emissions from aviation operations.
2. Elevates the sustainability of aviation by outperforming petroleum-based jet fuel in terms of economic, social, and environmental factors.

3. Enables drop-in jet fuel production from multiple feedstocks and conversion processes, so no changes are required in aircraft or engine fuel systems, distribution infrastructure, or storage facilities. As such, SAF can be mixed interchangeably (is fungible) with existing jet fuel.

Internationally, the International Civil Aviation Organization (ICAO) defines SAFs as renewable or waste-derived aviation fuels that meet the following sustainability criteria [10]:

1. Achieve net GHG emission reductions of at least 10% compared to the baseline life-cycle emission values for aviation fuel on a life-cycle basis.
2. Will not be made from biomass that is either obtained from land converted after January 1, 2008, that was primary forest, wetlands, or peat lands or contributes to degradation of the carbon stock in primary forests, wetlands, or peat lands as these lands all have high carbon stocks.
3. In the event of land use conversion after January 1, 2008, as defined based on the Intergovernmental Panel on Climate Change land categories, direct land-use change emissions will be calculated. If direct land-use change GHG emissions exceed the default "induced" land-use change value, the direct value will replace the default induced value.

It should be noted that the CAAFI definition does not state a specific target for GHG reduction relative to Jet A, and the ICAO definition specifies a target of 10% reduction relative to Jet A. The SAF Grand Challenge [6] targets a GHG reduction of 50% relative to Jet A.

2.2 Fuel and Chemical Properties of SAF

ASTM D1655 (*Standard Specification for Aviation Turbine Fuels*, ASTM 2019a) provides the specifications for jet fuel, and ASTM D4054 (*Standard Practice for Evaluation and Approval of New Aviation Turbine Fuels and Fuel Additives*, ASTM 2019b) defines the steps for qualification and approval of new aviation turbine fuels. ASTM D7566 (*Standard Specification for Aviation Turbine Fuels Containing Synthesized Hydrocarbons*, ASTM 2019c) defines the specification for SAF. An important point is that meeting all these specifications ensures that the fuel is fully fungible, which is critical for the market acceptance of SAF.

Due to the high operability and performance demands of jet fuel, the properties of SAF are very tightly specified. For the bulk fuel properties, the category requirements for SAF are (1) performance, (2) operability, and (3) drop-in. The precise specifications for all three bulk fuel categories are outlined in Holladay et al. [7]. In addition to bulk fuel properties, trace fuel properties are also important to be met for SAF. Where bulk fuel properties primarily address performance, operability, and drop-in aspects, trace fuel properties are associated with material and component

wear, corrosion, thermal stability, and conductivity. These properties are very important for engine and fuel system durability, reliability, and operability.

Like gasoline and diesel fuels, jet fuel is a blended mixture of several hundred different hydrocarbon molecules defined by boiling points. Jet fuel has an initial boiling point at atmospheric pressure of about 125 °C and a final boiling point of about 290 °C and ranges from approximately eight carbon atoms to 16. For comparison, gasoline has an initial boiling point at atmospheric pressure of about 35 °C and a final boiling point of about 200 °C, and diesel has an initial boiling point at atmospheric pressure of about 150 °C and a final boiling point of about 380 °C [7]. Molecules in gasoline fuel range from those containing four carbon atoms to those containing 12, and molecules in diesel fuel range from those containing eight carbon atoms to those containing 23.

Figure 2.2 depicts this graphically where jet fuel is the middle distillate product between gasoline and diesel. As shown in Fig. 2.2, the boiling point range of gasoline and jet fuel overlaps significantly, and there is an extensive overlap in the boiling point range between jet fuel and diesel. Refineries specifically exploit the overlap between jet fuel and diesel by selecting which product to favor, depending on market conditions and incentives.

If the higher portion of the diesel fraction (boiling point >290 °C) is to be converted to SAF, hydrocracking will be required. This will require hydrogen and higher-severity operations, thus increasing costs and decreasing carbon yields given that some portion of the diesel fraction will be over-cracked to the gasoline fraction. In general, renewable diesel has a higher market value and slightly more favorable policy incentives than SAF, so hydrocracking the higher end of the diesel fraction to SAF is generally not a good option.

As stated earlier, SAF properties are specified by bulk and trace fuel properties, but because these properties are highly correlated to the chemical properties, understanding the chemical composition of Jet A is important for the development of SAF. Figure 2.3 shows an average composition of Jet A, represented by four

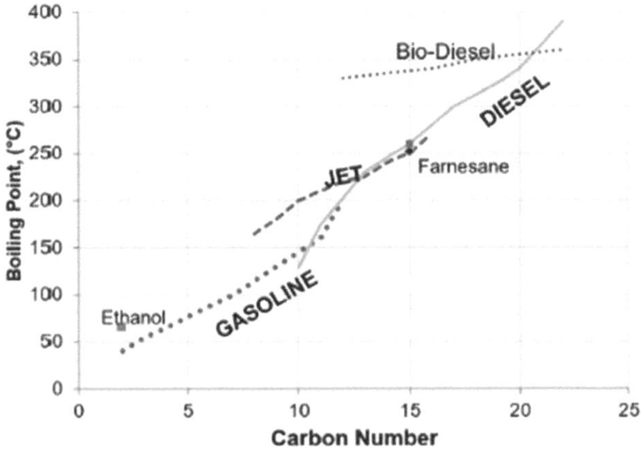

Fig. 2.2 Carbon numbers and boiling points for gasoline, jet, and diesel fuels [7]

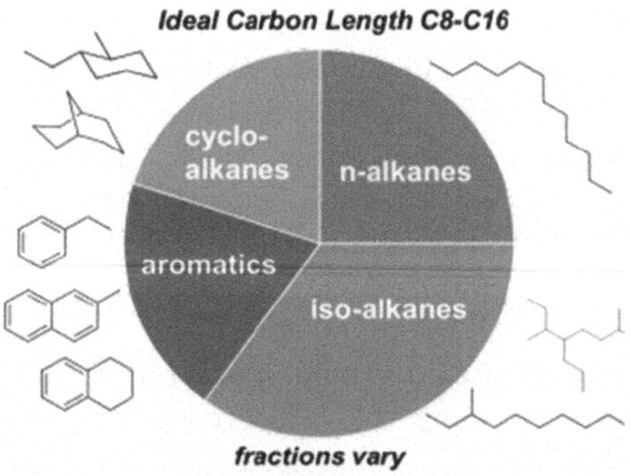

Fig. 2.3 Composition of average Jet A [7, 11]

families of hydrocarbons: aromatics (~18%), cycloalkanes (~32%), iso-alkanes (~30%), and n-alkanes (~20%) [7, 11].

The composition of Jet A is very relevant for the development and market introduction of SAF. As will be covered in subsequent sections of this chapter, different SAF production pathways tend to produce specific chemical families, and as covered earlier, because fuel properties are correlated to chemical composition, certainly properties could come out of balance by blending in too high a percentage of a certain chemical family. Thus, this could impede the market acceptance of certain SAF blendstocks on the pathways to high volumes of SAF.

A possible market driver for imparting a market pull for SAF is producing fuel with performance-advantaged properties or lower emissions. Specific energy and energy density improvements have immediate benefits by decreasing the volume/weight of fuel required, and this can be directly correlated to increases in payload capacity. N-alkanes and iso-alkanes impart high specific energy and higher energy density, and their concentration could be increased up to the limit of cold flow limitations. Aromatics, although required at some percentage to ensure O-ring swelling to prevent fuel leakages, have lower specific energy and contribute to higher soot emissions, so it is desirable to limit their content to the minimum required for sufficient O-ring swelling.

2.3 *Approved Pathways for SAF*

Most of the anticipated growth in hard-to-electrify transportation is attributed to increased air travel. Despite planned improvements in aircraft technology and operations, as a reference case, the FAA projects that domestic jet fuel demand will grow from 23 BGPY in 2019 to 34 BGPY by 2050 [12]. The EIA [13], US Environmental Protection Agency (EPA) [14], and FAA [12] estimate that nearly 70 BGPY of liquid transportation fuels will be needed in hard-to-electrify aviation, marine, rail, and off-road sectors by 2050. This initial analysis estimates optimistic biofuel production and GHG emission reduction potentials for hard-to-electrify sectors, providing

insight into meeting potential demands predicted by FAA, EPA, and EIA. Note that these are not projections but should be viewed as the upper bound. There are many factors that could limit this potential. Likewise, there are feedstocks and processes not considered that could increase this potential. Table 2.1 shows the eight ASTM-approved SAF pathways with blending levels from 5 to 50% [15].

SAF, or "bio-jet" as it was initially called, has been under development as an aviation decarbonization option for well over a decade. The FAA certifies commercial jet airplanes to operate on jet fuel that meets ASTM specifications. Any new fuel must meet ASTM specifications and be approved through the ASTM D4054 process [7]. To date, ASTM has approved seven pathways, listed in Annexes 1–7 of ASTM D7566 plus ASTM D1655 Annex A1, which allows 5% coprocessing of

Table 2.1 List of ASTM-approved SAF pathways

Conversion technology	ASTM name	Technical identity; feedstock	Commercial developer	Blending limitation (volume)
Fisher–Tropsch synthetic paraffinic kerosene (FT-SPK)	ASTM D7566 Annex 1	Thermochemical; biomass, coal, natural gas	Fulcrum Bioenergy, Red Rock Biofuels, SG Preston, Velocys, Sasol	50%
Hydroprocessed esters and fatty acids synthetic paraffinic kerosene (HEFA-SPK)	ASTM D7566 Annex 2	Chemical; triglyceride vegetable oil, cooked oils, grease, animal fats	World Energy, Honeywell UOP, Neste Oil	50%
Hydroprocessed fermented sugars to synthetic iso-paraffin (HFS-SIP)	ASTM D7566 Annex 3	Biochemical; cellulosic, starchy sugar	Amyris	10%
Fisher–Tropsch synthetic kerosene aromatics (FT-SKA)	ASTM D7566 Annex 4	Thermochemical; biomass, coal, natural gas	Sasol	50%
Alcohol-to-jet synthetic paraffinic kerosene (ATJ-SPK)	ASTM D7566 Annex 5	Biochemical; cellulosic ethanol or iso-butanol	Gevo, Honeywell UOP, LanzaTech, Byogy	50%
Catalytic hydrothermolysis jet (CHJ)	ASTM D7566 Annex 6	Chemical; vegetable oils such as soybean, jatropha, camelina oil	ARA and Chevron Lummus Global	50%
Hydrocarbon hydroprocessed esters and fatty acids synthetic paraffinic kerosene (HC-HEFA-SPK)	ASTM D7566 Annex 7	Biochemical; algae	IHI Corporation	10%
ATJ derivative starting with the mixed alcohols synthetic kerosene aromatics (ATJ-SKA)	D7566 Annex 8	Biochemical; mixed alcohols	Not available	Not available

fatty acid esters or Fischer–Tropsch (FT) hydrocarbon streams with conventional jet fuel [7, 16]. These ASTM D7566-22 approved pathways are as follows:

- Annex A1 (FT-SPK) was approved in June 2009 for up to a 50% blend with petroleum-derived jet fuel. FT-SPK is a mixture of iso- and *n*-alkanes derived from synthesis gas using the FT process. Syngas can be produced by reforming natural gas or by gasifying coal or biomass.
- Annex A2 (HEFA-SPK) was approved in July 2011 for up to a 50% blend with petroleum-derived jet fuel. The molecular composition of HEFA-SPK is similar to FT-SPK, consisting of iso- and *n*-alkanes. The alkanes are the product of hydrotreating esters and fatty acids from fats, oils, and greases and from oilseed crops or algae.
- Annex A3 (SIP, hydroprocessed fermented sugar to synthetic iso-paraffins) was approved in June 2014 for up to a 10% blend with petroleum-derived jet fuel. Unlike SPK from HEFA or FT, this is a single molecule, a 15-carbon hydrotreated sesquiterpene called farnesane, produced from the fermentation of sugars. Today, the fermentation is done commercially from sugar cane juice and is used in higher-value applications, most commonly in personal care products.
- Annex A5 (alcohol-to-jet [ATJ]-SPK) was approved in April 2016 for SPK from iso-butanol (30% blend with petroleum) and expanded in April 2018 for SPK from ethanol and for fuel blends of up to 50% with petroleum. ATJ-SPK consists of iso-alkanes of 8, 12, or 16 carbons when starting from iso-butanol. The iso-alkanes are highly branched and have lower DCNs than FT or HEFA, based on data from Gevo, Inc. The carbon number is broadened and the branching level can be significantly reduced, leading to a DCN similar to FT and HEFA when starting from ethanol.[1]
- Annex A6 (Applied Research Associates catalytic hydrothermolysis jet, or ARA CHJ) was approved in January 2020 as a 50% blend. The fuel is produced from lipids using a supercritical hydrothermal process, creating a blendstock that contains all four hydrocarbon families: *n*-, iso-, and cyclo-alkanes and aromatics.
- Annex A7 (HC-HEFA), synthesized paraffinic kerosene from hydroprocessed hydrocarbons, esters, and fatty acids, was approved in 2020 as a 10% blend. This is specifically for lipids from *B. braunii* algae that have been hydrocracking/hydroisomerizing to remove all oxygen and saturate double bonds. The product is rich in iso-alkanes. This is the first approval through the fast-track process.
- ASTM D1655-20b allows coprocessing of up to 5% mono-, di-, and triglycerides; free fatty acids; and fatty acid esters or up to 5% of FT hydrocarbons. Hydrocracking/hydrotreating and fractionation are required. No other coprocessing in refineries is allowed for jet fuel.
- **New, to be approved, Annex A8 (ATJ-SKA)** ATJ-SPK with aromatics, was submitted by Swedish Biofuels AB and passed the ballot in June 2023. The fuel

[1] Pacific Northwest National Laboratory internal data; see also alcohol-to-jet ASTM report, https://www.astm.org/Standards/D7566.htm.

is produced from a single or a combination of C2 to C5 alcohols and may be approved up to a 50% blend in 2023 [17].

Several fuels are somewhere in the D4054 approval pipeline. These fuels include SPK (*n*- and iso-alkanes), cycloalkanes, aromatics, and mixtures of the four hydrocarbon families. New SAF blendstock approvals are currently being pursued by companies in exploratory discussions (Global BioEnergies, POET, SBI BioEnergies, Vertimas, and GSR/GTI) and some companies in phase 1 (Shell, Byogy, and Swedish Biofuels), phase 1 review (Neste, for green diesel), or phase 2 (Virent, for aromatics).

3 Carbon-Based Feedstocks

3.1 Types of Lignocellulosic Biomass and R&D Needs

R&D needs for feedstocks can be categorized based on the type of feedstock. The high-level goal of feedstock R&D is to prepare the feedstock for processing to SAF and any other coproduct associated with that conversion pathway. This section will focus feedstocks on specific pathways, as well as the R&D needs to produce, collect, and transport.

Because of the commercial deployment of HEFA-based SAF in various airports across the world [9], HEFA has a technology readiness level (TRL) up to 9, a benchmark developed by the US National Aeronautics and Space Administration in the 1980s. This pathway is commonly referred to as the hydroprocessed esters and fatty acids (HEFA) process. The hydroprocessing typically involves two catalytic steps: (1) hydrodeoxygenation and (2) isomerization/hydrocracking into the finished SAF. However, the availability of triglyceride feedstock and competition with the production of RD (which has a high yield compared to SAF) are the major issues for HEFA-based technologies. Feedstocks in this category include vegetable oils (soybean, rapeseed, palm oil, etc.), waste oils (fats, oils, and greases [FOG]), algal oils, and pyrolysis oils. Soybean, rapeseed, and palm oils have been viewed as the major feedstocks for biodiesel and RD production [18].

Vegetable Oil Soybean oils account for 74% of the vegetable oils produced in the United States [19]. Soybean oil has been used extensively in the United States for biodiesel production, using 30.5% (in 2009) to 66.7% (in 2012) of the total soybean oil production [19–21]. The demand for soybean oil has been increasing in the past 10 years due to several factors, including increased production of biodiesel as well as RD. This increase in demand has led to fluctuating prices, reaching an all-time high in the early 2020s, but the USDA projects a price decrease in the mid-2020s, essentially remaining flat until about 2032–2033 [22].

Other vegetable oils produced in the United States in smaller quantities compared to soybean oil are canola, sunflower, and cottonseed oils. Rapeseed oil is the

main feedstock used for biodiesel production in Europe, with approximately 58,000 barrels produced per day [18]. Production of palm oil has increased recently, and 90% of that comes from Indonesia and Malaysia [18]. Biodiesel production has expanded based on the abundant palm oil resource in this area. However, the use of soybean, palm, and rapeseed oils as SAF feedstocks could lead to a large uncertainty in the amount of GHG emissions due to direct or indirect land-use change [18]. Palm oil biodiesel has been rejected by the EPA as not suitable for addition to the renewable fuel program due to high GHG emissions [23]. While DOE predicts lignocellulosic availability close to a billion tons across the United States, there is less information about the availability of triglyceride feedstock.

FOG FOG feedstocks are excellent sources of triglycerides that can be fairly straightforward to hydroprocess into SAF. Depending on the quality of the vegetable oil, the FOG pretreatment step may be required to eliminate or greatly lower the number of contaminants, which will improve the conversion catalyst life span as well as the process control for better product quality. FOG is a collective term to include animal byproducts and waste grease generated at food handling operations. Animal fats can be obtained from slaughterhouse and livestock farm waste. Used cooking oil can be collected from commercial and industrial cooking operations as well as trap/interceptor grease recovered from traps installed in the sewage lines of restaurants/food processing plants and wastewater treatment plants [24]. Production of FOG from all the various sources in the United States has fluctuated about 5% from an average of 6.3 million tons/year for about the past decade [25]. Production rates of FOG are expected to increase to eight million tons/year by mid-century, mostly driven by increases in livestock and poultry production. Prices for FOG have essentially doubled in the early 2020s, up to $1400 per ton for used cooking oil and $1800 per ton for inedible tallow. Projections are expected to continue to increase slightly and then remain constant for the foreseeable future. Increasing demand is mostly driven by increases in the usage of FOG for biofuel production.

There are no major R&D needs for the HEFA process using vegetable oils or FOG feedstocks because it is already a commercially proven process. An argument could be made that vegetable oil prices and FOG prices are too high for the overall economic viability of producing SAF, but this is more of an over-demand issue and logistical issue versus an R&D need.

Starch Starch to ethanol is a proven commercial technology, with the overwhelming majority of fuel production ethanol coming from corn grain. The biggest issue with using corn grain for ethanol production for the ATJ pathway for SAF is the relatively high carbon intensity of ethanol from corn grain compared to other renewable biomass and waste carbon feedstocks. Considerable progress has been made over the past several decades to reduce the carbon intensity of corn ethanol, and projections are for this trend to continue, so corn grain ethanol to SAF can be a contributor to SAF projection. Corn prices are a major factor in the cost of SAF from the ATJ pathway from corn grain, and they have fluctuated extensively since the early 2000s. Projections are for prices to somewhat stabilize over the next sev-

eral decades and increase in proportion to inflation [26]. Again, as is the case for vegetable oil and FOG production, corn grain production is a proven commercial process with more than a 100-year history, so there are really no R&D needs.

The major issue with ATJ is the supply of alcohol, which can be produced by fermentation or thermochemical routes. Renewable alcohols can be produced from feedstocks that contain fermentable sugars, such as sugar cane and beet, which are rich in sucrose [27], including starch and lignocellulosic biomass sugars. Feedstocks for producing alcohols are the most important consideration when considering economic feasibility. In the United States, corn grain is used as the principal feedstock for starch-derived alcohol production [27].

Lignocellulosic feedstock sugars are derived from the hydrolysis of hemicellulose and cellulose biomass, including agricultural residues and dedicated energy crops. The fundamental difference between the constituents of various feedstock is the length of the carbon chain. The components of lignocellulosic feedstocks are dominated by 5-carbon and 6-carbon elements (~60–80 wt %), followed by lignin (~30 wt %) [28]. Lignin, another feedstock option with less oxygen than carbohydrates, contains phenolic moieties that provide the aromatic components of SAF [29].

Agricultural Residues and Dedicated Energy Crops This subcategory refers to the biomass residual that is available for use after a food crop, in most cases a grain crop such as corn or wheat, has been harvested. Some amount of residue is necessary for soil health and erosion control, specific to the type of crop and geographical conditions [30]. The 2016 billion-ton study [31] provides a good summary of the number of agricultural residues projected to be available after accounting for soil health and erosion concerns at price points of ≤ $40, $60, and $80 per ton out to 2040. As expected, higher amounts are available at higher price points and in later years, with an upper limit of 214 million tons/year at ≤$80/ton in 2040. A roadmap was developed [32] for producing, collecting, and transporting agricultural residues to biorefineries for the production of fuels and other products.

According to DOE's billion-ton study [31], agricultural and forestry residues and dedicated energy crops are the largest potential sources of biomass feedstock for SAF production, with estimates ranging from 140 to 680 million metric tons of available feedstock. The R&D needs and the price of these feedstocks have been extensively studied, as summarized in [31]. An update of this important study is underway and is scheduled for publication in 2024. Among agricultural residues, corn stover is the most abundant, produced primarily in the Midwest region [33]. The yield ratio of corn grain to corn stover is 1:1, which is why the productivity of both feedstocks is also in a similar range [33, 34].

For the production of gas-derived fuels, the essential requirement for the feedstock is syngas with the proper ratio of the carbon and hydrogen contents [35]. A wide variety of municipal solid waste (MSW), construction waste, forestry residue, and dedicated energy crops, which contain abundant carbon and H_2, can be used for this purpose [35]. Although biomass is renewable, different types of biomass

feedstocks have different carbon contents. For example, woody biomass has a carbon content of approximately 50%. Grass crops or agricultural residues contain around 45% carbon [36]. The biomass-to-liquid process via Fischer–Tropsch synthesis is a thermochemical route that produces synthetic fuels from biomass. The choice of feedstock and pretreatment steps are closely related to the economic results.

MSW and Construction Waste Waste biomass from municipal waste and other sources can be a good source of biomass for SAF production. In the United States, the US EPA collects [37] and reports the amount of waste biomass available from MSW and other sources. EPA projections show that up to 290 million metric tons of waste biomass could be available from these sources. Prices can vary significantly from technically negative to much higher costs depending on the quality and how well the biomass is sorted. The challenges of sorting MSW for a good quality biomass feedstock for SAF production have been extensively covered for sorting MSW for biofuel production [38] and will not be discussed here. There are really no R&D needs associated with MSW, construction waste, and other sources of waste biomass beyond sorting and collection logistics.

Forestry Residues Forestry residues are also a high-volume feedstock for SAF production. Similar to agricultural residues, the billion-ton study [31] does a good job of capturing the amount of forestry residues available at price points of ≤ $40, $60, and $80 per ton out to 2040. The upper bound is 116 million tons at the ≤ $80 price point in 2040. The sustainable harvest of forestry residues and whole-tree biomass has been extensively studied [39] to identify the best harvesting practices at the lowest price point while positively contributing to forest health and biodiversity.

Dedicated Energy Crops This is the highest potential feedstock for SAF production. The billion-ton study shows a high potential of 729 million tons/year for herbaceous energy crops and 125 million tons/year for woody energy crops at the price point of ≤$80 per ton in 2040. The production, harvest, and collection of herbaceous energy crops [40] and woody energy [41] have been extensively studied to produce energy crops for SAF production at the lowest price point while accounting for important sustainability parameters such as soil health, biodiversity, and wildlife habitat.

Another renewable feedstock is biogas produced from anaerobic digestion of organic matter, such as animal manure, landfills, and wastewater, or from co-digestion of mixing liquid and solid biomass [42–44]. After purification by means of CO_2 adsorption, the methane product can serve as the gas feedstock for making syngas and then converting syngas to liquid fuels [44].

3.2 Type of Preprocessing Needed

The challenge with lignocellulosic biomass is the high oxygen content, which is removed as water, CO_2, or CO [45]. The chemistry of the removal process makes the conversion of lignocellulosic biomass to drop-in hydrocarbon fuel not only costly but also causes carbon losses [45] and hence low yields on a carbon basis. This fundamental chemistry challenge with lignocellulosic feedstocks is also the reason why triglyceride-based feedstocks with low oxygen content are more expensive due to their comparatively easier processability. However, as DOE's billion-ton update report suggests, the availability of triglyceride feedstock is much less than that of lignocellulosic feedstock for a future sustainable bioeconomy [31]. Table 2.2 enlists some of the major features of various types of feedstocks that have the potential to be part of a biomass-based SAF supply chain.

Moisture and ash contents also play a significant role in processing lignocellulosic feedstock into transportation fuel [51]. Due to the recalcitrant nature of the lignocellulosic feedstocks, pretreatment is required to condition the raw biomass for follow-on processing. Several pretreatment processes are available, including dilute sulfuric acid pretreatment, steam explosion, organosolv, hot water, and lime [52–56].

Pretreatment of triglycerides depends on the type of feedstock. Vegetable oil will not need any pretreatment before the conversion process, but used cooking oil or grease will probably need pretreatment in terms of contaminant removal, moisture content reduction, and free fatty acid content minimization [57].

In an algal biorefinery process, pretreatment is targeted to disrupt the cell walls to release the proteins, lipids, and carbohydrates. Acid hydrolysis, enzymatic hydrolysis, and thermal and mechanical pretreatment are some of the well-known pretreatment steps for algal biorefinery processes [58].

Productivity and fuel yield are two important indicators of how a potential feedstock can have a sustainable impact on biorefinery processes. Productivity indicates the average annual production of a feedstock in a specific region, which can depend on management such as site preparation, weed control, and fertilization [59]. Ochuodho et al. reported productivity of loblolly pine ranging from 1.6 to 8.5 dry MT per acre per year in Virginia for various management conditions [59].

Energy crops such as switchgrass can have a significant impact on SAF production and a productivity range of approximately 3.5–7 dry MT per acre per year [46]. The productivity of triglyceride-based feedstock is higher than that of lignocellulosic and starchy crops, as shown in Table 2.2. Vegetable oils such as soybean, rapeseed, canola, and jatropha oils are good candidates for their higher productivity, ranging from 18 to 636 gallons/acre [47].

On the other hand, algal oil is known for significantly higher productivity, ranging from 6000 to 15,000 gallons/acre [47]. The typical fuel yield for lignocellulosic feedstock ranges from 40 to 83 gal per dry MT [51, 60, 61], while triglyceride-based fuel yield is 170–216 gal per MT [49]. But even with high fuel yield, the actual availability or potential future availability of triglyceride feedstock is less than that of lignocellulosic feedstock (Table 2.2).

Table 2.2 High-level features of potential feedstocks toward SAF implementation

Category	Agricultural/forest residues/ energy crops	Starch crop	Vegetable oil/FOG	Algal oil	Municipal solid waste
Constituents	Cellulose, hemicellulose, lignin	Starch	Triglycerides	Triglycerides	Cellulose, hemicellulose, lignin
Preprocessing/ pretreatment	Particle size reduction, drying, solvent extraction, acid dilution	Particle size reduction	Oil extraction, degumming, filtration	Oil extraction	Sorting
Productivity (dry MT/ acre/year)	3–4.5/1.6–8.5/3.5–7 [46]	3–4.5 (corn grain) [26]	18–636 gal/acre [47]	6000 – 15,000 gal/ acre [47]	–
Fuel yield (gal per dry MT feed)	76–83/47	80 of corn grain [48]	170–216/210[49]	100 gal/dry MT feed	21.3
Value-added intermediates	Sugar mixture, levoglucosan, lignin	Distiller-dried grains	Propane	Protein	
Availability (million MT)	139–679	109	4.1 [19]/0.9		292 [37]
High oxygen content, low product yield, transportation		Always prioritized as a food crop, high water consumption	Lower availability	Large land area required	
Prices	$110–$125/MT [50]	$215/MT of corn [26]	700–1600/480		
Associated processes toward SAF	FT, pyrolysis, ATJ, direct sugar to hydrocarbon	Direct sugar to hydrocarbon	HEFA, esterification	HC-HEFA	FT, pyrolysis

DOE's billion-ton update provides a detailed analysis of the near-term availability of agricultural and forestry residues, which range from 130 to 760 million MT [31]. However, the historic availability of vegetable oils and FOG is lower. For example, the US Department of Agriculture (USDA) reports a soybean oil availability of 4.1 million MT in 2021 for biofuel production [19]. Municipal solid waste (MSW) has great potential to become feedstock for SAF production facilities. In the United States, the generation of MSW has increased by over 200% in the past 50 years. In 2018, EPA reported a total MSW generation of 292 million MT [37].

The price of lignocellulosic material is relatively uncertain compared to other well-known feedstocks such as soybean oil, FOG, and corn grain. In their billion-ton report [31], DOE adopts an ideal scenario of a price range between $40 and $100 per dry ton based on various levels of yield data. However, the International Civil Aviation Organization reports the current range of lignocellulosic prices between $110 and $125 per dry MT [50]. Currently, the demand for lignocellulosic biomass in biorefineries is unpredictable as it is unknown how much the refiners are willing to pay and because data on such factors as yield, moisture content, bulk density (and other physical properties), and transportation differ from one region to another.

The supply chain of vegetable oils and FOG is well-established due to the food market. The USDA provides a detailed level of pricing on both a quarterly and annual basis for this feedstock [19]. From Table 2.2, it is noteworthy that the prices of lignocellulosic feeds are cheaper than other established feedstocks.

3.3 Biofuel Production Potentials

A recent National Renewable Energy Laboratory (NREL) study [62] demonstrates the biofuel production potential to complementary to the petroleum-based fuels and mitigate GHG emissions in hard-to-electrify transportation sectors by 2050, using a variety of feedstocks.

In this study, dry feedstock and algae feedstock volumes are based on DOE's *2016 Billion-Ton Report* (BT16) 2030 base case [31]. DOE's 2017 *Biofuels and Bioproducts from Wet and Gaseous Waste Streams* report [63] and US Census Bureau projections of population growth are used to estimate volumes of wet waste. Starch-based, oilseed, and waste fat, oil, and grease (FOG) feedstock volumes are based on the Renewable Fuels Association and EIA projections of fuel ethanol, RD, and biodiesel production.

In the base case scenario, this study indicates 858 million tons of sustainable biomass supply [62]. And in an extended supply scenario, the study shows 1279 million tons of sustainable biomass supply [62]. The expanded case scenario includes both energy crops based on the BT16 2040 base case and algae feedstock volume based on DOE's 2017 Algae Harmonization Study [64]. Under the two biomass supply scenarios and coupled with the appropriate conversion pathway for each feedstock, this NREL study predicts that 55 and 80 billion gallons of gasoline

equivalent (GGE) per year of sustainable biofuel could be produced by 2050 [62]. Supplementing equivalent volumes of petroleum-based fuel with sustainable biofuel would result in 410 and 640 million metric tons (MT) of carbon dioxide equivalent (CO_2e) per year of life-cycle GHG reductions—a total 68%–73% reduction in GHG emissions over petroleum-based fuels [62].

4 Conversion Technologies

4.1 Thermochemical Conversion

Thermochemical conversion processes have the capability to deconstruct biomass very rapidly, on the order of seconds, to either a liquid or gaseous phase, which can then be further processed to produce hydrocarbon fuels or chemicals [65]. The rapid conversion reactions and continuous operation enable process intensification, thus lowering capital expenses [66]. Two primary thermochemical conversion processes are fast pyrolysis, which produces primarily a liquid product (bio-oil), and gasification, which produces a gas [67, 68].

The gasification process is more mature for making SAF as two pathways have already been approved. ASTM D7566 Annexes A1 and A4 have approved synthetic paraffinic kerosene (SPK) and synthetic paraffinic kerosene with aromatics (SPK/A), respectively, from the gasification–Fischer–Tropsch pathway for up to 50% blends with petroleum-derived jet fuel. In addition, ASTM D7566 Annex A5 has approved the alcohol-to-jet pathway, which was expanded in 2018, to include ethanol produced from syngas fermentation.

Fast pyrolysis pathways have not been approved by ASTM at the present time; however, they are included in this discussion because these pathways have technical merit and are rapidly being developed. A key challenge with fast pyrolysis pathways has been the difficulties associated with upgrading bio-oil or vapors for extended time on stream without catalyst deactivation [69, 70]. This problem is now being better understood, and several approaches are being investigated for bio-oil upgrading to produce hydrocarbon fuels. There is already a rich literature on R&D in fast pyrolysis [71]. The section below will therefore focus on technologies that are closer to commercial deployment.

4.1.1 Fast Pyrolysis

In the fast pyrolysis process, biomass is fed to a reactor heated nominally to 500 °C under an inert (oxygen-free) environment. Because oxygen is not present, the biomass does not ignite; rather, most of the macromolecular polymers in biomass thermally decompose within 1 or 2 s to a gas phase, most of which is then condensed to a liquid phase (bio-oil) [65]. There is some residual solid material, which does not get pyrolyzed (char), and a non-condensable light gas is also produced [71–74].

Table 2.3 provides approximate distributions of solid, liquid, and gaseous phases from biomass pyrolysis.

To produce SAF, the liquid "bio-oil" phase is the desired product, and its yield must be maximized. Maximum yield of bio-oil is achieved with high biomass heating rates (>1000 °C/s) and reactor temperature between 480 °C and 500 °C [65, 75, 76]. At temperatures below ~480 °C, the production of char increases, and the production of the gas and liquid phases decreases. At temperatures greater than 500 °C, the production of the gas phase increases, and the production of bio-oil and char decreases.

Biomass heating rates are determined by heat transfer, which is limited by (1) ambient temperature in the reactor, (2) heat transfer coefficient on the surface of the biomass particle, and (3) heat transfer through the particle. Ambient temperature in the reactor is maintained by heat transfer media such as sand for fast pyrolysis or a catalyst for in situ catalytic fast pyrolysis [77]. Heat transfer media is maintained at approximately 500 °C. High heat transfer coefficients on the biomass particle surface are achieved by maintaining high turbulence via the use of fluidized bed reactors [78, 79]. Heat transfer through the particle is limited by particle size and moisture. If the particle size is large, the heat transfer rate at the core of the particle is low [79], and if the moisture content is high [80], energy is used to evaporate the water rather than heating up the particle. Particle size is generally limited to 3 mm (1/8″), and moisture is limited to 5% to maintain high heat transfer rates that maximize bio-oil yield. Lower heat transfer rates favor char production.

Although externally heated bubbling fluid bed reactors are used widely in laboratories conducting fast pyrolysis R&D, these reactors have not been successful in commercial operations. One of the problems with using such reactors is char accumulation in the bed during operations. Char tends to act as a cracking catalyst, which reduces bio-oil yield and produces a higher yield of gas [81]. High ash content in the feedstock can make this worse as the ash tends to accumulate in the char, leading to increased cracking rates. Two pyrolysis reactor designs that are being scaled to commercial operation are the circulating fluid bed reactor and the rotating cone reactor.

The circulating fluidized-bed fast pyrolysis reactor design concept has been developed and commercialized by Ensyn [67, 82]. In this design, pyrolysis is conducted in a vertical tube fluidized bed reactor with a residence time of approximately 2 s. Sand is used as the heating medium. The char and sand are separated from the pyrolysis gases via multiple cyclones and routed to the regenerator, where the char is combusted to provide process heat. Fluidization of the reactor is achieved by recycling the non-condensable product gas.

The rotating cone reactor is being commercialized by BTG [67, 83]. In this design, pyrolysis is conducted in a rotating cone where sand flows radially outward

Table 2.3 Biomass pyrolysis products

Char	Bio-oil	Gas
15–25%	60–75%	10–20%

and upward because of the rotation of the cone. The sand and the char overflow the edge of the cone and are directed to a regenerator, where the char is combusted to provide process heat. The pyrolysis gas is directed to the condenser. One unique feature of the BTG system is that there is no fluidizing gas in the reactor; thus, the pyrolysis gas is undiluted.

Pyrolysis gas consists of entrained aerosols and char, condensable vapors, and non-condensable gas. Because char is entrained in the pyrolysis gas flow, it is important to minimize the time between pyrolysis and condensation because of char-catalyzed cracking reactions in the pyrolysis gas [81]. In commercial-scale systems, pyrolysis gas is condensed in spray condensers. There are many options for spray coolant, ranging from bio-oil to hydrocarbons, such as dodecane. Bio-oil is not miscible in hydrocarbons and therefore can be phase-separated easily in decanters if hydrocarbons are used as the coolant. Aerosol capture in pyrolysis systems has proven to be difficult using conventional demister systems; however, simple tube-type electrostatic precipitators worked well in continuous operation.

Fast pyrolysis bio-oil is a dark brown liquid with an elemental composition similar to biomass [84]. It is an emulsion of water-insoluble oxygenated hydrocarbons and aromatics along with water-soluble components from cellulose and hemicellulose decomposition, such as organic acids, aldehydes, alcohols, esters, and sugars, as well as solid char particulates. Typical properties of bio oil are shown in Table 2.4.

Upgrading of fast pyrolysis bio-oil via hydrotreatment had been difficult, with the main problem being rapid deactivation of the fixed bed hydrotreatment catalysts within about 100 h of time on stream (TOS). Recent progress has identified key issues that have resulted in hydrotreater catalyst deactivation. Fast pyrolysis bio-oil has been successfully hydrotreated for over 1200 h TOS, reported for the first time in the literature, after the following three root causes of hydrotreater catalyst deactivation were addressed [85]:

Table 2.4 Typical properties of fast Pyrolysis Bio-Oil [84]

Physical property	Typical value
Moisture content	25%
pH	2.5
Specific gravity	1.2
Elemental analysis	
C	56%
H	6%
O	38%
N	0%–0.1%
Higher heating value (HHV) as produced	17 MJ/kg
Viscosity	40–100 mpa s
Solids	0.1%
Vacuum distillation residue	< 50%

- Solid char particulate deposition on catalysts: Bio-oil was filtered through a 0.2-μm pressure filter to remove char particulate. This eliminated hydrotreater catalyst deactivation because of char particulate deposition.
- Catalyst deactivation because of metal poisoning: PPM levels of sodium, potassium, magnesium, and iron in the bio-oil can slowly poison hydrotreater catalysts. The concentration of these metals can be reduced significantly via ion-exchange slurry treatment.
- Coke formation due to aldehydes in the bio-oil: Aldehydes are extremely reactive components in bio-oil and are a root cause of high coking rates on hydrotreater catalysts. Aldehydes were removed via mild hydrotreatment on a Ru/TiO_2 catalyst. After the removal of solid char particles and metals, bio-oil could be processed on a Ru/TiO_2 catalyst to remove aldehydes and acids for several hundred hours without deactivation. In the original study [85], minor coke deposited on the Ru/TiO_2 catalyst was removed by oxidation, followed by reduction with hydrogen to maintain catalyst performance over 1200 h TOS.

After the removal of the char particulate, metal ions, and aldehydes, the mild hydrotreated bio-oil could be upgraded using conventional sulfided catalysts for over 1200 h without the need for regeneration [85].

The first two steps, bio-oil filtration and ion exchange treatment, are already commercial. There are several options to remove aldehydes from the bio-oil. The original Battelle study used custom-made Ru/TiO_2 catalysts; however, there are many other options for commercially available catalysts and other processes. One active area of research is catalytic fast pyrolysis (CFP) in which active components are removed via in situ or ex situ catalytic pyrolysis using Pt/TiO_2, ZSM-5, and Mo_2C [66, 86–89]. Another approach is solvent extraction, where proprietary solvent may be used to remove aldehydes and other reactive components to make the bio-oil suitable for hydrotreatment.

R&D Needs for Fast Pyrolysis

Mettler et al. [90] list complexities of pyrolysis that offer lower TRL R&D opportunities. These include:

- Multi-scale nature of biomass feedstock, i.e., these feedstocks exhibit a multi-scale (with variation over multiple orders of magnitude) structure, which varies during degradation, rendering fundamental prediction of properties and product distributions a challenge.
- Multiphase nature of pyrolysis, i.e., it has been shown that solid biomass, upon heating, decomposes through an intermediate liquid before forming volatile products. Although this intermediate liquid exists for short periods of time and some of it persists in aerosol, pyrolysis reactions include all three phases—solid char, liquid, and gas.

As a result, challenges include the need for comprehensive analytical and modeling techniques to better understand pyrolysis chemistry, degradation mechanisms of cellulose and lignin, mechanisms of intermediate liquid formation, aerosol formation and its impact on heat and mass transfer, and chemistry on the solid particle.

Other challenges include understanding char formation and the catalytic effect of char, and the inorganic (ash) that is enriched in the char phase.

For catalytic pyrolysis, some of the R&D challenges include [66] gaining operational knowledge with an integrated CFP system, understanding operation with realistic heterogeneous contaminated feedstock, and opportunities for colocating CFP systems with other industrial facilities such as pulp mills, petroleum refineries, and chemical plants to optimize synergies such as heat integration. In addition, catalysts may be improved to enhance yield to desired products and eliminate undesired products such as aldehydes. Generally, catalyst supports such as alumina are not resistant to water, which is generated when oxygenated species are reduced with hydrogen. There have been many studies done with hydrogen as a co-feed with biomass for CFP. This can be accomplished safely at the bench scale but can have safety issues at the commercial scale. It is advisable to design CFP systems that do not utilize hydrogen as a co-feed. Overall, process carbon efficiency is an extremely important parameter to maximize because that also maximizes yield and minimizes carbon intensity. Pyrolysis bio-oils contain multiple species, and with CFP, high-value coproducts can be produced [86, 91, 92]. There is a significant R&D need for separation technologies to extract high-value coproducts from bio-oil streams. Catalyst performance is dependent on pyrolysis gas composition, which, in turn, is dependent on feedstock. To increase the scale of production, feedstocks will have to be diversified; however, this implies that there may be significant variability in pyrolysis gas. Catalysts used in CFP must be tolerant of the variability in pyrolysis gas. R&D needs are to understand pyrolysis gas composition as a function of feedstock and then develop catalysts that would be targeted to the pyrolysis gas composition variability.

There are a number of engineering issues at high TRL as well. Some of these are listed below:

- Biomass particles vary in size and moisture, and as they are fed into a pyrolysis reactor at approximately 500 °C, the moisture flashes to steam, causing a short-term increase in gas volume and therefore pressure. This increase in pressure causes pyrolysis gas to flow backward into the biomass feeder. The internals of the biomass feeder are at a much lower temperature, and the pyrolysis gas condenses on the feeder screws and internals. Another cause of pressure fluctuation is the pulse-like flow of biomass into a reactor if screw feeders are used. Some manufacturers use multiple screw feeders to reduce the variability in biomass feed rate. Pyrolysis gas, which condenses on the internals of the feed system, makes these surfaces sticky, resulting in biomass accumulation and eventual plugging of the feed system. This is a significant operational issue for systems that must be operated continuously for extended periods.
- The material of construction for pyrolysis systems is mostly stainless steel. When pyrolysis exits the cyclone where the heating media such as sand is separated, it is transported in an insulated pipe to the condenser. One major problem that occurs in these pipes carrying olefin-rich pyrolysis gas is filamentous coke formation, thought to be catalyzed Fe, Ni, and Cr grains in the steel. Coke formation

does not appear to be significant in the pyrolysis reactor and cyclones because of high abrasion rates but can plug up the gas lines from the pyrolysis reactor to the condenser. The mechanism of coke formation appears to have some similarity to coke formation in ethylene crackers [86]. The commonly used workaround in commercial systems is to periodically introduce air to "burn the coke deposits"; however, this results in loss of production and very high thermal stresses on the transfer lines because of high temperatures from coke combustion. Additional research is necessary to understand the mechanisms of coke formation and develop solutions for commercial systems.

- The pyrolysis gas line from the reactor cyclone to the condenser is insulated, and its temperature is maintained close to the reactor temperature of approximately 500 °C. This gas line is physically connected to the condenser, which is at a much lower temperature, say approximately 50 °C. Due to conduction, the gas line temperature is lower in the region where it is connected to the condenser. Pyrolysis gas produced at approximately 500 °C condenses to a liquid phase as soon as it encounters a surface that is at a lower temperature; thus, condensation takes place inside the pyrolysis gas line at the condenser inlet. The liquid phase, which condenses inside the gas line, tends to polymerize and turns into a very thick viscous deposit that is very difficult to remove. Commercial systems use powerful plungers to keep the condenser inlet clean; however, these tend to be unreliable and cause operational problems. Additional research and engineering solutions are necessary to develop solutions for commercial systems.

4.1.2 Gasification

Gasification is a thermochemical deconstruction process conducted at a higher temperature than fast pyrolysis (~1000 °C versus ~500 °C) such that the gasification product is a non-condensable gas rather than a pyrolysis product, aerosol-laden condensable gas [93]. Another main difference between pyrolysis and gasification is that pyrolysis is endothermic and is conducted in an oxygen-free environment, whereas in gasification, air or pure oxygen (if nitrogen is not desired in the product syngas) is introduced along with the feed. Oxygen is produced via commercial air separation units, and since this is a mature technology, it will not be discussed in this section. The addition of an appropriate amount of oxygen and steam makes gasification a self-sustaining exothermic process. The product "syngas" from a gasifier would preferably consist of H_2 and CO; however, in actual operations, undesirable components such as CO_2, aromatics, phenols, CH_4, H_2S, COS, NH_3, aerosols, tars, soot, and char are also produced. There are several strategies to clean up the syngas, and these will be discussed below. The ratio of H_2 and CO can be adjusted via the water–gas shift (WGS) reaction [94].

Because gasification is a high-temperature, more intense process relative to pyrolysis, it is considered more "feedstock agnostic" relative to fast pyrolysis, which can only process lignocellulosic biomass. Gasification can process coal, woody biomass (softwood, hardwood, and waste wood), agricultural residues and

wastes (straw, maize, husks, and manure), and waste (MSW, RDF, black liquor, and organic waste) [95]. In commercial applications, gasification is being implemented to process MSW to methanol by Enerkem in Edmonton, Alberta, Canada and convert MSW to Fischer–Tropsch liquid at the Fulcrum Sierra Biofuels Plant in Reno, Nevada, United States. There are several gasification facilities planned and under construction for fuels [96].

There are several different types of gasifiers [67], which have been built at different scales:

Downdraft Gasifier Downdraft gasifier is a (mostly cylindrical) vessel in which both the biomass feed and the oxidant (generally air) are introduced at the top, and both streams flow downward concurrently. This device has the advantage that the biomass feed can be large (> 50 mm in characteristic dimensions) because the residence time in the unit is high. The biomass feed first undergoes drying as it moves downward; then, once the water has been mostly removed, temperatures rise and pyrolysis takes place. The pyrolysis gases, which include volatiles, tars, and char, ignite in a combustion zone below the pyrolysis zone and produce process heat. The char not being combusted moves further downward and creates a reduction zone. This zone is endothermic; the energy for reduction is provided by the combustion zone, but because reduction is endothermic, gas temperature drops. The exit gas temperature is between 600 °C and 650 °C. These are generally smaller units to produce combustible syngas, which can be used in engines as well as energy production.

Updraft Gasifier The updraft gasifier is a vessel in which the biomass feed is introduced at the top and flows downward, and the oxidant (usually air) is introduced at the bottom, flowing upward. As in the downdraft gasifier, this device also has the advantage that the biomass feed can be large (> 50 mm in characteristic dimensions) because the residence time in the unit is high. As in the downdraft gasifier, the biomass feed first undergoes drying as it moves downward; then, once the water has been mostly removed, temperatures rise and pyrolysis takes place. The pyrolysis gases, which include volatiles, tars, and char, ignite in a combustion zone below the pyrolysis zone and produce process heat. Because the oxidant is introduced from below, it burns the char phase, which is at the bottom, and the pyrolysis gases, creating high temperatures. The gases travel upward through the pyrolysis zone and the drying zone and thus carry with them tars, pyrolysis oils, and water. Tars tend to be a major problem in updraft gasifiers. This gas requires considerable cleanup before it can be used in downstream applications. Commercial examples of this type of gasifier include the Lurgi gasifier (for coal feedstock) as well as the Nexterra gasifier for biomass feedstock. These are also generally smaller units to produce combustible syngas, which can be used in engines as well as energy production.

Fluidized Bed Gasifier The fluidized gasifier is a vessel in which a bed of biomass feed, or a heat carrier such as sand, is fluidized by an upward flow of superheated steam, along with an oxidant. One major advantage of the fluidized bed design,

especially when a heating media such as sand is used, is that relatively large feedstock particles can be used, and because residence time and temperature is high, even "difficult" feedstock such as MSW can be gasified. Both the Enerkem gasifier in Edmonton and the Fulcrum gasifier in Reno utilize fluidized bed technology. In direct fluidized bed gasification, energy is provided through partial combustion of the feed or the intermediate pyrolysis gases, and in indirect gasification, energy is provided through heat exchangers that could be immersed in the bed, such as the TRI gasifier in the Fulcrum facility in Nevada and the MILENA gasifier. Direct gasification systems produce a medium-quality syngas relative to indirectly heated fluidized beds. Temperatures are between 925 °C and 1040 °C, generally below the ash fusion temperature to prevent slag formation. Another approach to providing heat is through the combustion of the char in a separate bed, such as in the Battelle/FERCO gasifier. Product gas emanating from the gasifier fluidized bed is passed through cyclones in the headspace above the bed, and solids are separated and returned to the bed. The product exits the reactor through the cyclones that can separate large particles, but tar formation is a problem in many fluidized bed gasifiers.

Transport Gasifier A transport gasifier is an upward moving "riser" bed in which biomass or biomass entrained in heating media such as sand is gasified. Superheated steam and oxidant are used as fluidizing gases. The solids disengage from the gas phase in a cyclone at the top of the riser and are redirected into a combustor, or back to the bottom of the riser. Transport gasifiers have a residence time on the order of a second and thus require feedstock to be ground to about less than ½ mm so that pyrolysis and gasification can take place. Temperatures in transport gasifiers are similar to those in fluidized bed gasifiers, between 925 °C and 1040 °C. The Valmet gasifier design is an example of a transport gasifier. Foster Wheeler and KBR also supply transport gasifiers. As in fluidized bed gasifiers, tar formation is still significant enough to be an issue with transport gasifiers.

Entrained Flow Gasifier The entrained flow gasifier is a vessel in which feed in the form of a slurry of finely pulverized particles is introduced with an oxidant in a downward co-flowing configuration. Water slurries work for energy-dense feeds such as coal but are not suitable for biomass feeds because of the lower energy density of biomass. For this reason, entrained flow gasifiers must be fed dry biomass [93]. Feeding solid biomass into hot reactors is nontrivial as pyrolysis may start taking place in the feed system, thus creating sticky deposits that can plug up feeders. The small feed particles are entrained with the flow and gasify rapidly, and temperatures in excess of 1000 °C are maintained, thus destroying most aerosols and tars. The residence time in an entrained flow gasifier is very short, on the order of a second. The gas is cooled via heat exchangers and exits at the bottom of the unit. High temperature in entrain flow gasifiers melts the ash into slag, which is removed via lock hoppers at the bottom, and high temperature also results in reduced tar formation. The units have been used successfully in large-scale (> 250 MW) power plants for coal but not for biomass feeds. Commercial suppliers for entrained flow gasifiers include Siemens, Lurgi, GE, Shell, MHI, and Uhde [93] (Table 2.5).

Syngas Cleanup and Conversion At the present time, fluidized bed gasifiers have been considered primarily for SAF production because they can be scaled and are more tolerant to variability in feedstock, such as MSW. As mentioned above, gas from fluidized bed gasifiers carries particulate and tar in addition to other heteroatoms depending on the feedstock. For methanol production and Fischer–Tropsch processes, heteroatom concentrations must be on the order of 10 ppb [93], and in addition, for downstream applications, the H_2 to CO ratio may also have to be adjusted to ~2. Some considerations for syngas cleanup include [93, 95, 97, 98]:

Gas Temperature Fluidized bed gasifiers operate at approximately 1000 °C, a temperature at which much of the ash, including alkali metals, is molten and thus cannot be removed via cyclones. The gas must be cooled first to below ~500 °C, a temperature at which the ash and metals solidify and can be removed.

Particulate Removal Once the gas is cooled, larger particles can be removed via cyclones, and candle filters may also be used for finer particles. Alkali metals tend to condense on particulates and are also removed along with the particulate.

Chloride Removal Biomass can contain chlorine, especially if it is grown in coastal areas. Chlorine forms HCl in the reducing environment of a gasifier, and because HCl is very corrosive, it must be removed via alkaline scrubbers as soon as the gas has been cooled. Chlorine can also reduce the melting temperature of ashes and make these difficult to remove. Residual particulate and sulfur, which appear as H_2S or COS, are also removed via scrubbers. Additional sulfur may be bound to and removed along with the tars.

Nitrogen Removal Nitrogen in wood is ~1%; however, in sludge, it may be as high as 9%. Nitrogen is converted to NH_3, HCN, or N_2. Nitrogen species may be removed via wet scrubbing with water or dilute acids.

Tar Removal Tar is produced in fluidized bed gasifiers and is very difficult to remove completely from the syngas. Tar particles can be partially removed via oxi-

Table 2.5 Gasifier configurations and operating conditions

	Downdraft	Updraft	Fluidized bed	Entrained flow	Transport
Operating temperature (°C)	425–675	425–675	925–1040	> 1100	950–1000
Fuel particle size (mm)	< 50	< 50	< 6	< 0.10	< 0.40
Fuel preparation	Drying	None	Grinding	Pulverization	Grinding
Residence time	High	High	Medium	Low	Low
Ash condition	Slag/fly ash	Slag/fly ash	Clinker agglomeration	Slag	Fly ash
Feed conditioning	Dry	Dry	Dry	Dry or slurry	Dry

dation, or by catalytic cracking. Catalytic cracking can be difficult because this requires high temperatures, and therefore must be done before scrubbing. Upstream of the scrubbing operation, the gas can contain particulate and metals, and these can deactivate the cracking catalyst. Other approaches include scrubbing in an oil wash (OLGA process) [93] and thermal destruction via plasma; however, plasma treatment may be cost-prohibitive.

For SAF, target intermediate products from syngas upgrading include methanol and Fischer–Tropsch liquids. For both these processes, the desired H_2 to CO ratio is approximately 2, which can be adjusted by the WGS reaction.

$$CO + H_2O \leftrightarrow CO_2 + H_2$$

The WGS reaction is at ~500 °C, and generally, Fe, Cu, Zu, or sulfided Co/Mo catalysts may be used if there is significant sulfur concentration in the gas. If sulfur is in the form of COS, it can be hydrolyzed to H_2S using sulfided catalysts. Other sorbent-based technologies may be used for removing sulfur from syngas [90].

Methanol and Fischer–Tropsch Synthesis

One of the major benefits of gasification is that it can be integrated into existing refineries. Two approaches that are currently being developed are the methanol and Fischer–Tropsch pathways [99]. The Fischer–Tropsch (FT) process produces an "FT wax," consisting of a wide linear hydrocarbon product distribution predominantly in the diesel carbon number range [99]. The methanol-to-olefins pathway uses methanol as an intermediate and produces a product consisting of paraffins and olefins, which can be upgraded to jet fuel carbon numbers.

Production of methanol can be done over a Cu-Zn/Al_2O_3 catalyst with high selectivity, and syngas conversion reactions are well-known:

$$2H_2 + CO \rightarrow CH_3OH$$

$$CO_2 + 3H_2 \rightarrow CH_3OH + H_2O$$

Major licensors of methanol synthesis plants include Air Liquide, Johnson Matthey, Mitsubishi, Methanex, and KBR [93].

Fischer–Tropsch synthesis can be done over Ru-promoted Fe or Co catalysts on silica, alumina, or titania supports and can produce a distribution of long chain ~C11–C20 alkanes and alkenes. Fe catalysts produce more alkenes than Co catalysts [93] and produce shorter chains at higher temperatures.

Both methanol synthesis and Fischer–Tropsch reactions are exothermic, and reactors must be designed to manage heat transfer and bed temperatures. Although fluidized bed reactors can be used, these result in mechanical attrition of the catalyst and are more suited for large-scale plants, and fixed bed reactors are more common for smaller plants with a catalyst life of approximately 4 years between change-outs [93].

R&D Needs for Gasification and Syngas Synthesis

- Technologies for dry biomass feeding, at scale, into entrained flow reactors still need to be developed.
- Feedstock preprocessing technologies, which may include heteroatom removal, cleanup, and size reduction, especially for challenging heterogeneous feed such as municipal solid waste, must be developed.
- Gasification kinetics need to be better understood, especially for heterogeneous feedstock.
- Dry biomass cannot be fed as a slurry; therefore, dry feeding approaches or slurrying with other renewable liquids need to be developed and de-risked.
- Gasifiers and gas processing trains at atmospheric pressure are very large and capital-intensive. At higher pressure, the equipment size can be reduced, but biomass feeding becomes very difficult and is an area of research.
- Material corrosion and degradation in the presence of oxygen and other heteroatoms such as chlorides are a problem at high temperatures.
- High thermal gradients in refractory line vessels degrade the refractory lining, and this requires additional research.
- Syngas cleanup, and especially tar cleanup, remains a major problem, both from a process complexity perspective and from a cost perspective.
- If syngas contains heteroatoms, there is a risk that acids are produced when the gas is cooled. These acids are corrosive and can cause operational problems in gas processing equipment.
- FT systems tend to be very sensitive to contaminants in syngas. Novel approaches to make FT catalytic reactors more robust against contaminants in syngas may reduce syngas cleanup requirements.
- Gasifiers and upgrading equipment, such as gas cleanup and FT reactors, are very capital-intensive. There may be an opportunity to integrate this equipment and operate at a higher pressure to achieve process intensification and reduced capital costs.

4.2 Biochemical Conversion

Biochemical conversion is loosely defined as conversion processes that utilize biological catalysts, such as enzymes and biological cells, for the conversion of renewable carbon to fuels and products. Another definition for biochemical conversion is using the carbohydrate portion of the biomass, cellulose, and hemicellulose for conversion to fuels where the lignin component can either be used for process heat and energy or valorized to higher-value fuels and products. Hence, by these definitions for biochemical conversion, the two approved SAF pathways per ASTM D7566 are Annex A5, ATJ-SPK, and Annex A3, hydroprocessed fermented sugars to synthetic isoparaffins. The states of these pathways are as follows.

4.2.1 Alcohol-to-Jet Synthetic Paraffinic Kerosene

ASTM D7566 Annex A5, ATJ-SPK, was approved in April 2016 for SPK from iso-butanol (30% blend with petroleum) and expanded in April 2018 for SPK from ethanol and for fuel blends up to 50% with petroleum. ATJ-SPK consists of iso-alkanes of 8, 12, or 16 carbons when starting from iso-butanol. The iso-alkanes are highly branched and have lower DCNs than FT or HEFA, based on data from Gevo, Inc. The carbon number is broadened, and the branching level can be significantly reduced, leading to a DCN similar to FT and HEFA when starting from ethanol [100].

Biobutanol or butanol production from biomass or other renewable carbon feedstocks is a well-researched and documented process [101]. Ethanol production from biomass is even more developed than biobutanol and is arguably an early-stage commercial technology. Numerous cellulosic ethanol plants have been built and commissioned in the United States and around the world, but many have been shut down for market reasons and some technical underperformance issues. In addition to cellulosic ethanol, the large volumes of corn ethanol available in the United States and sugar cane ethanol available in Brazil could be used as a starting feedstock. Using corn ethanol as a starting alcohol feedstock could present some challenges for meeting the SAF definition of >50% GHG reductions compared to conventional jet fuel. Sugar cane ethanol for the most part would be able to meet the >50%-GHG-reduction provision, but the availability of sugar cane ethanol for SAF production in the United States is quite low.

The process for converting both butanol and ethanol to SAF is well-researched [102]. Several companies, for example, Lanzajet and Vertimass, are actively working on taking the ATJ-SPK process to commercial application. This process shows very good promise for both near- and long-term deployment for the production of high volumes of SAF at comparatively promising economics when compared to other SAF production options, with projected mature plant economics approaching the $2/GGE [103]. These SAF fuels are approved for up to 50% blends, which is a good percentage, although the SAF markets have expressed interest in getting to 100% SAF blends.

R&D Needs for Alcohol to SAF

- Improve cost-effective biomass pretreatment technologies and de-risk solid feeding into the pretreatment reactor with heterogeneous feedstocks.
- Develop more efficient catalysts and better process economics for the conversion of alcohols to SAF. Several companies and organizations are actively working in this area and making good progress [104].
- Secure low-cost enzymes.

4.2.2 Hydroprocessed Fermented Sugars to Synthetic Isoparaffins

This pathway was approved in June 2014 for up to a 10% blend with petroleum-derived jet fuel. Unlike SPK from HEFA or FT, this is a single molecule, a 15-carbon hydrotreated sesquiterpene called farnesane, produced from the fermentation of sugars. Today, the fermentation is done commercially from sugar cane juice and is used in higher-value applications, most commonly in personal care.

Farnesane as a branched C15 saturated hydrocarbon has a low freezing point, good cold flow properties, and good energy density, so at up to a 10% blend, it is a good blendstock for SAF production. Theoretically, it could be produced in high volumes at low carbon intensities given that it can be produced from lignocellulosic biomass sugars. Most of the work on developing this SAF pathway has been done by private companies, most notably Amyris, so public cost data is very scarce and mostly speculative. Additionally, as stated earlier, there are large higher-value markets for farnesane that also address significant sustainability issues such as shark harvesting to produce squalene. So, at the time of writing, the future of this SAF pathway is not clear.

In addition to these ASTM-approved pathways, there are a number of promising biochemical pathways to SAF that show considerable promise. An overview of these pathways is as follows:

Carboxylic Acids Fermentation of sugars to carboxylic acids (primarily butyric acid), followed by catalytic upgrading of acids to SAF. This is a promising conversion pathway of lignocellulosic biomass via a sugar intermediate where the concentrated/clarified hydrolysate is anaerobically fermented to butyric acid. This process does involve a fairly complicated separation step to continuously remove the butyric acid from the fermentation broth. The butyric acid is concentrated and then catalytically upgraded through a ketonization step to primarily 4-hepanone, which is then upgraded by condensation to an oxygenated C14 when hydrotreated to remove the oxygen to an isomerized C14 hydrocarbon for use as an SAF blendstock.

R&D Needs for Carboxylic Acids to SAF

- Stain engineering and fermentation process to improve the butyric acid fermentation process and the separation step to continuously remove the butyric acid from the fermentation broth.
- Given that this pathway can utilize lignocellulosic biomass feedstocks, it has a very large production volume potential as a mature industry cost of projection potential in the mid $2/GGE. An important point for comparison purposes is that the product is a single C14 hydrocarbon molecule versus a range of SAF fuel range molecules, and hence, these very precise fuel properties of a single molecule could offer some benefits for tailoring fuel properties.
- Improve the separation process efficiency of carboxylic acid as well as reduce the separation process complexity to improve cost-effectiveness.

Modified BDO to SAF
Fermentation of sugars to 2,3-butanediol (BDO) followed by catalytic upgrading of BDO to SAF. In this SAF pathway, batch fermentation is initiated in the same vessels that were utilized for enzymatic hydrolysis after a cooling step. The BDO fermentation converts sugars to 2,3-BDO plus hydrogen. The fermentation is conducted on whole-slurry hydrolysate with the presence of 25% total solids. The fermentation broth goes through a lignin press to remove solids, and then filtration to remove fine particles and ion exchange to remove ionic/salt species. The aqueous BDO stream is then heated at elevated pressure and undergoes catalytic BDO upgrading, producing butene and minor byproducts. The butene product is oligomerized and hydrotreated to produce C8–C16 range paraffinic hydrocarbons suitable as an SAF blendstock.

R&D Needs for BDO to SAF

- Improve whole-slurry hydrolysate fermentation process to 2,3-BDO, which typically utilizes an engineered strain of *Zymomonas mobilis*.
- Separations research to improve the butene recovery and concentration step would also be beneficial. Volume potentials are very similar to any of the lignocellulosic SAF pathways, and cost projections are very good at < $2/GGE if the excess aldehyde can be captured and sold at market prices as a coproduct.

Furans Catalytic conversion of sugars to furans (furfural and hydroxymethylfurfural [HMF]) followed by condensation via reaction with methyl ethyl ketone (MEK) and then hydrotreating. This pathway starts with solids removal and concentration of C5 and C6 sugars in the hydrolysate, and then catalytic dehydration to produce furfural and HMF intermediates. The furfural and HMF intermediates are mixed with a homogeneous catalyst and an MEK/isobutyraldehyde mixture to undergo aldol condensation. The condensate intermediates are ultimately deoxygenated via hydrotreating to produce C14–C16 SAF range hydrocarbons. Because this is a lower TRL SAF pathway, there are many potentially impactful R&D needs that exist [105].

R&D Needs for Furan to SAF

- Recovery of the solvent before the aldol condensation reaction is a significant cost and energy use driver for the process. R&D to improve the performance of the solvent recovery would have a major impact on both the cost and GHG reduction potential of this SAF pathway.
- R&D to improve the aldol condensation reaction so that the homogeneous catalyst can be cost-effectively recovered and reused.
- Overall process R&D and process improvement to reduce energy demands and efficiency of the process to improve process economics and the GHG reduction profile. This is linked to more favorable aldol condensation temperature and residence time.
- This SAF production process shows good production potential and reasonably good economics of production at around $4/GGE without coproduct credits.

Aqueous Phase Reforming Catalytic conversion of sugars to SAF/fuels via aqueous phase reforming (APR), dehydration/oligomerization, and hydrotreating. This process starts with a clarification of the hydrolysate and then aqueous phase reforming/hydrodeoxygenation (APR/HDO) to deoxygenate the high-reactivity, highly oxygenated compounds, sugars, and other oxygenated organics to lower-reactivity and lower-oxygen-content molecules. Then, a dehydration–oligomerization condensation step is performed to condense products and increase carbon chain lengths of feed molecules (< 6) to that in the jet fuel range (approximately C9–C19). Finally, hydrotreating is performed to remove the remaining oxygen and produce SAF. Like the above processes, the production potential is very large via this pathway given that it utilizes the lignocellulosic feedstock base.

R&D Needs and Data Gaps for APR

- Most of the development of this pathway has been by private companies, most notably Virent, so public economic data is somewhat sparse, although some public data on the cost of production does exist [106]. This data does show favorable economics and very good GHG reduction values.
- Improve process conditions and catalyst for the APR step as well as process optimization to improve heat integration for further improvement in process economics and GHG reduction potentials.

5 Analysis

5.1 Techno-economic Analysis and Published Results

Techno-economic analysis (TEA) is a methodology utilized by the government, industry, and academia to quantify the impact of research discoveries and engineering advances on the economic viability of an integrated process. When effectively coupled with research and development (R&D) efforts, TEA is an important complementary tool for understanding and identifying key process attributes that can be improved to minimize overall production costs. The output of the TEA is the production cost. Typically, a minimum fuel selling price (MFSP) represents the minimum price that the hydrocarbon fuels must sell to generate a net present value of zero for a 10% internal rate of return on investment.

When performing TEA across a range of TRLs, different analysis strategies apply. For instance, early-stage TEA is used to validate an initial idea, while the development of more rigorous process designs and economic evaluations are performed with high-TRL technologies. This analysis is necessary to support R&D directions, identify production and cost drivers, and inform key decisions on cost-efficient strategies for bio-derived fuels and chemicals. Cost numbers are often used to map/track individual program accomplishments to evaluate the potential of technical success and allow the inclusion of multiple products and processes in the

aggregate net value estimate. The maturity of a conversion process and feedstock price impact the process economics most significantly.

For the FT pathway (Annex A1), the MFSP value is estimated to range from $3.7 to $7.6/GGE, depending on technology specifics, plant scales, and feedstock prices [99, 107, 108]. This corresponds to 39% jet fuel, 32% gasoline, 18% diesel, and 11% wax fraction on a mass basis. The overall yield of hydrocarbon fuel is estimated to be 57 GGE/dry ton considering all the hydrocarbon fuel products. The key cost drivers and parameters impacting FT pathway production costs are federal/state incentives, total capital investment, and feedstock prices. Product selection, which leads to the yield of jet fuel, is strongly dependent upon the operating temperature, partial pressure of syngas, and the applied catalyst.

For the HEFA pathway (Annex A3), MFSP value is estimated to range from $3.2 to $11.3/GGE, primarily due to wide ranges of HEFA feedstock prices, feedstock types, and plant scales [49, 109]. This corresponds to 68% jet fuel, 24% gasoline, 7% diesel, and 1% propane fraction on a mass basis. The overall yield of all HEFA feedstock-based processes is estimated to be 283 GGE/dry ton considering all the coproducts. The feedstock contributes a large portion (> 70%) to the HEFA production cost, while the use of hydrogen and the selection of a catalyst contribute to a significant part of the production costs.

The ATJ (Annex A5) processes' MFSP ranges from $4.7 to $13.6/GGE, depending on alcohol types, feedstock types and prices, and processing efficiencies [110, 111]. This corresponds to 71% jet fuel, 22% gasoline, and 7% C2 mix fraction on a mass basis. The overall yield of the process is estimated to be roughly 50 GGE/dry ton considering all the coproducts. The key cost drivers are federal/state incentives, total capital investment, feedstock prices, catalyst costs, hydrogen costs, and process on-stream factor.

5.2 *Life Cycle Assessment and Published Results*

The Department of the Treasury and Internal Revenue Service provided guidance on December 19, 2022 and December 15, 2023 regarding the sustainable aviation fuel (SAF) credits in Notice 2023–6, 2023–2 I.R.B. 328 [112]. Under § 40B(d)(1)(D) and (e), SAF must be certified to have a life-cycle GHG emission reduction of at least 50% compared to fossil jet fuel baselines. The amount of emission reduction generated by the use of SAF depends on its life-cycle emission value, expressed in terms of grams of CO_2 equivalent per megajoule (gCO_2e/MJ), so a 50% reduction is equivalent to 44.5 gCO_2e/MJ. The life cycle emissions are composed of core life cycle assessment emissions and induced land-use change (ILUC) emissions. ILUC emissions assessment accounts for these different effects by evaluating greenhouse gas released from the conversion of natural vegetation (forest and other natural land), soil organic carbon, oxidation of peatlands, and sequestered biomass [10]. CORSIA eligible fuel production may require some additional land to be used and generate land-use change GHG emissions [10, 113]. These could occur where the

new CORSIA eligible fuel production is taking place (direct land-use change) but also in other locations due to the displacement of crops (or animals) for which the land was previously used (indirect land-use change). Most ASTM-approved SAF pathways could meet the 50% GHG emission reduction threshold, except for pathways using first-generation feedstocks, such as corn grain. In that case, decarbonization strategies have been investigated extensively to realize the GHG emission reduction target by incorporating carbon capture storage, using renewable energy sources, and using green farming practices.

5.3 Local Air Quality Analysis

Facilities producing hydrocarbon fuels are considered chemical production facilities, which is one of the 28 listed source categories [114]. Thus, SAF facilities are considered chemical production facilities for air-permitting purposes and are subject to environmental laws, including complex air quality regulations that aim to protect and improve the quality of the air. The federal Clean Air Act requires states to regulate stationary and mobile sources pursuant to an EPA-approved state implementation plan for the implementation, maintenance, and enforcement of the National Ambient Air Quality Standards. More importantly, local air quality evaluation is a critical consideration for rolling our biorefineries successfully [115].

The National Ambient Air Quality Standards set ambient concentration limits for six air pollutants (ozone, PM, nitrogen oxides [NO_x], sulfur dioxide [SO_2], carbon monoxide [CO], and lead); this set is called "criteria air pollutants." The Clean Air Act established a New Source Review (NSR) [116] C program that requires an SAF facility to apply for a construction permit prior to the project being built or modified. There are three types of NSR permitting requirements, and biorefineries need to meet one or more of these:

- Prevention of Significant Deterioration (PSD) permits, which are applicable for new facilities in attainment status that are major for at least one pollutant.
- Nonattainment New Source Review (NNSR) permits, which are applicable to new major source facilities in a nonattainment area.
- Minor source permits.

These regulations limit the acceptable levels of various air pollutants emitted by sources. Federal regulations under the New Source Performance Standards (CFR Title 40, Part 60) and National Emission Standards for Hazardous Air Pollutants (CFR Title 40, Parts 61 and 63) include criteria that are used to determine the applicability of a regulation to specific unit operations or a whole facility [117].

Project developers are responsible for identifying needed permits, compiling the requisite information (e.g., type and magnitude of regulated pollutants to be emitted), and drafting the permit application. The permitting authority will conduct reviews and, if necessary, require changes or additional mitigation. For most permitting processes that have legally mandated timelines, the timeline starts either when

the application is received or when the agency deems the application complete [118]. Upon completion of the review process, the permitting agency approves or denies the permit.

6 Policies and Incentives

The SAF volumetric goal could potentially be achieved by licensing existing commercially ready conversion technologies (HEFA, alcohol-to-jet, and Fischer–Tropsch synthesis). Investments are on hold as the biofuel sector is waiting for policy incentives to materialize to verify economic payback. As a result, in 2023, biofuel investment was only slightly over \$1 billion, while the total global investment in low-carbon technology was over \$1.7 trillion [119]. This means continuing to enhance the market pull for SAF through aggressive credits and incentives to close the gap in price with conventional jet fuel.

Policies supporting the decarbonization of fuels broadly or aviation fuels more specifically establish demand- and supply-side pressure and support for the generation and utilization of renewable fuels. The 2021 proposed SAF Act would establish a low-carbon fuel standard of 20% by 2030 and 50% by 2050 [120, 121]. In response to the federal government's target for the production of 3 billion gallons of SAF annually by 2030, an aviation industry coalition recommends that the EPA include SAF technologies and corresponding feedstock categories in RFS policy [122]. The primary policies are RFS and low-carbon fuel standards (LCFSs, both federal and state), which set targets for renewable fuels and fuel carbon intensity respectively, with tax credits and incentives promoting biofuel production and use.

In the United States, the RFS policy, governed by the EPA, had a major update in August 2022 [123]. This update set volumetric requirements for biofuels broadly, including for cellulosic biofuel, biomass-based diesel, advanced biofuel, and total renewable fuel with respective volumetric requirements. Although the EPA has clearly defined fuel categories and corresponding feedstock categories, SAF does not have a separate volumetric requirement.

For federal-level LCFSs, the Inflation Reduction Act (IRA) of 2022 advances SAF in the United States by providing an SAF fuel credit for the sale or use of a qualified mixture [124]. Importantly, the definition established for SAF is in alignment with ASTM International Standards, excluding palm oil–based feedstocks, and in terms of a life-cycle greenhouse-gasemission reduction percentage of at least 50%.

Leading the way, California adopted an LCFS in 2011, updated in 2018, designed to reduce the GHG emissions from fuels in alignment with state commitments for reduced GHG emissions [125]. The policy specifies reduced carbon intensity benchmarks for increased decarbonization across different fuel types year over year through 2030, utilizing multiple replacement fuels [126]. California is a member of the Pacific Coast Collaborative, along with Oregon, Washington, British Columbia, and six major city governments as of 2016; the Collaborative establishes aligned

commitments and shared strategies to achieve reduced carbon emissions in the region [127, 128]. Oregon adopted an LCFS in 2016 with a goal to reduce the carbon intensity of transportation fuels and, as of 2019, includes an aviation fuel target of 10% by 2025 [129–131]. Similarly, in 2021, the state of Washington proposed a Clean Fuel Standard for transportation fuels (used in Washington), with a reduction goal of 20% below 2017 levels by 2038 [132, 133]. A member of the Pacific Coast Collaborative, British Columbia first adopted an LCFS in 2008 that reduces the carbon intensity of fuels used in the Canadian province, and this policy was modified in 2022 to include aviation fuel [134, 135]. While Canada has set forth a national Clean Fuel Standard influenced by California, British Columbia, and other members of the Pacific Coast Collaborative that reduces the carbon intensity of select fuels [136], it does not currently include aviation fuels [137]. Additional states are in the process of developing SAF mandates; for instance, New York State proposed a bill in 2022 that would require the use of SAF [138, 139].

Recommendations and Best Practices for SAF Mandates

The recent adoption of many SAF mandates demonstrates the increasing recognition of the significant GHG emissions from the aviation sector and the importance of decarbonizing aviation fuels to achieve broader goals for decreased GHG emissions. The design of the targets and timing of these policies reflects the use of mandates as a tool to achieve those larger goals, as stated within ReFuelEU [140]. However, in structuring SAF targets and timelines, policymakers should consider possible constraints on the availability of feedstocks and the capacity to achieve SAF [141]. Further recommendations call for policies to consider and mitigate adverse sustainability and land-use impacts [142, 143]. Such policy measures can reduce competition over crop-based biofuel between energy and food products and avoid negative indirect land-use impacts associated with certain feedstocks and practices. The consideration of innovative feedstocks and fuels in ReFuelEU policy [140] and the exclusion of select feedstocks associated with potentially negative sustainability impacts (such as palm oil–based fuels) in ReFuelEU [140] and as defined in the US Inflation Reduction Act of 2022 demonstrate the incorporation of these principles [124].

The use of LCFSs that include aviation fuel has been associated with increased biofuel production [144]. An analysis by the International Council on Clean Transportation found that while an LCFS could potentially be better suited to reduce GHG emissions and greater decarbonization compared to an RFS by specifically focusing on carbon intensity from fuel production to consumption, actions may be needed to control potential adverse sustainability impacts, including indirect land-use impacts [145]. Their study indicates that potentially shifting federal US policy from an RFS to an LCFS could require measures to mitigate a possible increased demand on crop-based biofuels and increased imports of fuels with poor sustainability (such as palm oil-based fuels), as previously discussed. The US Inflation Reduction Act of 2022 reflects some aspects of this shift by not only defining SAF to exclude palm oil–based components but also in terms of having a life-cycle GHG emission reduction of at least 50% [124].

7 Conclusion

In this chapter, we show that replacing fossil jet fuel with SAF is a reasonable near- to medium-term approach for reducing the GHG emissions needed to meet decarbonization goals for commercial air travel. We discussed the US interagency team (DOE, US Department of Transportation, and USDA)-issued *SAF Grand Challenge Roadmap* for accelerating the growth of the SAF market through collaboration between industries and governmental organizations. The development of multiple pathways that utilize various renewable carbon feedstocks will help eliminate competition for feedstocks from producing RD. Locating SAF biorefineries near airports can help alleviate the challenges associated with the United States's capacity-constrained fuel distribution systems. Federal- and/or state-level policies are essential to encourage more stable investments for producing SAF. SAF consists of four chemical families (*n*-alkanes, iso-alkanes, cycloalkanes, and aromatic hydrocarbons), and all ASTM-approved pathways in Table 2.1 produce a fraction of the required four chemical families. As a result, the approved pathways are limited to either 10% or 50% blending levels.

Thermochemical conversion pathways are attractive because they deconstruct biomass in seconds to produce an intermediate (bio-oil or syngas) for downstream processing into SAF. Perhaps, the most common thermochemical conversion processes are pyrolysis and gasification. Pyrolysis produces highly oxygenated bio-oil, comprising hundreds of molecules that cause it to be acidic, unstable, and immiscible with hydrocarbons plus inorganics, necessitating stabilization and hydrotreating into SAF. Catalysts are often added during pyrolysis, a process called CFP, to remove oxygen and generate a stable bio-oil for hydroprocessing into SAF. The SAF yields for CFP are low due to light gas and coke production, and this has limited the commercial adoption of the biomass CFP pathway. Gasification, on the other hand, produces syngas, but the production of tars and inorganic impurities remains a challenge and has also limited the broad adoption of this pathway for SAF. Biological conversion pathways are attractive in that they use biological catalysts to funnel the sugar portion of biomass into single molecules, like ethanol, BDO, butyric acid, etc., for downstream catalytic conversion into SAF. The biological pathways suffer from low fermentation yields, low SAF yields, high GHG emissions, and separation challenges.

Despite these challenges, some pathways have successfully produced SAF. As we show in Table 2.1, seven or eight ASTM D7566 pathways plus one ASTM D1655 pathway have been approved by ASTM but are limited to 10–50% blend levels with fossil jet fuel. The low blend limits are because the pathways produce a fraction of the four chemical families in the fossil jet fuel. For example, the ATJ-SPK (Annex A5) produces paraffins (alkanes), which are two of the four required chemical families in jet fuel. The HEFA-SPK pathway is the most mature pathway for making alkanes for SAF; however, refineries prefer producing RD. The HEFA-SPK pathway is also impacted by the limited supply of feedstock. These pathways produce hydrocarbon fuels with modeled MFSPs that are higher than the selling

price for fossil jet fuels. Federal/state incentives, feedstock costs, total capital investments, catalyst costs, and H_2 costs are the major cost drivers for the low TRL pathways, and feedstock cost is the major cost driver for the mature HEFA-SPK pathway. The SAF should have a life-cycle GHG emission reduction that is greater than or equal to 50% compared to fossil jet fuels. We highlight the importance of federal/state incentives in reducing costs to make SAF competitive with fossil jet fuel. Incentives that include SAF can help increase biofuel production and accelerate SAF market growth.

Acknowledgments This work was authored in part by the National Renewable Energy Laboratory, operated by Alliance for Sustainable Energy, LLC, for the US Department of Energy (DOE) under Contract No. DE-AC36-08GO28308. Funding was provided by the US Department of Energy Office of Energy Efficiency and Renewable Energy Bioenergy Technologies Office. The views expressed in the article do not necessarily represent the views of the DOE or the US government. The US government retains and the publisher, by accepting the article for publication, acknowledges that the US government retains a nonexclusive, paid-up, irrevocable, worldwide license to publish or reproduce the published form of this work or allow others to do so, for US government purposes.

References

1. U.S. Environmental Protection Agency, *Inventory of U.S. Greenhouse Gas Emissions and Sinks: 1990–2021. EPA 430-R-23-002.* https://www.epa.gov/ghgemissions/inventory-us-greenhouse-gas-emissions-and-sinks-1990-2021. Accessed 30 Nov 2024.
2. U. S. Department of Transportation. (2022). *FAA proposes requiring more efficient jets and turboprops to lower climate impacts*. In *U.S. Department of Transportation*. https://www.faa.gov/newsroom/faa-proposes-requiring-more-efficient-jets-andturboprops-lower-climate-impacts
3. U. S. Energy Information Administration. *Analysis & projections*. https://www.eia.gov/analysis/index.php. Accessed 18 Dec 2024.
4. U. S. Department of Transportation. *Bureau of Transportation Statistics*. https://www.bts.gov/. Accessed 30 Nov 2024.
5. Graver, B., Rutherford, D., & Zheng, S., (2020). *CO2 emissions from commercial aviation: 2013, 2018, and 2019*. ICCT The International Council on Clean Transportation (in en).
6. U. S. Department of Energy, U. S. Department of Transportation, U. S. Department of Agriculture, and U. S. Environmental Protection Agency, SAF Grand Challenge Roadmap. (2022). https://www.energy.gov/sites/default/files/2022-09/beto-saf-gcroadmap-report-sept-2022.pdf.
7. Holladay, J., Abdullah, Z., & Heyne, J., (2020). Sustainable aviation fuel: Review of technical pathways. https://www.energy.gov/sites/prod/files/2020/09/f78/beto-sust-aviation-fuel-sep-2020.pdf.
8. Government Accountability Office (GAO) Sustainable Aviation Fuel. *agencies should track progress toward ambitious federal goals* [Reissued with Revisions May 17, 2023], ed, 2023.
9. International Civil Aviation Organization (ICAO). *Sustainable aviation fuels guide*. [Online]. Available: https://www.icao.int/environmental-protection/knowledge-sharing/Docs/Sustainable%20Aviation%20Fuels%20Guide_vf.pdf
10. International Civil Aviation Organization (ICAO). (2022). *CORSIA default life cycle emission values for CORSIA eligible fuels*. [Online]. Available: https://www.icao.int/environmental-protection/CORSIA/Pages/CORSIA-Eligible-Fuels.aspx

11. Edwards, J. T., (2017). Reference jet fuels for combustion testing. In *55th AIAA Aerospace Sciences Meeting*, Grapevine, Texas, 2017/01/09 2017. https://doi.org/10.2514/6.2017-0146.
12. U. S. Federal Aviation Administration. *United States 2021 Aviation Climate Action Plan*. https://www.faa.gov/sites/faa.gov/files/2021-11/Aviation_Climate_Action_Plan.pdf. Aaccessed 30 Nov 2024).
13. U. S. Energy Information Administration. *Total energy: Production: Crude oil and lease condensate*. https://www.eia.gov/outlooks/aeo/data/browser/. Accessed 30 Nov 2024.
14. U. S. Environmental Protection Agency. *Fast Facts U.S.Transportation Sector Greenhouse Gas Emissions 1990–2019*. https://nepis.epa.gov/Exe/ZyPDF.cgi?Dockey=P10127TU.pdf. Accessed 30 Nov 2024.
15. International Civil Aviation Organization (ICAO). *ICAO global framework for aviation alternative fuels: Conversion processes*. [Online]. Available: https://www.icao.int/environmental-protection/GFAAF/Pages/Conversion-processes.aspx
16. U. S. Department of Energy an Office of Energy Efficiency an Renewable Energy. *Sustainable aviation fuel*. https://afdc.energy.gov/fuels/sustainable_aviation_fuel.html. Accessed 18 Dec 2024.
17. A. B. Swedish Biofuels. (2023). ASTM decision brings 100% SAF certification within reach. In *Cision PR Newswire*.
18. Stratton, R. W., Wong, H. M., & Hileman, J. I., Life cycle greenhouse gas emissions from alternative jet fuels. Massachusetts Institute of Technology, PARTNER-COE-2010-001 2010.
19. U. S. Department of Agriculture Economic Research Service. Oil Crops Yearbook (2023). https://www.ers.usda.gov/data-products/oil-crops-yearbook/oil-crops-yearbook/. Accessed 18 Dec 2024.
20. U.S. Energy Information Administration. *Monthly biodiesel production report*. http://www.eia.gov/biofuels/biodiesel/production/. Accessed 27 June 2013.
21. U.S. Department of Agriculture. *Oil crops yearbook*. http://usda.mannlib.cornell.edu/MannUsda/viewDocumentInfo.do?documentID=1290. Accessed 27 June 2013.
22. U. S. Department of Agriculture. *USDA agricultural projections to 2032 (Long-term projections report OCE-2023-1, Prepared by the Interagency Agricultural Projections Committee, 2023)*. https://www.usda.gov/sites/default/files/documents/USDA-Agricultural-Projections-to-2032.pdf Accessed 30 Nov 2024.
23. Bloomberg. *EPA rejects palm-oil based biodiesel for renewable fuels program*. http://www.bloomberg.com/news/2012-01-27/epa-rejects-palm-oil-based-biodiesel-for-renewable-fuels-program.html. Accessed 14 Dec 2012.
24. Milbrandt, A., Seiple, T., Heimiller, D., Skaggs, R., & Coleman, A. (2018). Wet waste-to-energy resources in the United States, (in en). *Resources, Conservation and Recycling, 137*, 32–47. https://doi.org/10.1016/j.resconrec.2018.05.023
25. R. Magazine. (2023). US Market Report. Render Mag. 52. Available: https://pubs.rendermagazine.com/2023-04/files/extfile/DownloadURL.pdf
26. NCGA. *World of Corn 2022*. https://ncga.com/world-of-corn#/. Accessed 10 May 2022.
27. Cardona, C. A., & Sanchez, O. J. (2007). Fuel ethanol production: Process design trends and integration opportunities. *Bioresource Technology, 98*(12), 2415–2457. https://doi.org/10.1016/j.biortech.2007.01.002
28. Maity, S. K. (2015, March). Opportunities, recent trends and challenges of integrated biorefinery: Part I. *Renewable & Sustainable Energy Reviews, 43*, 1427–1445. https://doi.org/10.1016/j.rser.2014.11.092
29. U. S. Department of Energy. "Conversion Technologies for Advanced Biofuels," Department of Energy, Office of the Biomass Program in the Office of Energy Efficiency and Renewable Energy, Conversion Technologies for Advanced Biofuels Workshop, December 2011.
30. Mirzaei, M., et al. (2021). Preliminary effects of crop residue management on soil quality and crop production under different soil management regimes in corn-wheat rotation systems. *Agronomy, 11*(2), 302.
31. M. H. Langholtz, B. J. Stokes, and L. M. Eaton, (2016). Billion-Ton report: Advancing domestic resources for a thriving bioeconomy, volume 1: Economic availability of feed-

stocks, Oak Ridge, [Online]. Available: https://www.srs.fs.usda.gov/pubs/books/2016/book_2016_abt_001.pdf
32. Hess, J. R., Foust, T. D., Hoskinson, R., & Thompson, D. (2003). *Roadmap for agriculture biomass feedstock supply in the United States*. US Department of Energy (USDOE).
33. National Corn Growers Association. *World of Corn 2021*. [Online]. Available: https://www.ncga.com/world-of-corn
34. Farm Energy. Corn Stover for Biofuel Production. [Online]. Available: https://farm-energy.extension.org/corn-stover-for-biofuel-production/
35. ASTM International. ASTM D975 – 12a Standard Specification for Diesel Fuel Oils. http://www.astm.org/Standards/D975.htm. Accessed 8 May 2013.
36. Antal, M. J., Allen, S. G., Dai, X., Shimizu, B., Tam, M. S., & Grønli, M. (2000). Attainment of the Theoretical Yield of Carbon from Biomass. *Industrial and Engineering Chemistry Research, 39*, 4024–4031.
37. U. S. Environmental Protection Agency. National overview: Facts and figures on materials, wastes and recycling. https://www.epa.gov/facts-and-figures-about-materials-waste-and-recycling/national-overview-facts-and-figures-materials. Accessed 18 Dec 2024.
38. Seiple, T., Jiang, Y., Snowden-Swan, L. J., Betzsold, N., Ramasamy, K. K., & Fuller, C. (2023). Cost-effective opportunities to produce sustainable aviation fuel from low-cost wastes in the US. *ACS Sustainable Chemistry & Engineering, 11*(33), 12326–12335.
39. Giuntoli, J., et al. (2022). The quest for sustainable forest bioenergy: win-win solutions for climate and biodiversity. *Renewable and Sustainable Energy Reviews, 159*, 112180.
40. Abreu, M., et al. (2022). Low indirect land use change (ILUC) energy crops to bioenergy and biofuels—A review. *Energies, 15*(12), 4348.
41. Schwerz, F., et al. (2020). Biomass and potential energy yield of perennial woody energy crops under reduced planting spacing. *Renewable Energy, 153*, 1238–1250.
42. Jingura, R. M., & Matengaifa, R. (2009). Optimization of biogas production by anaerobic digestion for sustainable energy development in Zimbabwe. *Renewable and Sustainable Energy Reviews, 13*(5), 1116–1120. https://doi.org/10.1016/j.rser.2007.06.015
43. Hessami, M.-A., Christensen, S., & Gani, R. (1996). Anaerobic digestion of household organic waste to produce biogas. *Renewable Energy, 9*(1–4), 954–957. https://doi.org/10.1016/0960-1481(96)88438-2
44. Lastella, G., Testa, C., Cornacchia, G., Notornicola, M., Voltasio, F., & Sharma, V. K. (2002). Anaerobic digestion of semi-solid organic waste: Biogas production and its purification. *Energy Conversion and Management, 43*(1), 63–75. https://doi.org/10.1016/S0196-8904(01)00011-5
45. Tanzil, A. H., Brandt, K., Wolcott, M., Zhang, X., & Garcia-Perez, M. (2021). Strategic assessment of sustainable aviation fuel production technologies: Yield improvement and cost reduction opportunities. *Biomass and Bioenergy, 145*, 105942. https://doi.org/10.1016/j.biombioe.2020.105942
46. Farm Energy. The Economics of Switchgrass for Biofuel. [Online]. Available: https://farm--energy.extension.org/the-economics-of-switchgrass-for-biofuel/
47. Farm Energy. Algae for biofuel production. [Online]. Available: https://farm-energy.extension.org/algae-for-biofuel-production/
48. Tao, L., Markham, J. N., Haq, Z., & Biddy, M. J. (2017). Techno-economic analysis for upgrading the biomass-derived ethanol-to-jet blendstocks. *Green Chemistry, 19*(4), 1082–1101.
49. Tao, L., Milbrandt, A., Zhang, Y. A., & Wang, W. C. (2017, November). Techno-economic and resource analysis of hydroprocessed renewable jet fuel. *Biotechnology for Biofuels, 10*, 261. https://doi.org/10.1186/s13068-017-0945-3
50. International Civil Aviation Organization. *ICAO Environment – SAF rules of thumb*. [Online]. Available: https://www.icao.int/environmental-protection/Pages/SAF_RULESOFTHUMB.aspx

51. Swanson, R. M., Satrio, J. A., Brown, R. C., Platon, A., & Hsu, D. D. (2010). *Techno-economic analysis of biofuels production based on gasification, NREL/TP-6A20-46587*. National Renewable Energy Laboratory.
52. Wyman, C. E., Dale, B. E., Elander, R. T., Holtzapple, M., Ladisch, M. R., & Lee, Y. Y. (2005, December). Coordinated development of leading biomass pretreatment technologies. *Bioresource Technology, 96*(18), 1959–1966. https://doi.org/10.1016/j.biortech.2005.01.010
53. Lloyd, T. A., & Wyman, C. E. (2005, December). Combined sugar yields for dilute sulfuric acid pretreatment of corn stover followed by enzymatic hydrolysis of the remaining solids. *Bioresource Technology, 96*(18), 1967–1977. https://doi.org/10.1016/j.biortech.2005.01.011
54. Sun, F., & Chen, H. Z. (2008, September). Organosolv pretreatment by crude glycerol from oleochemicals industry for enzymatic hydrolysis of wheat straw. *Bioresource Technology, 99*(13), 5474–5479. https://doi.org/10.1016/j.biortech.2007.11.001
55. Kim, Y., Hendrickson, R., Mosier, N. S., & Ladisch, M. R. (2009). Liquid hot water pretreatment of cellulosic biomass. *Biofuels: Methods and Protocols, 581*, 93–102. https://doi.org/10.1007/978-1-60761-214-8_7
56. Xu, J. L., Cheng, J. J., Sharma-Shivappa, R. R., & Burns, J. C. (2010, April). Lime pretreatment of switchgrass at mild temperatures for ethanol production. *Bioresource Technology, 101*(8), 2900–2903. https://doi.org/10.1016/j.biortech.2009.12.015
57. Goh, B. H. H., et al. (2020). Progress in utilisation of waste cooking oil for sustainable biodiesel and biojet fuel production. *Energy Conversion and Management, 223*, 113296. https://doi.org/10.1016/j.enconman.2020.113296
58. Kannah, R. Y., Kavitha, S., Banu, J. R., Sivashanmugam, P., Gunasekaran, M., & Kumar, G. (2021). A mini review of biochemical conversion of algal biorefinery. *Energy & Fuels, 35*(21), 16995–17007. https://doi.org/10.1021/acs.energyfuels.1c02294
59. Ochuodho, T. O., Alavalapati, J. R. R., Lal, P., Agyeman, D. A., Wolde, B., & Burli, P. (2019). Potential economic impacts of allocating more land for bioenergy biomass production in Virginia. *Forests, 10*(2), 159. [Online]. Available: https://www.mdpi.com/1999-4907/10/2/159
60. Davis, R., et al. (2015). *Process design and economics for the conversion of lignocellulosic biomass to hydrocarbons: Dilute-acid and enzymatic deconstruction of biomass to sugars and catalytic conversion of sugars to hydrocarbons*. National Renewable Energy Laboratory.
61. Jones, S., et al. (2013). *Process design and economics for the conversion of lignocellulosic biomass to hydrocarbon fuels: Fast pyrolysis and hydrotreating bio-oil pathway, PNNL-23053, NREL/TP-5100-61178*. Pacific Northwest National Laboratory/National Renewable Energy Laboratory.
62. Brown, C., & Tao, L., (2023). *Biofuel production and greenhouse gas reduction potential*. [Online]. Available: https://www.osti.gov/servlets/purl/2202642
63. Bioenergy Technologies Office, "biofuels and bioproducts from wet and gaseous waste streams: Challenges and Opportunities," 2017. [Online]. Available: https://energy.gov/sites/prod/files/2017/01/f34/biofuels_and_bioproducts_from_wet_and_gaseous_waste_streams_full_report_2.pdf
64. Davis, R. E., et al. (2017). Algae harmonization study: Evaluating the potential for future algal biofuel costs, sustainability, and resource assessment from harmonized modeling. *National Renewable Energy Lab..* (NREL), Golden, CO (United States), 2018.
65. Mohan, D., Pittman, C. U., Jr., & Steele, P. H. (2006). Pyrolysis of wood/biomass for bio-oil: A critical review. *Energy & Fuels, 20*(3), 848–889.
66. Wrasman, C. J., et al. (2023). Catalytic pyrolysis as a platform technology for supporting the circular carbon economy. *Nature Catalysis, 6*(7), 563–573.
67. Bridgwater, A. V. (2003). Renewable fuels and chemicals by thermal processing of biomass. *Chemical Engineering Journal, 91*(2–3), 87–102.
68. Bahng, M.-K., Mukarakate, C., Robichaud, D. J., & Nimlos, M. R. (2009). Current technologies for analysis of biomass thermochemical processing: A review. *Analytica Chimica Acta, 651*(2), 117–138.

69. Venderbosch, R. H., Ardiyanti, A. R., Wildschut, J., Oasmaa, A., & Heeres, H. (2010). Stabilization of biomass-derived pyrolysis oils. *Journal of Chemical Technology & Biotechnology, 85*(5), 674–686.
70. Han, Y., et al. (2019). Hydrotreatment of pyrolysis bio-oil: A review. *Fuel Processing Technology, 195*, 106140.
71. Wang, G., et al. (2020). A review of recent advances in biomass pyrolysis. *Energy & Fuels, 34*(12), 15557–15578.
72. Jahirul, M. I., Rasul, M. G., Chowdhury, A. A., & Ashwath, N. (2012). Biofuels production through biomass pyrolysis—a technological review. *Energies, 5*(12), 4952–5001.
73. Pecha, M. B., Arbelaez, J. I. M., Garcia-Perez, M., Chejne, F., & Ciesielski, P. N. (2019). Progress in understanding the four dominant intra-particle phenomena of lignocellulose pyrolysis: Chemical reactions, heat transfer, mass transfer, and phase change. *Green Chemistry, 21*(11), 2868–2898.
74. Sharma, A., Pareek, V., & Zhang, D. (2015). Biomass pyrolysis—A review of modelling, process parameters and catalytic studies. *Renewable and Sustainable Energy Reviews, 50*, 1081–1096.
75. Garcia-Perez, M., et al. (2008). Fast pyrolysis of oil mallee woody biomass: Effect of temperature on the yield and quality of pyrolysis products. *Industrial & Engineering Chemistry Research, 47*(6), 1846–1854.
76. Bridgwater, A. V., Meier, D., & Radlein, D. (1999). An overview of fast pyrolysis of biomass. *Organic Geochemistry, 30*(12), 1479–1493.
77. Paasikallio, V., Kalogiannis, K., Lappas, A., Lehto, J., & Lehtonen, J. (2017). Catalytic fast pyrolysis: Influencing bio-oil quality with the catalyst-to-biomass ratio. *Energy Technology, 5*(1), 94–103.
78. Ciesielski, P. N., et al. (2018). Advancing catalytic fast pyrolysis through integrated multiscale modeling and experimentation: Challenges, progress, and perspectives. *Wiley Interdisciplinary Reviews: Energy and Environment, 7*(4), e297.
79. Shen, J., Wang, X.-S., Garcia-Perez, M., Mourant, D., Rhodes, M. J., & Li, C.-Z. (2009). Effects of particle size on the fast pyrolysis of oil mallee woody biomass. *Fuel, 88*(10), 1810–1817.
80. Bryden, K. M., & Hagge, M. J. (2003). Modeling the combined impact of moisture and char shrinkage on the pyrolysis of a biomass particle☆. *Fuel, 82*(13), 1633–1644.
81. Fan, M., et al. (2022). Impact of biochar catalyst on pyrolysis of biomass of the same origin. *Journal of Environmental Chemical Engineering, 10*(5), 108546.
82. Ensyn. http://www.ensyn.com/overview.html. Accessed 30 Nov 2024.
83. BTG. https://www.btg-bioliquids.com/our-technology/. Accessed 30 Nov 2024.
84. Bridgwater, A. V. (2012). Review of fast pyrolysis of biomass and product upgrading. *Biomass and Bioenergy, 38*, 68–94.
85. Abdullah, Z., (2015). *Upgrading of biomass fast pyrolysis oil*. DOE Bioenergy Technologies Office (BETO) 2015 Project Peer Review. [Online]. Available: https://www.energy.gov/eere/bioenergy/articles/2015-peer-review-presentations-thermochemical-conversion
86. Yung, M. M., et al. (2023). Advancements and challenges in the production of low-carbon fuels via catalytic fast pyrolysis of biomass through refinery integration and co-product generation. *Green Chemistry, 25*(17), 6809–6822.
87. Mukarakate, C., et al. (2022). Accelerating catalyst development for biofuel production through multiscale catalytic fast pyrolysis of biomass over Mo2C. *Chem Catalysis, 2*(7), 1819–1831.
88. French, R. J., et al. (2021). Optimizing process conditions during catalytic fast pyrolysis of pine with Pt/TiO2—improving the viability of a multiple-fixed-bed configuration. *ACS Sustainable Chemistry & Engineering, 9*(3), 1235–1245.
89. Iisa, K., et al. (2020). Ga/ZSM-5 catalyst improves hydrocarbon yields and increases alkene selectivity during catalytic fast pyrolysis of biomass with co-fed hydrogen. *Green Chemistry, 22*(8), 2403–2418.

90. Mettler, M. S., Vlachos, D. G., & Dauenhauer, P. J. (2012). Top ten fundamental challenges of biomass pyrolysis for biofuels. *Energy & Environmental Science, 5*(7), 7797–7809.
91. Wilson, A. N., et al. (2019). Valorization of aqueous waste streams from thermochemical biorefineries. *Green Chemistry, 21*(15), 4217–4230.
92. Wilson, A. N., et al. (2017). Integrated biorefining: coproduction of renewable resol biopolymer for aqueous stream valorization. *ACS Sustainable Chemistry & Engineering, 5*(8), 6615–6625.
93. Long, A., Arnold, H., Brown, B., et al. (2022). *Guidelines/handbook for the design of modular gasification systems*. National energy Technology Laboratory, (NETL). US DOE, DOE/NETL-2022/3209.
94. National Energy Technology Laboratory (NETL). Gasopedia. https://netl.doe.gov/carbon-management/energy-systems/gasification/gasifipedia. Accessed 18 Dec 2024.
95. International Energy Agency Bioenergy Task 33. Gasification applications in existing infrastructures for production of sustainable value-added products, 2021. [Online]. Available: https://www.ieabioenergy.com/blog/publications/gasification-applications-in-existing-infrastructures-for-the-production-of-sustainable-value-added-products/
96. Hrbek, J. (2015). Status report on thermal gasification of biomass and waste 2019. *IEA Bioenergy Task, 33*, 1–125.
97. NETL. *Syngas Contaminant Removal and Conditioning*. [Online]. Available: https://netl.doe.gov/research/coal/energy-systems/gasification/gasifipedia/cleanup
98. Abdoulmoumine, N., Adhikari, S., Kulkarni, A., & Chattanathan, S. (2015). A review on biomass gasification syngas cleanup. *Applied Energy, 155*, 294–307.
99. Reinhard Rauch, X. K. (2021). IEA Bioenergy Task 33,"Gasification applications in existing infrastructures for production of sustainable value-added products", Case study 3: Integration of renewables into existing refineries. [Online].. Available: https://www.ieabioenergy.com/blog/publications/gasification-applications-in-existing-infrastructures-for-the-production-of-sustainable-value-added-products/
100. Brooks, K., et al. (2016). Low-carbon aviation fuel through the alcohol to jet pathway. In *Biofuels for aviation* (pp. 109–150). Elsevier.
101. Liu, Y., et al. (2022). A review on the promising fuel of the future–Biobutanol; the hindrances and future perspectives. *Fuel, 327*, 125166.
102. Goh, B. H. H., et al. (2022). Recent advancements in catalytic conversion pathways for synthetic jet fuel produced from bioresources. *Energy Conversion and Management, 251*, 114974.
103. Davis, R. E. et al., (2018). Process design and economics for the conversion of lignocellulosic biomass to hydrocarbon fuels and coproducts: 2018 biochemical design case update; biochemical deconstruction and conversion of biomass to fuels and products via integrated biorefinery pathways. National Renewable Energy Lab.(NREL). [Online]. Available: https://www.nrel.gov/docs/fy19osti/71949.pdf
104. Hannon, J. R., et al. (2020). Technoeconomic and life-cycle analysis of single-step catalytic conversion of wet ethanol into fungible fuel blendstocks. *Proceedings of the National Academy of Sciences, 117*(23), 12576–12583.
105. Klein, B., McNamara, I., Davis, R., Mittal, A., & Johnson, D. (2021). *Techno-economic assessment for the production of hydrocarbon fuels via catalytic upgrading of furans*. National Renewable Energy Lab.(NREL).
106. Davis, R. et al., (2015). Process design and economics for the conversion of lignocellulosic biomass to hydrocarbons: Dilute-acid and enzymatic deconstruction of biomass to sugars and catalytic conversion of sugars to hydrocarbons. National Renewable Energy Laboratory (NREL). NREL/TP-5100-62498. Accessed 19 May 2015, 18:41:42.
107. Atsonios, K., Kougioumtzis, M.-A., Panopoulos, K. D., & Kakaras, E. (2015). Alternative thermochemical routes for aviation biofuels via alcohols synthesis: Process modeling, techno-economic assessment and comparison. *Applied Energy, 138*, 346–366. https://doi.org/10.1016/j.apenergy.2014.10.056

108. Neuling, U., & Kaltschmitt, M. (2018). Techno-economic and environmental analysis of aviation biofuels. *Fuel Processing Technology, 171*, 54–69. https://doi.org/10.1016/j.fuproc.2017.09.022
109. Pearlson, M., Wollersheim, C., & Hileman, J. (2013). A techno-economic review of hydroprocessed renewable esters and fatty acids for jet fuel production. *Biofuels, Bioproducts and Biorefining, 7*(1), 89–96. https://doi.org/10.1002/bbb.1378
110. Diederichs, G. W., Ali Mandegari, M., Farzad, S., & Görgens, J. F. (2016). Techno-economic comparison of biojet fuel production from lignocellulose, vegetable oil and sugar cane juice. *Bioresource Technology, 216*, 331–339. https://doi.org/10.1016/j.biortech.2016.05.090
111. Yao, G., Staples, M. D., Malina, R., & Tyner, W. E. (2017). Stochastic techno-economic analysis of alcohol-to-jet fuel production. *Biotechnology for Biofuels, 10*(1), 18. https://doi.org/10.1186/s13068-017-0702-7
112. DOE News. U.S. Department of the Treasury, IRS Release Guidance to Drive American Innovation, Cut Aviation Sector Emissions. ed, 2023.
113. International Civil Aviation Organization (ICAO). CORSIA eligible fuels. [Online]. Available: https://www.icao.int/environmental-protection/CORSIA/Pages/CORSIA-Eligible-Fuels.aspx
114. Minnesota Pollution Control Agency. *Facts about determining applicability of new source review*. https://www.pca.state.mn.us/sites/default/files/aq4-25.pdf. Accessed 30 Nov 2024.
115. Inside Climate News. *Fulcrum bioenergy, aiming to produce 'Net-Zero' jet fuel from plastic waste, hits heavy turbulence*. ed, 2023, pp. https://insideclimatenews.org/news/20112023/fulcrum-gary-indiana-net-zero-jet-fuel/.
116. U. S. Environmental Protection Agency. *Nonattainment NSR Basic Information*. https://www.epa.gov/nsr/nonattainment-nsr-basic-information. Accessed 18 Dec 2024.
117. ECFR. *Code of Federal Regulations, 40 CFR Part 71 Subpart A – Operating Permits*. https://www.ecfr.gov/current/title-40/part-71/subpart-A. Accessed 30 Nov 2024.
118. U. S. Environmental Protection Agency. *Phosphoric acid supply chain—executive summary*. https://www.epa.gov/system/files/documents/2023-03/Phosphoric%20Acid%20Supply%20Chain%20Profile.pdf. Accessed 30 Nov 2024.
119. BloombergNEF. Over $50 Billion Flow to Climate-Tech Startups in a Stormy Year. ed, 2024.
120. The White House. *The Long-Term Strategy of the United States: Pathways to Net-Zero Greenhouse Gas Emissions by 2050*. https://www.whitehouse.gov/wp-content/uploads/2021/10/US-Long-Term-Strategy.pdf. Accessed 30 Nov 2024.
121. U.S. Congress. H.R.741 – 117th Congress (2021–2022), Sustainable Aviation Fuel Act. [Online]. Available: https://www.congress.gov/bill/117th-congress/house-bill/741
122. Hubbard, Dan. *Coalition calls on EPA to expand sustainable aviation fuel feedstocks in updated RFS*. [Online]. Available: https://nbaa.org/press-releases/coalition-calls-on-epa-to-expand-sustainable-aviation-fuel-feedstocks-in-updated-rfs/
123. U. S. Environmental Protection Agency. Final Volume Standards for 2020, 2021, and 2022 (last updated July 1, 2022). https://www.epa.gov/renewable-fuel-standard-program/final-volume-standards-2020-2021-and-2022. Aaccessed 30 Nov 2024.
124. U.S. Congress. *Inflation Reduction Act of 2022*. [Online]. Available: https://www.congress.gov/bill/117th-congress/house-bill/5376/text
125. California Air Resources Board. *Low Carbon Fuel Standard*. [Online]. Available: https://ww2.arb.ca.gov/our-work/programs/low-carbon-fuel-standard/about
126. California Air Resources Board. *Low carbon fuel standard basics*. [Online]. Available: https://ww2.arb.ca.gov/resources/documents/lcfs-basics
127. Pacific Coast Collaborative. *We have a bold vision for our future*. [Online]. Available: https://pacificcoastcollaborative.org/about/
128. Pacific Coast Collaborative. *Pacific Coast action plan on climate and energy*. [Online]. Available: https://46h83069gmc37jdhm425hbh3-wpengine.netdna-ssl.com/wp-content/uploads/2018/09/Pacific-Coast-Climate-Action-Plan.pdf

129. Witcover, J., & Murphy, C.. Oregon clean fuels program: A review and status update. [Online].. Available: https://doi.org/10.1177%2F0361198120972394
130. Oregon Department of Environmental Quality. Clean fuels program overview. [Online]. Available: https://www.oregon.gov/deq/ghgp/cfp/Pages/CFP-Overview.aspx
131. Oregon Secretary of State. Department of Environmental Quality: Chapter 340, Division 253, Oregon Clean Fuels Program. [Online]. Available: https://secure.sos.state.or.us/oard/displayDivisionRules.action?selectedDivision=1560
132. State of Washington Department of Ecology. Chapters 173–424 WAC and 173–455 WAC. [Online]. Available: https://ecology.wa.gov/Regulations-Permits/Laws-rules-rulemaking/Rulemaking/WAC-173-424-455
133. State of Washington Department of Ecology. *Clean Fuel Standard*. [Online]. Available: https://ecology.wa.gov/Air-Climate/Climate-change/Reducing-greenhouse-gases/Clean-Fuel-Standard
134. Legislative Assembly of British Columbia. *Low Carbon Fuels Act*. [Online]. Available: https://www.leg.bc.ca/parliamentary-business/legislation-debates-proceedings/42nd-parliament/3rd-session/bills/bills-with-hansard-debate
135. Government of British Columbia. *Renewable & Low Carbon Fuel Requirements Regulation: BC Low Carbon Standard*. [Online]. Available: https://www2.gov.bc.ca/gov/content/home
136. Government of Canada. *What are the clean fuel regulations?*. [Online]. Available: https://www.canada.ca/en/environment-climate-change/services/managing-pollution/energy-production/fuel-regulations/clean-fuel-regulations/about.html
137. Government of Canada. *Clean fuel regulations*. [Online]. Available: https://pollution-waste.canada.ca/environmental-protection-registry/regulations/view?Id=1170
138. The New York State Senate. *Assembly bill A9701*. [Online]. Available: https://www.nysenate.gov/legislation/bills/2021/A9701
139. LegiScan. *New York Assembly Bill A9701*. [Online]. Available: https://legiscan.com/NY/bill/A09701/2021
140. European Parliament. *Sustainable aviation fuels (ReFuelEU Aviation Initiative)*. [Online]. Available: https://www.europarl.europa.eu/doceo/document/TA-9-2022-0297_EN.html
141. Pavlenko, N.. (2021). An assessment of the policy options for driving sustainable aviation fuels in the European Union. *CCT briefing: An assessment of the policy options for driving sustainable aviation fuels in the EU. Accessed 6/2021.*
142. Mirolo, M.. *Making aviation fuel mandates sustainable*. [Online]. Available: https://www.transportenvironment.org/discover/making-aviation-fuel-mandates-sustainable/
143. Ng, K. S., Farooq, D., & Yang, A. (2021). Global biorenewable development strategies for sustainable aviation fuel production. *Renewable and Sustainable Energy Reviews, 150*, 111502.
144. Vimmerstedt, L. J., & Newes, E. K. (2017). *Effect of Additional Incentives for Aviation Biofuels: Results from the Biomass Scenario Model*. National Renewable Energy Lab.(NREL).
145. Pavlenko, N., Searle, S., & Christensen, A., (2022). Opportunities and risks for a national low-carbon fuel standard.

Chapter 3
Microalgae: Next-Generation Feedstock for Sustainable Aviation Fuel

Bidhu Bhusan Makut, Pritikrishna Majhi, Sanjeev Mishra, Mohammad Aslam, and Radhakanta Nag

1 Introduction

The substantial expansion of the aviation sector constitutes a significant concern regarding the impending energy crisis and environmental degradation. In 2019, the commercial aviation sector discharged approximately 900 million metric tons of CO_2, or about 2.5% of total CO_2 emissions, globally by human activities [1]. The aviation sector's carbon dioxide emissions are anticipated to rise to 3100 metric tons by 2050. The components of aircraft exhaust gases, such as CO_2, NO_x, SO_2, CO, and Unburned Hydrocarbon (UHC), contribute to global warming by degrading the ozone layer. To effectively mitigate the environmental repercussions associated with aviation, a holistic approach is imperative, which incorporates improvements in fuel efficiency alongside the exploration of carbon-neutral alternatives. Fuels derived from biomass exhibit significant promise in minimizing emissions while aligning with ecological objectives. In light of environmental considerations, multiple institutions, particularly the International Air Transport Association (IATA), have dedicated resources to promote the progression and implementation of

B. B. Makut (✉) · P. Majhi
Department of Biotechnology, Faculty of Health and Life Sciences, Chandigarh University, Unnao, Uttar Pradesh, India
e-mail: bidhu.l100107@culko.in

S. Mishra
Biochemical Conversion Division, Sardar Swaran Singh National Institute of Bio-Energy, Kapurthala, Punjab, India

M. Aslam
Department of Chemistry, National Institute of Technology Srinagar, Srinagar (J&K), India

R. Nag
Department of Microbiology, College of Basic Science and Humanities, Odisha University of Agriculture and Technology, Bhubaneswar, Odisha, India

M. Aslam et al. (eds.), *Sustainable Aviation Fuels*, Sustainable Aviation,
https://doi.org/10.1007/978-3-031-83721-0_3

sustainable aviation fuels (SAFs) [24]. The criteria for fuel sustainability must embrace (i) alignment with the operational capabilities of existing aircraft engines, (ii) the necessity for no modifications to the existing infrastructure, (iii) the production of minimal life cycle carbon emissions in comparison to conventional jet fuel, (iv) the safeguarding of food security and ecological integrity in relation to the cultivation of energy crops, and (v) the mitigation of deforestation as a significant factor [21].

Sustainable aviation fuel (SAF) produced from microalgae is garnering an increasing level of scholarly interest and possesses the potential to materialize as a noteworthy advancement with extensive global implications in the near future. At present, the oil extracted from microalgae is unsuitable for human consumption and poses no threat to food security. However, the costs associated with producing aviation fuel from microalgae remain considerably higher when juxtaposed with conventional fossil-based aviation fuel, thereby resulting in an escalation in the market price of the former. The production of aviation fuel from algal biomass feedstock necessitates a comprehensive evaluation of the prospective challenges. The methodology involved in producing aviation fuel from algal biomass, which includes the harvesting of microalgae, the processing pathways, and the attributes of the resultant aviation fuel, poses significant hurdles. Microalgae-derived aviation fuel, which exhibits similar physicochemical properties to conventional petroleum aviation fuel, can be synthesized through the meticulous identification and rectification of the existing challenges. The present chapter emphasizes various features relating to microalgae aviation fuel production, predominantly on the factors that have a significant impact on the yield and quality of microalgae-derived aviation fuel. The benefits and drawbacks of every route, such as hydro-processing, gas-to-liquid, and alcohol-to-aviation fuel, are delineated. A comprehensive analysis is conducted on the hurdles encountered in the production of aviation fuel derived from microalgae biomass, along with the strategies proposed to address these obstacles.

2 Microalgae as a Promising Feedstock for Sustainable Aviation Fuel (SAF)

Biomass-derived sustainable aviation fuel (SAF) is widely acknowledged as a crucial option for promoting a sustainable aviation industry. The ideal biomass feedstocks utilized in the production of SAF are agricultural commodities like soybeans, corn, maize, sugar beets, and sugarcane. These resources are categorized as first-generation biofuels. The diverse classifications of biofuels have engendered considerable debate due to their detrimental effects on food safety, universal food trade, water resource allocation, and deforestation [18]. Furthermore, second-generation biofuels, sourced from inedible oils like *Jatropha curcas*, *Pongamia pinnata*, and *Simarouba glauca*, together with lignocellulosic biomass and forest by-products, need significant land areas for production. Moreover, the production of

second-generation biofuels has challenges related to the advancement of effective technology for the commercial use of waste materials as feedstock for biofuel synthesis. Given the limitations of first- and second-generation biofuels, third-generation biofuels derived from microalgal biomass are being explored as a viable alternative to fossil fuels because of their distinctive properties. These attributes include [1] being effective as a suitable nonfood substitute for biofuel production; [3] having a lipid content exceeding that of terrestrial plants (between 40 and 80% of dry mass); (3) exhibiting a rapid reproductive and growth rate, with a doubling time of about 24 hours, enabling multiple harvests annually; [4] utilizing atmospheric carbon dioxide for growth; and [2] being free of sulfur, being nontoxic, and demonstrating biodegradability.

Furthermore, these organisms are recognized as potential options in biofuel generation because of their substantial concentrations of fatty acids and lipids. Fatty acids and lipids serve as the primary energy reserves for biofuel production. The predominant microalgal species, including *Nannochloropsis* sp., *Tetraselmis* sp., *Chlamydomonas* sp., and *Synechococcus* sp., are chiefly utilized in the production of biofuels, including biodiesel, bio-hydrogen, and bio-jet fuel [3]. The growing focus on SAFs generated from algae indicates their potential to become a substantial opportunity with universal insinuations in the upcoming years.

There is a significant focus on the advancement of biofuels for aviation to mitigate the effects of climate change and global warming. Moreover, this will contribute to the reduction of operational expenses within the aviation sector. Additionally, the substantial influence of petroleum-based aviation fuel prices and their volatility profoundly affect the operations of airlines. The fluctuations in fuel costs have far-reaching consequences on the economies of numerous airlines and countries. Moreover, the need for aviation fuel is growing alarmingly. The consumption of fuel by commercial airlines on a global scale exhibited a consistent annual increase from the year 2009 onward, ultimately peaking at a record-breaking 95 billion gallons in 2019. Nevertheless, the onset of the coronavirus pandemic led to a substantial decline in fuel consumption, plummeting to a mere 52 billion gallons in the year 2020. In the year 2021, there was a resurgence in fuel consumption by commercial airlines, with projections indicating a prospective escalation to 92 billion gallons in 2023 and 99 billion gallons in 2024 [6].

3 Implementation of Novel Strategies for the Conversion of Microalgae into SAF

There are several novel strategies adopted for the transformation of microalgal biomass into aviation fuel, which can be synthesized from triglycerides, sugars, and starch present in microalgal biomass. Basically, there are three conversion pathways for the generation of aviation fuel, which include (i) gas-to-aviation fuel, (ii)

oil-to-aviation fuel, and (iii) alcohol-to-aviation fuel. These methods pointed out in Fig. 3.1 can be used to produce microalgae aviation fuel.

3.1 Biomass Gasification Technology Using Fischer–Tropsch Synthesis

Gasification is a complex technological procedure that transforms various carbon sources (including natural gas, coal, and biomass) into a combination of gaseous components (such as CO, H_2, CO_2, and CH_4) by the application of high temperatures in a carefully controlled oxidizing environment (such as air, steam, or oxygen). Biomass gasification by Fischer–Tropsch (FT) synthesis is a viable method for generating sustainable liquid fuels, including synthetic gasoline, diesel, and aviation fuels. This procedure encompasses multiple critical phases, including biomass gasification, syngas conditioning, and the Fischer–Tropsch synthesis itself. Every phase is essential for guaranteeing the efficacy and financial feasibility of the biomass-to-liquid process.

In the biomass gasification process, syngas is generated from microalgal biomass at elevated temperatures (150–300 °C) and at high pressures (10–40 bars). The generated syngas yields a liquid hydrocarbon fuel composed of elongated alkane chains via the Fischer–Tropsch synthesis process, which transpires when it interacts with hydrogen in the presence of a catalyst, often iron, cobalt, ruthenium, or nickel. Nonetheless, the utilization of microalgal biomass for the production of aviation fuel carries significant implications for its commercial feasibility through the Fischer–Tropsch synthesis. The microalgal biomass is composed of a greater quantity of contaminants (such as nitrogen, chlorine, sulfur, tars, ashes, and alkalis) that

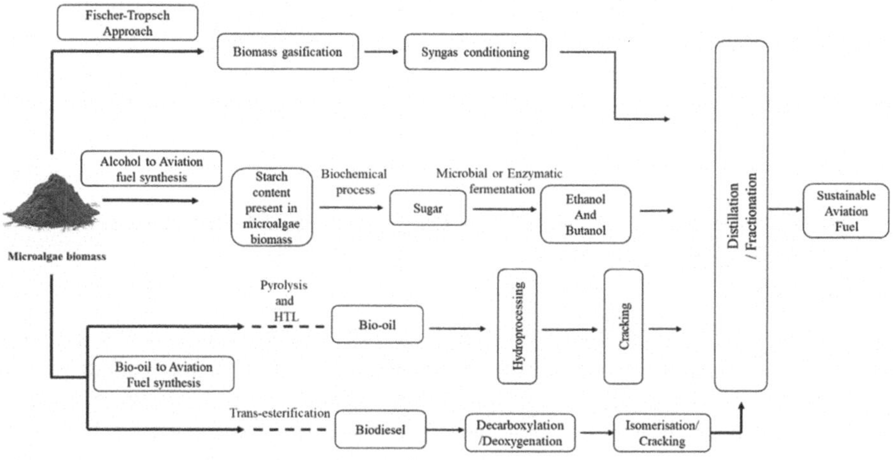

Fig. 3.1 Conversion processes to produce microalgae-based aviation fuel

are commonly present alongside carbon monoxide and hydrogen in the gas phase. The existence of these pollutants and the oxygen content in the biomass presents considerable hurdles since they adversely affect the effectiveness of Fischer–Tropsch catalysts. Consequently, the purification of syngas is the most economically onerous element of a Fischer–Tropsch (FT) synthesis facility. This phase constitutes around 60–70% of the total expense when employing natural gas as the feedstock. The purification of syngas is therefore a meticulous challenge to the successful commercialization of the biomass-to-liquid (BTL) process. The occurrence of pollutants in the syngas formed during the gasification process is unavoidable. Syngas contains several impurities, like particles; condensable tars; benzene, toluene, and xylenes (BTX); alkali compounds; hydrogen sulfide (H_2S); hydrochloric acid (HCl); ammonia (NH_3); and hydrogen cyanide (HCN). The catalysts employed in the Fischer–Tropsch (FT) reactor for converting syngas into liquid fuels have considerable sensitivity to impurities, especially sulfur and nitrogen compounds, which can permanently deactivate Fischer–Tropsch catalysts. The accumulation of alkaline metals and tars on the catalysts results in product contamination, whereas particulate matter causes reactor fouling. Therefore, thorough syngas purification is essential before it enters the Fischer–Tropsch (FT) reactor.

The first step in syngas conditioning is the exclusion of BTX and the condensable tars (larger hydrocarbons). BTX must be eliminated prior to the active carbon filters within the syngas purification system as active carbon has a propensity to adsorb BTX, necessitating frequent regeneration and consequently diminishing process reliability. Tars typically condense under the conventional operational conditions of the Fischer–Tropsch reactor, leading to fouling of downstream apparatus, deposition on surfaces, and infiltration into the pores of filters and sorbents. Hence, it is imperative that tars be extracted to a concentration that falls below the condensation threshold at the operational pressure of the Fischer–Tropsch (FT) reactor [10]. Following tar extraction, further impurities can be eliminated from the syngas using either the traditional "wet" low-temperature method or the "dry" high-temperature purification processes. The wet gas cleaning method is supported by empirical data and has attained considerable commercialization in large-scale coal gasification systems [25]. Ammonia (NH_3) is subsequently eliminated via a water washing apparatus, concomitantly extracting halides, while hydrogen sulfide (H_2S) is either absorbed or converted to elemental sulfur through the Claus process. In the concluding phase, the gas traverses through zinc oxide (ZnO) and activated carbon filtration systems, which effectively eradicate H_2S along with any residual trace contaminants. Following the gas purification process, the syngas produced from biomass must undergo conditioning to calibrate the H_2/CO ratio to the specifications necessary for the Fischer–Tropsch (FT) reactor. Conventional conditioning procedures encompass the steam reforming of methane and lighter hydrocarbons into carbon monoxide (CO) and hydrogen (H_2) utilizing a nickel-based catalyst, succeeded by a water–gas shift (WGS) reactor [12].

To generate hydrocarbons within the aviation fuel range (C9–C16), it is possible to initially produce heavy hydrocarbons (known as waxes) by conducting operations at reduced temperatures (230 °C). Subsequently, controlled cracking and

isomerization processes can be employed to convert these heavy hydrocarbons into aviation fuel components. Therefore, traditional petrochemical processes like hydrocracking, isomerization, and fractionation should be applied to the Fischer–Tropsch reactor to transform the molecular weight and configuration of the hydrocarbons to align with the jet fuel specifications. As the hydrocarbon oil obtained after this process does not contain aromatic hydrocarbon, this oil could be used as a blending mixture.

3.2 Transformation of Microalgal Bio-Oil to Aviation Fuel

The transformation of microalgal oil to aviation fuel is considered a promising way for sustainable aviation fuel (SAF) production, leveraging the high productivity and adaptability of microalgae. The bio-oil derived from microalgae is generated via thermochemical conversion mechanisms, specifically fast pyrolysis, and hydrothermal liquefaction, carried out at elevated temperatures (ranging from 300 °C to 700 °C) and pressures. Within the realm of thermochemical conversion processes, microalgal biomass yields biochar, bio-crude, and non-condensable gases (CH_4 and CO_2) predominantly in oxygen-deprived conditions.

The bio-crude oil acquired subsequent to the pyrolysis of microalgal biomass encompasses a variety of aliphatic and aromatic hydrocarbons, along with compounds holding oxygen and nitrogen. The presence of oxygenated compounds in bio-crude oil may lead to the generation of acidity through carboxylic acids, thereby triggering subsequent reactions such as oil polymerization in storage. This phenomenon could potentially elevate viscosity levels, consequently impeding flow characteristics when utilized in engines. Additionally, the inclusion of these compounds diminishes the calorific value of the oils, rendering them incompatible with blending with conventional fossil oils, whereas the nitrogenous compounds contaminate the catalyst and produce nitrogen oxides during combustion [16]. Hence, the exclusion of O_2 and N_2 components from bio-crude oil is crucial in order to optimize the hydrocarbon ratio, thereby yielding transportation fuels characterized by elevated heating values and reduced environmental footprint.

The upgarde of the microalgal bio-oil falls into two main categories: (i) Physical upgrading and (ii) chemical upgrading. Some methodologies for the physical upgrading of bio-oil to improve its quality encompass filtration, emulsification, and the incorporation of solvents. Filtration represents a straightforward and extensively utilized technique aimed at the removal of solid particulates and char residues from bio-oil. Various filters and membranes are employed to eradicate contaminants present within the oil [15]. A significant benefit of hot vapor filtration is its efficacy in substantially diminishing char and ash concentrations, thereby mitigating the occurrence of secondary reactions. Furthermore, this filtration technique enhances the combustion rate and lessens the latency associated with fuel ignition. Emulsification may be facilitated through the incorporation of diesel into the oil. The liquid microstructure undergoes modification upon the addition of ethanol or methanol, both of

which are polar solvents. This process facilitates the dissolution of components that exhibit lower solubility, ultimately contributing to the stabilization of the aging process. Additional outcomes of this methodology include a decrease in the viscosity of bio-oil and the achievement of homogenization [20].

The chemical upgradation of microalgal bio-oil involves various processes aimed at improving the quality of bio-oil derived from microalgae, making it more suitable as a renewable energy source. This upgradation typically focuses on enhancing the energy content, reducing heteroatom content, and increasing the yield of hydrocarbons. The process of catalytic pyrolysis using Fe-HZSM-5 has been shown to significantly improve the quality of bio-oil from microalgae. This method enhances the carbon content and higher heating value (HHV) of the bio-oil while reducing nitrogen-containing compounds by 50% and doubling the hydrocarbons compared to raw bio-crude [17]. In situ catalytic pyrolysis with ZSM-5 and MgO catalysts under nitrogen and carbon dioxide environments also improves bio-oil yield and quality. The MgO catalyst in a nitrogen environment yielded the highest bio-oil, with a high hydrocarbon content in the C7–C18 range [19].

In the cosolvent assisted hydrothermal liquefaction process, the use of cosolvents like alcohols and formic acid in hydrothermal liquefaction (HTL) increases bio-crude yield by 30–63% compared to water-only HTL. Formic acid, in particular, results in higher energy content and less viscous bio-crude [13]. Subsequent hydrodeoxygenation (HDO) of the bio-crude further enhances its energy content and reduces the oxygen and nitrogen content, making it more suitable for fuel applications. In the case of solvent-based upgrading, upgrading bio-crude in solvents such as methanol, ethanol, and acetone increases the HHV and reduces the O/C ratio. Ethanol upgrading, in particular, achieves the highest bio-oil yield of 82.8 wt % at 360 °C [23]. A staged hydro-upgrading process using tetralin and Pt/C catalyst effectively removes nitrogen, oxygen, and sulfur compounds from bio-oil. This method prevents secondary reactions and prolongs catalyst activity, resulting in high-quality liquid fuel [5].

3.3 Alcohol to Aviation Fuel

The starch constituent found within the microalgae cellular structure undergoes a transformation into sugar via a biochemical process, subsequently being converted into alcohol (e.g., ethanol and butanol) through the enzymatic fermentation activity of microorganisms. The alcohol obtained is then subjected to further processing aimed at the production of aviation fuel, a process involving a sequence of chemical transformations (namely, dehydration, oligomerization, hydrogenation, and distillation) [8].

Ethanol dehydration, for instance, leads to the generation of ethylene through the utilization of acidic catalysts such as silica–alumina, silicoaluminophosphates, zeolites, and heteropolyacids. The H-ZMS-5 zeolite catalyst demonstrates considerable potential as a catalyst for the dehydration process within the framework of industrial

applications. The olefin monomers, acquired post dehydration, undergo conversion into oligomers via the oligomerization process facilitated by employing homogeneous and heterogeneous catalysts. In the context of the catalytic oligomerization process, both homogeneous and heterogeneous catalysts, such as chromium diphosphine, Ziegler–Natta-type, and zeolite catalysts, have undergone extensive investigation. The parameters of temperature and pressure within the oligomerization process are manipulated within the ranges of 90 °C–350 °C and 89 bar–250 bar, respectively. A case in point is the transformation of ethylene into linear α-olefins by means of the widely utilized Ziegler–Natta catalysts [22]. The oligomerization process is pivotal for the generation of high-quality hydrocarbon fuels utilizing light olefins or alkene monomers. This procedure has been conducted in an organic solvent employing either homogeneous or heterogeneous catalysts; however, in alignment with the principles of green chemistry, the heterogeneous catalyst is deemed more environmentally benign owing to its enhanced recovery efficiency and convenience. The olefin product may undergo further upgrading to yield commercial products including acids, detergents, plasticizers, oil additives, and transportation fuels.

Harvey et al. [11] elucidated the synthesis of bio-jet fuel utilizing 2-ethyl-1-hexene as a precursor. 2-ethyl-1-hexene is recognized as a renewable feedstock that is derived from 1-butene through a dehydration reaction. The 2-ethyl-1-hexene underwent dimerization at a temperature of 116 °C for a duration of 2 h in the presence of montmorillonite K-10 and sulfated zirconia catalysts. The transformation of 2-ethyl-1-hexene resulted in the successful production of $C_{16}H_{32}$ hydrocarbon with an impressive yield of 90%. The physicochemical properties of the hydrocarbon, including a density of 0.78 g/mL and a freezing point below −60 °C, render it suitable for application as a bio-jet fuel.

4 Compliance for SAF

The predominant jet fuel, designated as Jet A-1, constitutes a complex mixture of hydrocarbons situated within the C9–C16 range. These hydrocarbons are classified into four distinct categories: linear alkanes; slightly branched alkanes, also referred to as iso-alkanes; cycloalkanes, including naphthenes; and aromatic compounds. Typically, alkanes (comprising linear, iso-, and cyclic forms) constitute approximately 80% of the total fuel composition, whereas aromatics, which include substituted benzenes and naphthalenes, represent the remaining fraction. The proportional representation of these four compound families is subject to variation contingent upon seasonal changes and geographical location. Jet fuels are subject to stringent specifications, particularly in comparison to fuels utilized for road transportation [7]. Thus, compliance with these regulations primarily emphasizes the physicochemical characteristics or parameters, which include properties such as cold flow behavior, energy density, energy content, kinematic viscosity, freezing point,

aromatic concentration, material compatibility, safety attributes, flash point, and thermal stability as per ASTM D1655-09.

The principal obstacle confronting producers of microalgae-derived biofuels lies in the formulation of an aviation fuel that adheres entirely to the rigorous standards established for conventional aviation fuels. The physical and chemical properties of SAFs from microalgae need to be carefully characterized before the fuels are certified for operation. The length of the carbon chain, along with the proportional composition of the four categories of hydrocarbons previously elucidated, is meticulously regulated to endow the jet fuel with the requisite characteristics. In this sense, the SAF must have high flash points to allow the fuel ignition for safe operation, a low freezing point that influences the pumping capacity of the fuel at lower temperatures, high energy density to minimize fuel storage room, good sealing properties to avoid fuel leakage on board, high thermal stability to prevent the decomposition of chemicals present in the fuel, and low viscosity to restart the engine at higher altitudes. The above physicochemical properties are dependable on the nature of hydrocarbons present in the fuel. Linear alkanes play a crucial role in enhancing the energy density of aviation fuel while concurrently augmenting its combustion efficiency as they are characterized by their propensity to combust with minimal emissions [2]. Despite the high energy density property of iso-alkanes, these specific hydrocarbons are predominantly incorporated into the aviation fuel to enhance its cold flow characteristics, due to their freezing points being significantly lower than those of their linear counterparts.

Cycloalkanes exhibit a lower hydrogen-to-carbon ratio in comparison to their linear and iso-alkane counterparts, consequently yielding reduced energy per unit mass. Notwithstanding, cycloalkanes are valued as constituents of aviation fuel due to their capacity to augment the fuel density while concurrently decreasing its freezing point. Aromatic compounds hold significant relevance within the composition of jet fuels. Although they adversely influence the combustion efficacy of the fuel, a requisite proportion of aromatics (for instance, 20%) is essential as they facilitate the swelling of elastomeric valves within fuel systems, thereby guaranteeing effective sealing [14].

The comparison in physicochemical properties between conventional aviation fuel and microalgae-derived aviation fuel is shown in Table 3.1. The high heating

Table 3.1 Physicochemical properties of conventional aviation fuel and microalgae-derived aviation fuel

Physicochemical properties	Jet A (ASTM D1655) [9]	Jet A-1 (ASTM D1655) [9]	Microalgal aviation fuel [4]
Heating value (MJ/kg)	42.8	42.8	44.0
Freezing point (°C)	−40	−47	−30
Flash point (°C)	38	38	68
Viscosity at −20°C	8	8	2.8
Density at 15°C	775–840	775–840	1380
Total sulfur (%)	0.30	0.30	0.27

value (HHV) of microalgae-derived aviation fuel (44 MJ/kg) is higher than that of conventional aviation fuel, which is suitable to be used in aircraft. The flash point of the microalgae-derived aviation fuel is very high (68 °C) compared to conventional aviation fuel, which indicates its lower risks of flammability and utility in hot environments. However, the presence of many dissolved particulate substances in the microalgae-derived aviation fuel causes its density to be higher [4]. In addition, the freezing point of microalgae-derived aviation fuel is comparatively higher than that of conventional aviation fuel, which leads to the crystallization of the fuel at high altitudes. To prevent crystallization, an anti-freezing substance could be added to decrease the freezing point of the microalgae aviation fuel [4]. The freezing temperature (−30 °C) and the specific density (1380 kg/m^3) present considerable obstacles for the advancement of microalgae-derived aviation fuel. In summary, microalgae aviation fuel exhibits significant challenges, including economic viability and environmental sustainability, when juxtaposed with conventional aviation fuel. The processes of upgrading, reforming, and incorporating additives will be imperative to achieve a microalgae-derived aviation fuel that fully complies with ASTM standards. The physicochemical properties of conventional aviation fuel and microalgae-derived aviation fuel are shown in Table 3.1.

5 Challenges and Prospects

Sustainable aviation fuels (SAFs) have attracted significant scrutiny, as demonstrated by the successful execution of test flights and the commercial applications conducted by numerous corporations. Microalgae present an economically viable alternative for the production of sustainable aviation fuels (SAFs) owing to their superior yield and reduced fuel expenses. Nevertheless, the scalability of this process necessitates the implementation of efficient cracking and isomerization methodologies. The expanding market for SAFs mandates rigorous sustainability criteria, with feasible feedstock alternatives such as waste cooking oil and microplastics. The SAF sector encounters obstacles including restricted feedstock availability, elevated production expenses, and diminished yields; however, the shift toward third-generation feedstock such as algae demonstrates considerable promise. However, there are certain challenges that hinder the utilization of microalgae-based biofuel toward its commercialization.

In addressing the biological challenges, the selection of strains is of paramount importance, given that each strain exhibits distinct metabolic accumulation characteristics, including variations in lipid composition and accumulation levels. In the context of SAF production, it is imperative that microalgae strains are chosen based on a multitude of criteria, which encompass lipid accumulation, growth velocity, ease of harvesting, resistance to contamination, optimal light intensity, elevated oxygen concentrations, and capacity to withstand extreme temperature conditions. Conversely, the genetic alteration of microalgal species represents a significant avenue that warrants investigation within the SAF fuel production sector. Consequently,

genetically modified microalgal species may serve as a viable feedstock in the synthesis of SAFs.

Chemical challenges are frequently encountered within the SAF sector when engaging in processes such as oil extraction (solvent extraction), transesterification, and the elimination of hazardous substances in water or wastewater employed for microalgae cultivation. Although these technologies have been effectively advanced, the energy consumption and production expenditures remain exceedingly high at the present stage. Furthermore, the formulation of appropriate catalysts to enhance the conversion efficiency of SAFs and diminish energy utilization constitutes a notably arduous endeavor. Therefore, the objective of the researchers is to develop catalysts that are both efficient and economically viable in order to enhance the sustainability of the aviation fuel sector.

6 Conclusion

The synthesis of microalgae-derived sustainable aviation fuel represents an exceedingly promising methodology to replace environmentally detrimental fossil fuels. The conversion of microalgal oil into sustainable aviation fuel has been effectively achieved through hydro-processing techniques, while alternative conversion methods such as syngas-to-aviation and alcohol-to-aviation processes are anticipated to emerge as significant future pathways for producing microalgae aviation fuel. However, the currently available chemical process to convert microalgal oil into aviation fuel is extremely expensive. A comprehensive optimization process and rigorous modeling are essential to reduce operational costs. The parameters of microalgae-derived aviation fuel, which include sulfur concentration (0.27 wt %), net heating value (44 MJ/kg), viscosity (2.8 mm^2/s), and flash point (68 °C), have been systematically aligned with ASTM standards. This configuration yields a higher heating value (HHV), enhanced fuel volatility (characterized by an elevated flash point), and reduced kinematic viscosity in comparison to alternative bio-jet fuel sources. Further investigation into the utilization of microalgae-derived aviation fuel as a viable alternative to conventional aviation fuels is imperative owing to its exceptional fuel characteristics and minimal greenhouse gas emissions.

References

1. Air Transport Action Group [ATAG] Aviation Benefits Beyond Borders. Geneva, Switzerland. (2014). Available online: http://aviationbenefits.org/media/26786/ATAG__AviationBenefits2014_FULL_LowRes.pdf
2. Brooks, K. P., Snowden-Swan, L. J., Jones, S. B., Butcher, M. G., Lee, G. S., Anderson, D. M., & Shonnard, D. (2016). Low-carbon aviation fuel through the alcohol to jet pathway. In *Biofuels for aviation* (pp. 109–150). Academic.

3. Bwapwa, J. K., Anandraj, A., & Trois, C. (2017). Possibilities for conversion of microalgae oil into aviation fuel: A review. *Renewable and Sustainable Energy Reviews, 80*, 1345–1354.
4. Bwapwa, J. K., Anandraj, A., & Trois, C. (2018). Microalgae processing for jet fuel production. *Biofuels, Bioproducts and Biorefining, 12*(4), 522–535.
5. Chen, L., Wang, Z. C., Chen, D., Yin, L. X., & Duan, P. G. (2021). Hydro-upgrading of algal bio-oil in tetralin for the production of high-quality liquid fuel: Process intensification. *Fuel Processing Technology, 224*, 107034.
6. Commercial airlines: worldwide fuel consumption 2024 – Statista. www.statista.com
7. Díaz-Pérez, M. A., & Serrano-Ruiz, J. C. (2020). Catalytic production of jet fuels from biomass. *Molecules, 25*(4), 802.
8. Gutiérrez-Antonio, C., Romero-Izquierdo, A. G., Gómez-Castro, F. I., Hernández, S., & Briones-Ramírez, A. (2016). Simultaneous energy integration and intensification of the hydrotreating process to produce biojet fuel from Jatropha curcas. *Chemical Engineering and Processing: Process Intensification, 110*, 134–145.
9. Gutiérrez-Antonio, C., Romero-Izquierdo, A. G., Castro, F. I. G., & Hernández, S. (2021). *Production processes of renewable aviation fuel: Present technologies and future trends*. Elsevier.
10. Hamelinck, C. N., Faaij, A. P., den Uil, H., & Boerrigter, H. (2004). Production of FT transportation fuels from biomass; technical options, process analysis and optimisation, and development potential. *Energy, 29*(11), 1743–1771.
11. Harvey, B. G., & Quintana, R. L. (2010). Synthesis of renewable jet and diesel fuels from 2-ethyl-1-hexene. *Energy & Environmental Science, 3*(3), 352–357.
12. Jahangiri, H., Lappas, A. A., Ouadi, M., & Heracleous, E. (2023). Production of biofuels via Fischer-Tropsch synthesis: Biomass-to-liquids. In *handbook of biofuels production* (pp. 449–509). Woodhead Publishing.
13. Jena, U., Eboibi, B. E., & Das, K. C. (2022). Co-solvent assisted hydrothermal liquefaction of algal biomass and biocrude upgrading. *Fuels, 3*(2), 326–341.
14. Kinder, J., & Rahmes, T. Evaluation of Bio-Derived Synthetic Paraffinic Kerosene (BioSPK). The Boeing Company; Sustainable Biofuels Research & Technology Program 2009. Available online: http://www.safug.org/assets/docs/biofuel-testing-summary.pdf. Accessed on 12 Feb 2020.
15. Kumar, R., & Strezov, V. (2021). Thermochemical production of bio-oil: A review of downstream processing technologies for bio-oil upgrading, production of hydrogen and high value-added products. *Renewable and Sustainable Energy Reviews, 135*, 110152.
16. Li, F., Srivatsa, S. C., & Bhattacharya, S. (2019). A review on catalytic pyrolysis of microalgae to high-quality bio-oil with low oxygeneous and nitrogenous compounds. *Renewable and Sustainable Energy Reviews, 108*, 481–497.
17. Li, H., Dong, Z., Wang, B., Wu, W., Cao, M., Zhang, Y., & Liu, Z. (2023). High-quality bio-oil production via catalytic pyrolysis of biocrude oil from hydrothermal liquefaction of microalgae Spirulina. *Journal of Analytical and Applied Pyrolysis, 173*, 106057.
18. Medipally, S. R., Yusoff, F. M., Banerjee, S., & Shariff, M. (2015). Microalgae as sustainable renewable energy feedstock for biofuel production. *BioMed Research International, 2015*.
19. Mo, L., Dai, H., Feng, L., Liu, B., Li, X., Chen, Y., & Khan, S. (2020). In-situ catalytic pyrolysis upgradation of microalgae into hydrocarbon rich bio-oil: Effects of nitrogen and carbon dioxide environment. *Bioresource Technology, 314*, 123758.
20. Sharifzadeh, M., Sadeqzadeh, M., Guo, M., Borhani, T. N., Konda, N. M., Garcia, M. C., & Shah, N. (2019). The multi-scale challenges of biomass fast pyrolysis and bio-oil upgrading: Review of the state of art and future research directions. *Progress in Energy and Combustion Science, 71*, 1–80.
21. Sharma, V., Hossain, A. K., Duraisamy, G., & Griffiths, G. (2023). Microalgal biodiesel: A challenging route toward a sustainable aviation fuel. *Fermentation, 9*(10), 907.
22. Weissermel, K., & Arpe, H. J. (2008). *Industrial organic chemistry*. Wiley.

23. Xu, D., Wei, N., Liang, Y., Wang, H., Liu, L., & Wang, S. (2021). Biocrude upgrading in different solvents after microalgae hydrothermal liquefaction. *Industrial & Engineering Chemistry Research, 60*(21), 7966–7974.
24. Zahid, I., Nazir, M. H., Chiang, K., Christo, F., & Ameen, M. (2024). Current outlook on sustainable feedstocks and processes for sustainable aviation fuel production. *Current Opinion in Green and Sustainable Chemistry*, 100959.
25. Zhang, Y., Hou, L., Tierney, J. W., & Wender, I. (2007). Addition of acetylene to the fischer–tropsch reaction. *Energy & Fuels, 21*(2), 640–645.

Chapter 4
Major Resources for Green Aviation Fuel Production, Challenges, and Opportunities

Paulo Gonçalves de Sousa Junior, Kaiany Moreira dos Santos, Francisco Izaias da Silva Aires, Isabelly Silveira Freitas, José Roberto de Matos Filho, Letícia Costa da Silva, Rodrigo da Silva Vieira, Patrick da Silva Sousa, Dayana Nascimento Dari, and José C. S. dos Santos

1 Introduction

The aviation industry has a significant influence on global commerce and transportation. It demands large quantities of fuel annually, resulting in the need to seek more renewable fuel sources that aim to reduce the environmental impacts caused by the burning of fossil fuels [1]. Due to this, the aviation industry has increased its interest in green fuels to reduce its ecological footprint [2]. These renewable fuels contribute significantly to the reduction of carbon emissions when compared to petroleum-derived fuels [3].

The large-scale production of these fuels is tied to the resources and availability of raw materials. The production of biofuels from biomass faces challenges in the availability of biomass from a single source and in large quantities, which include agricultural waste, forestry, or wood industry waste [4]. These raw materials produce biofuels, biodiesel, and bio-jet fuel. Algae, due to their rapid growth and high lipid content, present another promising source of biomass. Additionally, urban

P. G. de Sousa Junior · K. M. dos Santos · F. I. da Silva Aires · I. S. Freitas · J. R. de Matos Filho · L. C. da Silva · R. da Silva Vieira · D. N. Dari
Instituto de Engenharias e Desenvolvimento Sustentável, Universidade da Integração Internacional da Lusofonia Afro-Brasileira, Campus das Auroras, Redenção, Brazil

P. da Silva Sousa
Departamento de Engenharia Química, Universidade Federal do Ceará, Campus do Pici, Fortaleza, Brazil

J. C. S. dos Santos (✉)
Instituto de Engenharias e Desenvolvimento Sustentável, Universidade da Integração Internacional da Lusofonia Afro-Brasileira, Campus das Auroras, Redenção, Brazil

Departamento de Engenharia Química, Universidade Federal do Ceará, Campus do Pici, Fortaleza, Brazil
e-mail: jcs@unilab.edu.br

M. Aslam et al. (eds.), *Sustainable Aviation Fuels*, Sustainable Aviation, https://doi.org/10.1007/978-3-031-83721-0_4

waste, such as household waste and used cooking oil, can be converted into fuel, aligning with the principles of the circular economy and waste reduction [5].

However, there are several challenges to producing these green fuels and their allocation to the aviation industry. These include high production costs, challenges in the optimization and efficiency of green processes, and the availability and sustainability of raw materials. Sustainability is an essential point for the industry, and this material production impacts other agricultural areas, especially the competition between these energy crops and food production [6].

Despite the numerous challenges that transitioning from traditional fuels to biofuels presents, it has proven quite promising. It stands out in searching for opportunities for this transition, such as technological advancements that optimize fuel production [7]. Combining technology and the environment has also shown promise, such as using biocatalysts that reduce production time and increase the overall yield of the process. Partnerships between the public and private sectors and the increased interest in these biofuels by significant airlines have also helped drive investments in research, the development of new industrial technologies, and the commercialization of these bioproducts [8]. Thus, biofuels have attracted global attention for their importance in reducing the impacts caused by fuel burning. As a result, political measures are being taken to facilitate their development, such as subsidies, tax incentives, and even carbon pricing [9].

This bibliometric and literature analysis highlights the main topics regarding biofuels used by the aviation industry, key resources and raw materials, industry challenges, and prospects. A systematic search was conducted in the largest databases of articles and academic papers published in the past 10 years to detail the current conditions of biofuels for the aviation industry in the global scenario.

2 Methodology

A systematic analysis of scientific mapping was conducted following the methods of Nogueira et al. 2023 and collaborators [10]. Based on the research questions (RQs), the survey included studies from various regions, institutions, and authors. These studies were evaluated based on citations, publications, impact factors, and other relevant parameters. A clustering approach was also used to illustrate the trends and perspectives on biofuels produced through hydrothermal gasification [11].

2.1 Data Source

In this study, data for bibliometric analysis were obtained from the Web of Science (WoS) database, which many authors regard as an essential research resource in various fields of knowledge. The keywords "Aviation," "Resources," and "biofuel"

were used in the search. The initial search retrieved articles published between 2014 and 2024. Articles published in English were screened according to the Preferred Reporting Items for Systematic Reviews and Meta-Analyses (PRISMA). Papers that did not include all the necessary information on the platform, were presented at conferences, or were still in preprint format were excluded [12].

2.2 *Data Analysis*

The free software CiteSpace (version 6.1.R4 Basic) was used to create the bibliometric maps. CiteSpace supports the visualization and construction of bibliometric maps. The data obtained from WoS and compiled in CiteSpace enabled the creation of maps for journals, countries, institutions, authors, and keywords based on correlation data. Microsoft Excel (Microsoft Office 365®) spreadsheets were used for data analysis and plotting [13].

3 Results and Discussion

3.1 *Bibliometric Analysis*

3.1.1 Results of Publication: Overall

Research conducted in the Web of Science (WoS) database from 2014 to June 2024 resulted in 719 scientific productions and 15,956 citations. These data were selected according to the limitations of each topic for creating tables, graphs, and network maps. It is interesting to highlight that the United States stands out for the quantity of productions, with 202 published works and 5100 citations. The country appears as a central point in the network maps and is prominent in the topic of institutional contributions. However, in the journal analysis, the Netherlands' journals appear in greater quantity and have the highest impact factors (IFs), with numbers above 11.0 and having the most cited article. This may indicate that despite the United States being the region with the highest number of research and interest in the production of green aviation fuels, the quality of the Netherlands' works is also relevant. Similarly, when compared by the number of publications with their citations, publications from Belgium can be considered applicable to the study of the topic.

In Fig. 4.1, it is possible to observe the number of publications from 2014 to June 2024, highlighting a growing demand for research over the years. A slight decline in publications is noted in 2018 and 2019, with 55 and 56 publications, respectively. However, there was a continuous increase in the subsequent years, culminating in 2023 with 125 productions and 3465 citations. This represents a 594% increase in publications and a 38,400% increase in citations. Although 2024 presents 46 articles and 1562 citations so far, it is essential to consider that these data reflect only the

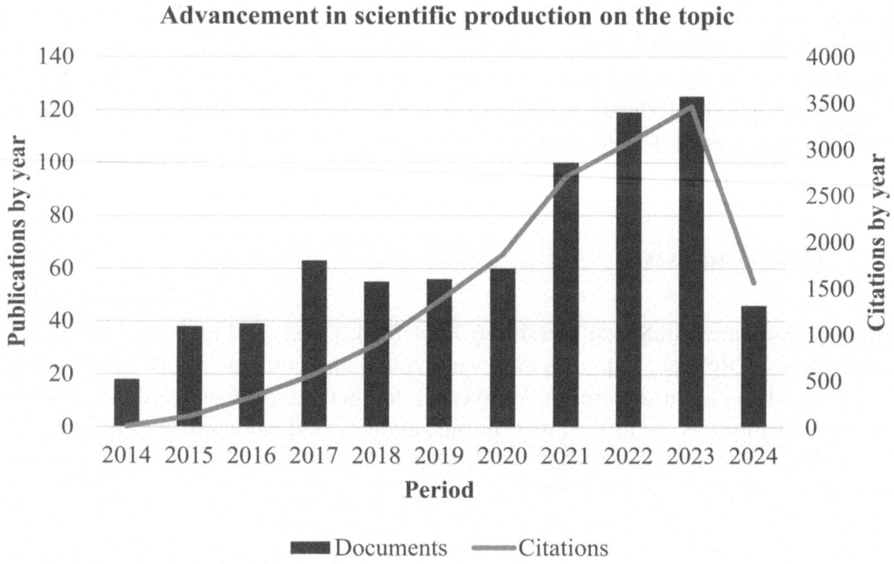

Fig. 4.1 Progress in the annual publication of scientific documents during the period analyzed, accompanied by the number of recurring citations

year's first half. Another relevant point is that the research was based exclusively on the Web of Science (WoS) database, and there are several other critical databases, such as Scopus, Science Direct, and SciFinder, among others.

3.1.2 Contributions of Countries/Regions

Regarding documents by region, as shown in Fig. 4.2a, the United States stands out with 202 publications, followed by China and Brazil, with 121 and 71 works, respectively. The United States's interest in the topic is evident, as is the growth of scientific production in China and Brazil. In Fig. 4.2b, the United States and China continue to lead, with 5100 and 3366 citations. Although Brazil ranks third in the number of publications, it does not maintain this position in citations; England appears in the third place with 1895 citations. Interestingly, despite Belgium having only 16 documents, it presents a significant level of citations relative to the number of publications, suggesting that Belgian works are few but relevant. Similarly, the works from England (63 works), although fewer in number compared to Brazil, prove to be more appropriate, with 1895 citations.

The network map in Fig. 4.3 details the partnership between the countries that contributed the most to publications from 2014 to June 2024. Out of 73 countries, only 21 met the minimum requirement of at least 11 documents. The United States (in red) stands out for its total link strength (TLS), with a value of 63, followed by England (in green) and China (also in green), both with a TLS of 43. Although the

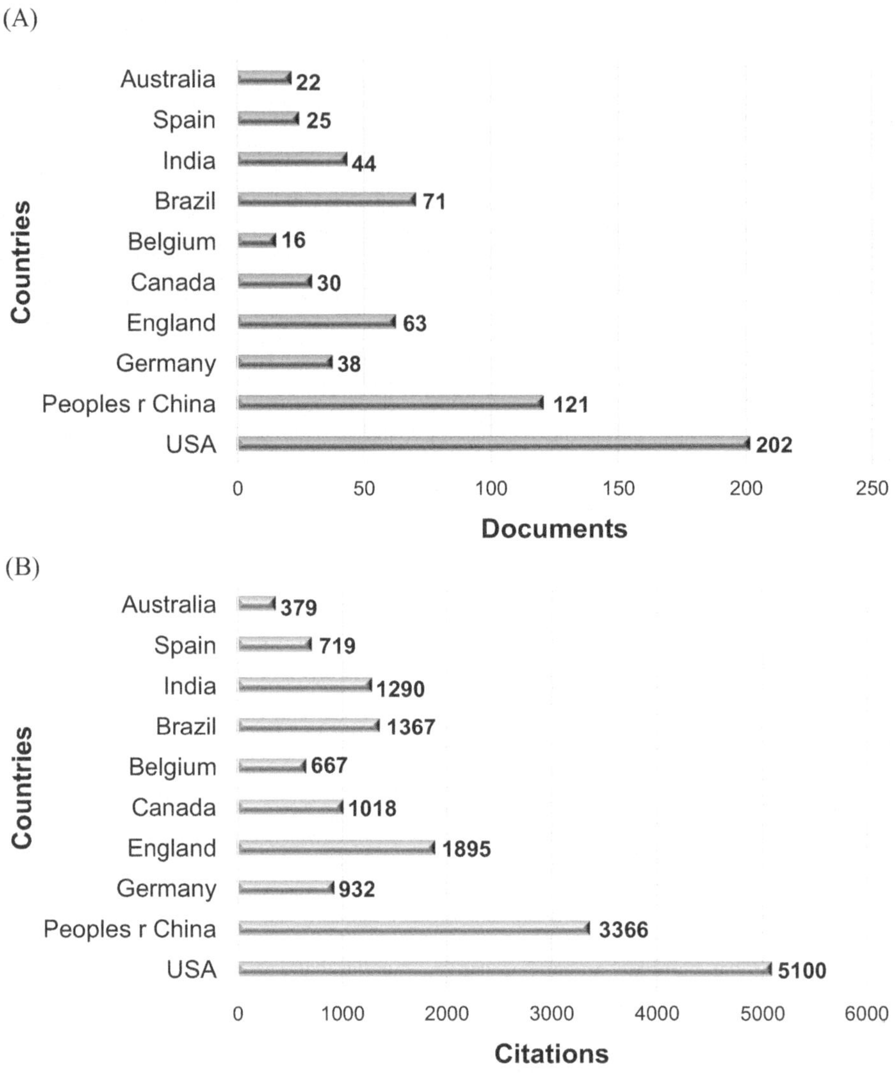

Fig. 4.2 (**a**) The ten countries with the highest quantities of scientific documents published within the range investigated. (**b**) The ten countries with the most citations for their published articles

United States and China are the two most prominent nodes, the central position of England (in green) on the network map indicates its involvement with all clusters and connections with all countries. Notably, the red cluster contributed the most, with 325 scientific productions and 8514 citations, followed by the green cluster, with 276 publications and 7766 citations, and the blue cluster, with 166 publications and 4166 citations.

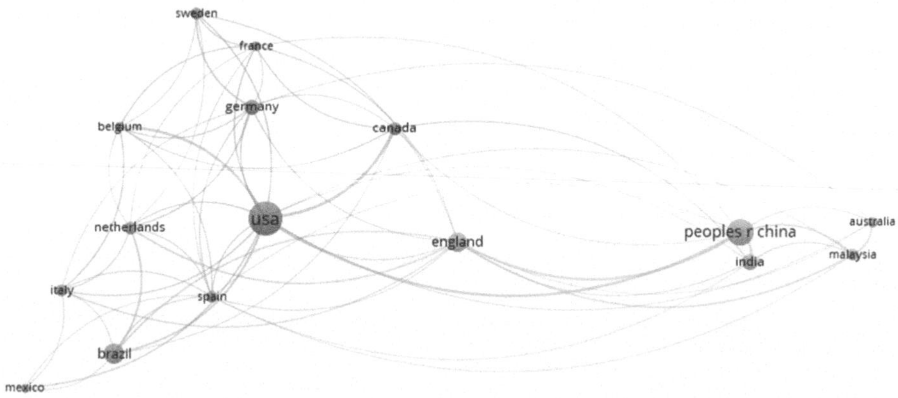

Fig. 4.3 Assessment of the partnership between the countries that contributed most to the publication of articles

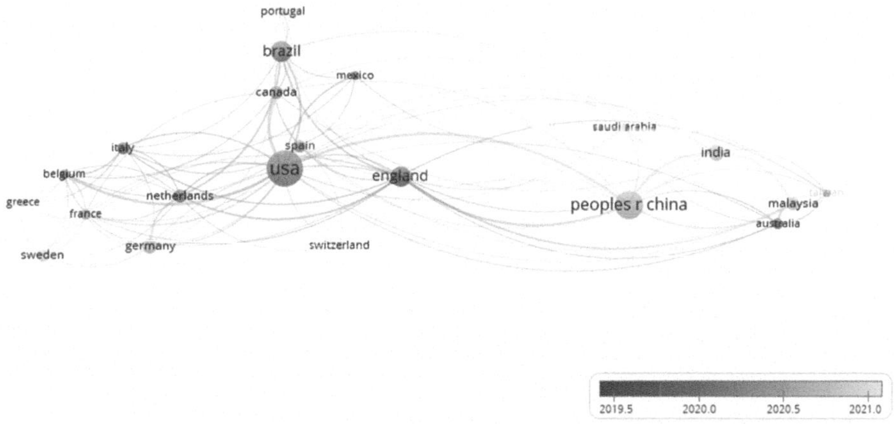

Fig. 4.4 Assessment of the partnership between the countries that contributed most to the publication of articles from 2019 to 2021

Visualizing the publications and correlating them with the period from 2019 to 2021, as shown in Fig. 4.4, it is easy to see why England is at the center of the map. Its purple color indicates that during the established period, it was one of the first countries to publish, thus possibly responsible for spreading interest to other regions, highlighting and increasing the overall contribution. Examples include Australia, Mexico, and Italy, dating from 2019. It is interesting to note that China, dating from May 2020, emerged more recently than the United States. This may indicate that despite the recent interest, China has shown rapid advancement and a remarkable interest in the topic. Additionally, there is emerging interest in some countries, such

as Greece, Sweden, Portugal, and Saudi Arabia, dating from 2021. These contributions denote that international collaboration and constant innovation are fundamental to the progress in producing sustainable biofuels for the aviation industry.

Table 4.1 highlights the top ten institutions that have published the highest number of documents related to the production of green fuels for the aviation industry. The institutions in the table come from various regions worldwide, emphasizing the global contribution to this research field. Additionally, it provides a detailed view of the number of publications per institution and their countries or regions of origin.

The analysis of Table 4.1 shows that besides dominating the top of the table, the United States is the most relevant country in terms of institutions, considering that out of the ten institutions mentioned, three are from the United States. Similarly, China has two out of the three institutions in the first and second places: Washington State University with 20 publications and the National Renewable Energy Lab (NREL) with 19 papers. This demonstrates their strong participation in research on the proposed topic. Purdue University, located in Denmark, follows below the United States, becoming third in the ranking, with 19 publications, indicating international collaboration in research. Regions such as Europe, Asia, and North America also show significant contributions, with institutions in the Netherlands, China, and Canada demonstrating a growing commitment to sustainable aviation fuel (SAF) research. Although these institutions rank high in the number of publications, the same does not apply to citations, where MIT leads with 876 citations, followed by the National Renewable Energy Lab, which remains in second place with 809 citations, and Tianjin University with 670 citations. This highlights that despite fewer publications, their work is highly relevant. In the case of the National Renewable Energy Lab, it demonstrates both quality and relevance, as well as a solid commitment to contribution.

Figure 4.5 presents a network map of the leading organizations contributing to publications. The connected colors demonstrate the correlation between these organizations. Three leading institutions stand out in publications and collaboration:

Table 4.1 Ranking of the ten institutions that published the most scientific documents on biofuel production for the aviation industry

Rank	Institution	Country/Region	Number of publications	Number of citations
1	Washington State University	United States	20	365
2	Natl Renewable Energy Lab	United States	19	809
3	Purdue University	Denmark	19	397
4	Beihang University	China	19	272
5	MIT	United States	17	876
6	Tianjin University	China	17	670
7	Chinese Academy of Sciences	China	16	578
8	Delft University Technol	Netherlands	12	166
9	University Utrecht	Netherlands	11	654
10	University Autonoma Queretaro	Mexico	11	300

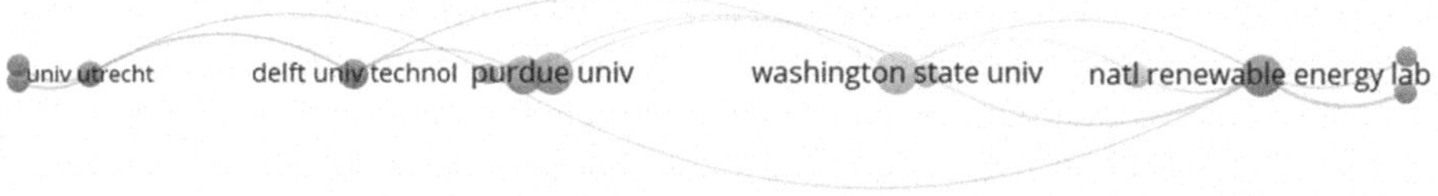

Fig. 4.5 Assessment of the partnership between the organizations that contributed most to the publication of articles

Washington State University, the National Renewable Energy Laboratory (NREL), and Purdue University. These institutions, which are also the first listed in Table 4.1, act as pillars of contribution and collaboration, serving as central connection points for other organizations. The partnerships between these institutions allow for a rapid exchange of knowledge and resources, accelerating the development of new technologies and practices in the discussed topic.

3.1.3 Contributions of Authors

Figure 4.6 presents a bar chart showing the ten authors with the highest number of scientific publications related to the theme, while Fig. 4.8 shows the chart titled "Authors with the highest number of citations." They highlight the number of documents published by researchers from various countries. In Fig. 4.6, these authors were selected with a minimum of four to six publications, and the most extensive bar represents the authors with the maximum number of publications, equivalent to 11 documents. Robert Malina and Xiangwen Zhang lead the ranking with 11 publications, both standing out as the leading researchers. Their research is likely of great relevance, given that they have a range of citations, 601 and 6546, respectively.

Following them, Ji-Jun Zou has ten publications and 597 citations, while Michael Wolcott, Kristin Brandt, Mark D. Staples, and Lun Pan follow closely, each with nine publications, as shown in Fig. 4.7. This suggests that, although some authors have fewer publications, their contributions are significant to the research on green aviation fuels, considering their high citations: 230, 213, 546, and 597, respectively. This indicates that their works have excellent visibility in the scientific community. The analysis of the most cited authors reveals that Robert Malina and Ji-Jun Zou are the leaders in impact, demonstrating that their works are highly influential and widely referenced by other researchers.

Complementing the other figures, Fig. 4.8 presents a network graph illustrating the connections between collaborating authors, highlighting those who contributed substantially to the scientific production of technologies and challenges in producing aviation biofuels. Each node represents an author, while the lines indicate their correlations, showing each node's cooperation on the subject. It is worth noting that besides having many authors with several publications, the green cluster is a group where all authors cooperate, totaling approximately 35 scientific productions.

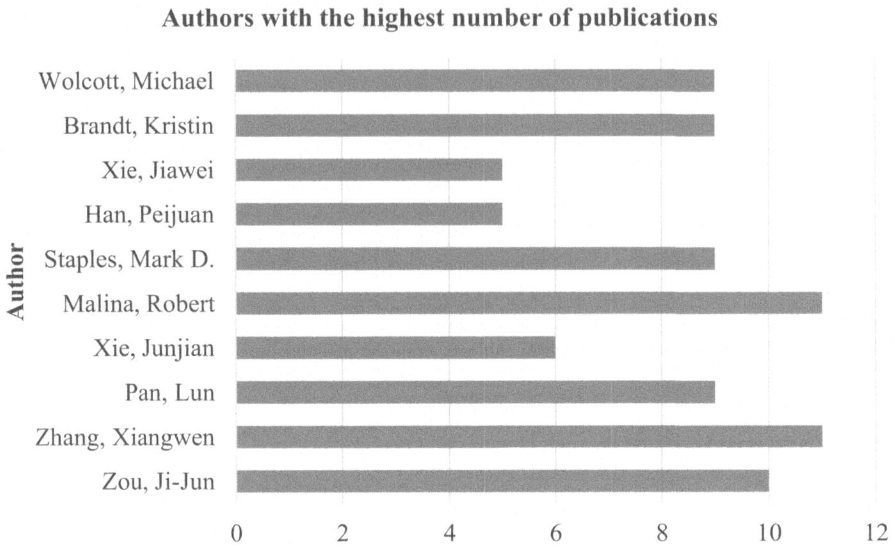

Fig. 4.6 List of ten authors who have published the most scientific works on aviation biofuels

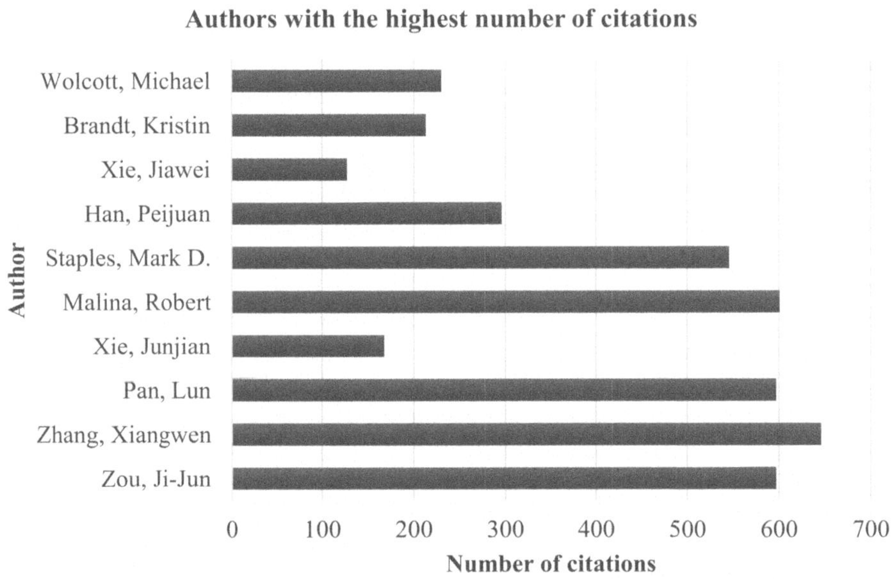

Fig. 4.7 Ranking of the ten most cited authors in scientific production on aviation biofuels

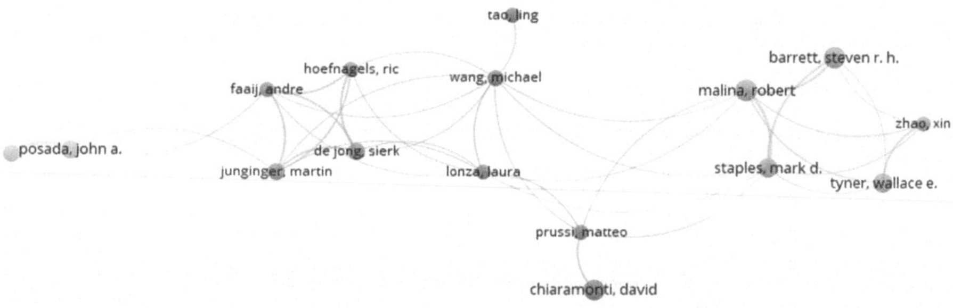

Fig. 4.8 Collaboration network between the authors who contributed most substantially to the scientific production of technologies and challenges in the production of biofuels for the aviation industry

3.1.4 Journal Analysis

Table 4.2 shows the scientific journals with the most published papers on producing green aviation fuels. It is of utmost importance to highlight that the Netherlands was the leading country in terms of influence in journals on the subject, appearing in the ranking eight times. This suggests that this country is an essential center for research on the proposed topic. *Renewable and Sustainable Energy Reviews* is the most cited journal, with 2057 citations, and has the highest impact factor (IF) of 11.889, standing out as a crucial reference on the subject. The journal *Journal of Cleaner Production and Renewable Energy* also has high impact factors of 11.533 and 11.446, respectively, indicating that besides the Netherlands being prominent in research, they also have the journals with the highest IF. Although the Netherlands predominantly dominates the table, there is also a significant contribution from the United States through the journal *Biofuels, Bioproducts, and Biorefining*, with 950 citations and an IF of 8.035, and from Switzerland, with the journal *Frontiers in Energy Research*, which has 251 citations and an IF of 8.4. The impact factors range from 3.252 (*Transportation Research Part D: Transport and Environment*) to 11.889 (*Renewable and Sustainable Energy Reviews*), showing the published research's variability in influence and reach.

Figure 4.9 illustrates a network graph on the collaboration between the most efficient journals for the scientific production of technologies and the challenges in producing green fuels for the aviation industry. Out of 222 journals, only 21 met the minimum parameters of eight documents. The color, size of the nodes, and thickness of the lines reflect different metrics, such as the number of publications and interconnection in the collaboration of various research areas. The largest cluster is the red area group, led by the journal *Fuel*, with 53 publications, 1019 citations, and a TLS of 255. Another significant group is the green one, centered on the journal *Renewable & Sustainable Energy Reviews*, with 40 publications, 2057 citations, and a TLS of 424, thus standing out as the main subject to be connected to the others. Following the relevance order by the size of the clusters, another subject of significant influence was *Biofuels, Bioproducts, & Biorefining*, in blue, with 34

Table 4.2 Scientific journals with the most published works on crude production

Rank	Journal title	Country	Citations	IF
1	*Renewable and Sustainable Energy Reviews*	Netherlands	2057	11.889
2	*Fuel*	Netherlands	1019	8.857
3	*Biofuels, Bioproducts and Biorefining*	United States	950	8.035
4	*Energy Conversion and Management*	Netherlands	635	5.276
5	*Applied Energy*	Netherlands	870	8.634
6	*Biomass and Bioenergy*	Netherlands	459	6.0
7	*Journal of Cleaner Production*	Netherlands	397	11.533
8	*Frontiers in Energy Research*	Switzerland	251	8.4
9	*Renewable Energy*	Netherlands	597	11.446
10	*Transportation Research Part d-Transportation*	Netherlands	392	3.252

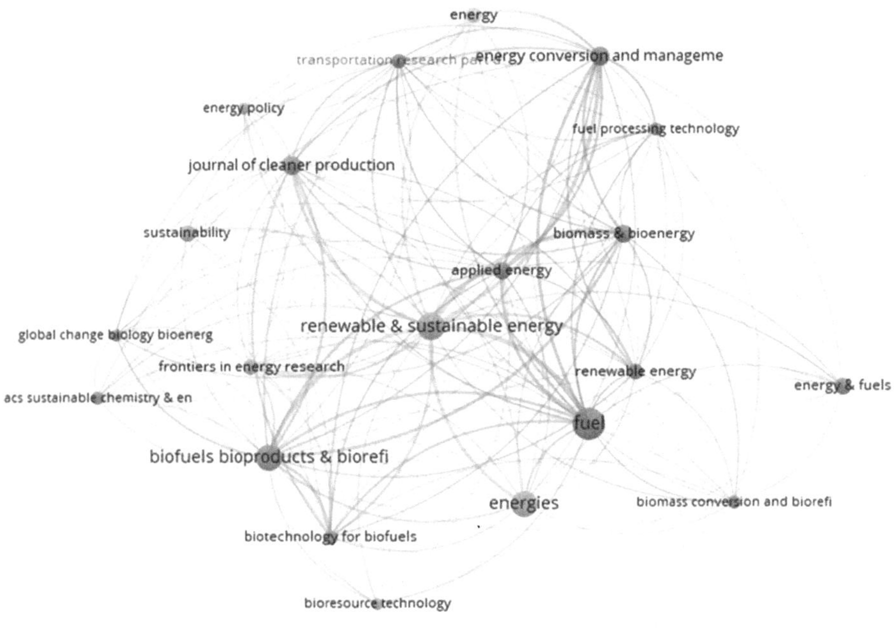

Fig. 4.9 Collaboration network between journals that contributed most substantially to the scientific production of technologies and challenges in the production of biofuels for the aviation industry

publications, 950 citations, and a TLS of 247. By color, line connection, and cluster size, the yellow area was less connected to other places than the others, with only the journal *Energy*, with 13 papers, 392 citations, and a TLS of 117.

3.1.5 Most Articles Cited

The ability to statistically analyze the publication of articles on a specific topic and the development of the scientific community in that research area is an essential feature of bibliometric analysis. Based on this, Table 4.3 presents the ten most cited articles on developing biofuel production technologies, including publication data and citation numbers.

Table 4.3 The ten most cited articles on developing bio-crude production technologies

Rank	Title	Journal	Year published	Citations	Reference
1	Bio-jet fuel conversion technologies	*Renewable and Sustainable Energy Reviews*	2016	310	[14]
2	Carbon-increasing catalytic strategies for upgrading biomass into energy-intensive fuels and chemicals	*ACS Catalysis*	2018	269	[15]
3	Aviation biofuel from renewable resources: Routes, opportunities and challenges	*Renewable and Sustainable Energy Reviews*	2015	244	[16]
4	Review on synthesis and properties of high-energy-density liquid fuels: Hydrocarbons, nanofluids and energetic ionic liquids	*Chemical Engineering Science*	2018	215	[17]
5	Biofuel blending reduces particle emissions from aircraft engines at cruise conditions	*Nature*	2017	199	[18]
6	The feasibility of short-term production strategies for renewable jet fuels—a comprehensive techno-economic comparison	*Biofuels, Bioproducts and Biorefining (Biofpr)*	2015	186	[19]
7	Life-cycle analysis of greenhouse gas emissions from renewable jet fuel production	*Biotechnology for Biofuels and Bioproducts*	2017	185	[20]
8	A review on the production processes of renewable jet fuel	*Renewable and Sustainable Energy Reviews*	2017	181	[21]
9	The compatibility of potential bioderived fuels with Jet A-1 aviation kerosene	*Applied Energy*	2014	181	[22]
10	Decarbonising ships, planes and trucks: An analysis of suitable low-carbon fuels for the maritime, aviation and haulage sectors	*Advances in Applied Energy*	2021	170	[23]

The most cited article is titled "Bio-jet fuel conversion technologies," published in 2016, with 310 citations. The work primarily discusses current technologies for producing renewable jet fuels, categorized by pathways such as alcohol-to-jet, oil-to-jet, gas-to-jet, and sugar-to-jet [14].

The second article is titled "Carbon-Increasing Catalytic Strategies for Upgrading Biomass into Energy-Intensive Fuels and Chemicals," published in 2018, with 269 citations. The authors mainly address the transformation of biomass into energy-intensive fuels and chemicals, the selective synthesis of long-chain oxygenates, and the effects of catalyst structure/type on catalytic performance [24].

The third document, titled "Aviation biofuel from renewable resources: Routes, opportunities and challenges," was published in 2015 and has 244 citations. It discusses the opportunities and challenges in developing alternative aviation fuels, the production process, raw materials used, and the most promising global projects [25].

The authors of the article "Review on synthesis and properties of high-energy-density liquid fuels: Hydrocarbons, nanofluids and energetic ionic liquids," published in 2018 with 215 citations, primarily discuss the synthesis and properties of high-energy-density liquid fuels. These include multi-cyclic fuels derived from fossil resources, strained and caged fuels with higher energy content, high-energy-density biofuels synthesized from biomass feedstock (such as terpenes, hemicellulose, cellulose, and lignin), nanofluid fuels with energetic nanoparticles, and hypergolic energetic ionic liquids [26].

The article titled "Biofuel blending reduces particle emissions from aircraft engines at cruise conditions," published in 2017 with 199 citations, primarily investigates the impact of biofuel blending on aerosol emissions during cruise conditions. It explores the potential use of biofuels in aviation to mitigate climate change and discusses the challenges and benefits of using biofuels for jet engines in the aviation sector.

3.1.6 Research Area

To better understand the research, it is essential to grasp its areas of study. With this objective in mind, Fig. 4.10 was created to quantify and investigate the main research areas studying green aviation fuels and their technologies. One thousand one hundred and twenty studies were filtered and divided into ten main areas. Energy Fuels leads with 395 studies, followed by Engineering Chemical, Green Sustainable Science Technology, Biotechnology Applied Microbiology, and Environmental Sciences, with 173, 130, 102, and 90 studies, respectively. Articles such as [27, 28] highlight the mentioned areas and demonstrate future studies and the creation of new technologies. Research areas like these, such as green fuel production, are essential for an eco-friendly aviation system.

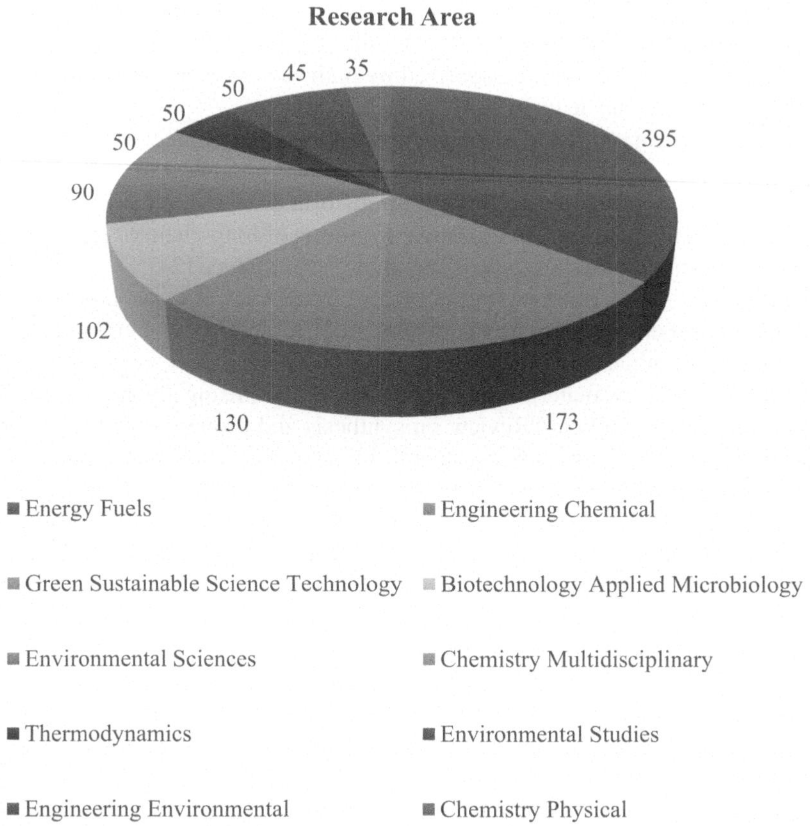

Fig. 4.10 Prominent research areas in the development of aviation biofuels

3.1.7 Keyword Analysis

This bibliometric study analyzed the most frequent keywords in the field of study. Table 4.4 presents the 20 most used keywords in scientific documents related to technologies for developing green aviation fuels, organized based on their total link strength (TLS). Among the top ten keywords, "Biofuels" stands out with 191 occurrences and a TLS of 1107, followed by "Biomass" and "Biofuel" with 121 and 139 occurrences and TLS of 881 and 806, respectively. In the second column, "Emissions," "Performance," and "Microalgae" are highlighted with 74, 58, and 50 occurrences and TLS of 426, 364, and 334, respectively.

Table 4.4 is based on a study that yielded 3448 keywords, further strengthening the research on green aviation fuels. It is interesting to note the high density of the words and their strong connections, highlighting that the research is correlated and addresses the meanings of the keywords. In terms of occurrences, the green, yellow, and blue groups stand out, featuring the words "Biofuels," "Biofuel," and "Biomass," where the highest density of links between clusters can be seen. It is worth

Table 4.4 Ranking of the 20 keywords with the highest occurrence in the scientific documents evaluated about biorefinery development technologies

Rank	Keyword	Frequency	Total link strength	Rank	Keyword	Frequency	Total link strength
1	Biofuels	191	1107	11	Emissions	74	426
2	Biomass	121	881	12	Performance	58	364
3	Biofuel	139	806	13	Microalgae	50	354
4	Biodiesel	95	607	14	Energy	49	334
5	Jet fuel	89	599	15	Sustainable aviation fuel	55	334
6	Aviation	97	580	16	Biodiesel production	44	310
7	Oil	74	509	17	Aviation biofuel	46	302
8	Diesel	69	507	18	Bio-oil	38	295
9	Conversion	74	505	19	Pyrolysis	38	289
10	Life-cycle assessment	63	485	20	Aviation fuel	43	284

mentioning that the most significant number of keywords belongs to the red cluster, with a total of 37 words. The total number of keywords and their various connections highlight the diverse research areas on the topic, which could potentially lead to the creation of new methods and technologies in the future.

Figure 4.11a, b highlight the emergence of keywords between 2019 and 2021. Although some words date back to 2019, such as "Fuels," "Bioenergy," and "Aviation biofuel," and others to 2020, like "Biofuel," "Biodiesel," and "Biomass," the appearance of new terms, such as "Combustion," "Sustainable aviation fuel," and "Green diesel," is evident. The large number of words in yellow, which emerged in 2021, reinforces the idea that more research is being conducted over the years, resulting in the development of new methods and technologies.

4 Feedstock Availability and Sustainability

Most aviation biofuels are derived from lipid feedstocks, especially vegetable oils, extracted from various biomass sources, many cultivated in agriculture [29, 30]. Biofuels derived from edible crops such as soybean, palm, corn, and sugar cane are first-generation biofuels [31]. Using first-generation raw materials has raised concerns about possible competition with food production, the use of large areas of land for cultivation, and significant greenhouse gas emissions [32]. However, the expectation is that the production of biofuels using these raw materials will continue to dominate the market for a long time. However, future demand will not be exclusively met by these types of raw materials as lignocellulosic waste and, more recently, biowaste and plastic waste are increasingly being used for the sustainable production of aviation biofuels [33, 34]. On the other hand, raw materials classified

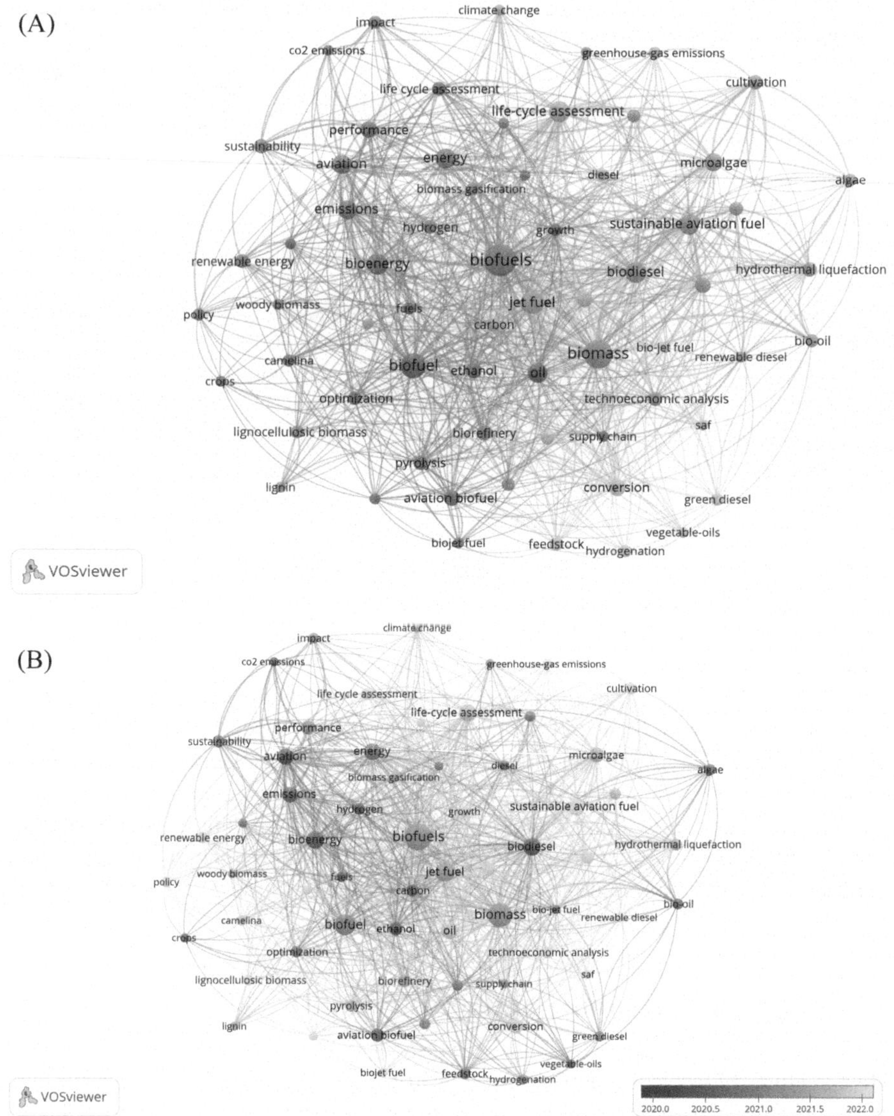

Fig. 4.11 (**a**) Connection network between the 20 prominent keywords in the evaluated scientific documents relating to crude production (**b**) Occurrence of the 20 recurring keywords in scientific works investigated about aviation biofuels over time

as second-generation are sources of nonfood biomass, such as municipal solid waste (MSW), forestry residues, agricultural residues, and other energy crops [35, 36]. They do not compete with food production and can replace first-generation raw materials due to their abundance, availability, and low cost, promoting the circular economy through waste recovery [37].

Lipid microalgae are an ideal feedstock for third-generation biofuels, such as aviation fuel, biodiesel, and biogasoline, surpassing conventional feedstocks [38]. The advantage of algae is that it can be cultivated using various water sources, including saline, brackish, and wastewater, and not just freshwater [39, 40]. Algae are converted into biofuels through pyrolysis and gasification processes, where biomass is broken down into liquids or synthetic gases, transforming them into aviation biofuel [41]. Although algae contribute to reducing greenhouse gas emissions and offer greater control and productivity, they have disadvantages such as low productivity, the need for large areas of land, expensive photobioreactors, and high-energy inputs [42].

In general, several raw materials derived from biomass, such as lignocellulosic materials, sugars and starch, oilseeds, and organic residues, are being widely used by the industry and biorefineries for the production of various final products, including biofuels [43, 44]. Due to its availability, abundance, and renewability, lignocellulosic biomass has become a preferred renewable raw material for producing alternative biofuels to fossil fuels [45].

In turn, plastics, especially plastic waste, are a potential, although nonrenewable, raw material source that can be converted into aviation biofuels through techniques such as pyrolysis and depolymerization [46]. Pyrolysis to convert plastics into oil faces several challenges, one of the main ones being the formation of wax composed of saturated and unsaturated hydrocarbons with linear and branched chains, which obstructs the reactor and other parts of the equipment [47, 48]. Plastic waste brings environmental benefits by eliminating the need for landfill disposal and reducing the demand for new fossil fuels. However, diversity in the types and compositions of plastics can complicate the consistent production of aviation biofuels using this waste [49].

5 Technological Advancements in Conversion Processes

Aviation biofuels can be produced through several conversion technologies, including catalytic lipid conversion, carbohydrate conversion, and lignocellulosic biomass, as well as the development of biorefinery concepts [50]. Among the most promising technologies in the near term to produce aviation biofuels are hydrogenated esters, fatty acids, and Fischer–Tropsch synthesis [51]. Renewable aviation fuels can be made from biomass and residual waste through gasification, syngas conditioning, and Fischer–Tropsch catalytic synthesis. However, energy efficiency and capital cost barriers still require advancements for widespread adoption [52].

Catalytic processes for green aviation fuel production include several critical steps, such as biomass pretreatment, carbohydrate hydrolysis, dehydration, and catalytic upgrading of platform chemicals [53]. Hydrothermal liquefaction of biomass at 340 °C with Fe catalyst, for example, produces crude oil with low production costs and a significant potential for reducing greenhouse gas (GHG) emissions compared to fossil aviation fuel [54]. Only esters and hydro-processed fatty acids (HEFA) are commercially available for biokerosene [55], highlighting the need for further technological development to expand the range of sustainable aviation fuels available.

Innovation in catalysts has also shown promising results. Nettle seed oil, for example, can be turned into sustainable aviation fuel, offering a viable alternative to fossil fuels [56]. Catalytic pyrolysis of coconut oil with Ni/SBA-15 catalysts can produce aviation fuel with a similar composition to drop-in fuels [57]. Furthermore, the Zeolite Y–ZSM-5 hybrid catalyst achieved the highest conversion of palm oil into aviation biofuel, with high alkane and low aromatic hydrocarbons [38]. Another study demonstrated that using a Pd/C catalyst in the conversion of palm kernel oil can produce green diesel and aviation fuel with up to 96% conversion efficiency, allowing up to 5% deoxygenation without compromising the properties of the cold fuel [46]. These advances highlight the potential of catalytic technologies in producing more sustainable and efficient aviation fuels.

Sustainable aviation fuels (SAFs) represent a significant step forward for green aviation [58], being able to replace fossil fuels without the need to modify the existing aircraft infrastructure [38]. Within this category, power-to-liquid (PtL) fuels stand out as SAFs, covering renewable source alternatives. The definition of SAF includes both bio-derived and synthetic fuels, and initiatives such as European Union Aviation Regulation (ReFuelEU) and IATA classify hydrogen-based PtL fuels as SAFs, which can be blended with conventional fuels and are suitable for long-haul flights [59]. The PtL method creates sustainable fuels that are chemically equivalent to regular aviation fuel, using water, CO_2, and renewable energy [60].

SAFs offer a promising alternative to power aircraft and reduce the dependence on Jet A-1 aviation fuel derived from fossils [61]. In addition to being carbon-neutral, they allow mixtures of up to 50% with approved drop-in fuels, reducing CO_2 emissions by up to 40%. When used in their pure form (100%) or in combination with hydrogen, SAFs can further reduce emissions [62]. They also demonstrated reductions in other emissions, such as NO_x, particles, and CO_2, depending on the fuel production routes [63].

Technological advances in the conversion processes of SAFs, such as esters and hydro-processed fatty acids (HEFA), have shown a significant reduction in ultrafine particle (PM) emissions compared to traditional fuels such as RP-3 and diesel. HEFA-SAF demonstrated nonvolatile particle (nvPM) emission rates up to 96% lower than diesel and 89% lower than RP-3, highlighting the efficiency and sustainability of these fuels [64]. Furthermore, the production of SAF from used cooking oil (UCO) has shown significant advances, with Ni-Mo/alumina–silica catalysts achieving high UCO conversion (86.69%) and selectivity for aviation hydrocarbons (74.79%) under moderate conditions, with stability for more than 100 h [65].

Table 4.5 Sustainable aviation fuel (SAF) candidates with high market potential, raw materials, and reduced greenhouse gas emissions

SAF candidates with high market potential	Feedstock	Reduction of GHG (greenhouse gas) (%)
Biofuels: Hydro-processed esters and fatty acids (HEFA)	Vegetable oils, residual lipids, and residues	84
Advanced biomass to liquids (ABtL)	Municipal solid waste, cellulosic cover crops, and agricultural and forestry residues	94
Power-to-liquids (PtLs)	Green hydrogen and a carbon source (e.g., biomass or CO_2)	94

The Fischer–Tropsch (FT) process, which converts syngas into fuel, and the power-to-liquid (PtL) process, which combines CO_2 with green hydrogen to produce SAFs, categorize these fuels into hydrocarbon groups such as synthetic paraffinic kerosene (SPK), isoparaffinic kerosene (IPK), and synthetic kerosene with aromatics (SKA) [36]. HEFA is highlighted as ready for commercial deployment (TRL 9 and FRL 9), despite higher costs, offering significant emission reductions (minimum 27%) [66].

Each of these SAF technologies is shown in Table 4.5 and highlights significant advances in reducing emissions and harnessing different feedstocks. HEFA is currently the most affordable and certified, while advanced biomass to liquids (ABtL) and PtL offer the potential for further emission reduction and the use of waste and innovative materials. Market expansion and certification of these emerging technologies will be crucial to their wider adoption and positive impact on aviation sustainability [67].

6 Economic Viability and Market Dynamics

The economic viability of green aviation fuels is a central issue for their large-scale adoption [68]. Production and infrastructure costs, government incentives, and the ability to compete with fossil fuels are determining factors [69, 70]. Currently, the production costs of aviation sector fuels vary depending on the technology and raw materials used [71, 72]. For example, biofuels produced via hydro-processed esters and fatty acids (HEFA) have a higher cost compared to conventional aviation kerosene [73, 74]. Similarly, synthetic fuels, produced through power-to-liquid (PtL) technology, and green hydrogen, obtained by electrolysis using renewable energy, also face cost challenges due to the need for advanced technological processes, infrastructure, and political incentives [75, 76].

The transition to green aviation fuels requires significant investments in production, storage, and distribution infrastructure [77, 78]. The existing infrastructure is predominantly designed for fossil fuels, requiring adaptations and new

developments. Creating biorefineries, green hydrogen production facilities, and carbon capture units are examples of necessary investments [79, 80]. These production costs can be amortized in the long term with increased demand and economies of scale [81]. Additionally, airlines can reduce operational costs in the long term by adopting more efficient and sustainable technologies [82].

Government policies play a crucial role in the economic viability of green aviation fuels [85]. Subsidies, carbon credits, and regulations incentivizing emission reduction can make these fuels more competitive [86]. For example, the European Union and the United States have already implemented policies to incentivize the production and use of biofuels [87]. Harmonizing international regulations is also essential to facilitate international trade of these fuels and accelerate a robust market [88]. Without these policies, green fuels may struggle to compete in price with fossil fuels, which could hinder their large-scale adoption. However, adopting green fuels contributes to reducing the demand for fossil fuels [89].

Consumer and regulatory pressure to reduce carbon emissions creates a growing demand for sustainable alternatives [90]. Airlines are setting ambitious carbon neutrality goals, driving the adoption of green fuels [91]. For example, companies like Delta Air Lines, ANA Group of Airlines, and Continental Airlines have announced significant commitments to reducing their carbon footprints [92]. However, market competitiveness is still challenged by the lower prices of fossil fuels and the volatility of prices for renewable raw materials [93]. The prospects for the economic viability of sustainable aviation fuels are promising, aiming to reduce environmental impacts, though they depend on technological advancements and innovations [94, 95]. Table 4.6 provides an overview of the key fuels influencing the economic viability and market dynamics of aviation fuels. Biofuels, synthetic fuels, and green hydrogen face similar challenges such as high production costs and the need for infrastructure investments. However, as mentioned earlier, each resource offers specific opportunities driven by technological advancements, incentive policies, and

Table 4.6 Main economic aspects of aviation fuels

Fuel type	Production costs	Major challenges	Future prospects	Reference
Biofuels	High, vary with raw material (e.g., vegetable oil, agricultural residues)	High cost, inadequate existing infrastructure	Cost reduction with economies of scale, increased demand	[96]
Synthetic fuels (e-fuels)	High, due to PtL technology and CO_2	High cost, scalability	Possible cost reduction with technological innovations	[97]
Green hydrogen	Very high, due to electrolysis and the use of renewable energy	High cost, need for new infrastructure	Cost reduction with technological advances, increased demand	[98]
Fossil fuels	Low, historically stable prices	Market dominance, concerns about environmental impacts	Stable prices, gradual efficiency improvements	[99]

strategic partnerships. Lower fossil fuel prices currently constrain market competitiveness, but increasing demand for sustainability and emission reduction opens promising pathways for the future of sustainable fuels.

7 Infrastructure Development and Integration

Worldwide, aviation is responsible for 2% of all human-caused carbon dioxide (CO_2) emissions and 12% of all CO_2 emissions from transportation. Given this, the search for alternative and sustainable solutions arises intending to mitigate environmental problems and climate change and promote sustainability driven by the development of biofuels for aviation [96, 97]. Therefore, technical and economic feasibility for expanding the use of these renewable fuels by investing in the infrastructure required for their use, production, and distribution becomes essential. In this way, integration with existing infrastructures and adaptation of regulatory policies that encourage its adoption are carried out [98].

Therefore, it becomes a challenge to encourage the competitiveness of bioproducts such as aviation fuels, and the construction of specialized biorefineries capable of large-scale operations is one of the promising alternatives that would help reduce production costs [99]. However, the obstacles to the integration of biofuels are not restricted only to production but also to distribution. While the storage of raw materials often minimally impacts the supply chain, energy-intensive facilities that prevent deterioration and preserve these materials can significantly contribute to costs and emissions [100]. Sophisticated mathematical models that consider the various storage and transportation factors can be used to optimize supply chains. These models consider the distribution of raw materials geographically, the type and location of production facilities, the appropriate storage facilities, the available modes of transportation, and the routes [101].

It is necessary to adapt infrastructures inherent to the transport and storage of renewable fuels. Storage tanks, pipeline systems, and distribution networks must be built or adapted to transport biofuels. Another aspect to be highlighted is the qualified preparation of airlines and airports to manage the use of biofuels as they will require new certifications and training for their employees [102].

Furthermore, it is worth mentioning the leading role of government policies in promoting the production, distribution, and use of renewable aviation fuels [103]. Therefore, there are reports of governments around the world developing standards that encourage the research, development, and adoption of biofuels through subsidies, tax incentives, and the establishment of targets to reduce greenhouse gas emissions, that is, stimulating market heating [99, 104]. The policy "Roadmap for carbon neutrality 2050" launched in 2019 by Portugal aims to eliminate carbon emissions in the aviation industry by 2050 [96]. Brazil has a "Brazilian National Biofuel Policy," also launched in 2019, which aims to reduce the intensity of carbon released through the decarbonization process across all types of fuels and establish a carbon credits market [99]. The European Union, through the "ReFuelEU Aviation

initiative" launched in 2021, aims to achieve a use rate of renewable aviation fuels of 2% by 2025 and sequentially increase it to 63% by 2050 [98]. In this way, international collaboration presents itself as a solution for harmonized policy practices aimed at standardizing sustainable techniques involved in the entire production chain.

8 Regulatory Support and International Collaboration

To comply with environmental policies, sustainable aviation fuels must meet specific sustainability criteria, including the reduction of greenhouse gas (GHG) emissions, consideration of land-use changes, and the avoidance of competition for raw materials and nutrients [105]. The safety, efficiency, and sustainability of airline operations depend on regulatory support for aviation fuels [61]. The rigorous standards set by aviation authorities, including the Federal Aviation Administration (FAA) in the United States and the European Aviation Safety Agency (EASA) in Europe, determine the type of fuel that is used in aircraft [106]. These regulations regulate the production, storage, and transportation of fuels, ensuring that they are free from contaminants and suitable for expected performance under different flight conditions [107].

The regulation also expands innovation in the aviation fuel sector. The focus on the development and certification of sustainable aviation fuels (SAFs) has increased in recent years. Aviation carbon emissions can be significantly reduced with these fuels produced from renewable sources. Regulatory authorities and the industry have worked together to create standards and processes that facilitate the adoption of these new fuels, ensuring that they meet the same performance and safety standards as traditional fuels [108].

Government incentive schemes must reduce costs and increase production at the same time to make SAFs commercially viable [93]. Norway was the first nation to mandate an SAF requirement. Norway wants SAF to be 0.5% of airlines' annual fuel and aims to reach 30% by 2030 [109]. The Biden administration in the United States has announced the goal of replacing all aviation fossil fuels with SAFs by 2050. The administration also plans to increase SAF production to 3 billion gallons per year by 2030, in collaboration with Airlines for America. The Sustainable Skies Act, passed by the US Congress in May 2021, supports this goal. In Australia, the Sustainable Aviation Fuels Alliance (SAFAANZ) of Australia and New Zealand put forward a proposal in 2022 that aimed to reduce aviation fuel emissions by 2.5% by 2025, 3% by 2030, and 10% by 2050 [110]. From 2025, the European Union Aviation Regulation (ReFuelEU) requires SAF mandates (quotas) of 2–70% for aviation fuel. The quota has gradually increased in 5-year increments [111].

International support is needed to make aviation fuel regulations uniform across the world. The International Civil Aviation Organization (ICAO), an agency of the United Nations, coordinates aviation fuel control efforts worldwide. ICAO collaborates with its member states to create international standards and recommendations

to improve aviation safety, efficiency, and sustainability. Due to the cooperation between regulatory authorities, international organizations, and the industry, significant advances have been achieved in the development of sustainable aviation fuels and compliance with international standards. To ensure a greener, safer future for global aviation, these cooperative efforts are essential [58].

9 Conclusion

Analyzing data from the Web of Science database spanning 2014 to June 2024, our study unveils compelling insights into biofuel research for the aviation sector. We reviewed a total of 719 scientific publications and 15,956 citations, revealing a vibrant field driven by global cooperation and technological advancements.

The United States emerges prominently with 202 publications and 5100 citations, showcasing its leadership in research volume. Meanwhile, the Netherlands excels in journal impact factors, boasting values exceeding 11.0 and producing highly cited articles, indicating their significant contributions to the field's scholarly impact.

Our analysis reveals a remarkable growth trajectory in publications, from a low of 55 in 2018 to a peak of 125 in 2023, marking a 594% increase. Citations surged by an astonishing 38,400% during the same period, underscoring heightened global interest and investment in sustainable aviation fuels (SAFs). While the United States and China lead in sheer volume, England plays a pivotal role in fostering extensive international collaborations, evident in its strategic network connections across research clusters.

Institutions such as Washington State University and the National Renewable Energy Laboratory (NREL) are at the forefront of this research, driving innovation and shaping the discourse in biofuel technologies. These institutions not only lead in publication output but also demonstrate significant influence through high citation rates, reinforcing their impact on advancing the field globally.

Scientific journals like *Renewable and Sustainable Energy Reviews* and *Biofuels, Bioproducts and Biorefining* serve as crucial platforms for disseminating research findings. Their high impact factors and substantial citation rates highlight the interdisciplinary nature of biofuel research, spanning energy, engineering, biotechnology, and environmental sciences.

Overall, our study underscores the growing importance of biofuels in mitigating aviation's environmental impact. It emphasizes the ongoing need for global collaboration, technological innovation, and supportive policies to advance sustainable aviation fuels from research to practical implementation.

Acknowledgments The authors would like to express their gratitude for the support that was provided through Programa de Pós Graduação em Engenharia Química at the Universidade Federal do Ceará and the Instituto de Engenharias e Desenvolvimento Sustentável (IEDS) at the Universidade da Integração Internacional da Lusofonia Afro-Brasileira (UNILAB). The authors

gratefully acknowledge the following Brazilian Agencies for Scientific and Technological Development: Fundação Cearense de Apoio ao Desenvolvimento Científico e Tecnológico (FUNCAP) (PS1-0186-00216.01.00/21; 31052.000417/2023; UNI-0210-00537.01.00/23), Conselho Nacional de Desenvolvimento Científico e Tecnológico (CNPq) (311062/2019-9; 440891/2020-5; 307454/2022-3), and Coordenação de Aperfeiçoamento de Ensino Superior (CAPES) (finance code 001).

Declaration of Competing Interests The authors declare that they have no known competing financial interests or personal relationships that could have appeared to influence the work reported in this paper.

Data Availability Data will be made available on request.

References

1. Su-ungkavatin, P., Tiruta-Barna, L., & Hamelin, L. (2023). Biofuels, electro fuels, electric or hydrogen?: A review of current and emerging sustainable aviation systems. *Progress in Energy and Combustion Science, 96*, 101073. https://doi.org/10.1016/j.pecs.2023.101073
2. He, X., Wang, N., Zhou, Q., et al. (2024). Smart aviation biofuel energy system coupling with machine learning technology. *Renewable and Sustainable Energy Reviews, 189*, 113914. https://doi.org/10.1016/j.rser.2023.113914
3. Anderson, B. J., Mueller, D. W., Hoard, S. A., et al. (2022). Social science applications in sustainable aviation biofuels research: Opportunities, challenges, and advancements. *Frontiers in Energy Research, 9*. https://doi.org/10.3389/fenrg.2021.771849
4. Qasem, N. A. A., Mourad, A., Abderrahmane, A., et al. (2024). A recent review of aviation fuels and sustainable aviation fuels. *Journal of Thermal Analysis and Calorimetry, 149*, 4287–4312. https://doi.org/10.1007/s10973-024-13027-5
5. Rony, Z. I., Mofijur, M., Hasan, M. M., et al. (2023). Unanswered issues on decarbonizing the aviation industry through the development of sustainable aviation fuel from microalgae. *Fuel, 334*, 126553. https://doi.org/10.1016/j.fuel.2022.126553
6. Chong, J. W., Chemmangattuvalappil, N. G., & Thangalazhy-Gopakumar, S. (2023). Aviation biofuels: Conversion routes and challenges. In *Sustainable technologies for the oil Palm Industry* (pp. 33–85). Springer Nature Singapore.
7. Khujamberdiev, R., & Cho, H. M. (2024). Biofuels in aviation: Exploring the impact of sustainable aviation fuels in aircraft engines. *Energies (Basel), 17*, 2650. https://doi.org/10.3390/en17112650
8. Baxter, G. (2020). The use of aviation biofuels as an airport environmental sustainability measure: The case of Oslo Gardermoen Airport. *MAD – Magazine of Aviation Development, 8*, 6–17. https://doi.org/10.14311/MAD.2020.01.01
9. Wei, H., Liu, W., Chen, X., et al. (2019). Renewable bio-jet fuel production for aviation: A review. *Fuel, 254*, 115599. https://doi.org/10.1016/j.fuel.2019.06.007
10. Nogueira, R. C., Neto, F. S., de Sousa Junior, P. G., et al. (2023). Research trends and perspectives on hydrothermal gasification in producing biofuels. *Energy Nexus, 10*, 100199. https://doi.org/10.1016/j.nexus.2023.100199
11. Rodrigues, A. F. S., da Silva, A. F., da Silva, F. L. B., et al. (2023). A scientometric analysis of research progress and trends in the design of laccase biocatalysts for the decolorization of synthetic dyes. *Process Biochemistry, 126*, 272–291. https://doi.org/10.1016/j.procbio.2023.01.014
12. de Castro, B. V., Leandro Fernandes Melo, R., Simão Neto, F., et al. (2024). Exploring the synergy of ionic liquids and lipase in sustainable enzymatic engineering. *Journal of Molecular Liquids, 399*, 124373. https://doi.org/10.1016/j.molliq.2024.124373

13. Dari, D. N., Freitas, I. S., Aires, F. I. S., et al. (2024). An updated review of recent applications and perspectives of hydrogen production from biomass by fermentation: A comprehensive analysis. *Biomass, 4*, 132–163. https://doi.org/10.3390/biomass4010007
14. Wang, W. C., & Tao, L. (2016). Bio-jet fuel conversion technologies. *Renewable and Sustainable Energy Reviews, 53*, 801–822. https://doi.org/10.1016/j.rser.2015.09.016
15. Li, H., Riisager, A., Saravanamurugan, S., et al. (2018). Carbon-increasing catalytic strategies for upgrading biomass into energy-intensive fuels and chemicals. *ACS Catalysis, 8*, 148–187. https://doi.org/10.1021/acscatal.7b02577
16. Kandaramath Hari, T., Yaakob, Z., & Binitha, N. N. (2015). Aviation biofuel from renewable resources: Routes, opportunities and challenges. *Renewable and Sustainable Energy Reviews, 42*, 1234–1244.
17. Zhang, X., Pan, L., Wang, L., & Zou, J. J. (2018). Review on synthesis and properties of high-energy-density liquid fuels: Hydrocarbons, nanofluids and energetic ionic liquids. *Chemical Engineering Science, 180*, 95–125.
18. Moore, R. H., Thornhill, K. L., Weinzierl, B., et al. (2017). Biofuel blending reduces particle emissions from aircraft engines at cruise conditions. *Nature, 543*, 411–415. https://doi.org/10.1038/nature21420
19. de Jong, S., Hoefnagels, R., Faaij, A., et al. (2015). The feasibility of short-term production strategies for renewable jet fuels – A comprehensive techno-economic comparison. *Biofuels, Bioproducts and Biorefining, 9*, 778–800. https://doi.org/10.1002/bbb.1613
20. De Jong, S., Antonissen, K., Hoefnagels, R., et al. (2017). Life-cycle analysis of greenhouse gas emissions from renewable jet fuel production. *Biotechnology for Biofuels, 10*, 64. https://doi.org/10.1186/s13068-017-0739-7
21. Gutiérrez-Antonio, C., Gómez-Castro, F. I., de Lira-Flores, J. A., & Hernández, S. (2017). A review on the production processes of renewable jet fuel. *Renewable and Sustainable Energy Reviews, 79*, 709–729.
22. Chuck, C. J., & Donnelly, J. (2014). The compatibility of potential bioderived fuels with Jet A-1 aviation kerosene. *Applied Energy, 118*, 83–91. https://doi.org/10.1016/j.apenergy.2013.12.019
23. Gray, N., McDonagh, S., O'Shea, R., et al. (2021). Decarbonising ships, planes and trucks: An analysis of suitable low-carbon fuels for the maritime, aviation and haulage sectors. *Advances in Applied Energy, 1*, 100008.
24. Lobo, P., Christie, S., Khandelwal, B., et al. (2015). Evaluation of non-volatile particulate matter emission characteristics of an aircraft auxiliary power unit with varying alternative jet fuel blend ratios. *Energy & Fuels, 29*, 7705–7711. https://doi.org/10.1021/acs.energyfuels.5b01758
25. Gao, Y., Li, H., Wei, G., & He, Y. (2015). Viscoelastic analysis of a sleeve based on the BP neural network. *Journal of Mechanical Science and Technology, 29*, 4621–4629. https://doi.org/10.1007/s12206-015-1007-z
26. Chen, Y., Yu, J., Li, L., et al. (2018). An empirical study of the impact of the air transportation industry energy conservation and emission reduction projects on the local economy in China. *International Journal of Environmental Research and Public Health, 15*, 812. https://doi.org/10.3390/ijerph15040812
27. Ackerman, A. S., Fridlind, A. M., Grandin, A., et al. (2015). High ice water content at low radar reflectivity near deep convection—Part 2: Evaluation of microphysical pathways in updraft parcel simulations. *Atmospheric Chemistry and Physics, 15*, 11729–11751. https://doi.org/10.5194/acp-15-11729-2015
28. Wang, M., Dewil, R., Maniatis, K., et al. (2019). Biomass-derived aviation fuels: Challenges and perspective. *Progress in Energy and Combustion Science, 74*, 31–49. https://doi.org/10.1016/j.pecs.2019.04.004
29. Shahabuddin, M., Alam, M. T., Krishna, B. B., et al. (2020). A review on the production of renewable aviation fuels from the gasification of biomass and residual wastes. *Bioresource Technology, 312*, 123596. https://doi.org/10.1016/j.biortech.2020.123596

30. Bond, J. Q., Upadhye, A. A., Olcay, H., et al. (2014). Production of renewable jet fuel range alkanes and commodity chemicals from integrated catalytic processing of biomass. *Energy & Environmental Science, 7*, 1500–1523. https://doi.org/10.1039/C3EE43846E
31. Tzanetis, K. F., Posada, J. A., & Ramirez, A. (2017). Analysis of biomass hydrothermal liquefaction and biocrude-oil upgrading for renewable jet fuel production: The impact of reaction conditions on production costs and GHG emissions performance. *Renewable Energy, 113*, 1388–1398. https://doi.org/10.1016/j.renene.2017.06.104
32. Neuling, U., & Kaltschmitt, M. (2015). Conversion routes for production of biokerosene—Status and assessment. *Biomass Conversion and Biorefinery, 5*, 367–385. https://doi.org/10.1007/s13399-014-0154-2
33. Gurkan Aydin, S., & Ozgen, A. (2023). Bio-based jet fuel production by transesterification of nettle seeds. *Engineering, Technology & Applied Science Research, 13*, 10116–10120. https://doi.org/10.48084/etasr.5556
34. Miro de Medeiros, A., de Sousa, C. K., Gundim de Macêdo, M. L., et al. (2022). Catalytic pyrolysis of coconut oil with Ni/SBA-15 for the production of bio jet fuel. *RSC Advances, 12*, 10163–10176. https://doi.org/10.1039/D2RA00866A
35. Basir, N. M., Jamil, N. A. M., & Hamdan, H. (2021). Conversion of jet biofuel range hydrocarbons from palm oil over zeolite hybrid catalyst. *Nanomaterials and Nanotechnology, 11*, 184798042098153. https://doi.org/10.1177/1847980420981536
36. de Sousa, F. P., Cardoso, C. C., & Pasa, V. M. D. (2016). Producing hydrocarbons for green diesel and jet fuel formulation from palm kernel fat over Pd/C. *Fuel Processing Technology, 143*, 35–42. https://doi.org/10.1016/j.fuproc.2015.10.024
37. Hebish, A. A., Zulkifli, N. W. M., Julkapli, N. M., & Patah, M. F. A. (2024). The impact of operating conditions on H2 free catalytic deoxygenation of triglycerides for bio-jet fuel synthesis: A step towards sustainable aviation. *Applied Catalysis O: Open, 190*, 206926. https://doi.org/10.1016/j.apcato.2024.206926
38. Ozkan, M., Narappa, A. B., Namboodiri, T., et al. (2024). Forging a sustainable sky: Unveiling the pillars of aviation e-fuel production for carbon emission circularity. *iScience, 27*, 109154. https://doi.org/10.1016/j.isci.2024.109154
39. Mueller, T., Winter, E., & Grote, U. (2024). Economic impacts of power-to-liquid fuels in aviation: A general equilibrium analysis of production and utilization in Germany. *Energy Conversion and Management: X, 23*, 100632. https://doi.org/10.1016/j.ecmx.2024.100632
40. Amhamed, A. I., Al, A. A. H., Le Page, L. M., & Alrebei OF. (2024). Alternative sustainable aviation fuel and energy (SAFE)- A review with selected simulation cases of study. *Energy Reports, 11*, 3317–3344. https://doi.org/10.1016/j.egyr.2024.03.002
41. Lau, J. I. C., Wang, Y. S., Ang, T., et al. (2024). Emerging technologies, policies and challenges toward implementing sustainable aviation fuel (SAF). *Biomass and Bioenergy, 186*, 107277. https://doi.org/10.1016/j.biombioe.2024.107277
42. Undavalli, V., Gbadamosi Olatunde, O. B., Boylu, R., et al. (2023). Recent advancements in sustainable aviation fuels. *Progress in Aerospace Sciences, 136*, 100876. https://doi.org/10.1016/j.paerosci.2022.100876
43. Xu, Z., Wang, M., Chang, L., et al. (2024). Assessing the particulate matter emission reduction characteristics of small turbofan engine fueled with 100% HEFA sustainable aviation fuel. *Science of the Total Environment, 945*, 174128. https://doi.org/10.1016/j.scitotenv.2024.174128
44. Racha, A., Kumar, L., Pai, S., et al. (2024). Highly selective hydrodeoxygenation catalyst for sustainable aviation fuel production from used cooking oil. *Catalysis Today, 442*, 114895. https://doi.org/10.1016/j.cattod.2024.114895
45. Poblador-Ibanez, J., & Nocivelli, L. (2024). Toward a real-fluid modeling framework for sustainable aviation fuels. *Fuel Communications, 18*, 100100. https://doi.org/10.1016/j.jfueco.2023.100100

46. Watson, M. J., Machado, P. G., da Silva, A. V., et al. (2024). Sustainable aviation fuel technologies, costs, emissions, policies, and markets: A critical review. *Journal of Cleaner Production, 449*, 141472. https://doi.org/10.1016/j.jclepro.2024.141472
47. Braun, M., Grimme, W., & Oesingmann, K. (2024). Pathway to net zero: Reviewing sustainable aviation fuels, environmental impacts and pricing. *Journal of Air Transport Management, 117*, 102580. https://doi.org/10.1016/j.jairtraman.2024.102580
48. Hayward, J. A., O'Connell, D. A., Raison, R. J., et al. (2015). The economics of producing sustainable aviation fuel: A regional case study in Queensland, Australia. *GCB Bioenergy, 7*, 497–511. https://doi.org/10.1111/gcbb.12159
49. Hannon, M., Gimpel, J., Tran, M., et al. (2010). Biofuels from algae: Challenges and potential. *Biofuels, 1*, 763–784. https://doi.org/10.4155/bfs.10.44
50. Moorthy, K., Patwa, N., Gupta, Y., et al. (2019). Breaking barriers in deployment of renewable energy. *Heliyon, 5*, 1–23. https://doi.org/10.1016/j.heliyon.2019
51. Braun-Unkhoff, M., & Riedel, U. (2015). Alternative fuels in aviation. *CEAS Aeronautical Journal, 6*, 83–93. https://doi.org/10.1007/s13272-014-0131-2
52. Kurzawska, P., & Jasiński, R. (2021). Overview of sustainable aviation fuels with emission characteristic and particles emission of the turbine engine fueled ATJ blends with different percentages of ATJ fuel. *Energies (Basel), 14*, 1–18. https://doi.org/10.3390/en14071858
53. Seymour, K., Held, M., Stolz, B., et al. (2024). Future costs of power-to-liquid sustainable aviation fuels produced from hybrid solar PV-wind plants in Europe. *Sustainable Energy & Fuels, 8*, 811–825. https://doi.org/10.1039/d3se00978e
54. Vásquez, M. C., Silva, E. E., & Castillo, E. F. (2017). Hydrotreatment of vegetable oils: A review of the technologies and its developments for jet biofuel production. *Biomass and Bioenergy, 105*, 197–206. https://doi.org/10.1016/j.biombioe.2017.07.008
55. Tao, L., Milbrandt, A., Zhang, Y., & Wang, W. C. (2017). Techno-economic and resource analysis of hydroprocessed renewable jet fuel. *Biotechnology for Biofuels, 10*, 1–16. https://doi.org/10.1186/s13068-017-0945-3
56. Ram, V., & Salkuti, S. R. (2023). An overview of major synthetic fuels. *Energies (Basel), 16*, 1–35. https://doi.org/10.3390/en16062834
57. Hassan, Q., Algburi, S., Sameen, A. Z., et al. (2024). Green hydrogen: A pathway to a sustainable energy future. *International Journal of Hydrogen Energy, 50*, 310–333. https://doi.org/10.1016/j.ijhydene.2023.08.321
58. Cabrera, E., & Melo de Sousa, J. M. (2022). Use of sustainable fuels in aviation—A review. *Energies (Basel), 15*, 1–23. https://doi.org/10.3390/en15072440
59. Nurdiawati, A., & Urban, F. (2022). Decarbonising the refinery sector: A socio-technical analysis of advanced biofuels, green hydrogen and carbon capture and storage developments in Sweden. *Energy Research and Social Science, 84*, 1–16. https://doi.org/10.1016/j.erss.2021.102358
60. Arias, A., Feijoo, G., & Moreira, M. T. (2023). Biorefineries as a driver for sustainability: Key aspects, actual development and future prospects. *Journal of Cleaner Production, 418*, 1–17. https://doi.org/10.1016/j.jclepro.2023.137925
61. Grimme, W. (2023). The introduction of sustainable aviation fuels—A discussion of challenges, options and alternatives. *Aerospace, 10*, 1–15. https://doi.org/10.3390/aerospace10030218
62. Lee, J., & Mo, J. (2011). Analysis of technological innovation and environmental performance improvement in aviation sector. *International Journal of Environmental Research and Public Health, 8*, 3777–3795. https://doi.org/10.3390/ijerph8093777
63. Kim, H., & Son, J. (2021). Analyzing the environmental efficiency of global airlines by continent for sustainability. *Sustainability (Switzerland), 13*, 1–16. https://doi.org/10.3390/su13031571
64. Thummala, V., & Hiremath, R. B. (2022). Green aviation in India: Airline's implementation for achieving sustainability. *Cleaner and Responsible Consumption, 7*, 1–11. https://doi.org/10.1016/j.clrc.2022.100082

65. Haksevenler, B. H. G., Temel, N. U., & Topal, Y. E. (2023). Civil aviation policies and practices in Turkey in a global context through sustainable aviation. *Global Challenges, 7*, 1–16. https://doi.org/10.1002/gch2.202300194
66. Hooda, S. K., & Yadav, S. (2023). Green finance for sustainable aviation: Stakeholder perspectives and systematic review. *International Journal of Professional Business Review, 8*, 1–45. https://doi.org/10.26668/businessreview/2023.v8i5.2085
67. Li, B., Geng, Y., Xia, X., & Qiao, D. (2021). The impact of government subsidies on the low-carbon supply chain based on carbon emission reduction level. *International Journal of Environmental Research and Public Health, 18*, 1–19. https://doi.org/10.3390/ijerph18147603
68. Wang, W., Wang, Y., Zhang, X., & Zhang, D. (2021). Effects of government subsidies on production and emissions reduction decisions under carbon tax regulation and consumer low-carbon awareness. *International Journal of Environmental Research and Public Health, 18*, 1–17. https://doi.org/10.3390/ijerph182010959
69. Ebadian, M., van Dyk, S., McMillan, J. D., & Saddler, J. (2020). Biofuels policies that have encouraged their production and use: An international perspective. *Energy Policy, 147*, 111906. https://doi.org/10.1016/j.enpol.2020.111906
70. Ziolkowska, J., Meyers, W. H., & Meyer, S. D. (2010). Targets and mandates: Lessons learned from eu and us biofuel enforcement mechanisms. *International Consortium on Applied Bioeconomy Research (ICABR), 14*, 1–36. https://doi.org/10.22004/ag.econ.188119
71. Asmelash, H. (2023). The role of international trade law in the energy transition. *The Journal of World Investment & Trade, 24*, 847–878. https://doi.org/10.1163/22119000-12340312
72. Foster, E., Contestabile, M., Blazquez, J., et al. (2017). The unstudied barriers to widespread renewable energy deployment: Fossil fuel price responses. *Energy Policy, 103*, 258–264. https://doi.org/10.1016/j.enpol.2016.12.050
73. Dalei, N. N., & Gupta, A. (2024). Adoption of renewable energy to phase down fossil fuel energy consumption and mitigate territorial emissions: Evidence from BRICS group countries using panel FGLS and panel GEE models. *Discover Sustainability, 5*, 1–19. https://doi.org/10.1007/s43621-024-00237-y
74. Yang, M., Chen, L., Wang, J., et al. (2023). Circular economy strategies for combating climate change and other environmental issues. *Environmental Chemistry Letters, 21*, 55–80. https://doi.org/10.1007/s10311-022-01499-6
75. Sim, J., & Kim, B. (2021). Regulatory versus consumer pressure and retailer responsibility for upstream pollution in a supply chain. *Omega (Westport), 101*, 102250. https://doi.org/10.1016/j.omega.2020.102250
76. Li, H., Hao, Y., Xie, C., et al. (2023). Emerging technologies and policies for carbon–neutral transportation. *International Journal of Transportation Science and Technology, 12*, 329–334. https://doi.org/10.1016/j.ijtst.2022.09.002
77. Reis, V., & Khammash, L. (2022). Chapter 12 – How can airports influence airline behavior to reduce carbon footprints? *The Air Transportation Industry, 1*, 273–294. https://doi.org/10.1016/B978-0-323-91522-9.00008-7
78. Takebayashi, M., & Yamaguchi, H. (2024). The Net Zero policy and its impact on the air transport network: Considering the modal competition between airlines and high-speed rail. *Transport Economics and Management, 2*, 122–129. https://doi.org/10.1016/j.team.2024.06.002
79. Minato, N., & Morimoto, R. (2017). Dynamically interdependent business model for airline–airport coexistence. *Journal of Air Transport Management, 64*, 161–172. https://doi.org/10.1016/j.jairtraman.2016.08.002
80. Fu, X., & Zhang, A. (2010). Effects of airport concession revenue sharing on airline competition and social welfare. *Journal of Transport Economics and Policy, 44*, 119–138.
81. Dixit, A., Choi, T.-M., Kumar, P., & Jakhar, S. K. (2024). Roles of reciprocity and fairness concerns in airline-airport systems with environmental considerations. *European Journal of Operational Research, 312*, 1011–1023. https://doi.org/10.1016/j.ejor.2023.07.016

82. Lemon, D. P. (2023). Delta Airlines–A carbon neutrality pact to 2050 and beyond a public policy white paper. *The Gettysburg Journal for Public Policy, 1*, 1–37.
83. Tian, J., Yu, L., Xue, R., et al. (2022). Global low-carbon energy transition in the post-COVID-19 era. *Applied Energy, 307*, 1–12. https://doi.org/10.1016/j.apenergy.2021.118205
84. Lv, Y. (2023). Transitioning to sustainable energy: Opportunities, challenges, and the potential of blockchain technology. *Frontiers in Energy Research, 11*, 1–20. https://doi.org/10.3389/fenrg.2023.1258044
85. Bauen, A., Bitossi, N., German, L., et al. (2020). Sustainable aviation fuels status, challenges and prospects of drop-in liquid fuels, hydrogen and electrification in aviation. *Johnson Matthey Technology Review, 64*, 263–278. https://doi.org/10.1595/205651320x15816756012040
86. Barke, A., Bley, T., Thies, C., et al. (2022). Are sustainable aviation fuels a viable option for decarbonizing air transport in Europe? An environmental and economic sustainability assessment. *Applied Sciences (Switzerland), 12*, 1–20. https://doi.org/10.3390/app12020597
87. Ambaye, T. G., Vaccari, M., Bonilla-Petriciolet, A., et al. (2021). Emerging technologies for biofuel production: A critical review on recent progress, challenges and perspectives. *Journal of Environmental Management, 290*, 1–16. https://doi.org/10.1016/j.jenvman.2021.112627
88. Balan, V. (2014). Current challenges in commercially producing biofuels from lignocellulosic biomass. *ISRN Biotechnology, 2014*, 1–31. https://doi.org/10.1155/2014/463074
89. Ruth, J. C., & Stephanopoulos, G. (2023). Synthetic fuels: What are they and where do they come from? *Current Opinion in Biotechnology, 81*, 102919. https://doi.org/10.1016/j.copbio.2023.102919
90. Shi, K., Guan, B., Zhuang, Z., et al. (2024). Perspectives and outlook of E-fuels: Production, cost effectiveness, and applications. *Energy & Fuels, 38*, 7665–7692. https://doi.org/10.1021/acs.energyfuels.4c00409
91. Agaton, C. B., Batac, K. I. T., & Reyes, E. M., Jr. (2022). Prospects and challenges for green hydrogen production and utilization in The Philippines. *International Journal of Hydrogen Energy, 47*, 17859–17870. https://doi.org/10.1016/j.ijhydene.2022.04.101
92. Chipangamate, N. S., & Nwaila, G. T. (2024). Assessment of challenges and strategies for driving energy transitions in emerging markets: A socio-technological systems perspective. *Energy Geoscience, 5*, 100257. https://doi.org/10.1016/j.engeos.2023.100257
93. Wang, J., & Azam, W. (2024). Natural resource scarcity, fossil fuel energy consumption, and total greenhouse gas emissions in top emitting countries. *Geoscience Frontiers, 15*, 101757. https://doi.org/10.1016/j.gsf.2023.101757
94. Nehring, R. (2009). Traversing the mountaintop: World fossil fuel production to 2050. *Philosophical Transactions of the Royal Society B: Biological Sciences, 364*, 3067–3079. https://doi.org/10.1098/rstb.2009.0170
95. Efthymiou, M., & Ryley, T. (2022). Governance and policy developments for sustainable aviation fuels. In *Sustainable alternatives for aviation fuels* (pp. 201–214). Elsevier.
96. Korkut, E., & Fowler, L. B. (2021). Regulatory and policy analysis of production, development and use of sustainable aviation fuels in the United States. *Frontiers in Energy Research, 9*. https://doi.org/10.3389/fenrg.2021.750514
97. Yang, J., Xin, Z., He, Q., et al. (2019). An overview on performance characteristics of bio-jet fuels. *Fuel, 237*, 916–936. https://doi.org/10.1016/j.fuel.2018.10.079
98. Zemanek, D., Champagne, P., & Mabee, W. (2020). Review of life-cycle greenhouse-gas emissions assessments of hydroprocessed renewable fuel (HEFA) from oilseeds. *Biofuels, Bioproducts and Biorefining, 14*, 935–949. https://doi.org/10.1002/bbb.2125
99. Santos, K., & Delina, L. (2021). Soaring sustainably: Promoting the uptake of sustainable aviation fuels during and post-pandemic. *Energy Research and Social Science, 77*, 102074. https://doi.org/10.1016/j.erss.2021.102074
100. Kohlman, M. (2020). *As jet fuel market craters, sustainable aviation fuel prepares for takeoff*. https://www.spglobal.com/platts/en/market-insights/blogs/agriculture/092320-as-jet-fuel-market-craters-sustainable-aviation-fuel-prepares-for-takeoff. Accessed 26 June 2024.

101. U.S. Congress. (2021). *H.R.3440 – Sustainable Skies Act*. https://www.nortonrosefulbright.com/en/knowledge/publications/f81d3390/new-regulatory-initiatives-supporting-sustainable-aviation-fuel. Accessed 26 June 2024.
102. Bioenergy Australia. (2022). *Bioenergy Australia's bridging the price gap SAF report*. https://cdn.revolutionise.com.au/cups/bioenergy/files/ephcnscv89wbfzjz.pdf. Accessed 26 June 2024.
103. ICAO. (2019). *Sustainable aviation takes significant step forward at ICAO*. https://www.icao.int/Newsroom/Pages/Sustainable-aviation-takes-significant-step-forward-at-ICAO.aspx. Accessed 26 June 2024.
104. Ansell, P. J. (2023). Review of sustainable energy carriers for aviation: Benefits, challenges, and future viability. *Progress in Aerospace Sciences, 141*, 100919. https://doi.org/10.1016/j.paerosci.2023.100919
105. Edwards, T., Moses, C., & Dryer, F. (2010). Evaluation of combustion performance of alternative aviation fuels. In *46th AIAA/ASME/SAE/ASEE Joint propulsion conference & exhibit*. American Institute of Aeronautics and Astronautics.
106. Rye, L., Blakey, S., & Wilson, C. W. (2010). Sustainability of supply or the planet: A review of potential drop-in alternative aviation fuels. *Energy & Environmental Science, 3*, 17–27. https://doi.org/10.1039/B918197K
107. Gan, C., Ma, Q., Bao, S., et al. (2023). Discussion of the standards system for sustainable aviation fuels: An aero-engine safety perspective. *Sustainability, 15*, 16905. https://doi.org/10.3390/su152416905
108. Kolosz, B. W., Luo, Y., Xu, B., et al. (2020). Life cycle environmental analysis of 'drop in' alternative aviation fuels: A review. *Sustainable Energy & Fuels, 4*, 3229–3263. https://doi.org/10.1039/C9SE00788A
109. Paschke, M., & Lutter, C. (2018). Regulatory framework of global aviation. In *Biokerosene* (pp. 59–92). Springer.
110. Scheelhaase, J., Maertens, S., & Grimme, W. (2019). Synthetic fuels in aviation – Current barriers and potential political measures. *Transportation Research Procedia, 43*, 21–30. https://doi.org/10.1016/j.trpro.2019.12.015
111. Detsios, N., Theodoraki, S., Maragoudaki, L., et al. (2023). Recent advances on alternative aviation fuels/pathways: A critical review. *Energies (Basel), 16*, 1904. https://doi.org/10.3390/en16041904

Chapter 5
Conversion Technologies for Green Aviation Fuels

Xiaoyu Lin, Quan Sophia He, and Jie Yang

1 Introduction

The global aviation industry is expected to continuously expand in the coming years due to the significant role of air transport in the modern world. The demand for jet fuels would increase accordingly. Jet fuel is a type of aviation fuel designed specifically for use in aircraft powered by gas-turbine engines [1]. Conventional jet fuel is produced from the distillation of crude oil ranging from 205 °C to 260 °C [2]. The composition of jet fuel is mainly C_8–C_{16} hydrocarbons, and the chemical components are alkanes, iso-alkanes, naphthenic or naphthenic derivatives, and aromatic compounds [3]. The content of each component has a direct relationship with the characteristics of the jet fuel. A high hydrogen–carbon ratio of alkanes could ensure the energy density of the fuel. Naphthenes help to reduce the freezing point, which is critical in high-altitude flying. The aromatics contribute to the lubricity to enhance the material compatibility and prevent leakage in the seals of some aircraft, but excessive aromatics will influence the cleanliness of the fuel, so the content of aromatics should be controlled in a reasonable range [2–4].

Besides these features, commercial jet fuels also have strict requirements for other characteristics, including sulfur content, density, flash point, viscosity, smoke point, and conductivity [5]. This kind of fuel must meet very stringent international specifications, which makes it more difficult to develop an alternative fuel for aviation turbines than for automobile applications [6]. In addition, fuel consumption is

X. Lin · J. Yang (✉)
Institute of Oceanography, College of Geography and Oceanography, Minjiang University, Fuzhou, China

Department of Engineering, Faculty of Agriculture, Dalhousie University, Truro, NS, Canada
e-mail: jie.yang@mju.edu.cn

Q. S. He
Department of Engineering, Faculty of Agriculture, Dalhousie University, Truro, NS, Canada

M. Aslam et al. (eds.), *Sustainable Aviation Fuels*, Sustainable Aviation,
https://doi.org/10.1007/978-3-031-83721-0_5

one of the biggest operating costs for the aviation industry [1]. According to the U.S. Department of Energy, jet fuel represents approximately 12% of total transportation fuels in the United States [7], and it could represent as much as 40% of the cost of flying a commercial airliner [8]. The aviation industry worldwide consumes approximately 1.5–1.7 billion barrels (47.25–53.55 billion gallons) of conventional jet fuel per year [1].

The large consumption of jet fuel also generates considerable greenhouse gas (GHG) emissions, which account for around 2% global CO_2 emissions every year. Growing awareness of GHG emissions and climate change has pushed the aviation industry to set an ambitious goal of achieving a 50% reduction in CO_2 emissions by 2050 based on 2005's level [9]. The International Air Transport Association (IATA) and the International Civil Aviation Organization (ICAO) have developed four pillars to achieve this goal: (1) technological improvements, (2) operational improvements, (3) measures based on the market, and (4) alternative jet fuel. For pillar (4), the most promising alternative to petroleum jet fuel is renewable jet fuels derived from biomass. Biomass-derived bio-jet fuel has become a key element in the aviation industry's strategy to reduce operating costs and, most importantly, to eliminate environmental impact. Researchers and industrial practitioners from the oil-refining industry, aviation industry, government, biofuel companies, agricultural organizations, and academia are actively working on the development of commercially viable and sustainable processes that can produce long-lasting renewable jet fuels with low production costs and low greenhouse emissions. To date, bio-jet fuels that meet ASTM specifications can be blended with commercially available petroleum jet fuel at a percentage lower than 50% and potentially could become a 100% drop-in replacement for current petroleum jet fuel.

2 Certified Pathways and Certification Timeline

ASTM D7566-20a has defined seven types of synthetic paraffinic kerosene (SPK) as blending components for conventional jet fuels (in a certain volume percentage) to make up bio-jet fuels. The production pathways of certified SPKs are presented in Table 5.1, and the timeline is provided in Fig. 5.1.

2.1 *Route 1: Gas-to-Jet (FT-SPK)*

FT technology has been extensively studied to convert syngas ($CO + H_2$) into desirable liquid fuels (Fig. 5.2). FT-SPK should contain a high percent of n−/iso-paraffins and a maximum of 15 wt.% cyclo-paraffins and be almost free of sulfur and aromatic content. The hydrocarbon distribution of synthetic jet fuel via FT technology could vary with different original feedstocks (such as natural gas, coal, and biomass) and hydroprocessing routes [9]. The major chemical compositions of FT-SPK

Table 5.1 Seven pathways approved in ASTM D7566-20a

Route	Pathway	SPK	Descriptions
1	Gas-to-jet	FT-SPK	Gasification to obtain syngas ($CO + H_2$); Fischer–Tropsch (FT) to synthesize paraffins and olefins, followed by hydroprocessing fractionation
2	Oil-to-jet	HEFA-SPK	Deoxygenation of mono-, di-, and triglycerides; free fatty acids; and fatty acid esters, followed by hydroprocessing and fractionation
3	Sugar-to-jet	SIP-SPK	Hydrolysis to obtain fermentable sugars; fermentation of sugars for farnesene production, followed by hydroprocessing and fractionation
4	Gas-to-jet	FT-SPK/A	The aromatic content is intentionally increased by adding alkylated and bio-based aromatics
5	Alcohol-to-jet	ATJ-SPK	Hydrolysis to obtain fermentable sugars; fermentation of sugars for iso-butanol and ethanol production, followed by dehydration, oligomerization, hydrogenation, and fractionation
6	Oil-to-jet	CHJ	Hydrothermal conversion of fatty acid esters and free fatty acids to synthesize kerosene-containing aromatic compounds, followed by hydroprocessing and fractionation
7	Biomass-to-jet	HH-HEFA-SPK	Deoxygenation of bio-derived hydrocarbons, fatty acid esters, and free fatty acids (algae and other bio-sources), followed by hydroprocessing and fractionation

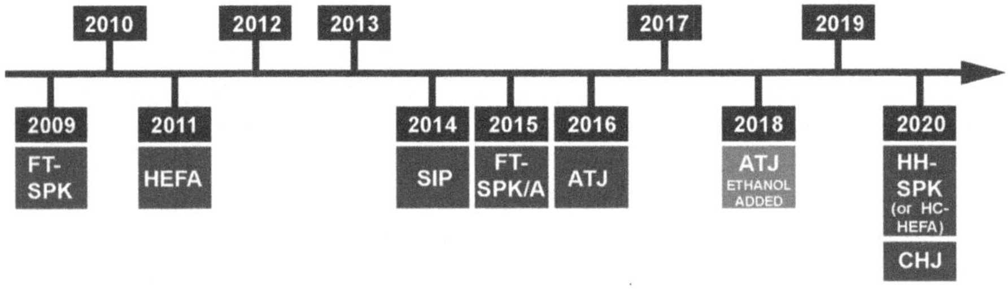

Fig. 5.1 The certification timeline for aviation biofuels. *FT-SPK* Fischer–Tropsch synthetic paraffinic kerosene, *HEFA* Hydroprocessed esters and fatty Acids, *SIP* Synthesized iso-paraffins, *FT-SPK/A* FT-SPK with aromatics, *ATJ* Alcohol-to-jet, *HH-SPK* Hydroprocessed hydrocarbons SPK, and *CHJ* Catalytic hydrothermolysis jet

and conventional jet fuels were very similar [10], with the exception of alkyl benzene content (2.73 wt.% and 14.05 wt.%, respectively). Up to 50 vol.% of FT-SPK is permitted to be blended with conventional jet fuel as specified in ASTM D7566.

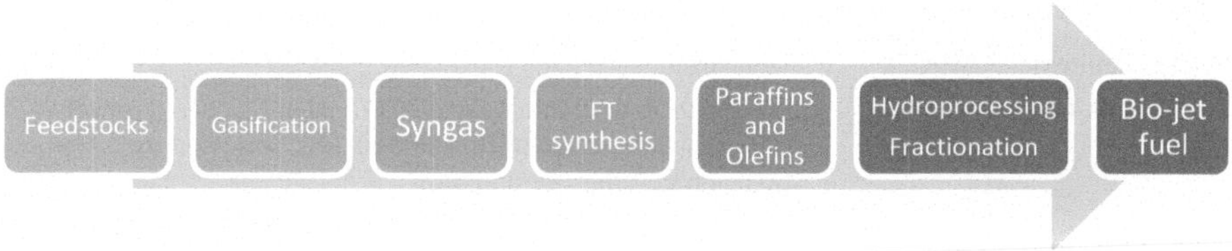

Fig. 5.2 The technical route of Fischer–Tropsch synthetic paraffinic kerosene (FT-SPK)

Many efforts have been contributed to studying the FT-SPK route. Monte et al. [11] explored the conversion of syngas into hydrocarbons within the kerosene range using catalytic dual beds formed by a potassium-cobalt-promoted iron and zeolites HZSM5 and HZSM12, increasing the selectivity toward C_9–C_{16} hydrocarbons. A triple-bed with Co-SiO_2, ZnO-SiO_2, and ZSM-5 catalysts demonstrated a promising route to enhance iso-paraffins in the FT synthesis reaction, and a selectivity of 40% for iso-paraffins was achieved, indicating the effectiveness of this system in producing desired hydrocarbon products [12]. Folkedahl et al. [13] explored the development of a process for converting coal and biomass into SPK, utilizing a supported iron-based FT catalyst in conjunction with a fluidized-bed gasification system. The produced liquid fuel nearly met the specifications for synthetic military-grade iso-paraffinic kerosene. Stadler et al. [14] investigated the impact of CO_2 presence in the gas feeds on the selectivity and reaction in a coupled FT/hydrocracking (FT-HC) process and found that the presence of CO_2 had a negligible impact on adsorption or reaction on the hydrocracking catalyst in the coupled FT-HC process. More detailed information of the FT-SPK route can be found in the review paper [15] that summarized the syngas production from biomass gasification and its conversion into fuels through the FT synthesis.

2.2 *Route 2: Oil-to-Jet (HEFA-SPK)*

Second to FT-SPK, hydroprocessed esters and fatty acids (HEFA) was certified as another blending component for fossil jet fuel in 2011. The chemical composition and carbon range of HEFA are generally similar to those of FT-SPK, indicating comparable physicochemical properties between FT-SPK and HEFA. HEFA, also

named "hydroprocessed renewable jet," is a bio-jet fuel produced from oils/lipids via hydroprocessing technologies, such as hydrodeoxygenation, hydrodecarboxylation, and hydroisomerization (Fig. 5.3). Various feedstock oils could be used for HEFA production, including conventional rapeseed oil in European countries and soybean and canola oils in North America, as well as nonconventional camelina oil, jatropha oil, waste cooking oil, algae oil, etc. These nonconventional oils usually contain a relatively high content of free fatty acids (FFA) that can be converted into HEFA via hydroprocessing treatments. Hydroisomerization or hydrocracking is needed to reduce the long carbon chain into the desirable carbon range for bio-jet fuel [9]. HEFA was proved for a blending ratio of 50 vol.% with petroleum jet fuels. Hydrotreatment of oleochemical/lipid feedstocks is currently the only technology that provides significant volumes (millions of liters per year) of bio-jet/sustainable aviation fuels.

The hydrotreatment process is generally divided into two stages. The first stage is to convert unsaturated fatty acids and/or triglycerides into saturated fatty acids, and then to n-paraffin by catalytic hydrogenation [16]. The second stage contains cracking and isomerization reactions [17]. Figure 5.4 shows reactions that might be involved and possible reaction pathways. Fractionation is the last step to obtain the desirable cut (kerosene) of synthesized paraffin hydrocarbon that fully complies with the specifications of HEFA.

Regarding the first stage of converting unsaturated fatty acids and triglycerides into straight-chain hydrocarbons, it involves hydrogenation, decarboxylation, decarbonylation, and deoxygenation, which occur in sequence and/or in parallel. The leading pathway, hydrogenation/hydrolysis of triglycerides, is combined with either β-elimination or γ-hydrogen transfer. During hydrogenation, stable saturated fatty acids are formed, no double bonds should be present in the final product, and no gaseous products are expected [19, 20]. After saturation, more hydrogen addition causes the breaking of the glycerol compound, forming propane. The resulting carboxylic acid group will be removed through three ways: hydrodeoxygenation, an exothermic reaction, which produces alkanes with the same number of carbons as the corresponding fatty acids and two moles of water; decarboxylation, which yields alkanes with one carbon atom less than the fatty acid chain and one mole of CO_2; and decarbonylation, which produces alkanes with one carbon atom less, as well as one mole of CO and one mole of water [21]. The catalyst employed included but was not limited to noble/transition metals Rh, Pt, Pd, Ni, Co, etc. [17, 22, 23]. The

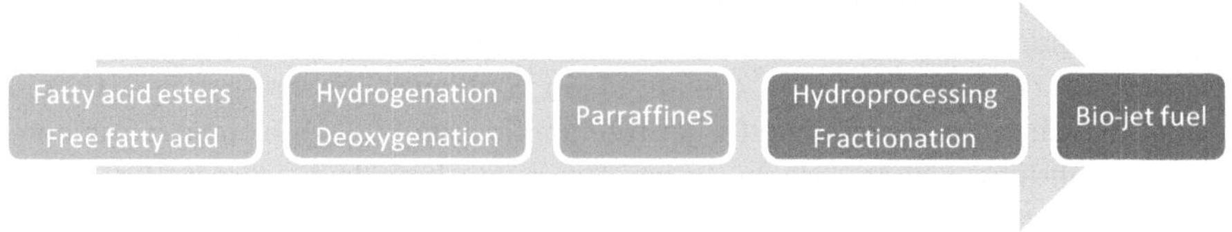

Fig. 5.3 The technical route of hydroprocessed ssters and fatty acids (HEFA) synthetic paraffinic kerosene (SPK)

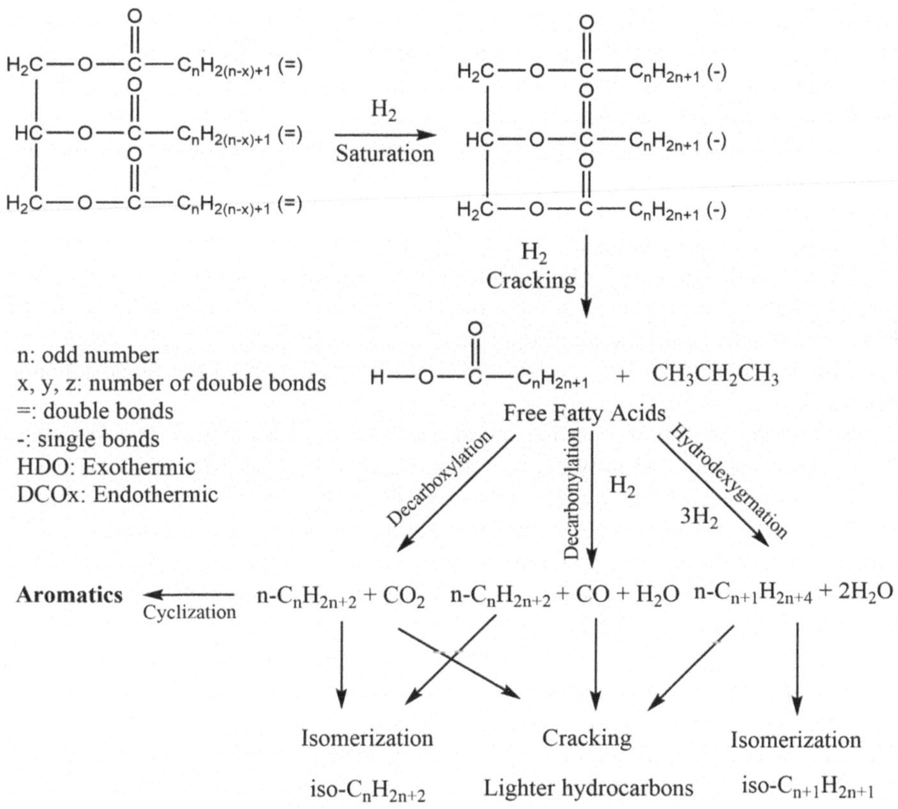

Fig. 5.4 Possible pathway reactions during the conversion of a triglyceride molecule under catalytic hydrotreatment. (Adapted from Veriansyah et al. [18] with permission)

reaction conditions also varied, including a temperature of 200–450 °C, a pressure of H_2 of 1.5–9 MPa, and a reaction time of 3–5 h [20]. Different types of reactors have been utilized, such as trickle bed reactors, batch reactors, and semi-batch reactors. The resulting yields were in the range of 55%–96%.

Cracking and isomerization are used to improve the properties of synthesized paraffin in the hydrotreating process. Hydrotreatment of triglycerides or fatty acids in the first stage generally results in straight-chain paraffinic hydrocarbons with a high cetane number, but these paraffinic oils may have compatibility issues with jet fuels due to their poor cold flow properties. It is therefore essential to improve their physicochemical properties by hydrocracking and hydroisomerization reactions. In the process of hydrocracking, straight-chain n-alkanes are broken into C_9–C_{15} chains, which are more suitable for jet fuel application [1, 24]. Isomerization converts straight-chain hydrocarbons (n-paraffins) into branched (iso-paraffin) structures, reducing the freezing point and thus improving the cold flow properties [24]. Cracking and isomerization reactions can be performed either concurrently or sequentially for the formation of desired hydrocarbons. The catalysts involved are

usually the noble or transition metals supported on activated carbon, Al_2O_3, and zeolite molecular sieves [25–27]. The reaction conditions and reactors used are comparable to the first stage of hydrotreatment. It is notable that a fractionation process is necessary to separate the resulting mixture to obtain jet-fuel-range hydrocarbons [17]. Based on the boiling point, jet-fuel-range fractions (120–235 °C) will be collected and subject to a full characterization. More detailed information on the HEFA route can be found in the review paper [28], which summarized the HEFA production that involves hydrodeoxygenation, hydrocracking, and hydroisomerization and highlighted the role of heterogeneous catalysts mediated by supported noble metals and bifunctional catalysts.

2.3 *Route 3: Sugar-to-Jet (SIP)*

Synthesized iso-paraffins (SIP) from hydroprocessing of biomass fermentation product (farnesene) was certified as one of the blending components (up to 10% by volume) for fossil jet fuel in 2014. Farnesene, a branched alkene with a chemical formula of $C_{15}H_{24}$, is typically derived from lignocellulosic and sugarcane biomass via fermentation in the presence of biocatalysts or microorganisms (Fig. 5.5). Similar to HEFA, the obtained farnesene needs to be further hydrogenated to break down the double bonds in the carbon chains and obtain desirable iso-paraffins. The specifications of SIP are largely different from those of FT-SPK and HEFA. This is mainly due to its unique chemical composition and carbon chain (a minimum of 97 wt.% farnesane that comprises a C_{15} hydrocarbon chain) [9].

As for farnesene production, various metabolic engineering techniques have been employed to improve the farnesene production efficiency. The study by [29, 30]) focused on improving α-farnesene production in the yeast *Pichia pastoris*. They achieved this through dual regulation of cytoplasm and peroxisomes. The research involved constructing a high-yielding α-farnesene strain, introducing the isopentenol utilization pathway into peroxisomes, and combining peroxisomal with cytoplasmic engineering. This resulted in a significant increase in α-farnesene production, reaching up to 2.56 g/L. The engineered strain of *Yarrowia lipolytica* was also used to generate α-farnesene [29, 30], which Erg12 was identified as a crucial limiting factor in α-farnesene production. Under optimized fermentation conditions, the engineered strain achieved an α-farnesene titer of 10.2 g/L and a yield of

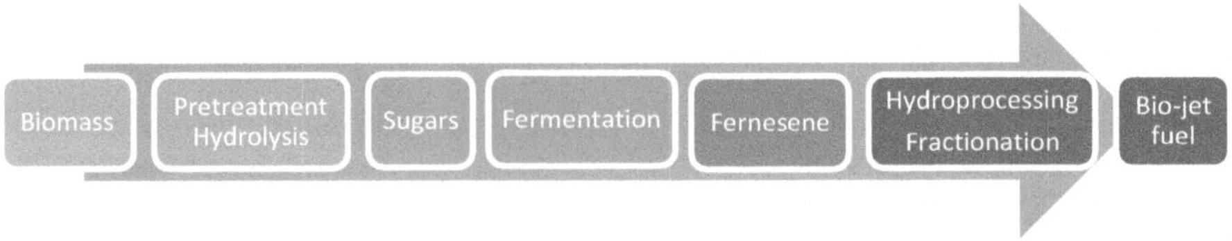

Fig. 5.5 The technical route of synthesized iso-paraffins (SIP) synthetic paraffinic kerosene (SPK)

0.1 g/g oleic acid. More detailed information for producing farnesene by using metabolic engineering techniques can be found in Liu et al. [31]. Additionally, the valorization of CO_2 to β-farnesene using *Rhodobacter sphaeroides* has been explored [32], and the attainment of high-yield β-farnesene production (234.08 mg/g CO_2) under autotrophic conditions has been reported, highlighting the potential of using *Rhodobacter sphaeroides* for the sustainable production of bio-jet fuel from CO_2.

The combustion properties of SIP have been evaluated as well. Goncalves et al. [33] explored the combustion of alternative jet fuels using reactive molecular dynamics. They focused on farnesane and its unsaturated variants, α-farnesene and β-farnesene, derived from sugarcane. The study revealed that lower activation energy was required for the combustion of α-farnesene and β-farnesene compared to farnesane, contributing to a better understanding of the combustion characteristics of bio-derived jet fuels. The key combustion properties of farnesane, such as ignition delay time and laminar burning velocity, were measured and compared with that of conventional Jet A-1 [34], and farnesane had a similar combustion behavior to conventional Jet A-1 fuel. Particularly noteworthy is that at temperatures below 1000 K, farnesane exhibited slightly shorter ignition delay times than Jet A-1. A reaction model was also developed to describe the combustion of farnesane, which was able to predict the measured combustion properties effectively, showing excellent agreement with experimental data [34]. The lower and higher flammability limits of SIP jet fuel have been explored in detail as well [35, 36].

2.4 Route 4: Gas-to-Jet (FT-SPK/A)

The specifications of FT-SPK/A are similar to those of FT-SPK and HEFA, with an exception in terms of aromatics content as well as a slightly different requirement in density (Fig. 5.6). The importance of adding a suitable amount of aromatics into n-paraffins has been identified. The study by Anuar et al. [37] investigated the swelling effects of nitrile O-rings in the presence of various fuels and aromatic species and found that in Jet-A1, nitrile O-rings exhibited swelling but none in SPK due to the absence of aromatics. For SPK blended with 4% of each of the two different aromatic species (totaling 8%), there was observable swelling in nitrile O-rings, especially in the mixture with tetralin and propylbenzene. A surrogate of 89 mol% n-decane and 11 mol% alkyl-substituted aromatics captured the major

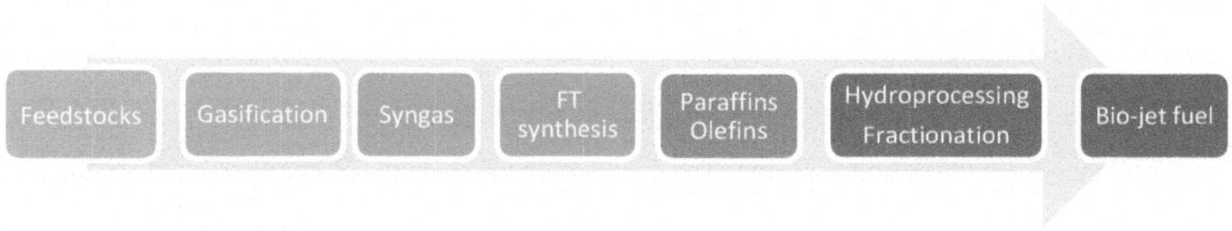

Fig. 5.6 The technical route of Fischer–Tropsch synthetic paraffinic kerosene with aromatics (FT-SPK/A)

characteristics of kerosene [9]. Due to the important role of aromatic content in aviation fuel, FT-SPK plus aromatics (FT-SPK/A) was added to the ASTM D7566 standard. The aromatics content in FT-SPK/A is intentionally increased to a maximum of 20 wt. %. Although FT-SPK/A has a relatively high aromatic content, it is still not satisfactory for direct use in aviation turbine engines unless it is blended with conventional jet fuel (up to 50 vol.%).

As for the production of aromatics, there are generally two routes; one is the traditional method that utilizes FT technology and corresponding catalysts as presented in Table 5.2. More details about the traditional method can be found in the comprehensive review contributed by Klerk et al. [38]. Many recent efforts have also been contributed to exploring the production of aromatics from biomass. The study by Bi et al. [39] focused on the transformation of lignin, a major component of biomass, into jet- and diesel-fuel-range hydrocarbons. They demonstrated a controllable process for converting lignin into C_8–C_{15} hydrocarbons. This involved catalytic depolymerization of lignin into C_6–C_8 low-carbon aromatic monomers, followed by the alkylation of these aromatics and the hydrogenation of the resulting C_8–C_{15} aromatics to produce cycloparaffins. Also, the synthesized biofuels met the main technical requirements of conventional jet fuels in terms of the heat of combustion, freezing point, and viscosity. The lignin-derived products compared favorably with standard commercial and military jet fuels, demonstrating their potential as sustainable alternatives.

The synthesis of bio-aromatics (benzene, toluene, and xylenes (BTX)) from black liquors using catalytic pyrolysis has been reported [40], and up to 7 wt.% of BTX based on the organic fraction of the black liquors was achieved. This was accomplished using a Ga-modified HZSM-5 catalyst at temperatures ranging from 500 to 600 °C. They also stated that from an industrial perspective, the ex situ pyrolysis of black liquor is preferred as it prevents catalyst deactivation by inorganics present in the black liquor feed. The study by Huang et al. [41] focused on enhancing the production of bio-aromatics (BTX) from lignin through a combined approach. The study used a combined approach of torrefaction deoxygenation pretreatment (TDP) and shape-selective catalytic fast pyrolysis (SS-CFP), employing bifunctional catalysts (metal-modified HZSM-5). The TDP process effectively removed the oxygen element from lignin, with a maximum oxygen removal efficiency of 22.27% at 300 °C. This reduction in oxygen content resulted in a significant decrease of unnecessary oxygenates in bio-oil. The use of metal-modified

Table 5.2 Aromatization processes that could be considered for producing aromatics from Fischer–Tropsch products [38]

Catalyst class	Examples	Typical feed	Comments on typical use
Monofunctional Pt	PtK/L-zeolite	C6–C8 alkanes	BTX petrochemicals
Pt/acid support	PtCl/alumina	C7–C10 hydrocarbons	Catalytic naphtha reforming
Non-Pt metal/acid support	Ga/HZSM-5	C3–C5 hydrocarbons	BTX petrochemicals
Acidic material	H-Y, HZSM-5	C3 and heavier	Fluid catalytic cracking

Note: BTX denotes benzene, toluene, and xylenes

HZSM-5 catalysts (specifically Ga/HZSM-5, Zn/HZSM-5, and Ga-Zn/HZSM-5) promoted the formation of bio-BTX. Among these, Zn/HZSM-5 showed the highest selective yield of bio-BTX due to enhanced deoxygenation reactions of oxygenates and aromatization of olefins. The combined approach of TDP and SS-CFP significantly improved the selective yield of bio-BTX, reaching a maximum value of 65.19%. This yield was substantially higher than those achieved using either the TDP (33.84%) or SS-CFP (47.36%) approaches alone. Tian et al. [42] also reported that the integrated TDP and CFP with ZSM-5 and MoO_3 significantly improved the selectivity of bio-BTX, achieving a high selectivity of 70.69% and a yield of 51.07%.

The production of aromatics from bio-oils derived from rice husks has been investigated [43], which implemented a combined catalytic process for converting bio-oils into aromatics. This process comprises two sequential steps: reforming and upgrading. During the reforming step, hydrogen is generated in situ from methanol and low-weight molecules present in the bio-oil. Structured nickel nets are used both as the reforming catalyst and as a heating grid to gasify high-boiling-point components of the bio-oil. The study reports the highest yield of aromatics at 53.7% and the lowest coke deposition at 7.1% when using Co–Ru–Zr/silica with Pt-ZSM-5 in the upgrading reactor. This suggests that the combined catalytic process effectively enhances the quality of the product while minimizing undesirable byproducts like coke. The study by Pan et al. [44] focused on the selective production of monocyclic aromatic hydrocarbons from agricultural waste, specifically wheat straw, for aviation fuel using Ni/ZSM-5 catalysts. The research showed that ZSM-5 zeolite catalysts with different Si/Al ratios and Ni loading amounts were effective in reducing phenolic components and significantly increasing the relative content of target product aromatics. The best catalytic properties for aromatic production were observed with ZSM-5 having a Si/Al ratio of 5:18. Loading 10 wt.% of Ni onto this ZSM-5 catalyst further enhanced its performance, achieving the highest relative concentration of aromatics at 67%, of which the ratio of monocyclic aromatic hydrocarbon was 55%. Lok et al. [45] focused on the research and development of Ga- and Zn-promoted ZSM-5 catalysts for producing bio-aromatics through the thermochemical conversion of biomass and waste. They identified key process parameters for aromatization, such as temperature (400–550 °C) and space velocity (0.5–1.0 h^{-1}), and emphasized the bifunctional nature of Ga/ZSM-5 and the importance of achieving an optimal Ga loading for each specific application.

2.5 Route 5: Alcohol-to-Jet (ATJ-SPK)

ATJ-SPK was recently certified as a blending component (up to 50 vol%) for conventional jet fuels. The use of iso-butanol and ethanol for ATJ-SPK production is allowed, and using all C_2–C_5 alcohols for ATJ-SPK production is the ultimate goal. ATJ-SPK starting from alcohols is usually processed through dehydration, oligomerization, hydrogenation, and fractionation (Fig. 5.7). Unfortunately, the chemical compositions of ATJ-SPK are rarely available [9]. ATJ aims to use C_2–C_5 alcohols

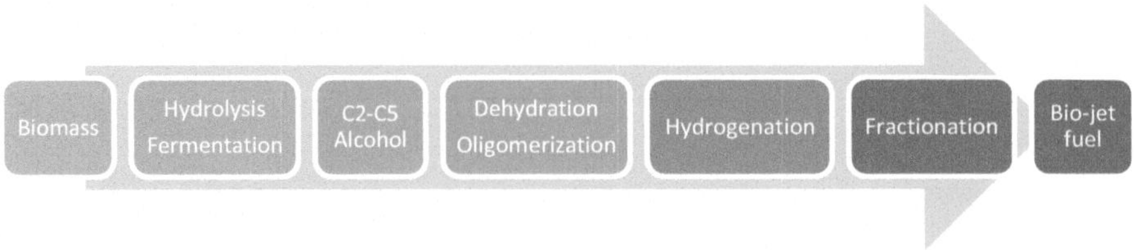

Fig. 5.7 The technical route of alcohol-to-jet (ATJ) synthetic paraffinic kerosene (SPK)

for the production of ATJ-SPK, and iso-butanol (C_4) was approved in 2015, while ethanol (C_2) was certified in 2018. Up to 50 vol.% blending of ATJ-SPK is permitted.

After obtaining the alcohols, the conversion into ATJ fuel involves an initial dehydration of alcohols to yield alkenes with an equivalent number of carbon atoms as illustrated in Fig. 5.8. This dehydration process employs a range of heterogeneous catalysts, including zeolites, silico-aluminum-phosphates, and acids. This process results in the formation of water, which is subsequently separated through techniques such as distillation, liquid–liquid separation, and the use of molecular sieves. Following dehydration, the resultant alkenes, along with short-chain olefins, undergo oligomerization. This step produces unsaturated compounds with higher molecular weights. Oligomerization is facilitated by a variety of heterogeneous or homogeneous catalysts, which could induce isomerization and cracking as well. Subsequently, hydrogen gas is introduced, facilitating the hydrogenation of olefins over solid catalysts into various paraffins, including normal, iso-, and cyclic forms. In the final stage, the resultant hydrocarbons undergo distillation, yielding products like the kerosene fraction.

However, ATJ-SPK presents certain limitations, particularly in blends with conventional jet fuel, due to its lack of aromatic hydrocarbons. This deficiency results in reduced fuel density, occasionally falling below acceptable limits, and alters the fractional composition. While high aromatic content in conventional jet fuel is generally deemed undesirable, its complete absence is not permissible, primarily due to its role in maintaining seal integrity in fuel systems. A minimum aromatic content of approximately 8 vol.% is recommended to prevent seal shrinkage and potential system damage [37]. In response to these challenges, the development of ATJ-SKA technology is underway. The pivotal distinction between ATJ-SKA and ATJ-SPK lies in the inclusion of a specific concentration of aromatic hydrocarbons in the fuel composition.

The combustion properties of ATJ-SPK have been explored as well. Richter et al. [47] focused on understanding the combustion properties of ATJ by analyzing laminar burning velocities and ignition delay times and highlighted its potential as a sustainable aviation fuel. With the aim to assess the environmental impact of different fuel types in aviation [48], particle number emissions were chosen as the indicator. The ATJ-SPK blend was found to significantly reduce both total and nonvolatile particle number emissions by up to 97% compared to Jet A-1 fuel. This reduction is attributed to the lower aromatic and sulfur content and higher hydrogen content in

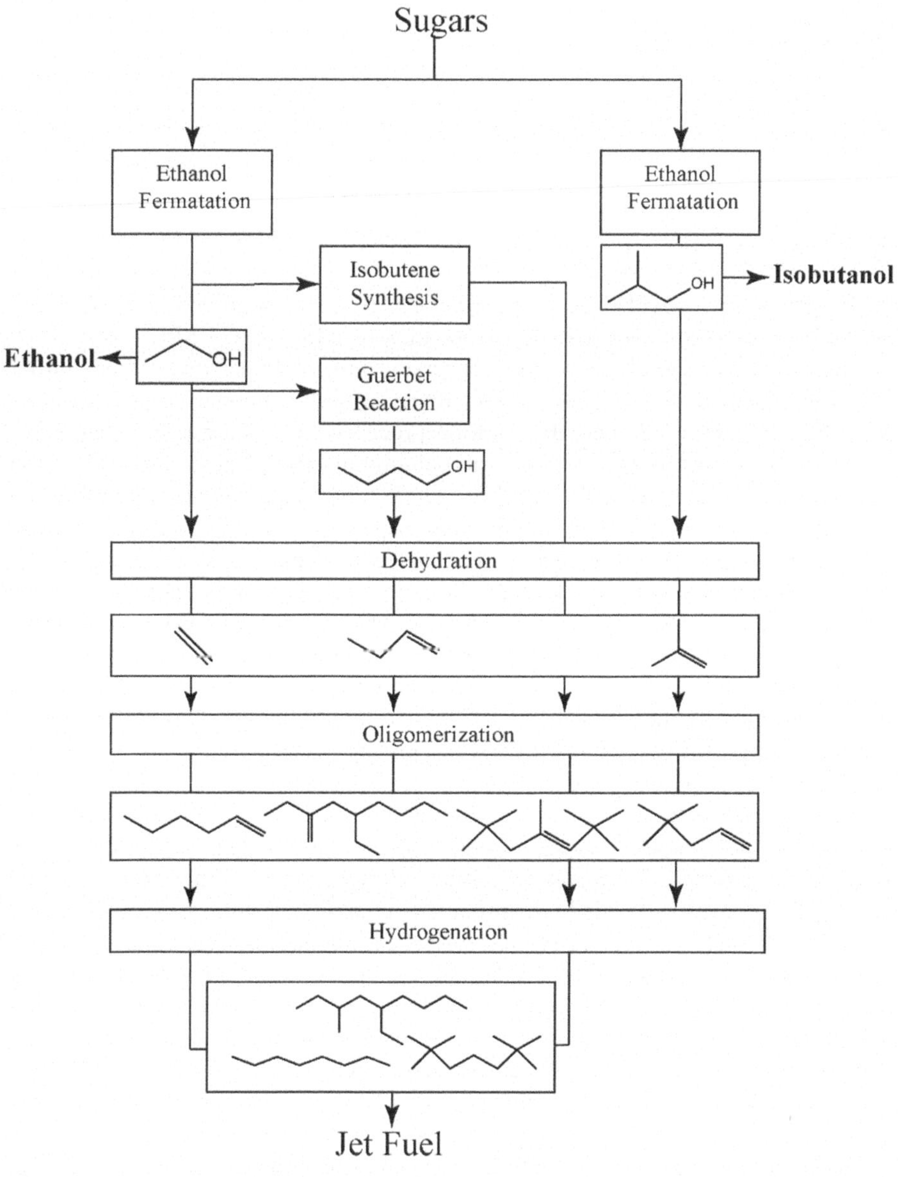

Fig. 5.8 The core catalytic steps of alcohol-to-jet process. [46]

the ATJ fuel [48]. The study provides evidence that alternative fuels like ATJ can substantially reduce particle emissions from aircraft, potentially leading to a lower environmental impact from aviation activities. Han et al. [49] present a well-to-wake analysis of ethanol-to-jet and reported that corn- and corn stover–based ATJ produced greenhouse gas emissions 16% and 73% lower than petroleum jet, respectively.

In addition to the environmental advantages, the techno-economic analysis of ATJ-SPK production has been carried out as well. Geleynse et al. [46] focused on the ATJ-SPK conversion pathway using ethanol or iso-butanol as intermediates and found that using iso-butanol offers a 34% lower conversion cost for the catalytic upgrading process compared to ethanol. However, the cost of alcohol production contributes more than 80% of the total cost at the refinery, indicating that feedstock and alcohol production costs are major factors in the overall process economics. They then examined the integration of ATJ-SPK conversion technology with a conventional kraft pulp mill [50] and identified the potential reductions in energy usage compared to traditional pulp production and once again highlighted the necessity of maintaining a high yield throughout the conversion process. Romero-Izquierdo et al. [51] focused on intensifying the ATJ process for producing renewable aviation fuel using bioethanol from lignocellulosic wastes and claimed a reduction in energy requirements by 5.31% in the intensified process compared to the conventional process. They also reported an overall decrease in the total annual cost and CO_2 emissions by 4.83% and 4.99%, respectively, for the intensified process. Voß et al. [52] evaluated the pathway for producing aviation fuel from lignocellulosic biomass via ethanol, which showed an overall carbon efficiency of 28.1%.

2.6 *Route 6: Oil-to-Jet (CHJ)*

Catalytic hydrothermolysis jet (CHJ) is a synthetic blending component that comprises essentially n-paraffin, cycloparaffin, iso-paraffin, and aromatic compounds. It is composed of hydroprocessed synthesized kerosene-containing aromatics that are wholly derived from the hydrothermal conversion of fatty acid esters and free fatty acids (Fig. 5.9). Van Dyk et al. [53] assessed three direct thermochemical liquefaction technologies, fast pyrolysis, catalytic fast pyrolysis, and hydrothermal liquefaction, for their potential to produce "biocrudes," which were subsequently upgraded to drop-in biofuels by either dedicated hydrotreatment or co-processed hydrotreatment. They showed that the production of bio-jet fuels based on direct thermochemical liquefaction of biocrudes, followed by hydrotreating, has considerable

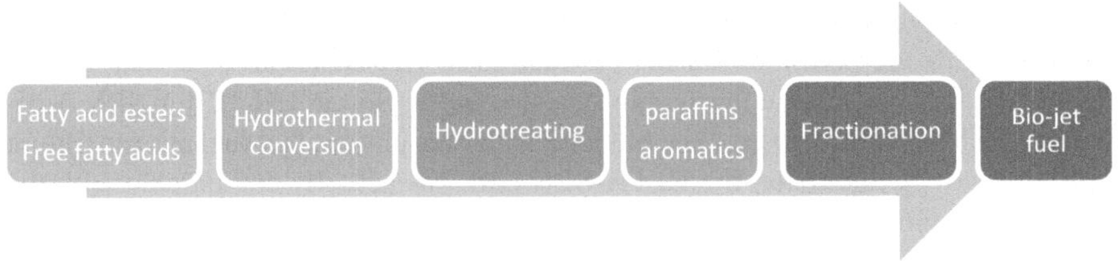

Fig. 5.9 The technical route of catalytic hydrothermolysis jet (CHJ) synthetic paraffinic kerosene (SPK)

potential. Based on the jet fuel certified to specifications in ASTM D7655, up to 50% by volume of CHJ can be blended with conventional jet fuels.

Li et al. [54] focused on producing bio-jet fuel from corncob using hydrothermal decomposition and catalytic hydrogenation and achieved a 51% yield of bio-jet-fuel-range hydrocarbons (C_8–C_{15}) from the total mass of oxygenated precursor. About 31% of the carbon and 47.6% of the energy from the carbohydrate content in corncob were recovered in bio-jet fuel. A pilot study on the production of aviation fuel from catalytic conversion of corn stover was carried out [55], which involved several units: steam stripping, hydrolysis, condensation, hydrogenation, hydrodeoxygenation, and upgrading. The process successfully produced aviation fuel with a 10.6% mass yield based on dry corn stover. The final oil product predominantly contained iso-paraffins (50.4% selectivity), contributing to a high heating value of 45.5 MJ/kg. Wang et al. [56] provided a comprehensive overview of the catalytic conversion of lignocellulosic biomass to aviation fuel. They explored various methods for converting cellulose, hemicellulose, and lignin, the primary components of lignocellulosic biomass, into hydrocarbons suitable for aviation fuel. The review paper also highlighted the challenges in biomass conversion, such as the need for hydrothermally stable catalysts and efficient hydrodeoxygenation conversion processes.

The techno-economic analysis of CHJ has been conducted as well. Eswaran et al. [57] focused on evaluating the economic viability of CHJ production from four different feedstocks: carinata oil, soybean oil, yellow grease, and brown grease, and found that without incentives, the minimum selling prices (MSPs) were \$1.50/L, \$1.32/L, \$1.19/L, and \$1.00/L, respectively. The MSP was more sensitive to the yield of CHJ and feedstock costs than to capital and supplementary costs. Also, the CH process was compared with other technologies like HEFA, highlighting the efficiency and carbon utilization advantages of CH. Similar to Eswaran et al. [57], an operation expense of \$1.18 per liter of bio-jet fuel was reported by Li et al. [54], including various costs like feedstock, catalysts, and hydrogen.

2.7 Route 7: Oil-to-Jet (HH-HEFA-SPK)

Based on ASTM D7566-20a, SKP produced from bio-derived hydroprocessed hydrocarbons (HH), fatty acid esters, and free fatty acids can be added as a synthetic blending component in jet fuels (Fig. 5.10). All bio-derived hydrocarbons, esters, and fatty acids are permitted for the production of HH-HEFA-SPK as long as sufficient test data is available for the biological sources. So far, HH-HEFA-SPK certified in ASTM is only allowed to be blended with conventional jet fuel at a maximum of 10 vol.%. For this route, the HEFA-SPK part was actually comparable with the one certified earlier in 2011. The only difference is that the source of fatty acids or fatty acid esters should originate from third-generation biomass, such as microalgae.

The HH-SPK is relatively new, and limited efforts have been contributed. Kang et al. [58] developed a novel process for synthesizing high-density bio-jet fuel,

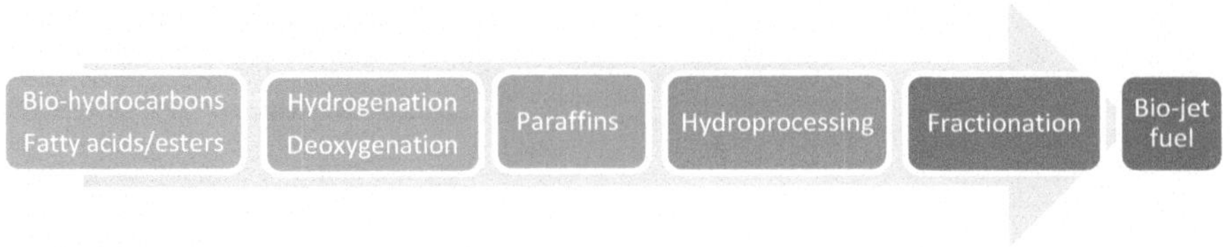

Fig. 5.10 The technical route of hydroprocessed hydrocarbons (HH) synthetic paraffinic kerosene (SPK)

involving a one-pot Mannich–Michael reaction followed by hydrodeoxygenation over alkali-treated zeolite catalysts. The Mannich–Michael reaction utilized cyclopentanone and formaldehyde derived from biomass as the starting materials. The synergistic interaction of pyrrolidine with acetic acid demonstrated high catalytic activity in the one-pot Mannich–Michael reaction, and the use of alkali-treated zeolites combined with Pd/C proved effective for the hydrodeoxygenation of condensate products. Overall, this process achieved a high yield of 92.3% [58]. The final hydrodeoxygenation products had a high density of 0.906 g/cm^3 and a high heating value of 40.5 MJ/L, making them suitable for use as jet-fuel components. Some recent efforts have been contributed to conduct cumulative exergy analysis of lignin-to-bio-jet fuel through aqueous-phase conversion [59] and claimed that the external-supplied hydrogen scenario led to the highest yield and efficiency. Zhang et al. [60] reported that the highest bio-jet fuel yield and overall energy and exergy efficiencies were observed in the case where lignin is directly converted to bio-jet fuel.

3 Readiness Level of Each Pathway

Evaluating the readiness of these routes for aviation biofuel production is crucial due to varied factors such as technological maturity, economic viability, environmental impact, and resource availability. It is essential to identify which methods are feasible and cost-effective while ensuring environmental sustainability. Additionally, the compatibility of each route with existing aviation infrastructure, compliance with policy and regulatory frameworks, and the potential for scalability and reliability are key considerations. This assessment aids in discerning the most promising and innovative routes, guiding future research and investments toward sustainable aviation fuel solutions. Such comprehensive evaluation is fundamental for the successful transition of the aviation industry toward more sustainable practices. Mawhood et al. [61] mapped development and manufacturing efforts for six renewable jet fuel production pathways expected to reach commercialization (Fig. 5.11). They found that HEFA was the most developed pathway. It has fueled the most testing flights and is produced at three commercial-scale facilities. FT-derived fuels are moving toward the start-up of the first commercial facilities although widespread deployment seems unlikely under current market conditions.

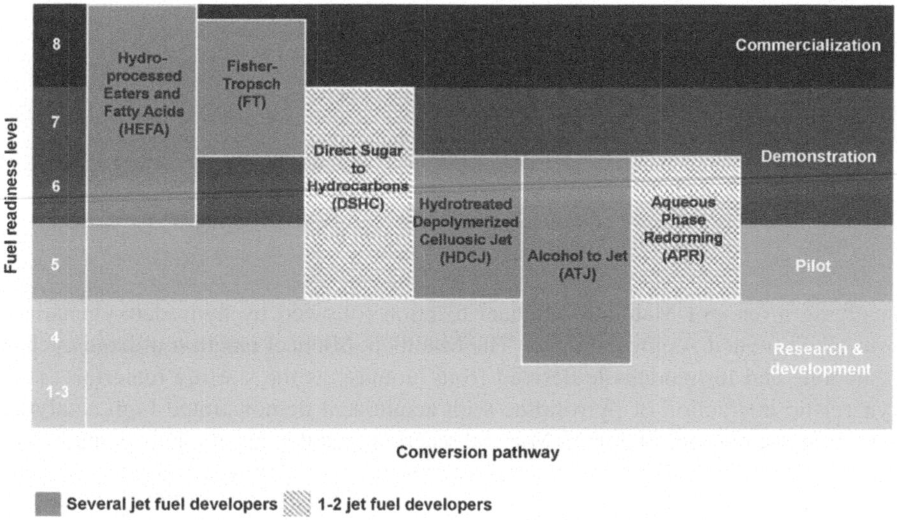

Fig. 5.11 Current fuel readiness levels of various conversion technologies. [61]

The direct sugars to hydrocarbons conversion pathway is being championed by Amyris and Total in Brazil but has not yet been demonstrated at a large scale.

Pereira et al. [8] then evaluated and compared the financial viability of six key bio-jet fuel production pathways using appropriate biomass feedstocks. The pathways were analyzed from both technical and financial perspectives, utilizing a common discounted cash flow approach and Monte Carlo analysis, considering internal (e.g., scale-up to commercial-scale) and external (e.g., oil price) uncertainties. The HEFA technology with oil feedstock had the most promising financial results. The next most attractive pathway was catalytic hydrothermolysis, which had favorable financial performance but only under a high oil price scenario. Pyrolysis and gasification with FT synthesis presented high financial risk under an oil price range of \$50–\$93 per barrel and low technical development scenarios, whereas the ATJ and direct-fermentation-to-jet technologies were found to be unlikely to achieve the minimum attractive rate of return for any of the scenarios. Wei et al. [5] conducted such research and suggested that HEFA and FT-SPK could be the most promising technologies for bio-jet fuel production in the near term, considering feedstock availability, technology state, environmental impact, and economic assessment. They stated that both these technologies had a wide range of feedstock resources, mature technology, and low CO_2 emissions.

Recently, the paper by Chong and Ng [62] presents a comprehensive analysis of aviation biofuel production, focusing on various limitations. The study introduces a holistic sustainability index encompassing energy, water, food, and governance aspects. This index is applied to assess the potential impact of aviation biofuel production on 154 countries using different conversion methods. Key findings include the identification of Finland as the top-ranked country in sustainability, highlighting

the balance between energy, food, water security, and governance. The study also discusses the economic viability of different aviation biofuel production methods (oil-to-jet (OTJ), ethanol-to-jet (ETJ), and gas-to-jet (GTJ)) under varying crude oil prices as illustrated in Fig. 5.12. The OTJ method, exemplified by HEFA, is a technologically advanced bio-jet fuel production method. Despite its capability to profit under various crude oil price–production cost scenarios, OTJ faces limitations in feedstock availability. It is estimated that only 35.15 billion liters of feedstock can be sustainably diverted from food sources to bio-jet fuels without impacting food supplies. Additionally, certain crude oil price–production cost combinations might necessitate government subsidies and supportive policies for viability. Currently, ETJ is not financially viable under the prevailing market price for crude oil. Due to the established global bioethanol industry, it is unlikely that ETJ would receive significant governmental subsidies as the required amount would be excessively high. However, if crude oil prices rise above USD 165 per barrel, ETJ becomes immediately viable and offers a higher feedstock availability ceiling of 94.04 billion liters. The GTJ method, a syngas-based approach, boasts the highest feedstock availability compared to OTJ and ETJ, with the potential to produce 704.32 billion liters of aviation biofuel. This method can utilize any organic biomass, including agricultural wastes and nonedible feedstocks, significantly reducing feedstock costs. Economically viable, even at current crude oil prices, with minimal or no government assistance, GTJ's major limitation is the FT method's low jet-fuel-range selectivity in larger-scale productions.

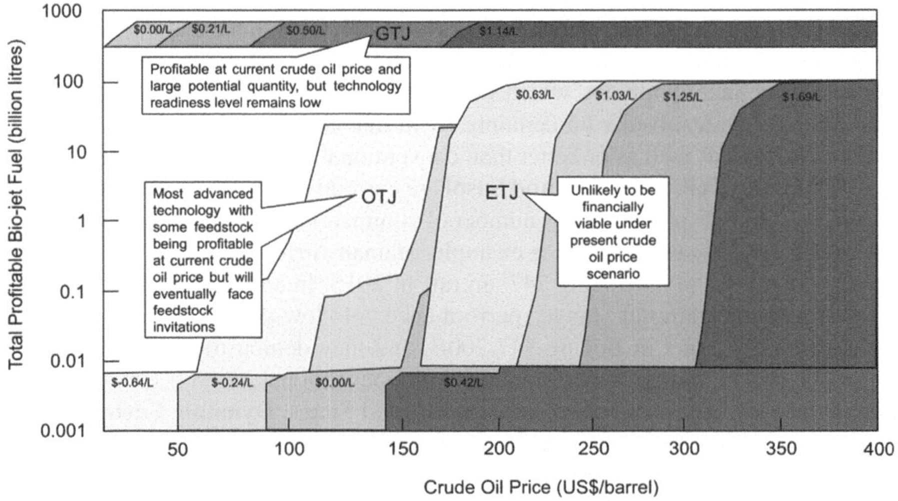

Fig. 5.12 Profitable price points of aviation biofuels by production technologies. The production methods of oil-to-jet (OTJ), ethanol-to-jet (ETJ), and gas-to-jet (GTJ) are shown in shades of green, blue, and red, respectively. (Adapted from Chong and Ng [62] with permission)

4 Involvement of Airlines Worldwide

Implications of the commercialization of alternative aviation fuels are mostly associated with the prevailing conditions of the oil market and the current jet fuel procurement mechanisms. While Jet A-1 as the dominant jet fuel type has been on sale in the aviation market since the early 1970s, alternative fuels have to compete with a highly integrated supply chain that is satisfactorily working with airlines for more than four decades [63]. A number of companies have invested in developing technologies for the production of bio-jet fuels. For instance, **Amyris**, whose SIP fuel was used in an experimental flight in 2012, is the only commercial bio-jet fuel producer with 13 million gallons (49.2 million liters) of capacity per year as of September 2015. However, plans to expand its plant in Brotas, Brazil, have shifted to the production of chemicals for the cosmetic industry in 2016.

Another potential producer, **Gevo**, has been selling its patented ATJ fuel derived from iso-butanol produced in its demonstration biorefinery in Texas to the US military and flew its first commercial flight test with Alaska Airlines in November of 2016. Lufthansa, with European Commission support, also tested the bio-jet fuels produced by **Gevo** to obtain the necessary approval for the commercial use in aviation of its ATJ fuel. In addition, from July to December 2011, Lufthansa conducted an in-service evaluation of HEFA bio-jet fuels, flying an Airbus A321 in commercial operations between Hamburg and Frankfurt using bio-kerosene on one engine. During this evaluation, aircraft and engine were extensively monitored, with engine-conditioning data downlinked and compared to the performance of the reference engine. Fuel quality was repeatedly tested and critical parameters were monitored. After the evaluation, fuel-bearing parts were removed from the bio-kerosene engine and compared to those of the reference engine, and the fuel tanks and their condition were accessed. The effect of the fuel on the fueling equipment used during the evaluation was also analyzed, and one observation was made concerning a possible impact on fueling equipment, which was probably not specific to bio-kerosene but could not yet be resolved with certainty. With that exception, HEFA-based bio-jet fuels performed as well as or better than conventional kerosene [64].

UOP Honeywell, SkyNRG, and Sasol are currently the world's leading companies in bio-jet fuels production. A number of commercial flight tests for bio-jet fuels have been carried out as well; for example, Hainan Airline tested the HEFA from waste cooking oil in its Boeing 747 aircraft in 2015. In addition, Blakey et al. [4] investigated the aircraft range performance of low-density/high-net-heat-of-combustion SPK fuel in Boeing 747-200B, and they demonstrated that SPK fuel had a comparable range of performance with conventional jet fuels (Jet A-1). The first commercial-scale biofuel plant in the United States, **Dynamic Fuels**, a 50/50 joint venture between Syntroleum and Tyson Foods, achieved a production of 65 million gallons per year. **Neste Oil**, one of the largest producers of renewable diesel, with plants in Finland, the Netherlands, and Singapore, totaling 670 million gallons per year, plans to have its fuel qualified as a jet-fuel-blending component by ASTM soon [8].

5 Conclusions and Recommendations

This chapter underscores the critical importance of developing green and sustainable aviation biofuels to mitigate the aviation industry's carbon footprint. It provides an in-depth analysis of various ASTM-certified bio-jet fuel conversion technologies, emphasizing their unique characteristics, economic feasibility, and environmental impacts. The analysis highlights the economic viability of HEFA and FT-SPK while noting the current financial challenges of ATJ. The necessity of incorporating aromatics in SPK for infrastructure compatibility is also underlined. Additionally, the chapter discusses the real-world application and testing of these fuels, showcasing their potential to revolutionize the aviation sector into a more environmentally sustainable domain. While this chapter highlights the current state and potential of various sustainable aviation biofuel technologies, future research should focus on enhancing the economic viability and environmental performance of these technologies, particularly under varying market conditions. The industrial sector is encouraged to invest in advanced research for improving conversion efficiencies and reducing costs, especially for less developed technologies like ATJ, FT-SPK/A, CHJ, etc. Collaborative efforts between the industry, academia, and governments are essential for overcoming technical challenges and establishing robust, sustainable aviation fuel supply chains. These future endeavors will be pivotal in realizing the full potential of green aviation fuels in reducing the industry's environmental impact.

References

1. Wang, W.-C., & Tao, L. (2016). Bio-jet fuel conversion technologies. *Renewable and Sustainable Energy Reviews, 53*, 801–822. https://doi.org/10.1016/j.rser.2015.09.016
2. Liu, G., Yan, B., & Chen, G. (2013). Technical review on jet fuel production. *Renewable and Sustainable Energy Reviews, 25*, 59–70. https://doi.org/10.1016/j.rser.2013.03.025
3. Hileman, J. I., & Stratton, R. W. (2014). Alternative jet fuel feasibility. *Transport Policy, Air Transportation and the Environment, 34*, 52–62. https://doi.org/10.1016/j.tranpol.2014.02.018
4. Blakey, S., Wilson, C. W., Farmery, M., & Midgley, R. (2011). Fuel effects on range versus payload for modern jet aircraft. *Aeronautical Journal, 115*, 627–634. https://doi.org/10.1017/S000192400000631X
5. Wei, H., Liu, W., Chen, X., Yang, Q., Li, J., & Chen, H. (2019). Renewable bio-jet fuel production for aviation: A review. *Fuel, 254*, 115599. https://doi.org/10.1016/j.fuel.2019.06.007
6. Wang, J., Bi, P., Zhang, Y., Xue, H., Jiang, P., Wu, X., Liu, J., Wang, T., & Li, Q. (2015). Preparation of jet fuel range hydrocarbons by catalytic transformation of bio-oil derived from fast pyrolysis of straw stalk. *Energy, 86*, 488–499. https://doi.org/10.1016/j.energy.2015.04.053
7. Hansen, S., Mirkouei, A., & Diaz, L. A. (2020). A comprehensive state-of-technology review for upgrading bio-oil to renewable or blended hydrocarbon fuels. *Renewable and Sustainable Energy Reviews, 118*, 109548. https://doi.org/10.1016/j.rser.2019.109548
8. Pereira, L. G., MacLean, H. L., & Saville, B. A. (2017). Financial analyses of potential biojet fuel production technologies. *Biofuels, Bioproducts and Biorefining, 11*, 665–681. https://doi.org/10.1002/bbb.1775

9. Yang, J., Xin, Z., He, Q. S., Corscadden, K., & Niu, H. (2019). An overview on performance characteristics of bio-jet fuels. *Fuel, 237*, 916–936. https://doi.org/10.1016/j.fuel.2018.10.079
10. van der Westhuizen, R., Ajam, M., De Coning, P., Beens, J., de Villiers, A., & Sandra, P. (2011). Comprehensive two-dimensional gas chromatography for the analysis of synthetic and crude-derived jet fuels. *Journal of Chromatography. A, 1218*, 4478–4486. https://doi.org/10.1016/j.chroma.2011.05.009
11. del Monte, D. M., Vizcaíno, A. J., Dufour, J., & Martos, C. (2021). Potential pathways for syngas transformation towards kerosene range hydrocarbons in a dual Fischer-Tropsch-zeolite bed. *International Journal of Energy Research, 46*, 5280–5287. https://doi.org/10.1002/er.7460
12. Zhang, G., Amoo, C. C., Li, M., Wang, J., Lu, C., Lu, P., Xing, C., Gao, X., Yang, R., & Tsubaki, N. (2019). Rational design of syngas to isoparaffins reaction route over additive dehydrogenation catalyst in a triple-bed system. *Catalysis Communications, 131*, 105799. https://doi.org/10.1016/J.CATCOM.2019.105799
13. Folkedahl, B. C., Snyder, A. C., Strege, J. R., & Bjorgaard, S. J. (2011). Process development and demonstration of coal and biomass indirect liquefaction to synthetic iso-paraffinic kerosene. *Fuel Processing Technology, 92*, 1939–1945. https://doi.org/10.1016/j.fuproc.2011.05.013
14. Stadler, T., Bertin-Mente, B., Dittmeyer, R., Brübach, L., Böltken, T., & Pfeifer, P. (2022). Influence of CO 2 -Rich Syngas on the selectivity to C 10 –C 14 in a coupled Fischer-Tropsch/Hydrocracking process. *Chemie Ingenieur Technik, 94*, 289. https://doi.org/10.1002/cite.202100172
15. Santos, R. G., & Alencar, A. C. (2020). Biomass-derived syngas production via gasification process and its catalytic conversion into fuels by Fischer Tropsch synthesis: A review. *International Journal of Hydrogen Energy, 45*, 18114–18132. https://doi.org/10.1016/J.IJHYDENE.2019.07.133
16. Morgan, T., Santillan-Jimenez, E., Harman-Ware, A. E., Ji, Y., Grubb, D., & Crocker, M. (2012). Catalytic deoxygenation of triglycerides to hydrocarbons over supported nickel catalysts. *Chemical Engineering Journal, 189–190*, 346–355. https://doi.org/10.1016/j.cej.2012.02.027
17. Galadima, A., & Muraza, O. (2015). Catalytic upgrading of vegetable oils into jet fuels range hydrocarbons using heterogeneous catalysts: A review. *Journal of Industrial and Engineering Chemistry, 29*, 12–23. https://doi.org/10.1016/j.jiec.2015.03.030
18. Veriansyah, B., Han, J. Y., Kim, S. K., Hong, S.-A., Kim, Y. J., Lim, J. S., Shu, Y.-W., Oh, S.-G., & Kim, J. (2012). Production of renewable diesel by hydroprocessing of soybean oil: Effect of catalysts. *Fuel, 94*, 578–585. https://doi.org/10.1016/j.fuel.2011.10.057
19. Sugami, Y., Minami, E., & Saka, S. (2016). Renewable diesel production from rapeseed oil with hydrothermal hydrogenation and subsequent decarboxylation. *Fuel, 166*, 376–381. https://doi.org/10.1016/j.fuel.2015.10.117
20. Vásquez, M. C., Silva, E. E., & Castillo, E. F. (2017). Hydrotreatment of vegetable oils: A review of the technologies and its developments for jet biofuel production. *Biomass and Bioenergy, 105*, 197–206. https://doi.org/10.1016/j.biombioe.2017.07.008
21. Hermida, L., Abdullah, A. Z., & Mohamed, A. R. (2015). Deoxygenation of fatty acid to produce diesel-like hydrocarbons: A review of process conditions, reaction kinetics and mechanism. *Renewable and Sustainable Energy Reviews, 42*, 1223–1233. https://doi.org/10.1016/j.rser.2014.10.099
22. Hari, T. K., & Yaakob, Z. (2015). Production of diesel fuel by the hydrotreatment of jatropha oil derived fatty acid methyl esters over γ-Al2O3 and SiO2 supported NiCo bimetallic catalysts. *Reaction Kinetics, Mechanisms and Catalysis, 116*, 131–145. https://doi.org/10.1007/s11144-015-0874-8
23. Wang, H., Yan, S., Salley, S. O., & Simon Ng, K. Y. (2013). Support effects on hydrotreating of soybean oil over NiMo carbide catalyst. *Fuel, 111*, 81–87. https://doi.org/10.1016/j.fuel.2013.04.066
24. Khan, S., Kay Lup, A. N., Qureshi, K. M., Abnisa, F., Wan Daud, W. M. A., & Patah, M. F. A. (2019). A review on deoxygenation of triglycerides for jet fuel range hydrocar-

bons. *Journal of Analytical and Applied Pyrolysis, 140*, 1–24. https://doi.org/10.1016/j.jaap.2019.03.005
25. Fu, J., Lu, X., & Savage, P. E. (2011). Hydrothermal Decarboxylation and Hydrogenation of Fatty Acids over Pt/C. *ChemSusChem, 4*, 481–486. https://doi.org/10.1002/cssc.201000370
26. Jeong, C. M., Park, G. W., Choi, J., Kang, J. W., Kim, S. M., Lee, W.-H., Woo, S. I., & Chang, H. N. (2011). Steam reforming of volatile fatty acids (VFAs) over supported Pt/Al2O3 catalysts. *International Journal of Hydrogen Energy, 36*, 7505–7515. https://doi.org/10.1016/j.ijhydene.2011.03.126
27. Lu, M., Liu, X., Li, Y., Nie, Y., Lu, X., Deng, D., Xie, Q., & Ji, J. (2016). Hydrocracking of bio-alkanes over Pt/Al-MCM-41 mesoporous molecular sieves for bio-jet fuel production. *Journal of Renewable and Sustainable Energy, 8*, 053103. https://doi.org/10.1063/1.4962561
28. Monteiro, R. R. C., dos Santos, I. A., Arcanjo, M. R. A., Cavalcante, C. L., de Luna, F. M. T., Fernandez-Lafuente, R., & Vieira, R. S. (2022). Production of jet biofuels by catalytic hydro-processing of esters and fatty acids: A review. *Catalysts, 12*, 237. https://doi.org/10.3390/catal12020237
29. Liu, H., Chen, S.-L., Xu, J.-Z., & Zhang, W.-G. (2021a). Dual Regulation of Cytoplasm and Peroxisomes for Improved A-Farnesene Production in Recombinant Pichia pastoris. *ACS Synthetic Biology, 10*, 1563–1573. https://doi.org/10.1021/acssynbio.1c00186
30. Liu, Y., Wang, Z., Cui, Z., Qi, Q., & Hou, J. (2021b). α-Farnesene production from lipid by engineered Yarrowia lipolytica. *Bioresources and Bioprocessing, 8*, 78. https://doi.org/10.1186/s40643-021-00431-0
31. Liu, Y., Wang, Z., Cui, Z., Qi, Q., & Hou, J. (2022). Progress and perspectives for microbial production of farnesene. *Bioresource Technology, 347*, 126682. https://doi.org/10.1016/j.biortech.2022.126682
32. Lee, S., Rim Lee, Y., Lee, W.-H., Youn Lee, S., Moon, M., Woo Park, G., Min, K., Lee, J., & Lee, J.-S. (2022). Valorization of CO2 to β-farnesene in Rhodobacter sphaeroides. *Bioresource Technology, 363*, 127955. https://doi.org/10.1016/j.biortech.2022.127955
33. Goncalves, R. F. B., Iha, B. K. V., Rocco, J. A. F. F., & Kuznetsov, A. E. (2022). Reactive molecular dynamics of pyrolysis and combustion of alternative jet fuels: A ReaxFF study. *Fuel, 310*, 122157. https://doi.org/10.1016/j.fuel.2021.122157
34. Richter, S., Kathrotia, T., Naumann, C., Kick, T., Slavinskaya, N., Braun-Unkhoff, M., & Riedel, U. (2018). Experimental and modeling study of farnesane. *Fuel, 215*, 22–29. https://doi.org/10.1016/j.fuel.2017.10.117
35. Barbosa, J. A., Coronado, C. J. R., De Andrade, J. C., Tuna, C. E., Silva, M. H., Carvalho Junior, J. A., & Mendiburu, A. Z. (2022). Experimental determination of upper flammability limits of synthesized iso-paraffins (SIP), Jet fuel and their mixtures with air at atmospheric and sub-atmospheric pressures. *Process Safety and Environment Protection, 160*, 102–115. https://doi.org/10.1016/j.psep.2022.02.016
36. Barbosa, J. A., Coronado, C. J. R., Tuna, C. E., Silva, M. H., Mendiburu, A. Z., Carvalho Junior, J. A., & De Andrade, J. C. (2021). Experimental determination of lower flammability limits of Synthesized Iso-Paraffins (SIP), jet fuel and mixtures at atmospheric and reduced pressures with air. *Fire Safety Journal, 121*, 103276. https://doi.org/10.1016/j.firesaf.2021.103276
37. Anuar, A., Undavalli, V. K., Khandelwal, B., & Blakey, S. (2021). Effect of fuels, aromatics and preparation methods on seal swell. *Aeronautical Journal, 125*, 1542–1565. https://doi.org/10.1017/aer.2021.25
38. Klerk, A., Chauhan, G., Halmenschlager, C., Link, F., Montoya Sánchez, N., Gartley, B., El-Sayed, H. E. M., Sehdev, R., & Lehoux, R. (2023). Sustainable aviation fuel: Pathways to fully formulated synthetic jet fuel via Fischer–Tropsch synthesis. *Energy Science & Engineering, 12*(2), 394–409. https://doi.org/10.1002/ese3.1379
39. Bi, P., Wang, J., Zhang, Y., Jiang, P., Wu, X., Liu, J., Xue, H., Wang, T., & Li, Q. (2015). From lignin to cycloparaffins and aromatics: Directional synthesis of jet and diesel fuel range biofuels using biomass. *Bioresource Technology, 183*, 10–17. https://doi.org/10.1016/j.biortech.2015.02.023

40. Heeres, A., Schenk, N., Muizebelt, I., Blees, R., De Waele, B., Zeeuw, A.-J., Meyer, N., Carr, R., Wilbers, E., & Heeres, H. J. (2018). Synthesis of bio-aromatics from black liquors using catalytic pyrolysis. *ACS Sustainable Chemistry & Engineering, 6*, 3472–3480. https://doi.org/10.1021/acssuschemeng.7b03728
41. Huang, M., Ma, Z., Zhou, B., Yang, Y., & Chen, D. (2020). Enhancement of the production of bio-aromatics from renewable lignin by combined approach of torrefaction deoxygenation pretreatment and shape selective catalytic fast pyrolysis using metal modified zeolites. *Bioresource Technology, 301*, 122754. https://doi.org/10.1016/j.biortech.2020.122754
42. Tian, H., Chen, L., Huang, Z., Cheng, S., & Yang, Y. (2022). Increasing the bio-aromatics yield in the biomass pyrolysis oils by the integration of torrefaction deoxygenation pretreatment and catalytic fast pyrolysis with a dual catalyst system. *Renewable Energy, 187*, 561–571. https://doi.org/10.1016/j.renene.2022.01.096
43. Li, F., Yuan, Y., Huang, Z., Chen, B., & Wang, F. (2015). Sustainable production of aromatics from bio-oils through combined catalytic upgrading with in situ generated hydrogen. *Applied Catalysis B: Environmental, 165*, 547–554. https://doi.org/10.1016/j.apcatb.2014.10.050
44. Pan, H., Zhou, X., Xie, S., Du, Z., Li, G., Zhang, C., Luo, Y., & Zhang, X. (2022). Selective production of monocyclic aromatic hydrocarbon from agricultural waste wheat straw for aviation fuel using Ni/ZSM-5 catalyst. *Biomass and Bioenergy, 165*, 106592. https://doi.org/10.1016/j.biombioe.2022.106592
45. Lok, C. M., Van Doorn, J., & Aranda Almansa, G. (2019). Promoted ZSM-5 catalysts for the production of bio-aromatics, a review. *Renewable and Sustainable Energy Reviews, 113*, 109248. https://doi.org/10.1016/j.rser.2019.109248
46. Geleynse, S., Brandt, K., Garcia-Perez, M., Wolcott, M., & Zhang, X. (2018). The alcohol-to-jet conversion pathway for drop-in biofuels: Techno-economic evaluation. *ChemSusChem, 11*, 3728–3741. https://doi.org/10.1002/cssc.201801690
47. Richter, S., Kukkadapu, G., Westbrook, C. K., Braun-Unkhoff, M., Naumann, C., Köhler, M., & Riedel, U. (2022). A combined experimental and modeling study of combustion properties of an isoparaffinic alcohol-to-jet fuel. *Combustion and Flame, 240*, 111994. https://doi.org/10.1016/j.combustflame.2022.111994
48. Tran, S., Brown, A., & Olfert, J. S. (2020). Comparison of particle number emissions from in-flight aircraft fueled with Jet A1, JP-5 and an alcohol-to-jet fuel blend. *Energy & Fuels, 34*, 7218–7222. https://doi.org/10.1021/acs.energyfuels.0c00260
49. Han, J., Tao, L., & Wang, M. (2017). Well-to-wake analysis of ethanol-to-jet and sugar-to-jet pathways. *Biotechnology for Biofuels, 10*, 21. https://doi.org/10.1186/s13068-017-0698-z
50. Geleynse, S., Jiang, Z., Brandt, K., Garcia-Perez, M., Wolcott, M., & Zhang, X. (2020). Pulp mill integration with alcohol-to-jet conversion technology. *Fuel Processing Technology, 201*, 106338. https://doi.org/10.1016/j.fuproc.2020.106338
51. Romero-Izquierdo, A. G., Gómez-Castro, F. I., Gutiérrez-Antonio, C., Hernández, S., & Errico, M. (2021). Intensification of the alcohol-to-jet process to produce renewable aviation fuel. *Chemical Engineering and Processing: Process Intensification, 160*, 108270. https://doi.org/10.1016/j.cep.2020.108270
52. Voß, S., Bube, S., & Kaltschmitt, M. (2023). Aviation fuel production pathways from lignocellulosic biomass via alcohol intermediates—A technical analysis. *Fuel Communications, 17*, 100093. https://doi.org/10.1016/j.jfueco.2023.100093
53. van Dyk, S., Su, J., Ebadian, M., O'Connor, D., Lakeman, M., & Saddler, J. J. (2019). Potential yields and emission reductions of biojet fuels produced via hydrotreatment of biocrudes produced through direct thermochemical liquefaction. *Biotechnology for Biofuels, 12*, 281. https://doi.org/10.1186/s13068-019-1625-2
54. Li, Y., Zhao, C., Chen, L., Zhang, X., Zhang, Q., Wang, T., Qiu, S., Tan, J., Li, K., Wang, C., & Ma, L. (2018). Production of bio-jet fuel from corncob by hydrothermal decomposition and catalytic hydrogenation: Lab analysis of process and techno-economics of a pilot-scale facility. *Applied Energy, 227*, 128–136. https://doi.org/10.1016/j.apenergy.2017.07.133

55. Liu, Y., Chen, L., Chen, Y., Zhang, X., Liu, J., Liu, Q., Li, Y., Wang, C., Zhang, Q., & Ma, L. (2023). Pilot study on production of aviation fuel from catalytic conversion of corn stover. *Bioresource Technology, 372*, 128653. https://doi.org/10.1016/j.biortech.2023.128653
56. Wang, H., Yang, B., Zhang, Q., & Zhu, W. (2020). Catalytic routes for the conversion of lignocellulosic biomass to aviation fuel range hydrocarbons. *Renewable and Sustainable Energy Reviews, 120*, 109612. https://doi.org/10.1016/j.rser.2019.109612
57. Eswaran, S., Subramaniam, S., Geleynse, S., Brandt, K., Wolcott, M., & Zhang, X. (2021). Techno-economic analysis of catalytic hydrothermolysis pathway for jet fuel production. *Renewable and Sustainable Energy Reviews, 151*, 111516. https://doi.org/10.1016/j.rser.2021.111516
58. Kang, Q., Zhang, X., Cui, Y., Zhao, X., Wang, W., Zhang, J., Zhang, Q., & Ma, L. (2023). Synthesis of high-density bio-jet fuel by one-pot Mannich-Michael reaction and subsequent hydrodeoxygenation over alkali-treated zeolite catalysts. *Fuel, 349*, 128706. https://doi.org/10.1016/j.fuel.2023.128706
59. Zhang, Q., Xiao, J., & Hao, J. (2023). Cumulative exergy analysis of lignocellulosic biomass to bio-jet fuel through aqueous-phase conversion with different lignin conversion pathways. *Energy, 265*, 126301. https://doi.org/10.1016/j.energy.2022.126301
60. Zhang, Q., Xiao, J., Hao, J., Wu, Q., & Song, G. (2022). Energy and exergy analyses of bio-jet fuel production from full components in lignocellulosic biomass via aqueous-phase conversion. *Applied Thermal Engineering, 201*, 117723. https://doi.org/10.1016/j.applthermaleng.2021.117723
61. Mawhood, R., Gazis, E., de Jong, S., Hoefnagels, R., & Slade, R. (2016). Production pathways for renewable jet fuel: a review of commercialization status and future prospects: Renewable jet fuel technologies: commercialisation status and future prospects. *Biofuels, Bioproducts and Biorefining, 10*, 462–484. https://doi.org/10.1002/bbb.1644
62. Chong, C. T., & Ng, J.-H. (2023). Limitations to sustainable renewable jet fuels production attributed to cost than energy-water-food resource availability. *Nature Communications, 14*, 8156. https://doi.org/10.1038/s41467-023-44049-6
63. Buse, J. (2018). Market Commercialization of Alternative Aviation Fuels. In M. Kaltschmitt & U. Neuling (Eds.), *Biokerosene: Status and Prospects* (pp. 741–758). Springer. https://doi.org/10.1007/978-3-662-53065-8_29
64. Zschocke, A. (2018). Lufthansa experiences using biokerosene. In M. Kaltschmitt & U. Neuling (Eds.), *Biokerosene: Status and prospects* (pp. 725–740). Springer. https://doi.org/10.1007/978-3-662-53065-8_28

Chapter 6
Hydroprocessing of Liquid Biomass Fractions to Drop-in Green Aviation Fuel

Elias Martinez-Hernandez and Diego Valencia

1 Introduction

Hydroprocessing or hydrotreating (HDT) of biomass fractions has been a possible transformation process because it is very mature in the petroleum-based industry [1–8]. HDT takes place at refinery operating conditions such as higher temperatures and pressures in the presence of a heterogeneous catalyst [9]. The focus of HDT is the elimination of targeted heteroatoms in the structure of organic molecules that can be undesirable in commercial products. For instance, the elimination of oxygen in biomass-derived feedstocks could lead to drop-in green fuels. This is possible because the organic products would be very similar to those produced from fossil fuels, so there will be no need of any changes in the existing infrastructure or engines for the green transition of transportation sectors [10]. However, several aspects of producing drop-in fuels using HDT such as green diesel and aviation fuels need to be addressed through research and development.

The main aspects of the HDT of liquid biomass fractions into aviation fuel that are discussed in this chapter are represented in Fig. 6.1.

E. Martinez-Hernandez · D. Valencia (✉)
Dirección de Investigación en Transformación de Hidrocarburos, Instituto Mexicano del Petróleo, Mexico City, Mexico
e-mail: dvalencia@imp.mx

M. Aslam et al. (eds.), *Sustainable Aviation Fuels*, Sustainable Aviation,
https://doi.org/10.1007/978-3-031-83721-0_6

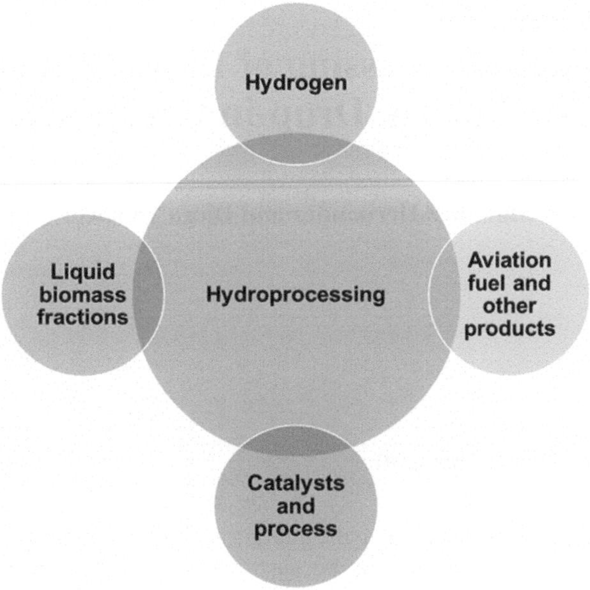

Fig. 6.1 The main aspects of hydroprocessing of liquid biomass fractions into aviation fuel

2 Liquid Biomass Fractions for Hydroprocessing

The experience gained from the hydrotreating of liquid crude oil fractions in refineries has led to the quick adoption of current hydroprocessing technologies for aviation fuel production from renewable raw materials. In general, hydroprocessing requires a liquid biomass fraction as a feedstock. Biomass can be processed into liquid fractions through different methods as shown in Fig. 6.2.

- Fischer–Tropsch (FT) wax: Lignocellulosic biomass can be gasified into syngas ($CO+H_2$). The syngas is then used in a Fischer–Tropsch synthesis process to generate hydrocarbons. Depending on the catalyst and process technologies, aviation-fuel-range hydrocarbons can be directly generated, but a heavy wax fraction is coproduced. This FT wax fraction can then be a feedstock for hydroprocessing to generate lighter and isomerized hydrocarbons of the aviation fuel range [11, 12].
- Bio-oil: Thermochemical processing of biomass using pyrolysis or hydrothermal liquefaction yields a liquid bio-oil fraction, which can serve as a feedstock for hydroprocessing into aviation fuel [13–16]. In the pyrolysis process, biomass is subject to thermal decomposition at temperatures between 400 and 800 °C, producing bio-oil, biochar, and gas. In the hydrothermal liquefaction process, the biomass is subject to high pressures (40–200 bar) and medium temperatures (300–400 °C) in water at supercritical or near-supercritical conditions. This process also yields bio-oil, biochar, and gas, but, unlike pyrolysis, it has the advantage that wet biomass can be used.

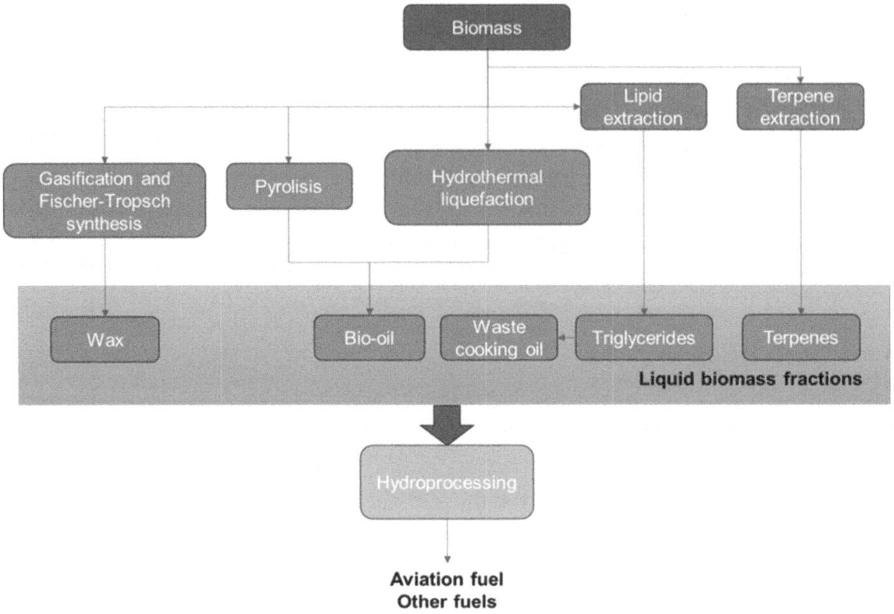

Fig. 6.2 Sources and processes to obtain liquid biomass fractions for hydroprocessing into aviation fuel

- Triglycerides: Triglycerides are esters formed by one glycerol and three fatty acids, which can be directly extracted from oil-rich biomass such as seeds and animal fat, which are usually known as first-generation feedstocks. New-generation feedstocks include algae oil and yeast oils from natural or genetically engineered organisms. The triglycerides are hydroprocessed to generate hydrocarbons, which can then be hydroisomerized to obtain aviation-fuel-range compounds; this is currently the most widely adopted technology for aviation fuel, and it is known as the hydroprocessed esters and fatty acids (HEFA) process [17, 18].
- Waste cooking oil (WCO): This is a secondary source of liquid biomass fraction, which is the residual vegetable oil from households, restaurants, and food shops. After cleaning and pretreatment to favor the process and protect the catalysts, the WCO can be used as a feedstock for hydroprocessing into aviation fuel in a similar process as with the virgin oils [19].
- Terpenes: These compounds are naturally found as components in resins and essential oils produced by plants, bacteria, yeast, and other microorganisms. Due to their unsaturated hydrocarbon nature, these are being researched as a potential feedstock for aviation fuel. Some of the more common terpenes include limonene found in citrus fruits, pinene, and, more recently, farnesene, bisabolene, and others that are of particular interest for aviation fuel production.

The liquid biomass fractions' origin and chemical nature represent a challenge for their catalytic hydroprocessing. FT wax, for example, requires only existing

technologies as it is a heavy hydrocarbon mixture. The most challenging are the bio-oil feedstocks due to their varying compositions, instability, and high oxygen content. Triglycerides and waste cooking oil are now the most common feedstocks in commercial applications, but new catalysts must be developed to remove oxygen, decrease hydrogen consumption and corrosion promoted by by-products, optimize aviation fuel yield, and decrease process complexity. Another challenge is moving from first-generation vegetable oils to triglycerides obtained in a more sustainable manner from microalgae or engineered yeast.

3 Catalytic Processes

3.1 Hydrodeoxygenation

Organic molecules are necessary for the satisfaction of societal needs, no matter how they are processed, synthesized, or isolated. Biomass-derived products can be used immediately as perfumes, key starting materials in pharmaceutics, polymers, or other substances. However, some applications that can be satisfied from biomass-derived products would require chemical transformation, not only physical separation or purification processes. Therefore, biomass transformation is a milestone in contemporary times, especially for producing intermediates or commodities for several uses. For instance, some products need to be produced towards chemical transformation such as biofuels, key intermediates towards hydrogenation or alkylation reactions. The development of new catalysts, utilization of polar solvents or making new infrastructure, and human resources capacitation are some aspects to continue transitioning to renewable energy in the present time.

Hydrodeoxygenation (HDO) is a catalytic process that belongs to the hydrotreating or hydroprocessing processes. These processes were developed in the petroleum industry for the removal of specific heteroatoms within the crude oils to meet environmental regulations or prevent damage to machinery. HDO takes place at higher temperatures and pressures in the presence of a heterogeneous catalyst. The main purpose of HDO is the elimination of oxygen atoms within the organic compounds (Scheme 6.1).

$$\text{OrgComp-O} + 2\text{H}_2 = \text{OrgComp-H}_2 + \text{H}_2\text{O} \tag{6.1}$$

where OrgComp means an organic compound present in the feedstock. HDO was used for treating biomass feedstocks with similar conditions and catalytic systems to the ones used in petroleum feedstocks. It was an initial step, but the chemical characteristics of each feedstock are very different. In the case of petroleum feedstocks, the oxygen content is very low compared to this atom concentration in biomass feedstocks. Table 6.1 showed a comparison between physical and chemical characteristics.

Table 6.1 Physical and chemical characteristics of crude oil and bio-oil

Property	Crude oil	Bio-oil
Water content (wt. %)	0.1	15–30
pH	–	2.8–3.8
Density (kg/L)	0.86	1.05–1.25
Viscosity (cP)	180	40–100
C (wt. %)	83–86	55–65
O (wt. %)	<1	28–40
H (wt. %)	11–14	5–7
S (wt. %)	<4	<0.05
N (wt. %)	<1	<0.4
H/C	1.5–2.0	0.9–1.5
O/C	~0	0.3–0.5

Water and oxygen atoms are present in high concentrations in bio-oils, which requires a pretreatment for dehydration prior to HDO. Dehydration processes applied to biomass result in the elimination of water that could damage the chemical reactors and inhibit catalytic performance. However, oxygen content in biomass will produce water as a by-product in the HDO process. This water produced in the HDO process is still a challenge in petroleum-based infrastructure or catalytic materials.

The most common catalysts applied to HDO of biomass are based on metal particles supported on metal oxides, carbon, or zeolitic frameworks. The use of the catalyst depends on the organic compounds that need to be eliminated. Then it is necessary a correct selection for using the existing materials or a proper design of a new catalyst for specific targeted molecules. The adaptation of the heterogeneous catalysts for the HDO of biomass feedstocks was the first step in facing this challenge. However, those materials were synthesized for other very different chemical processes [20, 21]. This is important since those catalytic materials also require specific operational conditions for their performance. Hydrotreating of petroleum-based feedstock takes place at higher temperatures and pressures, and those conditions need stainless steel and thicker reactors. On the other hand, heat and mass transfer in the petroleum-based feedstock deal with a very specific reactor design. As a result, the poor performance of the adapted catalytic materials takes place within reactors and operational conditions without any specific design for biomass feedstocks.

In principle, biomass feedstocks do not require higher temperatures or pressures as operating conditions. Water, CO, or CO_2 as products under higher temperatures or pressures might be in the vapor phase, and such a mixture has acidic properties that can damage the infrastructure designed for petroleum-based applications. Some improvements in the catalytic materials are focused on the use of low metal loading and avoiding the utilization of noble metals; this new generation of catalysts would be in the benefit of improving the conversion of biomass feedstocks to desired products [22–24].

Molecules from biomass can be converted into diesel fuels. For instance, HDO of palm oil can produce linear alkanes in the range of diesel. Once highly active catalysts and operational conditions are properly tuned, the production of diesel-range hydrocarbons can be achieved. However, the production of bio-jet fuel is not simple. A main reason is that the jet fuel requires ramifications in the alkane chain, which is not obtained in naturally occurring biomass feedstocks. Some approaches have been explored in recent years to produce green aviation fuel from triacylglycerides (TAGs). In general, there are two main processes: (i) alkylation of small molecules or (ii) hydroisomerization of alkanes obtained after HDO.

(i) *HDO + Hydroisomerization*

Hydrotreating of vegetable oils for fuels is of paramount importance. This method is based on mostly a two-stage process. The first stage is the conversion of vegetable oils into an oxygen-free C_{15}–C_{18} mixture of hydrocarbons. This is achieved toward the HDO process where ideally hydrocarbons, water, and CO or CO_2 gases are produced. Therefore, a separation scheme is needed for the immiscible products (hydrocarbons and water). The mixture of hydrocarbons could be used as fuels in the range of diesel or gasoline, but aviation biofuels need an additional chemical process. The second stage within this scheme is a hydroisomerization reaction [25, 26]. This step is required to produce molecules with alkyl ramifications that could meet the physicochemical properties of jet fuels. Nowadays, three technologies for the processing of vegetable oils into biofuel components, developed by UOP, Neste Oil, and Axens, are being commercialized [27].

The first step in this approach is the HDO of lipids from vegetable oils or animal fats, which mainly consist of molecules such as TAGs. The HDO of those molecules forms propane and fatty acids as the desired products. This is possible in the presence of highly active and selective heterogeneous catalysts, but several issues should be overcome since the commercial catalysts are deactivated due to the formation of coke at the surface of the catalysts. This C-O bond scission from a big molecule to form fatty acids is complicated due to the size of the molecules and the complexity of bonds such as the C=C bond in certain fatty acids and the partial HDO (Fig. 6.3).

The fatty acids produced within the HDO of TAGs within the same operational conditions and the same catalyst would react for the complete HDO. The purpose of the HDO for producing green fuels is the complete elimination of oxygen atoms.

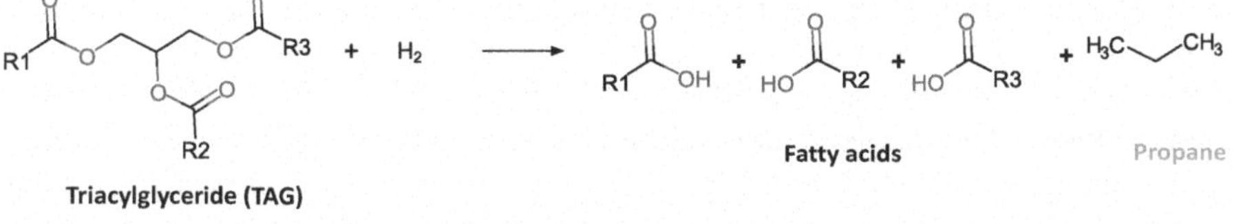

Fig. 6.3 Hydrodeoxygenation of triacylglycerides to form fatty acids, R1, R2, or R3, are linear hydrocarbon chains ranging from C_4–C_{22} with the possibility of C=C in different positions. The chemical reactions take place over heterogeneous catalysts

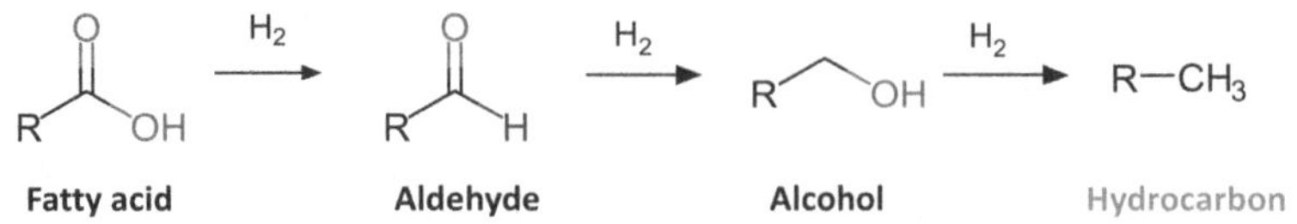

Fig. 6.4 Hydrodeoxygenation of fatty acids, R, is a linear hydrocarbon chain ranging from C_4–C_{22} with the possibility of C=C in different positions. The chemical reactions take place over heterogeneous catalysts

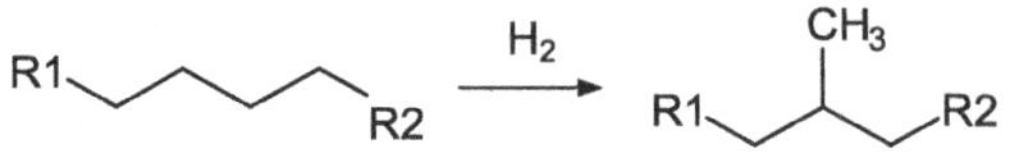

Fig. 6.5 Hydroisomerization of fatty acids to obtain drop-in aviation fuel

The chemical reaction under the H_2 atmosphere to produce hydrocarbons, which can be used as drop-in sustainable fuels, is depicted in Fig. 6.4.

The HDO of fatty acids produces linear hydrocarbons that could be used as drop-in diesel with a high cetane number. Some other chemical reactions take place under the HDO of TAGs, mostly decarbonylation or decarboxylation process. The preference of those chemical processes depends on the catalyst, biomass feedstock composition, and operational conditions. After total HDO, the hydrocarbon products need a subsequent chemical process, namely, hydroisomerization, to produce drop-in aviation fuel. This process takes place with a different catalyst and under dried conditions, meaning that water should be eliminated for a proper catalytic performance [28–30]. Zeolite-based (Si_xAl_y) materials that were developed for the isomerization of gasoline-range fuels have been adapted and utilized [31–33]. The presence of small amounts of water produces a partial or total deactivation of the active phase of the catalytic materials. The chemical reactions in the hydroisomerization process are shown in Fig. 6.5 [34–36].

(ii) *Alkylation + HDO*

Alternatives to produce bio-jet fuel in the HDO process can be achieved through molecules such as 2-methylfuran (2-MF) [37, 38]. This molecule is obtained from hemicellulose, which contains xylose that is dehydrated to furfural. The selective hydrogenation of furfural can produce 2-MF. This molecule has been employed to produce molecules in the range of bio-jet fuel toward alkylation reactions [39–42]. After alkylation reactions, high-carbon-number molecules are produced that can be treated within the HDO process to produce chemical mixtures in the range of aviation biofuels. These molecules are converted to bio-jet fuel over a Ru/SiO_2–Al_2O_3 heterogeneous catalyst producing 70% of bio-aviation fuel.

2-MF can react with furfural to form fuel precursors such as $C_{15}H_{14}O_3$. The formation of this precursor is achieved by chemical reactions in the presence of heterogeneous catalysts. The first step is the condensation of both molecules to form the intermediate product, and this will react with another molecule of 2-MF to form the fuel precursor with three furan-like molecules. The reactions are depicted in Fig. 6.6.

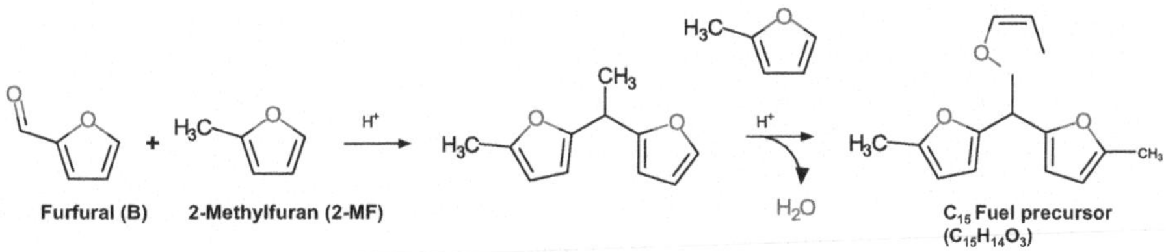

Fig. 6.6 Condensation reactions of 2-methylfuran (2-MF) and furfural (B) to form the C_{15} fuel precursor. The chemical reactions take place over heterogeneous catalysts

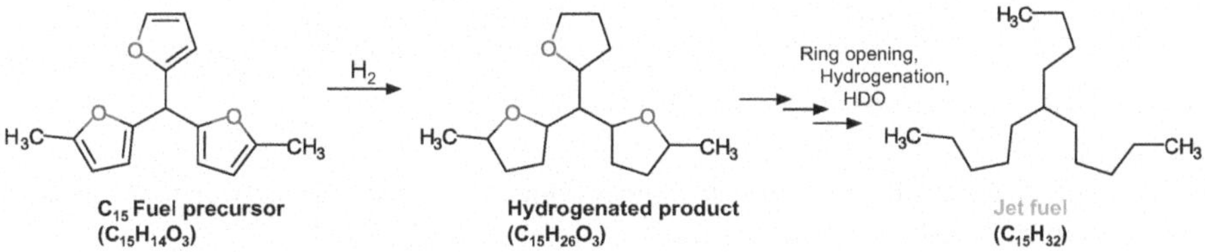

Fig. 6.7 Conversion of the C_{15} fuel precursor into the jet fuel molecule. The chemical reactions take place over heterogeneous catalysts

The molecule $C_{15}H_{14}O_3$ is a fuel precursor that should be further transformed in a complex scheme of chemical reactions that take place in the presence of a different heterogeneous catalyst that is able to promote hydrogenation, ring opening, and HDO processes to convert all the intermediates into jet fuel molecules such as $C_{15}H_{32}$. The main chemical reactions are shown in Fig. 6.7. This process allows to form a molecule with alkyl substitution that meets the physicochemical properties of the jet fuel from biomass feedstocks [43].

The formation of this jet fuel molecule ($C_{15}H_{32}$) requires two different types of catalysts and then two separate processes. The condensation reaction takes place in the presence of basic oxides or cation exchange resins as catalysts because the condensation reactions would form C-C bonds. Once the $C_{15}H_{14}O_3$ is formed, it should be separated and transformed with a different catalyst capable of hydrogenation and breaking of the C-O bond within cyclic molecules. The process is very complex, and the formation of several intermediate molecules is unavoidable, but it is very promising for producing a mixture of hydrocarbons that meets the jet fuel characteristics.

The 2-MF molecule is versatile, and it can also react with other molecules such as mesityl oxide ($C_6H_{10}O$) to form another jet fuel molecule ($C_{12}H_{24}$) with properties capable of being used as a jet fuel [44]. The change in the reactant can produce another molecule, with a cyclic structure, which can fulfill the requirements in the jet engines. The main chemical reactions of this process are shown in Fig. 6.8.

The alkylation or condensation reaction of 2-MF is possible due to the aromatic characteristics of this molecule and the formation of C-C bonds, but it does not occur naturally; it is made possible by an acid catalyst. In some cases, mineral acids (HCl and H_2SO_4, among others) can be used, or some heterogeneous resin-based

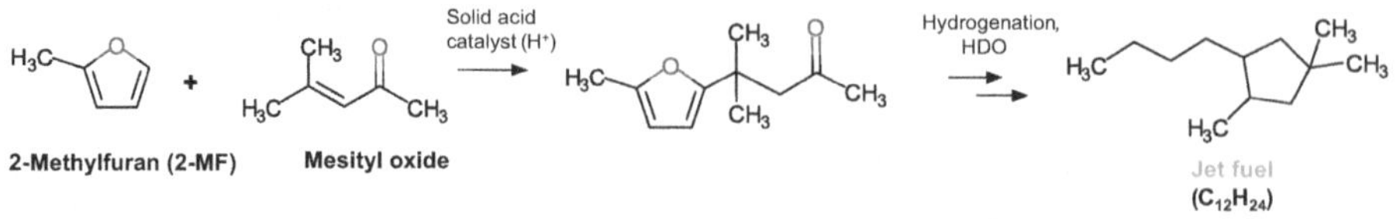

Fig. 6.8 Formation of $C_{12}H_{24}$ molecule as a jet fuel precursor, using 2-MF from biomass feedstocks. The chemical reactions take place over heterogeneous catalysts

(Nafion-212 and Amberlyst-15, among others) catalysts may perform better. The adduct molecule formed needs to be processed in a second process with a different heterogeneous catalyst and with other operating chemical conditions. This product should undergo hydrogenation and HDO process with a specific catalyst. The catalysts used in the step can be supported materials using Ni as an active phase.

Cyclopentane or cyclohexane derivatives exhibit physicochemical properties similar to drop-in aviation fuels; this is an alternative for isomerization of linear hydrocarbons described in the processes above. Cyclic molecules are present in several biomass feedstocks and are more adequate for coupling biological strategies with chemical processes.

3.2 Hydrogenation of Terpenes

Oligomerization of small molecules has been achieved since ancient times by cellules or microorganisms. The production of some chemicals has been achieved industrially by using microbial synthesis. Some molecules such as farnesene, an acyclic volatile sesquiterpene present in apple peels, are attractive commodities in present times. Farnesene is the main constituent of several plant essential oils, and it is a renewable resource. Because the content of farnesene in plants is relatively low and plant growth is significantly affected by season, regional climate, and other factors, extracting farnesene from plants for mass-market use is challenging. Microbial synthesis is an ideal alternative for farnesene production because it overcomes the disadvantages of natural extraction and chemical synthesis. Several microorganisms, including *Saccharomyces cerevisiae* and *Escherichia coli*, have been extensively engineered to produce terpenes [45, 46].

Apart from HDO process coupled with alkylation or hydroisomerization, hydrogenation processes are also envisaged as promising alternatives to overcome the complexity of coupling two or more processes, including chemical transformation and/or physical separation. In this line, hydrogenation processes of C=C bonds within cyclic structures can produce molecules with the desired properties to be used for green aviation fuels. The possibility of developing a new and green view about hydrogenation processes considers other biomass feedstocks typically considered waste in the society.

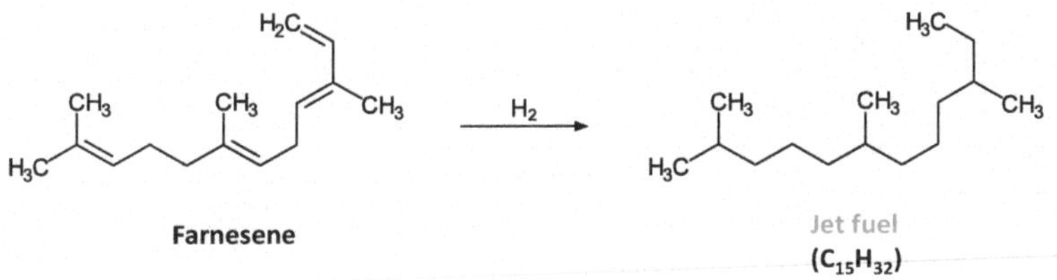

Fig. 6.9 Hydrogenation of farnesene to produce sustainable aviation fuel. The chemical reactions take place over heterogeneous catalysts

The use of other bio-based molecules such as terpenes is very promising. Terpenes are a valuable family of compounds to produce fuels and fine chemicals. Those molecules are central in the perfume, food, and pharmaceutical industries [47, 48]. As partial replacement to fossil fuel–based routes, plant-based chemicals are increasingly being evaluated as chemical building blocks per se or to other materials. Terpenes are classified based on the number of isoprene units ($H_2C{=}C(Me)CH{=}CH_2$) present in their structure. They can be hemiterpenes (C_5H_8), monoterpenes ($(C_5H_8)_2$), sesquiterpenes ($(C_5H_8)_3$), etc.

Farnesene is a family of isomers of sesquiterpenes that have differences in the chirality or position of the C=C bonds in the alkyl chains. This is another attractive molecule produced by living species, and it has the potential to be used as a precursor to drop-in aviation fuel. Similar to limonene, it is an organic molecule that requires hydrogenation to get the appropriate molecules for its use as SAF in the current aviation infrastructure. The chemical process from farnesene is shown in Fig. 6.9. Hydrogenation of farnesene produces a saturated molecule with a composition of $C_{15}H_{32}$ that satisfies the physicochemical properties for use as an SAF. In this line, this molecule is envisaged as a drop-in green aviation fuel.

Limonene is present in agro-industrial and municipal waste that is envisaged to further valorization by chemical transformations [49–51]. On the other hand, biosynthetic strategies are also developed to get higher quantities of limonene by using microorganisms. Hydrogenation of limonene to produce p-menthane, a sustainable aviation fuel, is possible with highly active heterogeneous catalysts [52]. This product can be used as a drop-in aviation fuel. In this line, the hydrogenation of this molecule can be achieved by only one step, namely hydrogenation (Fig. 6.10) over heterogeneous catalysts [53]. Compared to the previous processes, the production of SAF with limonene as a starting material requires only the hydrogenation of two C=C bonds, and there are no acidic or polar by-products.

Terpenes as biomass feedstocks for producing drop-in green aviation fuels are attractive because they require only the hydrogenation of the C=C bonds. Compared to other processes, the hydrogenation of terpenes produces SAF in one step without very complex chemical processes or using catalysts incompatible with the by-products. The economic value of terpenes as feedstock might be the limitation to producing SAFs. However, new developments in synthetic biology and heterogeneous catalysts might overcome such limitations [54]. The synthesis of SAF toward hydrogenation opens the door to more sustainable production.

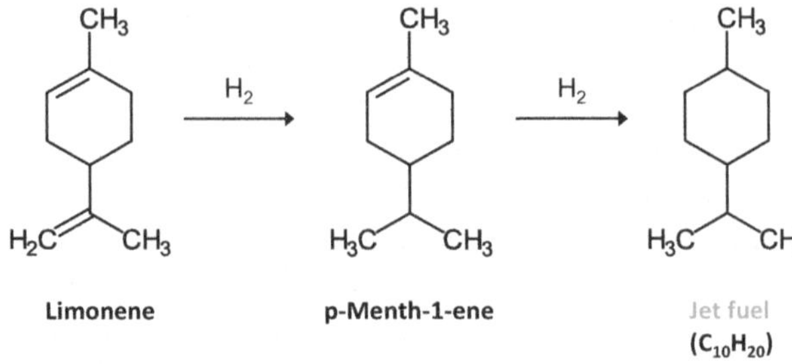

Fig. 6.10 Hydrogenation of limonene to produce sustainable aviation fuel. The chemical reactions take place over heterogeneous catalysts

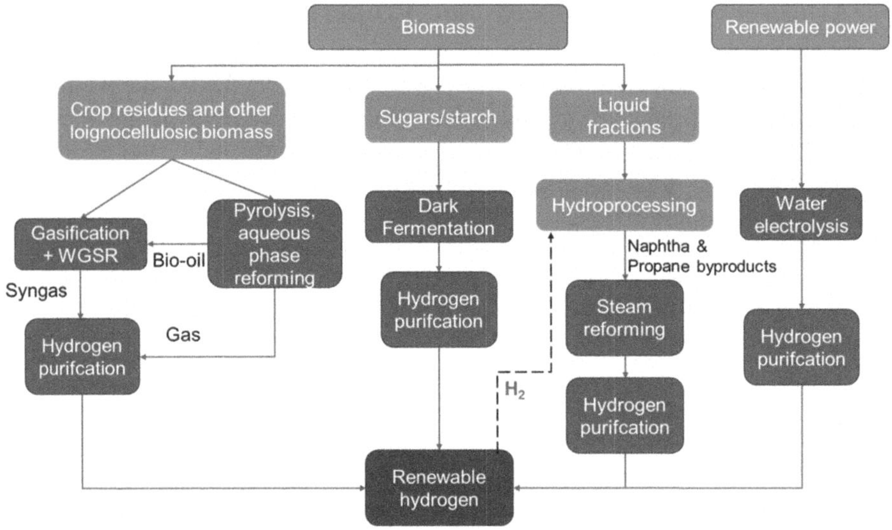

Fig. 6.11 Hydrogen sources for the hydroprocessing of liquid biomass fractions into aviation fuel

4 Hydrogen

In addition to the catalysts, sourcing hydrogen is a key aspect in the hydroprocessing of liquid biomass fractions. One of the first options is to decrease hydrogen demand by using more effective catalysts and process conditions. Hydrogen is currently under the spotlight for sustainability standards due to the high energy and carbon intensity of conventional production from the steam reforming process using fossil feedstocks (natural gas and naphtha mainly). Several standards and regulations for green and low-carbon hydrogen sources are being developed around the world. Due to the strict sustainability criteria demanded for biofuels such as aviation fuel, the producers should take care of the hydrogen sourcing. Figure 6.11 shows different hydrogen sources for the hydroprocessing of liquid fractions to aviation fuel. One strategy to ensure that aviation fuel remains renewable and sustainable is to use renewable energy for hydrogen production. However, this is still at a high cost while continuous and on-site production is preferable from an

operational and logistics perspective. Another strategy can be on-site generation through the steam reforming of hydroprocessed by-products. For example, the propane or naphtha fraction from the HEFA process could be used for steam reforming to generate hydrogen, thus making full utilization of the liquid fraction input. Another potential source of hydrogen is the gasification of the residual biomass such as seed husks and hulls or other agricultural residues. In the case of the FT wax option, part of the syngas can be purified into hydrogen for the process.

5 Aviation Fuel Components and Properties

Although hydrocarbons for aviation fuel can be produced by hydroprocessing, the final aviation fuel formulation can be critical for its use in commercial airplanes and approval by the standard ASTM D7566. The fuel must ultimately meet all the Jet A-1 specifications. Some key specifications are the cold-flow and low-temperature properties such as the freezing point. The maximum freezing point determined according to the ASTM D5972 should be −47 °C (226 K). Figure 6.12 shows the freezing points of some of the bio-jet fuel components obtained from a HEFA process. It is important to note how the isomers are required to decrease the freezing point of the mixture [18]. The freezing points of α− and β−pinene are also

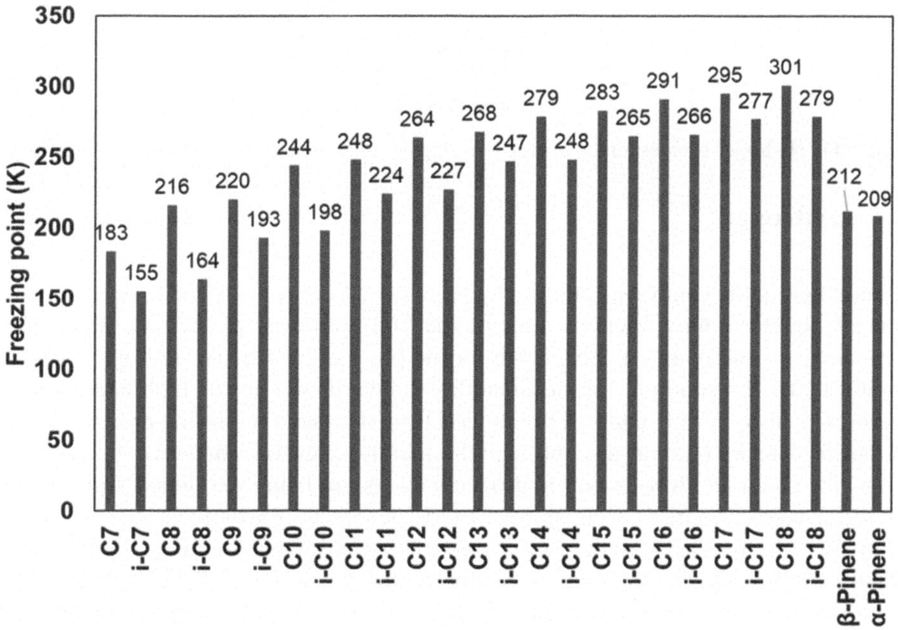

Fig. 6.12 Freezing points of representative bio-jet fuel components from a HEFA process and the terpenes, α-Pinene and β-Pinene

displayed, showing that these terpenes have freezing points, which are low enough for aviation fuel blends.

In addition, some aromatic content (8–25% volume) is desirable according to ASTM D1319. Most of the technologies currently generate an aromatic lean fuel, but researchers need to find ways to create all the required synthetic aviation fuel components including n-alkanes, iso-alkanes, cycloalkanes, and aromatics normally found in petroleum-derived jet fuel to make a fully drop-in blend. Thus, new catalysts and process conditions are required to achieve the specifications and address fuel formulation challenges.

6 Conclusions and Prospects

The chapter presents the general aspects of the production of aviation fuel from the hydroprocessing of liquid biomass fractions. There are several liquid streams that can be extracted from biomass and can be used as a feedstock for hydrocarbon molecules with potential as drop-in aviation fuel components. However, there are remaining challenges in converting these fractions due to the presence of oxygen or the need to improve the properties to comply with the industry standards of aviation fuels. In this respect, further research is envisioned to find inexpensive and more effective catalytic materials, which are also more sustainable and specifically designed to treat biomass liquid fractions. For example, ultralow loading of non-noble metals or metal oxides with resistance to acidic or water environments to achieve higher conversion to desired products.

Processes that incorporate the conversion of wastes and by-products or renewable power for hydrogen production are also required to close the carbon loop and produce fully renewable, drop-in aviation fuel components and reduce carbon footprint and other environmental impacts. This will require a combination of process technologies currently under development such as gasification, pyrolysis, dark fermentation, and the more conventional steam reforming of hydroprocessing by-products such as bionaphtha and propane. The strategies discussed in this chapter will be key in the decarbonization pathway of aviation fuels and the aviation sector.

References

1. Furimsky, E. (2000). Catalytic hydrodeoxygenation. *Applied Catalysis A: General, 199*, 147–190.
2. Bu, Q., et al. (2012). A review of catalytic hydrodeoxygenation of lignin-derived phenols from biomass pyrolysis. *Bioresource Technology, 124*, 470–477.
3. Grange, P., & Vanhaeren, X. (1997). Hydrotreating catalysts, an old story with new challenges. *Catalysis Today, 36*, 375–391.
4. Sullivan, R. F., Boduszynski, M. M., & Fetzer, J. C. (1989). Molecular transformations in hydrotreating and hydrocracking. *Energy & Fuels, 3*, 603–612.

5. Topsøe, H., Clausen, B. S., Topsøe, N.-Y., & Zeuthen, P. (1989). Progress in the design of hydrotreating catalysts based on fundamental molecular insight. In D. L. Trimm, S. Akashah, M. Absi-Halabi, & A. Bishara (Eds.), *Studies in surface science and catalysis* (Vol. 53, pp. 77–102). Elsevier.
6. Wang, H., Male, J., & Wang, Y. (2013). Recent advances in hydrotreating of pyrolysis bio-oil and its oxygen-containing model compounds. *ACS Catalysis, 3*, 1047–1070.
7. Gandarias, I., et al. (2008). From biomass to fuels: Hydrotreating of oxygenated compounds. *International Journal of Hydrogen Energy, 33*, 3485–3488.
8. Zacher, A. H., Olarte, M. V., Santosa, D. M., Elliott, D. C., & Jones, S. B. (2014). A review and perspective of recent bio-oil hydrotreating research. *Green Chemistry, 16*, 491–515.
9. Gates, B. C. (2015). From catalyst preparation toward catalyst synthesis. *Journal of Catalysis, 328*, 72–74.
10. Chen, Y. K., Lin, C. H., & Wang, W. C. (2020). The conversion of biomass into renewable jet fuel. *Energy, 201*, 117655.
11. de Klerk, A., et al. (2024). Sustainable aviation fuel: Pathways to fully formulated synthetic jet fuel via Fischer–Tropsch synthesis. *Energy Science & Engineering, 12*, 394–409.
12. Shahabuddin, M., Alam, M. T., Krishna, B. B., Bhaskar, T., & Perkins, G. (2020). A review on the production of renewable aviation fuels from the gasification of biomass and residual wastes. *Bioresource Technology, 312*, 123596.
13. Guo, Y., Yeh, T., Song, W., Xu, D., & Wang, S. (2015). A review of bio-oil production from hydrothermal liquefaction of algae. *Renewable and Sustainable Energy Reviews, 48*, 776–790.
14. Ramirez, J. A., Brown, R. J., & Rainey, T. J. (2015). A review of hydrothermal liquefaction bio-crude properties and prospects for upgrading to transportation fuels. *Energies, 8*, 6765–6794.
15. Isahak, W. N. R. W., Hisham, M. W. M., Yarmo, M. A., & Yun Hin, T. Y. (2012). A review on bio-oil production from biomass by using pyrolysis method. *Renewable and Sustainable Energy Reviews, 16*, 5910–5923.
16. Kumar, R., & Strezov, V. (2021). Thermochemical production of bio-oil: A review of downstream processing technologies for bio-oil upgrading, production of hydrogen and high value-added products. *Renewable and Sustainable Energy Reviews, 135*, 110152.
17. Monteiro, R. R. C., et al. (2022). Production of jet biofuels by catalytic hydroprocessing of esters and fatty acids: A review. *Catalysts, 12*, 237.
18. Martinez-Hernandez, E., Ramírez-Verduzco, L. F., Amezcua-Allieri, M. A., & Aburto, J. (2019). Process simulation and techno-economic analysis of bio-jet fuel and green diesel production — Minimum selling prices. *Chemical Engineering Research and Design, 146*, 60–70.
19. El-Araby, R., Abdelkader, E., El Diwani, G., & Hawash, S. I. (2020). Bio-aviation fuel via catalytic hydrocracking of waste cooking oils. *Bulletin of the National Research Centre, 44*, 1.
20. Valencia, D., et al. (2019). Paving the way towards green catalytic materials for green fuels: Impact of chemical species on Mo-based catalysts for hydrodeoxygenation. *RSC Advances, 9*, 18292–18301.
21. Díaz-García, L., et al. (2019). Tuning redox and chemical characteristics of Mo-based catalysts for bioenergy applications – The case of catalysts supported on TiO2 or ZrO2. *Materials Today Communications, 20*, 100543.
22. Valencia, D., Zenteno, C., Gómora-Herrera, D., Gómez-Pérez, M. T., & Aburto, J. (2020). Ni-based heterogeneous catalysts for the transformation of fatty acids into higher yields of O-free hydrocarbons. *Green Chemistry, 22*, 3470–3480.
23. Valencia, D., et al. (2020). Ultra-low loading of Ni in catalysts supported on mesoporous SiO<inf>2</inf> and their performance in hydrodeoxygenation of palmitic acid. *New Journal of Chemistry, 44*, 2435.
24. Valencia, D., Díaz, L., Ramírez-Verduzco, L. F., Amezcua-Allieri, M. A., & Aburto, J. (2019). MoO3-based catalysts supported on SiO2 and their performance in hydrodeoxygenation. *Materials Letters, 251*, 226–229.
25. Horáček, J., Tišler, Z., Rubáš, V., & Kubička, D. (2014). HDO catalysts for triglycerides conversion into pyrolysis and isomerization feedstock. *Fuel, 121*, 57–64.

26. Zhang, Z., Cheng, J., Zhu, Y., Guo, H., & Yang, W. (2020). Jet fuel range hydrocarbons production through competitive pathways of hydrocracking and isomerization over HPW-Ni/MCM-41 catalyst. *Fuel, 269*, 117465.
27. Zhang, B., et al. (2018). Recent developments in commercial processes for refining biofeedstocks to renewable diesel. *Bioenergy Research, 11*, 689–702.
28. Sun, J., et al. (2020). Supported heteropolyacids catalysts for the selective hydrocracking and isomerization of n-C16 to produce jet fuel. *Applied Catalysis A: General, 598*, 117556.
29. Główka, M., et al. (2024). Sustainable aviation fuel – Comprehensive study on highly selective isomerization route towards HEFA based bioadditives. *Renewable Energy, 220*, 119696.
30. Misra, P., Alvarez-Majmutov, A., & Chen, J. (2023). Isomerization catalysts and technologies for biorefining: Opportunities for producing sustainable aviation fuels. *Fuel, 351*. Preprint at https://doi.org/10.1016/j.fuel.2023.128994
31. Kang, J., et al. (2012). Hydrocracking and hydroisomerization of n-hexadecane, n-octacosane and fischer-tropsch wax over a Pt/SiO2-Al2O3 catalyst. *Catal Letters, 142*, 1295–1305.
32. Park, K.-C., & Ihm, S.-K. (2000). Comparison of Pt/Zeolite catalysts for n-Hexadecane Hydroisomerization. *Applied Catalysis A: General, 203*, 201.
33. Parmar, S., et al. (2015). Hydroisomerization of n-hexadecane over Pt/ZSM-22 framework: Effect of divalent cation exchange. *Journal of Molecular Catalysis A: Chemical, 404–405*, 47–56.
34. Romero, D., Rigutto, M., & Hensen, E. J. M. (2024). Mechanistic aspects of n-hexadecane hydroconversion: Impact of di-branched isomers on the cracked products distribution. *Fuel, 358*, 130264.
35. Batalha, N., Pinard, L., Bouchy, C., Guillon, E., & Guisnet, M. (2013). N-Hexadecane hydroisomerization over Pt-HBEA catalysts. Quantification and effect of the intimacy between metal and protonic sites. *Journal of Catalysis, 307*, 122–131.
36. Meng, J., et al. (2022). Hydroisomerization of n-hexadecane over Pt/ZSM-48 catalysts: Effects of metal-acid balance and crystal morphology. *Microporous and Mesoporous Materials, 330*, 111637.
37. Ravi, K., Lee, A. F., Wilson, K., & Biradar, A. V. (2024). Catalytic valorization of sugarcane bagasse to chemicals and aviation fuel precursors. *ACS Sustainable Chemistry & Engineering, 12*, 1632–1644.
38. Ahmad, F. B., Kalam, M. A., Zhang, Z., & Masjuki, H. H. (2022). Sustainable production of furan-based oxygenated fuel additives from pentose-rich biomass residues. *Energy Conversion and Management: X, 14*, 100222.
39. Xie, J., et al. (2019). Synthesis of high-density liquid fuel via Diels-Alder reaction of dicyclopentadiene and lignocellulose-derived 2-methylfuran. *Catalysis Today, 319*, 139–144.
40. Kunamalla, A., & Maity, S. K. (2023). Production of green jet fuel from furanics via hydroxyalkylation-alkylation over mesoporous MoO3-ZrO2 and hydrodeoxygenation over Co/γ-Al2O3: Role of calcination temperature and MoO3 content in MoO3-ZrO2. *Fuel, 332*, 125977.
41. Nie, G., et al. (2021). Co-conversion of lignocellulosic derivatives to jet fuel blending by an efficient hydrophobic acid resin. *Applied Catalysis. B, Environmental, 292*, 120181.
42. Xie, J., et al. (2018). Synthesis of high-density and low-freezing-point jet fuel using lignocellulose-derived isophorone and furanic aldehydes. *Sustainable Energy & Fuels, 2*, 1863–1869.
43. Kunamalla, A., Shrirame, B. S., & Maity, S. K. (2022). Production of jet fuel-range hydrocarbon biofuel by hydroxyalkylation-alkylation of furfural with 2-methylfuran and hydrodeoxygenation of C15fuel precursor over a Ni/γ-Al2O3catalyst: A reaction mechanism. *Energy Advances*, 99–112. https://doi.org/10.1039/d1ya00078k
44. Li, S., et al. (2016). Synthesis of jet fuel range branched cycloalkanes with mesityl oxide and 2-methylfuran from lignocellulose. *Scientific Reports, 6*, 32379.

45. Wu, J., Cheng, S., Cao, J., Qiao, J., & Zhao, G.-R. (2019). Systematic optimization of limonene production in engineered Escherichia coli. *Journal of Agricultural and Food Chemistry, 67*, 7087–7097.
46. Zhang, X., et al. (2021). Combinatorial engineering of Saccharomyces cerevisiae for improving limonene production. *Biochemical Engineering Journal, 176*, 108155.
47. Rubulotta, G., & Quadrelli, E. A. (2019). Terpenes: A valuable family of compounds for the production of fine chemicals. In *Studies in surface science and catalysis* (Vol. 178, pp. 215–229). Elsevier.
48. Ravichandran, C., Badgujar, P. C., Gundev, P., & Upadhyay, A. (2018). Review of toxicological assessment of d-limonene, a food and cosmetics additive. *Food and Chemical Toxicology, 120*, 668–680.
49. Ozturk, B., Winterburn, J., & Gonzalez-Miquel, M. (2019). Orange peel waste valorisation through limonene extraction using bio-based solvents. *Biochemical Engineering Journal, 151*, 107298.
50. Bogel-Łukasik, E., et al. (2008). Limonene hydrogenation in high-pressure CO2: Effect of hydrogen pressure. *Journal of Supercritical Fluids, 45*, 225–230.
51. Santiago, B., Moreira, M. T., Feijoo, G., & González-García, S. (2020). Identification of environmental aspects of citrus waste valorization into D-limonene from a biorefinery approach. *Biomass and Bioenergy, 143*, 105844.
52. Rubulotta, G., et al. (2017). Highly selective hydrogenation of R-(+)-Limonene to (+)-p-1-Menthene in batch and continuous flow reactors. *ACS Sustainable Chemistry & Engineering, 5*, 3762–3767.
53. Valencia, D., Martinez-Hernandez, E., García, A., & Aburto, J. (2024). Sustainable hydrogenation of limonene to value-added products using Cu–Ni catalysts supported on KIT-5. *Journal of Cleaner Production, 434*, 140356.
54. Serrano-Ruiz, J. C., Ramos-Fernández, E. V., & Sepúlveda-Escribano, A. (2012). From biodiesel and bioethanol to liquid hydrocarbon fuels: New hydrotreating and advanced microbial technologies. *Energy & Environmental Science, 5*, 5638–5652.

Chapter 7
Sustainable Aviation Fuel Production: Step Towards a More Prosperous and Energy-Independent Future

Prerna Tripathi and Abhay Kumar Choubey

1 Introduction

Sustainable aviation fuel (SAF) is recognised and regarded as a biofuel of the next generation for aircraft propulsion. At present, SAF stands as the sole sustainable substitute for conventional jet fuel, offering a critical remedy for the mitigation of greenhouse gas emissions in the aviation industry. As a result of apprehensions regarding climate change and rising emissions, scientific communities have begun to give decarbonisation considerable attention. Green transport is an initiative to promote sustainable economic development while ensuring that the local and global environments are not compromised. The objective of a green transport system is to offer the public a mode of transport that consumes less energy and generates less carbon dioxide (CO_2) and other detrimental pollutants. Aircraft and automobile travel continue to experience the most rapid expansion among the main modes of transportation, particularly due to a demand increase that has multiplied by a significant factor in the developing nations of Asia and Africa. Aviation currently accounts for a negligible 2–3% of global CO_2 emissions [1]. Nonetheless, its contribution to the overall transportation sector is merely 9%. As per the estimation, the present yearly CO_2 emissions from commercial aviation may reach 670 million tonnes by 2025, but they could soar to 1.2–1.5 billion tonnes, if air travel increases by a factor of three by that year. Without swift action to reduce fuel burn through technological and operational innovations and the exploration and implementation of alternatives to jet-kerosene, air travel is poised to become a significant source of greenhouse gas emissions, with projections showing that it will account for 40% of total passenger-kilometres travelled by 2050 [2]. Therefore, the development of alternative aviation

P. Tripathi · A. K. Choubey (✉)
Department of Sciences and Humanities, Rajiv Gandhi Institute of Petroleum Technology, Jais, Amethi, India
e-mail: achoubey@rgipt.ac.in

M. Aslam et al. (eds.), *Sustainable Aviation Fuels*, Sustainable Aviation,
https://doi.org/10.1007/978-3-031-83721-0_7

fuels is primarily inspired by environmental concerns and the escalating price of petroleum [3]. Additional considerations, which should not be overlooked, include the consistent operation of the aircraft engine, environmental impacts and cost escalation [4]. The utilisation of biofuels is motivated by several factors including the availability of renewable resources, reduced reliance on fossil fuel sources and low greenhouse gas emissions. The production of bio-aviation fuel has the potential to generate economic benefits, particularly in developing nations where non-cultivable land is scarce; thus, it can facilitate the supply chain [5]. Alternative aviation fuels are required to have certain characteristics such as having good cold flow capabilities, being thermally stable and having a low freezing point [6]. It is imperative that the fuel be suitably matched for the current configuration of the aircraft engine. The energy crops that are utilised as the production source should not pose a threat to the ecosystem and the production of food, and they should also not be harmful to the environment and should not result in the destruction of forests [7]. In a nutshell, aviation biofuels are sustainable from ecological, economic and social perspectives. Furthermore, despite the fact that bio-jet fuels are good alternatives to conventional aviation fuels, there are a great deal of obstacles and problems that need to be conquered in order to attain this goal. Extremely difficult challenges include the collection of feedstocks, the manufacturing route, the characteristics of the fuel that is generated and the characterisation of the fuel itself. This chapter focuses on the prospects and obstacles involved in the manufacturing of aviation biofuel using sustainable raw materials. Renewable hydrocarbon fuel is compared to alternative fuels such as biohydrogen, biodiesel, bio-alcohol and biomethane. This article also covers the features of the feedstock, the production methods, the industry's commitments and partnerships and the field's future prospects and suggestions.

2 Feedstocks for Biomass-Derived SAF

Producing sustainable aviation fuel from a variety of biomass feedstock is possible. Fuel crops, agricultural and forestry by-products, used cooking oil, straw, wood scraps and municipal solid waste (MSW) all fall under this category. Aeroplanes cannot use every kind of biofuel [8]. For instance, biodiesel cannot be used in aircraft because it has a lower energy density and a greater freezing point compared to Jet A fuel [9]. Studies with internal combustion engines have indicated possibilities for some bio-alcohols such as bioethanol. Still, additional sustainability and emission standards must be satisfied before a biofuel can be called an SAF. There are four generations of feedstocks that are suitable for SAF: first-generation (1-G), second-generation (2-G), third-generation (3-G) and fourth-generation (4-G). The various conversion processes necessitate distinct pre-treatment procedures, and each group has its own unique characteristics. The length of this study does not provide a comprehensive explanation of all potential biofuel feedstocks; nonetheless, a selection of the most promising materials for SAF production is provided below.

2.1 First-Generation Feedstocks

Food crops that may be consumed including sugarcane, corn, oil palm, sugar beets and oil-seed crops, such as camelina, rapeseed, soybean and sunflower, are classified under the 1-G category [10]. The fats or oils obtained from these crops can be readily transformed into jet fuel using the well-established hydroprocessed esters and fatty acids (HEFA) process. The manufacture of biofuels from oil feedstocks gives rise to sustainability concerns, including the adverse impact on food prices and alterations in local ecosystems, as a result of the utilisation of pesticides and other substances [11]. The main disadvantages of using 1-G feedstocks are water shortages and eutrophication, which are caused by the high water and nutrient demands of most food crops. An additional primary obstacle in 1-G feedstock production is the rivalry for land, water and energy resources with food production [12]. Expanding forestland through deforestation, resulting in the loss of biodiversity, appears to be an unfavourable choice [13].

Oil palm needs to be farmed and processed in a sustainable way for it to be considered a 'good' feedstock alternative for bio-aviation fuel generation. Plantations can have less of an adverse effect if farmers use optimal agronomic practices to increase oil production while decreasing the amount of resources they use. To integrate bio-alcohol fuel (BAF) production with food production from 1-G feedstocks, which could be technically viable, it is necessary to systematically examine and sustainably apply the policies and management in place [14].

2.2 Second-Generation Feedstocks

Materials that are rich in sugar or oil are known as 2-G feedstocks. A common oilseed energy crop, such as jatropha (*Jatropha curcas*), contains 30–40% of the seed weight, while castor bean (*Ricinus communis*) often has 50–60% oil [15]. In order to increase BAF production, Molefe et al. (2019) suggested hydrocracking the oils of castor beans and jatropha [16]. There are still many unanswered questions about the viability of using jatropha and castor bean as BAF raw materials, and the markets for these products are not yet developed [15]. A few of these issues include a lack of seed production, vulnerability to viruses and pests like rodents and insects and the high expense and difficulty of effective oil extraction techniques [17]. Even though soil salinity has an effect on agricultural productivity, advances in synthetic biology and genetic modification may provide a solution that leads to higher yields in the future [18]. Various types of energy crops, including grass and wood, have been suggested as potential raw materials for the production of bio-alcohol fuels through thermochemical and/or biochemical processes. Waste biomass may be more suitable as feedstocks compared to energy crops due to their lack of land requirements, minimal revenue potential and less water footprints when compared to cultivated crops [19]. Waste biomass, such as lignocellulosic by-products from

various activities like planting, harvesting, logging and subsequent processing, are being increasingly considered as feedstock by BAF developers due to their low cost [20]. Existing technologies can already transform lignocellulosic wastes into various jet fuel precursors such as syngas, pyrolysis oil, ethanol and butanol [21].

2.3 Third-Generation Feedstocks

Algae are highly intriguing since they lack food value yet offer significant yields with minimal land needs and relatively inexpensive costs [22]. Algae, unlike second-generation fuels, have higher growth rates and need less land surface compared to other feedstocks [23]. Microalgae are single-celled organisms that possess exceptional efficiency in photosynthesis and the ability to fix carbon [24]. Microalgae have the ability to increase their biomass by two times within a 24-hour period, and their growth rate can be as fast as 3.5 hours during the exponential development phase [25]. As a feedstock, they offer both high productivity and a ready supply of fatty acids that can be easily converted to BAF using HEFA [26]. Therefore, microalgae are highly esteemed for their potential to produce biofuels on a global scale. However, challenges in the areas of cultivation, harvesting and oil extraction technologies, as well as significant negative effects on the environment hinder their commercialisation [27]. Although current algal technologies will not yet be economically feasible within the next decade, it is crucial to continue researching supply chain studies and generating reliable data for microalgae-based biofuels namely algae as a possible source of biokerosene and diesel [28].

2.4 Fourth-Generation Feedstocks

Non-living resources and organisms with genetic modifications are classified jointly as fourth-generation (4-G) feedstocks [29]. Genetically modified organisms such as cyanobacteria, yeast, microalgae and fungus have been artificially transformed to increase their oil and/or sugar production. However, these modifications are still in the early stages of study. Further research is required to assess the health and environmental hazards associated with these organisms as well as to develop ways for containing and minimising any possible dangers when they are introduced into the global supply chain [30]. Non-biological feedstocks like CO_2, water, renewable electricity and sunlight have the potential to be a more environmentally friendly choice, particularly when utilising flue gases from industrial units [22]. One method is power-to-liquid, which entails using a renewable electricity-powered electrolyser to split water into hydrogen and oxygen. The hydrogen is then mixed with CO_2/CO to form BAF [22].

3 Alternative Fuels for Aviation

All renewable bioresources can be used to make green aviation fuels. The characteristics of alternative aviation fuels include (i) diminished greenhouse gas emissions, (ii) utilisation of resources that are sustainable, (iii) compatibility with traditional fuels and (iv) sustainability as well as green combustion. The ongoing development of sustainable fuels encompasses the following:

3.1 Hydroprocessed Renewable Jet (HRJ) Fuels

Hydroprocessed renewable jet fuels, also known as hydroprocessed esters and fatty acids (HEFA), are derived from various sources viz. animal fats, vegetable oils, waste lube, algae-derived oil and bio-oil. The main by-products of this process are propane and water [30]. Hydroprocessed regenerative aviation fuels are biofuels with a high energy content that can be used as fuels without the need for blending. They offer a significant advantage by reducing the emission of greenhouse gases [31]. The fuels have demonstrated stability for storage and a high resistance to microbial growth [32]. The hydroprocessed sustainable green fuels are compatible with traditional aircraft engines and do not pose any fuel quality concerns. The combustion of fuel results in a complete absence of ash residue. HRJ fuels are well-suited for flights at greater heights due to their superior cold flow properties [31].

3.2 Fischer–Tropsch Fuels

Fischer–Tropsch fuels, often known as FT fuels, are hydrocarbon fuels that are synthesised through the catalytic conversion of synthesis gas (CO and H_2) [33]. Various types of biomass feedstock can be utilised for the production of synthesis gas [34]. The FT fuels are often characterised by their clean combustion properties and great economic value [35]. The FT fuels possess the following characteristics: non-toxicity, absence of nitrogen oxide emissions, high cetane number, lowered particle emissions and minimal sulphur and aromatic content, among others. Furthermore, the process of fuel combustion is devoid of hydrocarbons and carbon dioxide [36]. The attributes of FT fuels are primarily influenced by the operating circumstances rather than the specific feedstock used [37]. These fuels have a lower energy density, resulting in less power and fuel efficiency [35]. The lack of sulphur in the fuel leads to a decrease in lubricity, which is an additional problem [37].

3.3 Biodiesel

Biodiesel refers to the alkyl esters of fatty acids, which are primarily derived from vegetable oils and animal fats. This fuel is created through the process of transesterification. Biodiesel as a fuel demonstrates the benefits of being sustainable, having a high ignition point, possessing an adequate energy balance and being easily mixed with petroleum fuels. The product is free of sulphur and exhibits significantly reduced greenhouse gas emissions [38]. Furthermore, it is worth mentioning that biodiesel fuels are free from aromatic compounds resulting in no water, soil or air pollution. Additionally, they are non-toxic in nature [39]. However, biodiesel fuel does not meet the requirements for use as an aviation fuel. Biodiesel has a significantly lower energy density when compared to conventional jet fuels [40]. Biodiesel that acts as an aviation fuel does not require additional engine modification or infrastructure. However, its efficiency is not particularly high [41]. Biodiesel has a significantly higher freezing point in comparison to oil-based fuel for aircraft, rendering it unsuitable for flights at higher elevations. The stability of biodiesel is reduced by the oxidation of unsaturated sites, which is caused by the presence of polyunsaturated and unsaturated fatty acids [36]. Due to the presence of ester groups, biodiesel exhibits some degree of polarity, leading to the formation of emulsion and making water separation challenging [42]. Insufficient availability of feedstock and unfavourable economics pose additional challenges for biodiesel fuel.

3.4 Liquid Biohydrogen and Biomethane

Research is currently underway to explore the viability of using liquid hydrogen as a potential substitute for traditional jet fuel. Biohydrogen is generated from various biomass resources using thermal and biochemical processes [43]. Liquid hydrogen is known for its higher energy production per weight when compared to traditional fuels for aircraft. However, it does come with the drawback of requiring a larger volume for storage [44]. The burning of liquid hydrogen fuels results in significantly lower greenhouse gas emissions in comparison to jet fuels derived from petroleum [45]. An issue arises when attempting to utilise liquid hydrogen fuel in a traditional aircraft engine as it requires modifications to the engine [46]. One potential application of liquid methane is as a fuel source for cryogenic aircraft. By utilising liquid methane fuel, it is possible to reduce carbon dioxide emissions by approximately 25%. Engine design poses a significant challenge in the process of commercialising liquid methane fuels. The combustion of liquid methane fuel releases methane, a significant greenhouse gas [36].

3.5 Bio-Alcohols

Bio-alcohols are generated through the process of fermentation of starch and sugar or by the catalytic conversion of biogas. Ethanol and butanol are commonly employed as fuel alternatives. Aviation purposes do not allow for the use of ethanol due to its volatile nature, low point of flash and low energy density [47]. The case of butanol exhibits a high degree of similarity. Alcohol as a propellant for aircraft requires a specialised delivery infrastructure and storage system [8]. Alcohol cannot be blended into traditional fuels due to its unfavourable fuel properties, leading to potential aviation complications.

4 Production Pathways

The utilisation of biomass as a resource for producing liquid fuel for transportation has made significant progress in the past decade and continues to hold great potential for future enhancements [48, 49]. The regulations governing aviation fuel are highly rigorous, encompassing specifications, quality standards and safety measures. In the field of aviation, the use of biofuel has the potential to address two key issues: reducing reliance on fossil fuels and minimising emissions throughout the life cycle [50]. This is primarily due to the carbon neutrality of biomass. Two primary methods for biofuel production are the biochemical route and the thermochemical pathway. Biofuel is typically produced using enzymes and microorganisms in the biochemical route. In the thermochemical route, biofuel can be obtained by converting the synthesis gas produced through pyrolysis or gasification technologies [51]. Some common examples include Fischer–Tropsch process, oil-to-jet, sugar-to-jet and alcohol-to-jet [52]. In the field of research, there is a current emphasis on the synthesis of jet-fuel-range hydrocarbons using platform compounds derived from lignocellulose [53]. Pre-treatment plays a crucial role in the production of bio-jet fuels, considering the diverse characteristics of biomass feedstock including moisture content, density and energy content. Torrefaction and pyrolysis are two primary thermal pre-treatment techniques [54]. Fast pyrolysis produces a high-energy-density fuel, pyrolysis oil, by thermally decomposing biomass feedstock at 500 °C temperature without oxygen. Different techniques can transform the intermediate products into bio-jet fuels. This section discusses four recognised bio-jet fuel generation routes: Fischer–Tropsch (FT) synthesis (3.1), hydroprocessed esters and fatty acids (HEFA) (3.2), alcohol-to-jet (ATJ) (3.3) and hydroprocessing of fermented sugars (HFS) (3.4). Aviation sector challenges and possibilities follow.

4.1 Fischer–Tropsch Process (FT)

Fischer–Tropsch process (FT) transforms synthesis gas into liquid hydrocarbon fuels. A typical FT process comprises feedstock prior treatments, biomass combustion, gas conditioning, acid gas elimination, FT synthesis and another method called synthetic crude refining. Before working on gasification, the feedstock is dried first and then processed to minimise the size of particles along with humidity. Biomass is gasified at a later stage. Factors such as gasifying substance biomass species, particle size, temperature, heating rate, operational pressure, equivalency ratio, reactor architecture and catalyst addition affect synthesis gas yields and its composition [55]. Gasification is possible using the FT process in the presence of high-purity oxygen and steam at a high temperature of 1300 °C. Gasification reactors consist of moving-bed, fluid-bed and entrained flow systems. A direct-quench synthesis gas cooling system near the gasifier could help in eliminating ash and tars [56]. An acid gas removal system removes CO_2, H_2S and sulphide from syngas after gasification. Eliminating CO_2 and H_2S could improve downstream synthesis kinetics and costs and prevent catalyst toxicity. The gas conditioning system adjusts CO and H_2 proportions by water–gas reaction with clean gas. The H_2/CO proportion is crucial to FT. The CO and H_2–mixed gas enters the FT reactor for FT synthesis. Reforming could recirculate unaltered syngas and some FT gas to the FT reactor. The liquid products should be processed to make different fuels. The excess gas can be used to generate electricity. The selectivity of the catalyst determines the target hydrocarbons. Fe, Co, Ni and Ru are frequent FT catalysts [57]. Ru's expensive price limits its application, but its catalytic activity and selectivity are found to be good [58]. Fe and Co are the most common industrial catalysts. Fe-based catalysts are short-lived yet have a high space-time yield. Co-based catalysts generate carbon chains, have low oxygen content and resist carbon deposition [59]. FT fuel is sulphur- and nitrogen-free and has a high specific energy, thermal stability and low aviation emissions [60]. However, low aromatic concentration and energy density make the fuel inefficient and expensive to produce.

4.2 Hydrogenated Esters and Fatty Acids (HEFA)

Triglycerides, cooking oils, both saturated and unsaturated fatty acids in vegetable oils and animal fats are hydrotreated by HEFA to make bio-jet fuels. It usually takes two steps. First, unsaturated fatty acids and triglycerides undergo catalytic hydrogenation to produce saturated fatty acids. Triglycerides undergo a β-hydrogen elimination reaction to produce fatty acids [61]. By hydrodeoxygenation and decarboxylation, saturated fatty acids become C15–C18 straight-chain alkanes [62]. H_2O, propane, CO and CO_2 are co-products. Due to the deactivation of catalysts by poisoning, cracking species and manufacturing costs, precious metals supported with zeolites or oxides were replaced by transition metals similar to Co, Ni and Mo or their reinforced bimetallic composites [63]. Deoxygenated straight-chain paraffinic compounds are selectively hydrocracked and isomerised to yield highly branched liquid

fuels in the next step. Pt, Ni or other similar noble metals held by activated carbon or zeolite molecular sieves are common catalysts [64]. In flight engines, HEFA's high-energy biofuels can be used without mixing. The fuel has high thermal stability, high cetane number, good cold flow and low releases from the exhaust system, but its low aromatic content causes low lubricity and leakage [65].

4.3 Hydroprocessing of Fermented Sugars (HFS)

This process utilises a biochemical conversion mechanism, wherein a biological platform (such as genetically engineered yeast or microorganisms) is employed to transform sugars into hydrocarbons. More specifically, it results in the production of a C15 alkene known as farnesene, which has four double bonds, through the process of fermentation. Subsequently, hydroprocessing is employed to transform it into the corresponding alkane, farnesane, which is then subjected to distillation in order to get aviation fuel. The finished product may additionally contain residual farnesene and olefins (partially hydrogenated farnesene). The current primary source of feedstock for this process is sugar cane, while other cellulose-based sugars can serve as viable alternatives to mitigate conflicts between food, feed and fuel. An inherent drawback of this method is the substantial operational expenses linked to the requisite processing procedures. However, it is anticipated that when the supply chain expands, these expenses will be much diminished [66, 67].

4.4 Alcohol-to-Jet (ATJ)

Dehydration, oligomerisation, hydroprocessing and distillation can make biofuels from alcohols such as methanol, ethanol or higher alcohols [68]. Commercial biomass-to-jet fuel conversion always uses ethanol, butanol and *iso*-butanol. Jet-fuel hydrocarbons are produced from alcohols using an upgrading process having four steps. After the dehydration of alcohol, olefinic compounds are oligomerised along with catalysts to yield middle distillate. Hydrogenation of the intermediate distillates gives jet-fuel hydrocarbons. The last step is adopted as the distillation [69]. Typical ethanol dehydration catalysts used are Al_2O_3, transition metal oxides, zeolite and heteropolyacid [70].

4.5 Catalytic Hydrothermolysis

In this process, fatty acid esters and free fatty acids are converted into sustainable aviation fuel through hydrothermolysis engaging catalysts, followed by a combination of hydrotreatment, hydrocracking or hydroisomerisation and fractionation. In the catalytic hydrothermolysis process, the maximum blend is 50 per cent.

4.6 Co-processing

Co-processing is the simultaneous processing of renewable feedstocks such as vegetable oils or animal fats with crude petroleum. The resulting fuel comprises renewable content that can help accelerate the aviation industry's energy transition. When these fuels are employed to power aircraft, the resulting greenhouse gas emissions come down in comparison to existing petroleum fuels. Co-processing is considered to be the smart and viable way, which could bring lower emissions rapidly.

4.7 Hydroprocessed Hydrocarbons, Esters and Fatty Acids

In this process, lipid feedstocks like plant or algae oils, animal fats or cooking oils are deoxygenated first and then hydroprocessed to yield hydrocarbon fuel as a blending component. The maximum blend ratio is 10 per cent.

5 Challenges

Despite the high demand for alternative aviation biofuels, several obstacles remain in fulfilling the targets. In alternative jet fuel development, the important difficulties faced are the pricing difference between bio- and conventional jet fuels and commercialisation and sustainability funds [71]. Economic and environmental challenges consisting of land–water utilisation, greenhouse gas and particle emissions and fuel–food competition are the main obstacles. Sustainability is difficult because feedstock availability and fuel production routes have social, economic and environmental implications. Afforestation difficulties caused by increased demand for bio-jet fuels influence soil fertility, biodiversity and atmospheric CO_2. Crop planting for biofuel generation will diminish soil quality and water availability and generate food scarcity. Fertilisers and pesticides pollute water and soil [71]. The supply process struggles with product quality and mixing for efficiency. Fuel infrastructure, marketing, feedstock and fuel storage, by-product marketing laws and certifications are also key challenges. Feedstock issues include bulk availability at a cheap cost and limits for large-scale uses. For aeroplane testing, a lot of fuel and feedstock are needed. Renewable and traditional jet fuels must be compatible. Unlike conventional jet fuels, bio-aviation fuels must have a low freezing point, high auto-ignition temperature and no sulphur or aromatics [71]. There are further issues with renewable fuel introduction in aviation at a commercial scale, certification, engine and quality testing and aircraft use [72]. Fuel must be resistant to several operational conditions, operate well and be safe. Thermal and storage stability are key aviation biofuel challenges. Low-temperature features, combustion and the presence of metals and micro-nutrients also matter [73].

6 Future Prospects of Bio-Jet Fuel

Bio-jet fuel comparable to aviation fuel has the capability to substitute the conventional hydrocarbon-based aviation fuel used in aircraft and deal with the fast-growing market of aviation while dropping greenhouse gas emissions. Apart from the technological and environmental features, the competition is between aviation biofuel and road transport biofuel markets, which should not be ignored. There is a rising interest in several countries in adopting blending directives and ambitious targets for the SAF. In promoting the development and uptake of the proposed SAF, a few recommendations are found like:

- Unless there is a clear case for feedstock substitution based on cost savings and increased greenhouse gas savings, lignocellulosic biomass and waste feedstocks should be prioritised above food and feed crops.
- Research and development efforts should prioritise non-HEFA routes like FT, ATJ and HFS to enhance cost-effectiveness through optimisation, integration and scaling-up of technologies.
- A preferred SAF feedstock should avoid competing with the existing road biofuel market and land use applicable for food production. It is important to avoid using food or feed crops, nevertheless, and think about the sustainability standards of the feedstock. A blending mandate may not be the most effective policy tool, but a target to reduce the intensity of greenhouse gas emissions may be.
- To encourage greater adoption of SAF, carbon tax schemes should include incentives for SAF adoption in addition to penalties for greenhouse gas emissions. Additional research should consider the effects and rivalry between the aviation and road transport industries.

7 Conclusions

Aviation fuels made from sustainable bioresources could replace petroleum-based fuels. Alternative jet fuels like HRJ and FT fuels may replace regular jet fuels. The main downsides of biodiesel and bio-alcohol as aviation fuels are their weaknesses. Though liquid hydrogen and liquid methane fuels are ideal for use, their high manufacturing expenditures and incompatibility with existing aviation engines mark them as less desirable. Growing government interests of the country concerned and international interest can boost scaling, commercialisation and supply chain infrastructural facilities. However, increasing feedstock complexity may hinder larger-scale implementations and undermine long-term sustainability aims. Synthetic fuels may lead decarbonisation once technology constraints are overcome. Biofuel research and marketing are still necessary to prove economies of scale and reduce emissions significantly. Long-term, hydrogen and battery-powered aircraft can reduce fossil dependence. However, without revolutionary advances in this field, such options will be restricted to short-range flights. Finally, scientific and political collaborations are necessary to speed up future progress.

References

1. www.iea.org
2. Agarwal, R. K. (2012). *Review of technologies to achieve sustainable (Green) aviation*. Intechopen.
3. Faaij, A., & van Dijk, M. (2012). *White paper on sustainable jet fuel*. SkyNRG (SkyNRG).
4. Hemighaus, G., Boval, T., Bosley, C., Organ, R., Lind, J., Brouette, R., Thompson, T., Lynch, J., & Jones, J. (2006). *Alternative jet fuels, a supplement to Chevron's aviation fuels technical review*. Chevron Corporation.
5. Air Transport Action Group. (2011). *Beginner's guide to aviation biofuels* (2nd ed.). Air Transport Action Group.
6. Rekoske, J. (2010). *Biofuels: Challenges & opportunities*. Asia Pacific Clean Energy Summit and Expo.
7. Hari, K., Thushara, Y., Zahira, & Binitha, N. N. (2015). Aviation biofuel from renewable resources: Routes, opportunities and challenges. *Renewable and Sustainable Energy Reviews, 42*, 1234–1244.
8. Air Transport Action Group. (2017). *Beginner's guide to sustainable aviation fuel*. Air Transport Action Group.
9. Adin, M. S., Altun, S., & Adin, M. S. (2021). Effect of using bioethanol as fuel on start-up and warm-up exhaust emissions from a diesel power generator. *International Journal of Ambient Energy, 43*, 1–7.
10. Lee, R. A., & Lavoi, J. M. (2013). From first- to third-generation biofuels: Challenges of producing a commodity from a biomass of increasing complexity. *Animal Frontiers, 3*, 6–11.
11. Air Transport Action Group. (2011). *Beginner's guide to aviation biofuels*. Air Transport Action Group.
12. Moioli, E., Salvati, F., Chiesa, M., Siecha, R. T., Manenti, F., Laio, F., et al. (2018). Analysis of the current world biofuel production under a water–food–energy nexus perspective. *Advances in Water Resources, 121*, 22–31.
13. Keles, D., Choumert-Nkolo, J., Motel, P. C., & Kéré, E. N. (2018). Does the expansion of biofuels encroach on the forest? *Journal of Forest Economics, 33*(1), 75–82.
14. Sims, R., Flammini, A., Puri, S., & Bracco, S. (2015). *Opportunities for agrifood chains to become energy-smart* (pp. 1–198). Food and Agriculture Organization & United States Agency for International Development.
15. Heinrich, G. (2018). Jatropha curcas L. – An alternative oil crop. In M. Kaltschmitt & U. Neuling (Eds.), *Biokerosene: Status and prospects* (pp. 237–257). Springer. Biotechnol. Biofuels 10, 1–16.
16. Molefe, M., Nkazi, D., & Mukaya, H. E. (2019). Method selection for biojet and biogasoline fuel production from Castor oil: A review. *Energy & Fuels, 33*(7), 5918–5932.
17. Moniruzzaman, M., Yaakob, Z., Shahinuzzaman, M., Khatun, R., & Islam, A. K. M. A. (2017). Jatropha biofuel industry: The challenges. In E. Jacob-Lopes & L. Q. Zepka (Eds.), *Frontiers in bioenergy and biofuels*. IntechOpen.
18. Hendricks, R. C., & Bushnell, D. M. (2008, 17 February). Halophytes energy feedstocks: Back to our roots. In *Proceedings of the 12th international symposium on transport phenomena and dynamics of rotating machinery, Honolulu, HI, USA*.
19. Caicedo, M., Barros, J., & Ordás, B. (2016). Redefining agricultural residues as bioenergy feedstocks. *Materials, 9*, 635.
20. Staples, M. D., Malina, R., Suresh, P., Hileman, J. I., & Barrett, S. R. H. (2018). Aviation CO_2 emissions reductions from the use of alternative jet fuels. *Energy Policy, 114*(C), 342–354.
21. Pandiyan, K., Singh, A., Singh, S., Saxena, A. K., & Nain, L. (2019). Technological interventions for utilization of crop residues and weedy biomass for second generation bio-ethanol production. *Renewable Energy, 132*(C), 723–741.
22. Mi, Z., & Sick, V. (2020). Taking a shortcut: Direct power-to-X conversion. *Frontiers in Energy Research, 8*, 153.

23. Lee, R. A., & Lavoie, J.-M. (2013). From first- to third- generation biofuels: Challenges of producing a commodity from a biomass of increasing complexity. *Animal Frontiers, 3*, 6–11.
24. Yujie, S., Song, K., Jhang, P., Yuquing, S., Cheng, X., & Chen, X. (2017). Progress of microalgae biofuel's commercialization. *Renewable and Sustainable Energy Reviews, 74*(C), 402–411.
25. Chisti, Y. (2007). Biodiesel from microalgae. *Biotechnology Advances, 25*(3), 294–306.
26. Tao, L., Milbrandt, A., Zhang, Y., & Wang, W. C. (2017). Techno-economic and resource analysis of hydroprocessed renewable jet fuel. *Biotechnology for Biofuels, 10*, 1–16.
27. Doshi, A., Pascoe, S., Coglan, L., & Rainey, T. J. (2016). Economic and policy issues in the production of algae-based biofuels: A review. *Renewable and Sustainable Energy Reviews, 64*, 329–337.
28. Behrendt, D., Schreiber, C., Pfaff, C., et al. (2018). Algae as a potential source of biokerosene and diesel-opportunities and challenges. In M. Kaltschmitt & U. Neuling (Eds.), *Biokerosene* (pp. 303–324). Springer.
29. Alalwan, H. A., Alminshid, A. H., & Aljaafari, H. A. S. (2019). Promising evolution of biofuel generations. Subject review. *Renewable Energy Focus, 28*, 127–139.
30. Arvidsson, R., Persson, S., Froling, M., & Svanstrom, M. (2011). Life cycle assessment of hydrotreated vegetable oil from rape, oil palm and Jatropha. *Journal of Cleaner Production, 19*, 129–137.
31. Gong, D. Y., Kaario, O., Tilli, A., Larmi, M., & Tanner, F. X. (2010). *A computational investigation of hydrotreated vegetable oil sprays using RANS and a modified version of the RNG kε model in OpenFOAM*. SAE World Congress.
32. Aatola, H., Larmi, M., Sarjovaara, T., & Mikkonen, S. (2008). *Hydrotreated vegetable oil (HVO) as a renewable diesel fuel: Trade-off between NOx, particulate emission and fuel consumption of a heavy duty engine*. SAE International.
33. Commercial case business evaluation of Dynamotive's fast pyrolysis and upgrading processes. Dynamotive Energy Systems Corporation.
34. Veringa, H. J. Advanced techniques for generation of energy from biomass and waste. ECN biomass. https://www.ecn.nl/fileadmin/ecn/units/bio/Overig/ pdf/Biomassa_voordelen.pdf
35. Clean alternative fuels: Fischer–Tropsch, United States Environmental Protection Agency EPA420-F-00-036, 2002.
36. Saynor, B., Bauen, A., & Leach, M. (2003). *The potential for renewable energy sources in aviation*. PRESAV final report.
37. Kreutz, T. G., Larson, E. D., Liu, G., & Williams, R. H. (2008). Fischer–Tropsch fuels from coal and biomass. In *25th annual international Pittsburgh coal conference*.
38. Moser, B. R. (2010). Camelina (Camelina sativa L.) oil as a biofuels feedstock: Golden opportunity or false hope. *Lipid Technology, 22*, 270–273.
39. Acute toxicity of biodiesel to freshwater and marine organisms. (1996). *Development of rapeseed biodiesel for use in high speed diesel engines* (pp. 117–131). Progress report.
40. Roberts, W. L. (2008). Bio jet fuels. In *The fifth international biofuels conference*.
41. Wardle, D. A. (2003). Global sale of green air travel supported using biodiesel. *Renewable and Sustainable Energy Reviews, 7*, 64–71.
42. Hemighaus, G., Boval, T., Bosley, C., Organ, R., Lind, J., Brouette, R., et al. (2006). *Alternative jet fuels, a supplement to Chevron's aviation fuels technical review*. Chevron Corporation.
43. Balat, M., & Balat, M. (2009). Political, economic and environmental impacts of biomass-based hydrogen. *International Journal of Hydrogen Energy, 34*(9), 3589–3603.
44. Faab, R. (2001). Cryoplane: Flugzeuge mit Wasserstoffantrieb. Accessed 13 April 2006.
45. Birkenstock, W. (1998). Hydrogen aircraft fuel research plans. Flug Revue, 66 (September).
46. Airbus Deutschland GmbH, liquid hydrogen fuelled aircraft-system analysis: final technical report (Publishable Version). (2003). Accessed 13 April 2006.
47. Hileman, J. I., Ortiz, D. S., Bartis, J. T., Wong, H. M., Donohoo, P. E., & Weiss, M. A. (2009). *Near-term feasibility of alternative jet fuels*. RAND Infrastruct Saf Environ.
48. Elia, J. A., Baliban, R. C., Floudas, C. A., Gurau, B., Weingarten, M. B., & Klotz, S. D. (2013). Hardwood biomass to gasoline, diesel, and jet fuel: 2. Supply chain optimization framework for a network of thermochemical refineries. *Energy & Fuels, 27*(8), 4325–4352.

49. Tan, E. C. D., Marker, T. L., & Roberts, M. J. (2014). Direct production of gasoline and diesel fuels from biomass via integrated hydropyrolysis and hydroconversion process-a techno- economic analysis. *Environmental Progress & Sustainable Energy, 33*(2), 609–617.
50. Augusdinata, D. B., Zhao, F., Ileleji, K., & DeLaurentis, D. (2011). Life cycle assessment of potential biojet fuel production in the United States. *Environmental Science & Technology, 45*(21), 9133–9143.
51. Slims, R. E. H., Mabee, W., Saddler, J. N., & Taylor, M. (2010). An overview of second generation biofuel technologies. *Bioresource Technology, 101*(6), 1570–1580.
52. Wang, W.-C., & Tao, L. (2016). Bio-jet fuel conversion technologies. *Renewable and Sustainable Energy Reviews, 53*, 801–822.
53. Xia, Q.-N., Cuan, Q., Liu, X.-H., Gong, X.-Q., Guan-Zhong, L., & Wang, Y.-Q. (2014). Pd/NbOPO4 multifucntional catalysts for the direct production of liquid alkanes for Aldol Adducts of Furans. *Angewandte Chemie International Ed, 53*(37), 9755–9760.
54. Güell, B. M., Bugge, M., Kempegowda, R. S., George, A., & Paap, S. M. (2012). *Benchmark of conversion and production technologies for synthetic biofuels for aviation*. SINTEF Energy Research.
55. Nor Afzanizam, B. S., Mohammad Nazri, B. M. J., Chongcheng, T., & Johan, N. (2014). A review of biomass gasification technology to produce syngas. *American-Eurasian Journal of Sustainable Agriculture, 8*(7), 69.
56. Kreutz, T. G., Larson, E. D., Liu, G., & Williams, R. H. (2008). Fischer-Tropsch fuels from coal and biomass. In *25th annual international Pittsburgh coal conference* (Vol. 29, No. 2. 10). Princeton University.
57. Vannice, M. A., & Garten, R. L. (1980). Influence of the support on the catalytic behavior of ruthenium in CO/H/sub 2/synthesis reactions. *Journal of Catalysis, 63*(1), 255–260.
58. Steynberg, A. P., Dry, M. E., Davis, B., & Breman, B. (2004). Fischer Tropsch reactors. Studies in surface science. *Catalysis., 152*, 64–195.
59. Lugue, R., De la Osa, A. R., Campelo, J. M., Romero, A. A., Valverde, J. A., & Sanchez, P. (2012). Design and development of catalysts for Biomass-to-Liquid-Fischer–Tropsch (BTL-FT) processes for biofuels production. *Energy & Environmental Science, 5*(1), 5186–5202.
60. Fox, E. B., Liu, Z.-W., & Liu, Z.-T. (2013). Ultraclean fuels production and utilization for the twenty-first century: Advances troward sustainable transportation fuels. *Energy & Fuels, 27*(11), 6335–6338.
61. Morgan, T., Santillan-Jimenez, E., Harman-Ware, E. A., Ji, Y., Grubb, D., & Crocker, M. (2012). Catalytic deoxygenation of triglycerides to hydrocarbons over supported nickel catalysts. *Chemical Engineering Journal, 189–190*(2), 346–355.
62. Yang, Y., Wang, Q., Zhang, X., Wang, L., & Li, G. (2013). Hydrotreating of C18 fatty acids to hydrocarbons on sulphided NiW/SiO_2–Al_2O3. *Fuel Processing Technology, 116*, 165.
63. Hari, T. K., & Yaakob, Z. (2015). Production of diesel fuel by the hydrotreatment of jatropha oil derived fatty acid methyl esters over γ-Al2O3 and SiO_2 supported Ni Co bimetallic catalysts. *Reaction Kinetics, Mechanisms, and Catalysis, 116*(1), 1–15.
64. Jeong, C. M., Park, G. W., Choi, J.-d.-r., Kang, J. W., Kim, S. M., Lee, W. H., Woo, S. L., & Chang, H. N. (2011). Steel reforming of volatile fatty acids (VFAs) over supported Pt/Al2O3 catalysts. *International Journal of Hydrogen Energy, 36*(13), 7505–7515.
65. Hari, T. K., Yakoob, Z., & Binitha, N. N. (2015). Aviation biofuel from renewable resources: Routes, opportunities and challenges. *Renewable and Sustainable Energy Reviews, 42*, 1234–1244.
66. International Air Transport Association. (2020). *Fact sheet 2: Sustainable aviation fuel: Technical certification*. International Air Transport Association.
67. Directorate-General for Energy. (2017). *High biofuel blends in aviation*. European Commission.
68. Chiaramonti, D., Prussi, M., Buffi, M., & Tacconi, D. (2014). Sustainable bio kerosene: Process routes and industrial demonstration activities in aviation biofuels. *Applied Energy, 136*, 767–774.

69. Serrano-Ruiz, J. C., Ramos-Fernández, E. V., & Sepúlveda-Escribano, A. (2012). From biodiesel and bioethanol to liquid hydrocarbon fuels: New hydrotreating and advanced microbial technologies. *Energy & Environmental Science, 5*(2), 5638.
70. Zhang, M., & Yu, Y. (2013). Dehydration of ethanol to ethylene. *Industrial and Engineering Chemistry Research, 52*(28), 9505–9514.
71. Hari, T. K., Yaakob, Z., & Binitha, N. N. (2015). Aviation biofuel from renewable resources: Routes, opportunities and challenges. *Renewable and Sustainable Energy Reviews, 42*, 1234–1244.
72. Pandey P. K. (2011). *Opportunities in bio-fuels for aviation*. National seminar on bioenergy Solutions.
73. Holmgren, J. (2008). Bio aviation fuel. In *World biofuels markets congress*.

Chapter 8
Development of Advanced Sustainable Processes for Aviation Fuel Production

Jorge Aburto and Arick Castillo-Landero

1 Introduction

Nowadays, the development of chemical processes involves a new challenge based on implementing strategies that not only include economic aspects, as has been considered for several years but also consider sustainability and safety aspects. All this is to reach a balance between these three spheres and lead from a conventional to a perspective sustainable industry. Based on this new need, a sustainable process is defined as the combination of different schemes of reaction and separation tools to satisfy a social need while contributing to economic growth, environmental protection, and social progress [1].

One example is the aviation industry, which is expected to grow to approximately 10 billion passengers by 2050 resulting in a carbon footprint of 1.8 gigatonnes of carbon. Therefore, different challenges were defined in the aviation industry, such as the net zero commitment which implies a total mitigation of 21.2 gigatonnes of carbon emitted from 2020 to 2050. The Air Transport Association (IATA) proposed to work with companies that can eliminate inefficiencies in air traffic management by governments and air navigation providers, create new technologies for the development of aircraft and aircraft engines, promote the use of low-carbon fuels by airport operators and airlines [2]. On the other hand, there is an area of opportunity where process engineers have an important role to play, which consists of designing new processes or bioprocesses with technology that implements biomass as a raw material and developing advanced processes from a sustainable approach and thus bringing the new chemical industry into a globalised and sustainable world.

J. Aburto · A. Castillo-Landero (✉)
Energy Efficiency and Biofuels Research Division, Instituto Mexicano del Petróleo, Mexico City, Mexico
e-mail: acastillo@imp.mx

M. Aslam et al. (eds.), *Sustainable Aviation Fuels*, Sustainable Aviation,
https://doi.org/10.1007/978-3-031-83721-0_8

As mentioned above, one of the goals is the production of renewable fuels, such as the use of biomass for the product may be phased out in the future [3]. Considering that a complete elimination of petroleum is not attractive nor realistic for the energy industry, since such an industry enjoys the scale of economy and biomass is widely geographically dispersed and competes with food and feed needs. Despite this, research and contributions are made in this line of research in order to contribute to the international targets for decarbonisation of the energy sector. Hence, it needs a strategy in the design of new chemical plants with strong commitments to developing a cleaner industry with economic benefits and social growth.

For sustainable aviation fuel (SAF) production, the ASTM has approved eight different routes to produce synthetic paraffins kerosene (SPK) or SAF such as: hydroprocessing of esters and fatty acids (HEFA-SPK) [4], Fischer-Tropsch (FT-SPK) [5, 6], alcohol to jet (ATJ-SPK) [7], isoparaffins from hydroprocessed fermented sugars (SIP) [8], Fischer-Tropsch plus aromatics (FT-SPK/A), catalytic hydrothermolysis jet (CHJ-SPK) and paraffinic kerosene produced from bioderived hydroprocessed hydrocarbons (HC-HEFA-SPK) [9]. There are other prototype technologies without ASTM approval for use as aqueous phase reforming (APR) [10] and hydrotreated depolymerised cellulosic jet (HDCJ) [11].

Hence, the hydroprocessed synthetic paraffinic kerosene and Fischer Tropsch (FT-SPK) route were certified in 2009 up to a blended content of 50% vol/vol. The feedstock includes renewable biomass such as agricultural and forestry waste, wood and energy crops, municipal solid waste, and non-renewable such as coal and natural gas. It is a thermochemical process based on the gasification of the raw material, followed by the synthesis of hydrocarbons from CO and H_2 through the Fischer Tropsch general reaction:

$$(2n+1)H_2 + nCO \rightarrow C_nH_{2n+2} + nH_2O, \tag{8.1}$$

The isoparaffin route synthesised from the hydroprocessing of fermentable sugars (SIP-SPK) was certified in 2014 at a maximum mixture of 10% vol/vol. The raw material consists of sugars from any source that are fermented by yeast to farnesene (C15 molecule), which is hydroprocessed to farnesane:

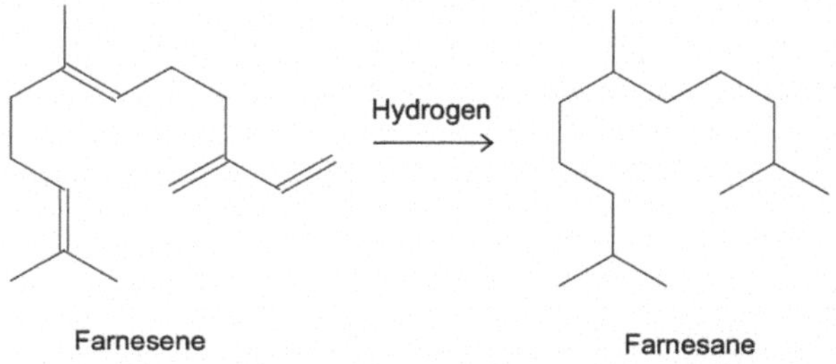

(8.2)

Now, the paraffinic kerosene route synthesised with aromatics (FT-SPK/A) contemplates the alkylation of light aromatics from coal, although biomass is also considered a raw material. This route was certified in 2015 for a mixture of up to 50%

vol/vol. The raw material is the same as in FT-SPK, and the alkylated aromatics, mainly benzene, are added to the biobased SPK.

Another relevant route, that had initiated commercial production in 2024 by LanzaTech, is the alcohol-to-jet (ATJ-SPK) synthetic paraffinic kerosene route certified in 2016 using an isobutane intermediate; and in 2018 using ethanol as an intermediary. This fuel can be blended with fossil jet fuel up to 50% vol/vol. The route considers the technology developed by LanzaTech, which involves the anaerobic fermentation of CO_2 from combustion gases to ethanol, dehydrated to ethylene, followed by oligomerization and hydrogenation to obtain linear and branched hydrocarbons.

The catalytic hydrothermolysis jet pathway (CHJ-SPK) was certified in 2010. It is based on the use of fatty acid esters as well as free fatty acids and a hydrothermal liquefaction technology. Mixtures of up to 50% are allowed. The product contains paraffin, isoparaffin, cycloparaffin and aromatic compounds that are in the boiling range of diesel and jet fuel, which is why fractionation is required to obtain each fuel separately.

The synthetic paraffinic kerosene route from bio-derived hydroprocessed hydrocarbons (HC-HEFA-SPK) was certified in 2020. The source of the approved raw material is *Botryococcus braunii*, a species of microalgae, which produces biobased lipids. The mixture is allowed up to 10% vol/vol and was the first aviation biofuel certified by express.

Finally, the alcohol-to-jet synthetic paraffinic kerosene route with the addition of aromatics (ATJ-SKA) may be blended up to 50% vol/vol. The raw material is the same used for the ATJ-SPK route where an aromatic product is added to the ATJ stream which includes dehydration, aromatization, hydrogenation and fractionation.

This chapter will further discuss some advances in process design using the HEFA route as well as the type of biomass that we have analysed for biojet fuel production.

2 Hydroprocessing of Esters and Fatty Acids (HEFA) Process

The American Society for Testing and Materials (ASTM) considers the HEFA process to be the most established route for producing SAF [4] (Table 8.1). For instance, the HEFA process has a level of 8–9, meaning that this technology has been completed and is for full commercial deployment. Neste Oil developed the first commercial-scale refinery using the HEFA process with a capacity of 170,000 tonnes per year in 2007 [12].

In general, the HEFA process consists of three main stages as shown in Fig. 8.1. Each stage has characteristics ranging from the transformation of the biomass to the

Table 8.1 Technology readiness level of technologies for SAF [9, 13]

<table>
<tr><th colspan="6">Technology readiness level</th></tr>
<tr><td>1–3</td><td>4</td><td>5</td><td>6–7</td><td>8</td><td>9</td></tr>
<tr><td></td><td></td><td></td><td colspan="2">Gasification & Fischer-Tropsch</td><td></td></tr>
<tr><td></td><td></td><td></td><td colspan="2">Sasol, Fulkrum/Marathon</td><td></td></tr>
<tr><td></td><td></td><td></td><td></td><td colspan="2">HEFA</td></tr>
<tr><td></td><td></td><td></td><td></td><td colspan="2">NESTE, ENI, TOTAL/AXENS, IMP, Etc.</td></tr>
<tr><td colspan="3"><SIP</td><td></td><td></td><td></td></tr>
<tr><td colspan="3">Amvris</td><td></td><td></td><td></td></tr>
<tr><td></td><td></td><td></td><td colspan="2">SPK/A</td><td></td></tr>
<tr><td></td><td></td><td></td><td colspan="2">Virent</td><td></td></tr>
<tr><td></td><td></td><td></td><td colspan="3">ATJ</td></tr>
<tr><td></td><td></td><td></td><td colspan="3">Lanzatech, Swedish biofuels, Biogy</td></tr>
<tr><td></td><td></td><td></td><td colspan="2">CHJ</td><td></td></tr>
<tr><td></td><td></td><td></td><td colspan="2">Reliance, ARA-Chevron</td><td></td></tr>
<tr><td colspan="3">HC-HEFA-SPK</td><td></td><td></td><td></td></tr>
<tr><td colspan="3">IHI Corporation</td><td></td><td></td><td></td></tr>
<tr><td></td><td></td><td></td><td colspan="2">ATJ-SKA</td><td></td></tr>
<tr><td></td><td></td><td></td><td colspan="2">Swedish biofuels, Vertex</td><td></td></tr>
</table>

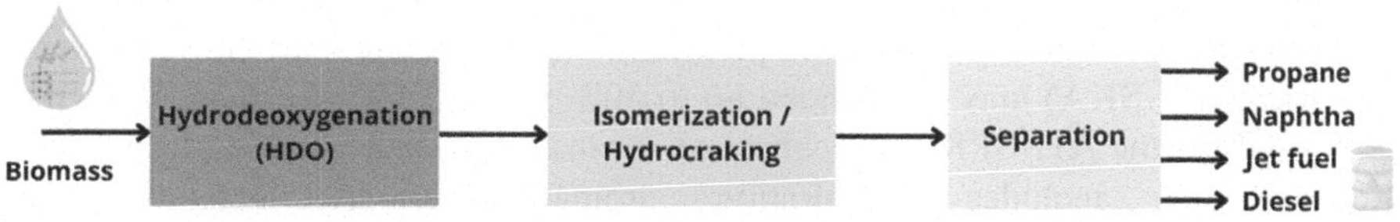

Fig. 8.1 General scheme of the HEFA process to produce jet fuel, diesel, naphtha and propane

production of the resulting products. The main activity of each stage is described below [14].

- *Hydrodeoxygenation (HDO) stage*. This consists of reacting the biomass through a mechanism based on hydrogenation, decarbonylation, saturation, hydrocracking, depropylation and hydrodeoxygenation reactions. The main objective of this stage is to remove oxygen from the triacylglycerides and produce a mixture of linear hydrocarbons such as propane and paraffin.
- *Isomerisation/cracking stage*. The mixture obtained in the previous stage is cracked and isomerised to obtain straight-chain hydrocarbons such as propane, and fractions boiling at the temperature range of naphtha, jet fuel and diesel.
- *Separation*. The straight-chain hydrocarbon mixture containing propane, naphtha, jet fuel and diesel is separated by selecting different unitary separation operations.

As mentioned above, the HEFA process can produce biojet fuel by hydroprocessing biomass oils such as animal, vegetable, used cooking, pyrolysis and algal

Table 8.2 Some examples of different types of biomass used in the HEFA process

Biomass	Description
Castor oil	The first step consists of HDO reaction of the vegetable oil in a catalytic fixed-bed microreactor. In the second stage, hydrocracking and isomerisation is carried out with Ni/SAPO-11 and Pt/SAPO-11 as catalyst. The results obtained show an alkane yield of 91.6% [16]
Coconut oil	Coconut oil yields a biofuel with a chain length distribution like that of fossil fuels during hydroprocessing stage and therefore no cracking or fractionation is necessary. They used a HDO reactor over the sulfided NiMo/Al_2O_3 followed by a hydroisomerisation (HIS) reactor over Pt/SAPO-11 catalyst [17]
Jatropha oil	The process was divided into six stages: shelling of the fruit, combustion of the husk, extraction of the oil, pyrolysis of the press cake, upgrading of the oil and separation of the product. Compared to the oil scenario used as a reference in this work, capital costs are 36% more expensive and operating costs are 13% cheaper [6]
Palm oil	The process consists of four steps. The first is a hydrodeoxygenation (HDO) section where a reactor is used to remove oxygen, followed by gas and water separation where the remaining water is removed from the paraffin mixture. The third stage is the isomerisation/hydrocracking zone to crack and isomerise the straight-chain hydrocarbons into hydrocarbons of jet fuel hydrocarbons range. Finally, the separation section when the products obtained in a train of separation units. The results show that the process is profitable at an IRR of 10% when the capacity is 75,000 bbl/year [14, 18]

[3, 15]. Table 8.2 shows some alternative oils used to produce biojet fuel using this technology.

3 Advanced Process to Produce Biojet Fuel

Nowadays, there is a lot of hard work in designing processes that meet sustainable characteristics, therefore there are different strategies that can be implemented such as integration of mass, energy or properties [1], process optimisation [19–21], and process intensification [22–26]. These techniques have involved the development of new methodologies to create processes but where the difficult part has been to include more than one response variable in the analysis. We illustrate below the latter by the development of an intensified and integrated process using palm oil through the HEFA process.

3.1 Conventional Process to Produce Biojet Fuel from Palm Oil via HEFA Process

The proposed conventional design represents the production of biojet fuel using palm oil as biomass feedstock (Fig. 8.2). The flowsheet has the capacity to transform 29,720.6 tonnes/year of palm oil to produce 11,437 tonnes/year of bio jet fuel

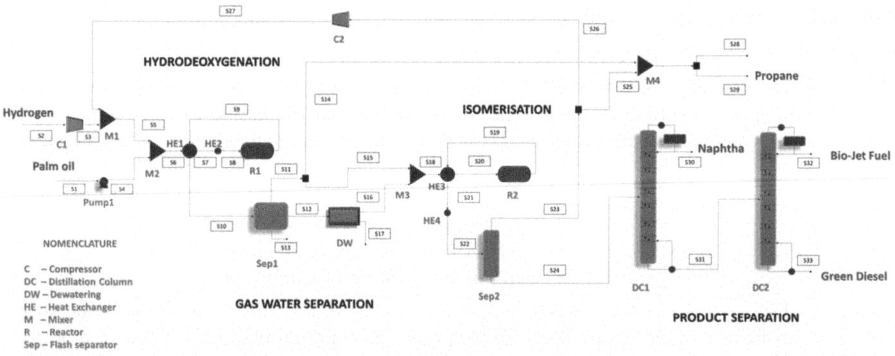

Fig. 8.2 Conventional scheme for the bio jet fuel production from palm oil via the HEFA process. (Reprinted (adapted) with permission from Castillo-Landero, A., Dominguillo-Ramirez, D., Aburto, J., Sadhukhan, J., & Martinez-Hernandez, E. (2023). Improving the Economic, Environmental, and Safety Performance of Bio-Jet Fuel Production through Process Intensification and Integration Using a Modularity Approach. ACS Sustainable Chemistry & Engineering, 11(2), 660-669. Copyright 2023 American Chemical Society)

Table 8.3 Reactions and conversion data present at the hydrodeoxygenation section

Reactions		Conversion (%)
Depropanation	$C_nH_{(2n-6m-1)})$ $(COOH)_3 + 3H_2 \rightarrow 3C_nH_{2n+1}COOH + C_3H_8$	100
DO	$C_nH_{2n+1}COOH + (m+3)$ $H_2 \rightarrow C_{n+1}H_{2n+4} + 2H_2O$	60.1
DCx	$C_nH_{2n+1} + mH_2 \rightarrow C_nH_{2n+2} + CO_2$	36.1
DC	$C_nH_{(2n+1-2m)}COOH + (m+1)$ $H_2 \rightarrow C_nH_{2n+2} + CO + H_2O$	3.8

Where *n* is the number of carbons in the aliphatic chain and m is the number of double bonds in the fatty acid molecule

and includes the three stages shown in Fig. 8.1 with the addition of the gas water separation stage in order to obtain the hydrocarbon mixture to carry out the isomerization, cracking reaction and followed by product and co-product separation stage.

The first stage represents the sections of hydrodeoxygenation and gas/water separation. The palm oil is fed into a reactor with H_2, in which the R1 reactor is used to remove oxygen from triacylglycerides. The reactions taking place are saturation, hydrocracking, depropanation, hydrodeoxygenation (DO), decarboxylation (DCx) and decarbonylation (DC) and their conversions are shown in Table 8.3 [14]. Then, H_2O, CO, and CO_2 are separated to get a mixture of linear hydrocarbons or paraffins.

Next, the mixture of linear hydrocarbons is fed into a reactor where isomerisation and hydrocracking reactions take place (Table 8.4) [14]. This reaction step is needed to achieve a low freezing point jet fuel to comply with the ASTM standard [9, 27]

Table 8.4 Reactions present at the isomerisation section

Reactions		*n* value
Hydrocraking	$C_nH_{2n+2} + H_2 \rightarrow C_{n-3}H_{2n-4} + C_3H_8$	13–18
Hydrocraking	$C_nH_{2n+2} + H_2 \rightarrow C_{n-8}H_{2n-14} + C_8H_{18}$	17, 18
Hydrocraking	$C_nH_{2n+2} + H_2 \rightarrow C_{n-7}H_{2n-12} + C_7H_{16}$	17, 18
Hydrocraking	$C_nH_{2n+2} + H_2 \rightarrow 2C_{n/2}H_{n/2+2}$	16, 18
Isomerisation	$C_nH_{2n+2} \rightarrow \text{iso} - C_nH_{2n+2}$	16, 18

Finally, the resulting hydrocarbon mixture is fed into a flash tank to separate the propane from the liquids, which are then sent to a train of two conventional distillation columns, and the naphtha is collected in the distillate stream of the first column, and the bio jet fuel and green diesel in the distillate and bottom streams respectively of the second column.

3.2 *Intensified and Integrated Process to Produce Biojet Fuel from Palm Oil via HEFA Process*

One of the alternatives to implement advanced designs is to integrate hybrid equipment, which has the capacity to do more than one task in a single piece of equipment, unlike a conventional design where each activity is done separately, i.e., if you have a mixture of three components, the separation would be carried out in at least two pieces of equipment. In a design with hybrid equipment, a single piece of equipment capable of obtaining the three components of the mixture could be designed as shown in Fig. 8.3 [28]. Among the tools for adding hybrid equipment is process intensification, which is defined as reducing inventory through smaller equipment, improving the reaction zone and minimising the consumption of energy and raw materials [29].

An example of this definition is presented in Fig. 8.3 which shows the process of joining two separation columns and for the separation of a three-component mixture. The integration of these two columns consists of removing the cooling and heating units from the first column and making a thermal coupling by harnessing the heat from the second column through the internal flows that keep the two columns connected as shown in Fig. 8.3b. The result of this coupling is a partition wall column where the left-hand column is the pre-fractionator and the right-hand column is the main column (see Fig. 8.3c). With this it is possible to obtain the three components through the three zones of the main column which is the distillate, sidestream, and bottom recovery sections into a divided-wall column (DWC). The importance of using this technology lies in the fact that it eliminates heating and cooling units, thus reducing the number of utilities used and having a positive impact on economic and environmental aspects [30–33].

Different methodologies have been proposed to obtain intensified designs with different objectives, which consider economic, environmental, safety and control

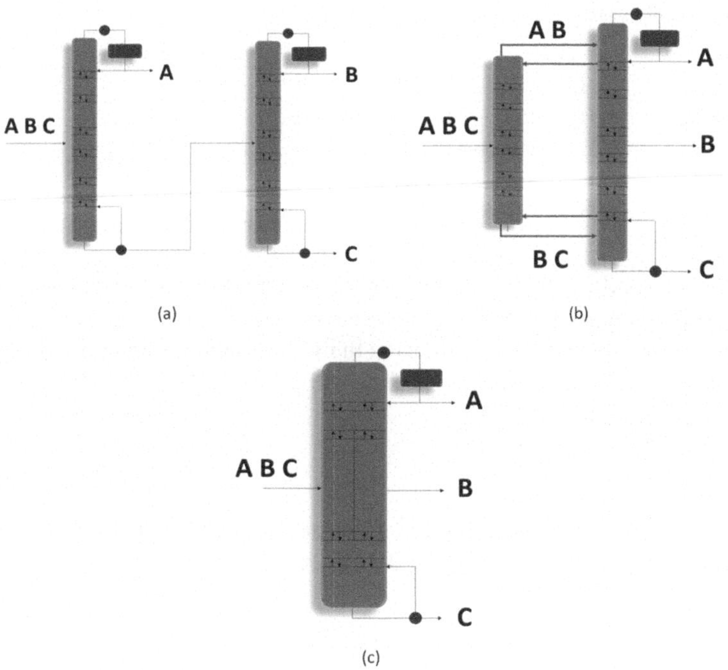

Fig. 8.3 Representative diagram of converting two conventional columns into one divided-wall column

criteria on an individual or multi-objective basis [34]. Among these methodologies are those based on the study of the phenomena involved in the equipment in a flow diagram, formal optimisation modelling, process improvement through supply chain optimisation and advanced model-based control [35, 36]. Recently, a methodology described intensified designs based on grouping teams by modules that have in common some criteria to improve [18].

For the conventional design in Fig. 8.2, the methodology of intensifying processes through modularity criteria was implemented to find a balance between economic/environmental aspects with inherent safety criteria. Since it has been proven that the fully intensified process is not always the most inherently safe. Hence, a DWC column is added to the design instead of the two distillation columns proposed in the conventional design (Fig. 8.4).

Once the reduction in the amount of equipment in the product separation zone has been implemented, an energy integration strategy was taken into account using the Pinch Point method, which consists of designing a network of heat exchangers in order to make use of the heat from the currents using the second law of thermodynamics as a basis. Figure 8.8 shows the heat integration problem, in which H1–H6 hot streams and C1–C3 cold streams were selected, and Table 8.5 shows the stream data for the intensified process (Fig. 8.5).

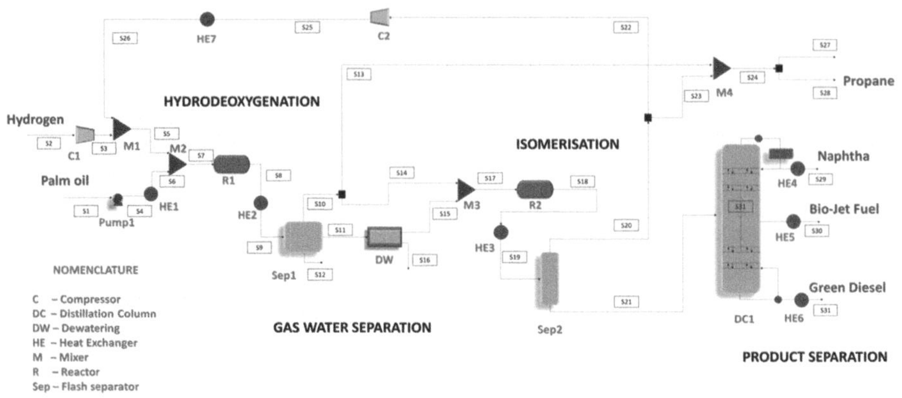

Fig. 8.4 Intensified bio jet fuel production flowsheet from palm oil via HEFA process. (Reprinted (adapted) with permission from Castillo-Landero, A., Dominguillo-Ramirez, D., Aburto, J., Sadhukhan, J., & Martinez-Hernandez, E. (2023). Improving the Economic, Environmental, and Safety Performance of Bio-Jet Fuel Production through Process Intensification and Integration Using a Modularity Approach. ACS Sustainable Chemistry & Engineering, 11(2), 660-669. Copyright 2023 American Chemical Society)

Table 8.5 Stream data of the intensified process

Stream	Flow rate × specific heat (Kcal/h−°C)	Supply temperature (°C)	Target temperature (°C)
H1	2386.7	310.0	250.0
H2	1945.5	276.8	98.0
H3	341.6	120.7	99.0
H4	308.0	104.5	25.0
H5	729.2	176.5	25.0
H6	431.0	288.1	25.0
C1	562.9	27.4	310.0
C2	127.1	173.1	310.0
C3	174.4	284.0	288.0

Considering a $\Delta T_{\min} = 10\,^{\circ}C$, Fig. 8.6a shows a cascade diagram for intensified process following the Pich Point method and Fig. 8.6b is the revised cascade diagram where the Pinch Point is located. Based on this result, the amount of needed heat satisfies the minimum heat quantity which is 651 kW for heating and 8.0 kW for cooling (Fig. 8.6b). Figure 8.7 shows one feasible solution in a grid diagram where the match streams represent the heat exchangers in this solution. The demand for heating and cooling services was reduced by 76 % and 95 %, respectively, compared to the conventional design shown in Fig. 8.2 [18].

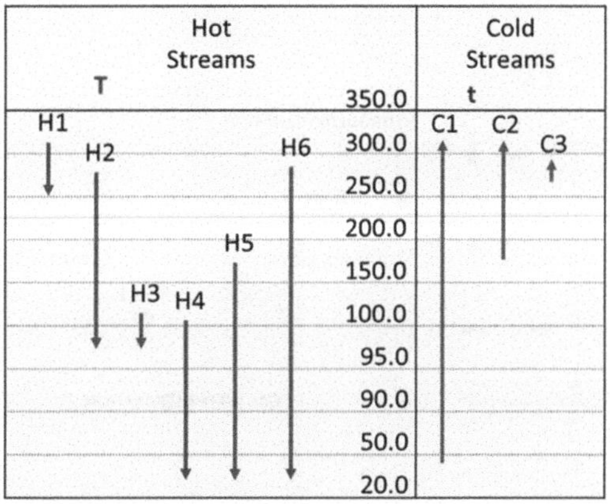

Fig. 8.5 Schematic representation of the heat-integration problem to solve the intensified process

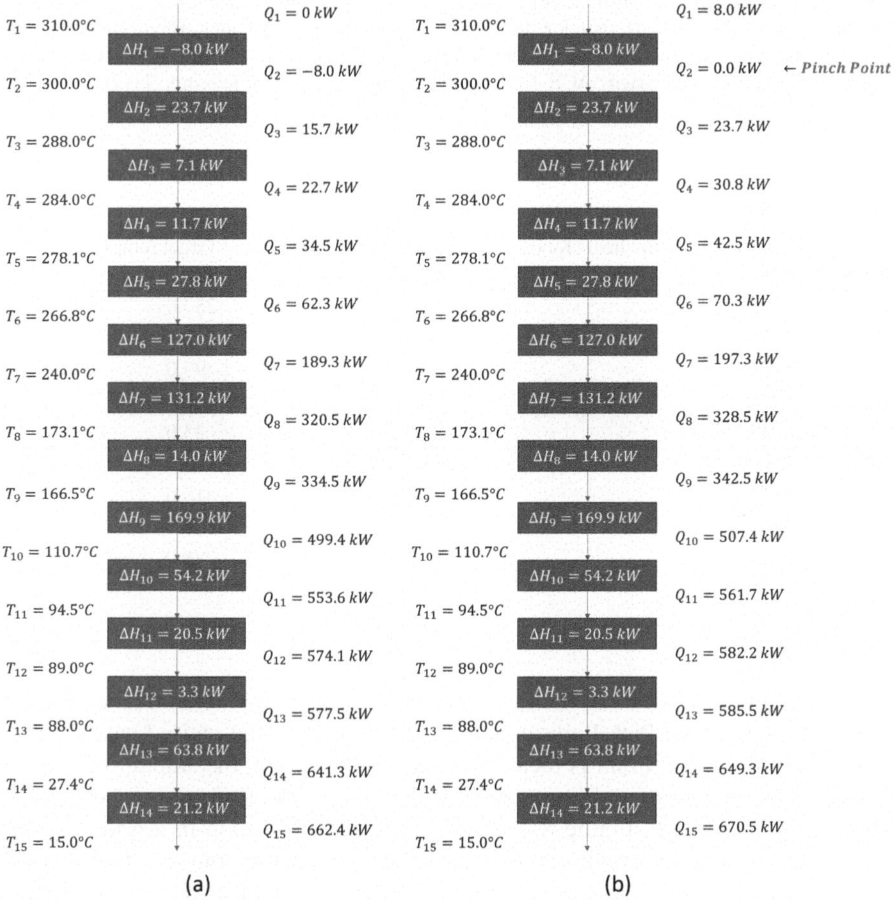

Fig. 8.6 Cascade diagram calculations for the heat-integration problem of a) heating and b) cooling of the intensified process to produce SAF

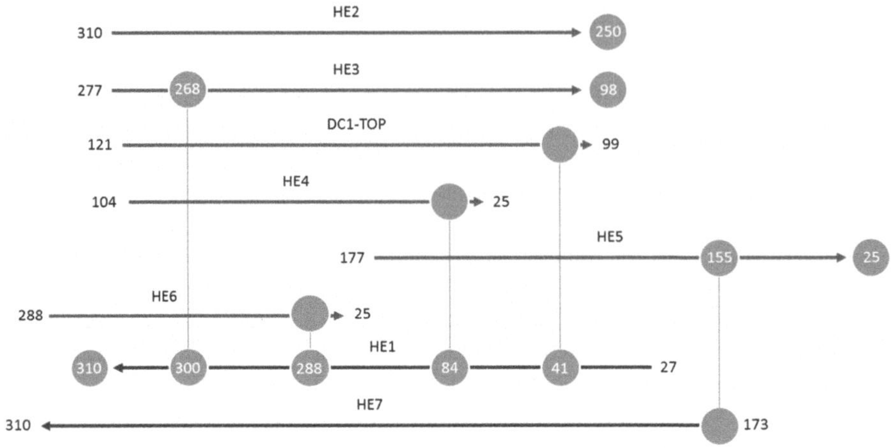

Fig. 8.7 Heat integration result of the intensified process. (Reprinted (adapted) with permission from Castillo-Landero, A., Dominguillo-Ramirez, D., Aburto, J., Sadhukhan, J., & Martinez-Hernandez, E. (2023). Improving the Economic, Environmental, and Safety Performance of Bio-Jet Fuel Production through Process Intensification and Integration Using a Modularity Approach. ACS Sustainable Chemistry & Engineering, 11(2), 660-669. Copyright 2023 American Chemical Society)

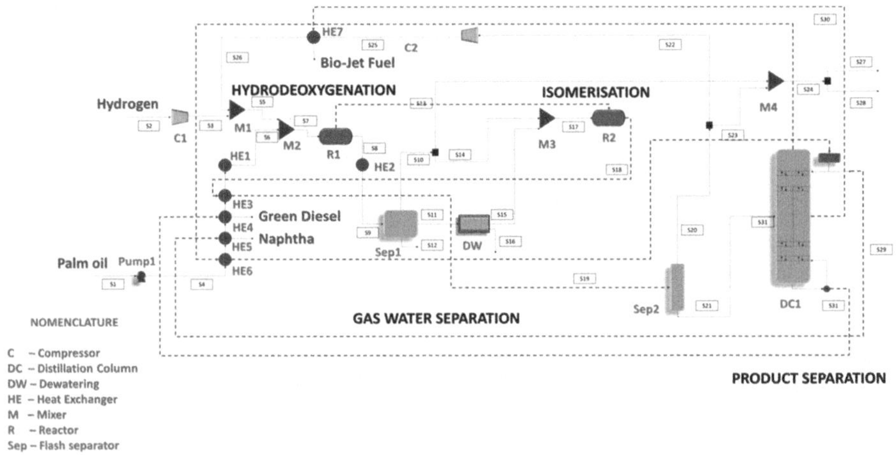

Fig. 8.8 Intensified and heat-integrated SAF production flowsheet from palm oil via the HEFA process. (Adapted with permission from Castillo-Landero, A., Dominguillo-Ramirez, D., Aburto, J., Sadhukhan, J., & Martinez-Hernandez, E. (2023). Improving the Economic, Environmental, and Safety Performance of Bio-Jet Fuel Production through Process Intensification and Integration Using a Modularity Approach. ACS Sustainable Chemistry & Engineering, 11(2), 660-669. Copyright 2023 American Chemical Society)

4 Results

We analyse the techno-economic performance and environmental impacts of the intensified and heat-integrated process through a techno-economic analysis [37] and life cycle analysis (LCA) using the OpenLCA software based on the CML-IA reference method and the energy allocation method for impact assessment [38]. The database used was Ecoinvent 3.8, considered the largest transparent database of unit processes [39]. Finally, the inherent safety metric used was the Fire and Explosion Damage Index (FEDI) because it requires in its calculation data that can be obtained at an early stage of the design since it is based on a rigorous simulation without considering the conditions to which the process is exposed once it is built. Within these considerations are the operating and design variables [29, 40].

Table 8.6 shows the values obtained for the economic, environmental, and inherent safety metrics considered for comparison. An improvement over the conventional design can be seen in the intensified and integrated process. In economic terms, there is a 20% increase in return on investment (ROI) and a 70% reduction in utility costs. In environmental aspects, global warming potential decreases by 30%, human toxicity by 90%, acidification by 42%, and eutrophication by 41%. With respect to photochemical oxidation, there is no significant change as it represents the sum of the impacts coming from the whole process and gases such as CO_2 and CO generated, which come mainly from the reaction in reactor R1, which was not

Table 8.6 Comparison of economic, environmental, and inherent safety indicators between the conventional design and the advanced designs

Indicator	Conventional design	Intensified design	Intensified and Integrated design
ROI (%)	8.9	9.0	11.0
Utility Cost ($/y)	443,790	433,227	144,876
Global warming potential (kg CO_2 eq./GJ)	2.84	2.62	2.00
Human toxicity (kg 1,4-DB eq./GJ)	1.24×10^{-2}	9.30×10^{-3}	2.64×10^{-3}
Acidification (kg SO_2 eq./GJ)	2.51×10^{-4}	2.22×10^{-4}	1.46×10^{-4}
Eutrophication (kg PO_4 eq./GJ)	6.60×10^{-5}	5.87×10^{-5}	3.89×10^{-5}
Photochemical oxidation (kg C_2H_2 eq./GJ)	2.66×10^{-3}	2.66×10^{-3}	2.64×10^{-3}
Aquatic ecotoxicity (kg 1,4-DB eq./GJ)	1.98×10^{-5}	1.65×10^{-5}	7.40×10^{-6}
FEDI – Column (C1) (1/m)	91	105 (C1 + C2)	105 (C1 + C2)
FEDI – Column (C2) (1/m)	123	–	–

modified in this work. With respect to the FEDI, a change can be seen in the product separation zone, as the conventional design considered an FEDI value of 123, modifying it to 105, which tells us that the process is inherently safer [18].

Acknowledgements This work was supported by a Newton Fund Impact Scheme (NFIS) grant, ID 540821111. The grant is funded by the UK Department for Business, Energy and Industrial Strategy and delivered by the British Council. For further information, please visit www.newtonfund.ac.uk. The authors would also like to acknowledge Project Y.62001 (IMP) for financial support. Arick Castillo acknowledges the financial support and Dr Elías Martínez-Hernández for his advice during the postdoctoral stay.

Author Contributions All authors contributed to the study's conception and design. Material preparation, data collection and analysis were performed by Jorge Aburto and Arick Castillo-Landero. The first draft of the manuscript was written by Arick Castillo-Landero and all authors commented on previous versions of the manuscript. All authors read and approved the final manuscript.

Declarations The authors have no competing interests to declare that are relevant to the content of this article.

Jorge Aburto is an editorial board member for the book title 'Sustainable Aviation Fuels: Recent Advances and Future Challenges' and was not involved in the editorial review, or the decision to publish, this article. All authors declare that there are no competing interests.

References

1. El-Halwagi, M. M. (2017). *Sustainable design through process integration: Fundamentals and applications to industrial pollution prevention, resource conservation, and profitability enhancement*. Butterworth-Heinemann.
2. Association IAT. (2021). *IATA annual review 2021*.
3. Ng, K. S., Farooq, D., & Yang, A. (2021). Global biorenewable development strategies for sustainable aviation fuel production. *Renewable and Sustainable Energy Reviews, 150*, 111502.
4. Monteiro, R. R. C., dos Santos, I. A., Arcanjo, M. R. A., et al. (2022). Production of jet biofuels by catalytic hydroprocessing of esters and fatty acids: A review. *Catalysts, 12*, 237.
5. Lundberg, S., Fagerström, A., & Johansson, K. (2020). *Life cycle assessment of gasification-based Fischer-Tropsch bio jet fuel production*
6. Wang, W.-C. (2016). Techno-economic analysis of a bio-refinery process for producing Hydro-processed Renewable Jet fuel from Jatropha. *Renew Energy, 95*, 63–73.
7. Yao, G., Staples, M. D., Malina, R., & Tyner, W. E. (2017). Stochastic techno-economic analysis of alcohol-to-jet fuel production. *Biotechnol Biofuels, 10*, 1–13.
8. Serrano-Ruiz, J. C., & Dumesic, J. A. (2011). Catalytic routes for the conversion of biomass into liquid hydrocarbon transportation fuels. *Energy & Environmental Science, 4*, 83–99.
9. ASTM A. (2016). *Standard specification for aviation turbine fuel containing synthesized hydrocarbons*. 2023.
10. Pipitone, G., Zoppi, G., Pirone, R., & Bensaid, S. (2023). Sustainable aviation fuel production using in-situ hydrogen supply via aqueous phase reforming: A techno-economic and life-cycle greenhouse gas emissions assessment. *Journal of Cleaner Production, 418*, 138141.
11. Mawhood, R. K., Gazis, E., Hoefnagels, R., et al. (2015). *Technological and commercial maturity of aviation biofuels: Emerging options to produce jet from lignocellulosic biomass*.
12. Bezergianni, S., & Dimitriadis, A. (2013). Comparison between different types of renewable diesel. *Renewable and Sustainable Energy Reviews, 21*, 110–116.

13. Panoutsou, C., & Maniatis, K. (2021). *Sustainable biomass availability in the EU, to 2050*. Brussels, Belgium.
14. Martinez-Hernandez, E., Ramírez-Verduzco, L. F., Amezcua-Allieri, M. A., & Aburto, J. (2019). Process simulation and techno-economic analysis of bio-jet fuel and green diesel production—Minimum selling prices. *Chemical Engineering Research and Design, 146*, 60–70.
15. Starck, L., Pidol, L., Jeuland, N., et al. (2016). Production of hydroprocessed esters and fatty acids (HEFA)–optimisation of process yield. *Oil & Gas Science and Technology–Revue d'IFP Energies Nouvelles, 71*, 10.
16. Liu, S., Zhu, Q., Guan, Q., et al. (2015). Bio-aviation fuel production from hydroprocessing castor oil promoted by the nickel-based bifunctional catalysts. *Bioresource Technology, 183*, 93–100.
17. Brandão, R. D., de Freitas Júnior, A. M., Oliveira, S. C., et al. (2021). The conversion of coconut oil into hydrocarbons within the chain length range of jet fuel. *Biomass Conversion and Biorefinery, 11*, 837–847.
18. Castillo-Landero, A., Aburto, J., Sadhukhan, J., & Martinez-Hernandez, E. (2022). A process modularity approach for chemical process intensification and inherently safer design. *Process Safety and Environmental Protection*.
19. Wang, S., Lu, X., Li, X. X., & Li, W. D. (2015). A systematic approach of process planning and scheduling optimization for sustainable machining. *Journal of Cleaner Production, 87*, 914–929.
20. El-halwagi, M. M., Foo, D. C. Y., & Tan, R. G. R. (2012). *Recent advances in sustainable process design and optimization* (With Cd-rom).
21. Halim, I., & Srinivasan, R. (2011). A knowledge-based simulation-optimization framework and system for sustainable process operations. *Computers & Chemical Engineering, 35*, 92–105.
22. Bielenberg, J., & Palou-Rivera, I. (2019). The RAPID Manufacturing Institute–Reenergizing US efforts in process intensification and modular chemical processing. *Chemical Engineering and Processing-Process Intensification, 138*, 49–54.
23. Ponce-Ortega, J. M., Al-Thubaiti, M. M., & El-Halwagi, M. M. (2012). Process intensification: New understanding and systematic approach. *Chemical Engineering and Processing: Process Intensification, 53*, 63–75.
24. Babi, D. K., Holtbruegge, J., Lutze, P., et al. (2014). Sustainable process synthesis-intensification. In *Computer aided chemical engineering* (pp. 255–260). Elsevier.
25. Lutze, P., Babi, D. K., Woodley, J. M., & Gani, R. (2013). Phenomena based methodology for process synthesis incorporating process intensification. *Industrial & Engineering Chemistry Research, 52*, 7127–7144.
26. Castillo-Landero, A., Jiménez-Gutiérrez, A., & Gani, R. (2018). Intensification methodology to minimize the number of pieces of equipment and its application to a process to produce dioxolane products. *Industrial & Engineering Chemistry Research, 57*, 9810–9820.
27. International A. (2009). *Standard specification for aviation turbine fuels*.
28. Wolff, E. A., & Skogestad, S. (1995). Operation of integrated three-product (Petlyuk) distillation columns. *Industrial & Engineering Chemistry Research, 34*, 2094–2103.
29. Castillo-Landero, A., Ortiz-Espinoza, A. P., & Jiménez-Gutiérrez, A. (2018). A process intensification methodology including economic, sustainability, and safety considerations. *Industrial & Engineering Chemistry Research, 58*, 6080–6092.
30. Kaibel, B. (2014). Dividing-wall columns. In *Distillation* (pp. 183–199). Elsevier.
31. Aurangzeb, M., & Jana, A. K. (2016). Dividing wall column: Improving thermal efficiency, energy savings and economic performance. *Applied Thermal Engineering, 106*, 1033–1041.
32. Errico, M., Tola, G., Rong, B.-G., et al. (2009). Energy saving and capital cost evaluation in distillation column sequences with a divided wall column. *Chemical Engineering Research and Design, 87*, 1649–1657.
33. Pandit, S. R., & Jana, A. K. (2022). Transforming conventional distillation sequence to dividing wall column: Minimizing cost, energy usage and environmental impact through genetic algorithm. *Separation and Purification Technology, 297*, 121437.

34. Sitter, S., Chen, Q., & Grossmann, I. E. (2019). An overview of process intensification methods. *Current Opinion in Chemical Engineering, 25*, 87–94.
35. Lutze, P., Gani, R., & Woodley, J. M. (2010). Process intensification: A perspective on process synthesis. *Chemical Engineering and Processing: Process Intensification, 49*, 547–558.
36. Pistikopoulos, E. N., Tian, Y., & Bindlish, R. (2021). Operability and control in process intensification and modular design: Challenges and opportunities. *AIChE Journal, 67*, e17204.
37. Peters, M. S., Timmerhaus, K. D., & West, R. E. (2003). *Plant design and economics for chemical engineers*. McGraw-Hill.
38. Ciroth, A. (2007). ICT for environment in life cycle applications openLCA – A new open source software for life cycle assessment. *The International Journal of Life Cycle Assessment, 12*, 209–210.
39. Wernet, G., Bauer, C., Steubing, B., et al. (2016). The ecoinvent database version 3 (Part I): Overview and methodology. *The International Journal of Life Cycle Assessment, 21*, 1218–1230.
40. Khan, F. I., & Abbasi, S. A. (1998). Multivariate hazard identification and ranking system. *Process Safety Progress, 17*, 157–170.

Chapter 9
Integrated System for Biojet Fuel Production

Edwin Santiago Rios Escalante, Pedro Teixeira Lacava, and João Andrade de Carvalho Júnior

1 Introduction

The aeronautical industry aims to reduce GHG emissions by up to 50% by 2050 (compared to 2005 levels) [1]. The strong dependence on oil-based jet fuel and the growing environmental impacts have encouraged the search for new alternatives in the aviation sector [2]. Thus, biojet fuels are an attractive option since they can be used without problems in currently operating aircraft engines. However, an adequate evaluation of the components of the biojet fuel supply chain is necessary, that is, raw materials (cultivation, harvest, storage, and transport), production routes, storage, and transport [2].

Although biojet fuel is a very attractive option to be used in air transport, there are still several obstacles to its implementation, such as inadequate government support and industrial commitment, unreliable raw material supply, commercial uncertainty of the conversion pathway, and lack of certification of the supply chain [3]. In addition, the biggest challenge for biojet fuel production is mainly economic, since the cost of this fuel is not competitive against fossil jet fuel. Fuel costs represent about 30% of airlines operating costs. The weighting of the cost/benefit ratio is a prerequisite for changing the energy matrix [4]. On the other hand, the type of biomass to choose is also another challenge, since certification of sustainable production of biojet fuel is required under which social and environmental requirements must be met [4].

The ASTM D7566 standard is the organization responsible for certifying the so-called synthesized paraffinic kerosene (SPK) to be applied in the aviation sector

E. S. R. Escalante (✉) · J. A. de Carvalho Júnior
São Paulo State University (UNESP), Faculty of Engineering, Department of Chemistry and Energy, Guaratinguetá, SP, Brazil

P. T. Lacava
Aeronautics Institute of Technology, São Jose dos Campos, Brazil

M. Aslam et al. (eds.), *Sustainable Aviation Fuels*, Sustainable Aviation, https://doi.org/10.1007/978-3-031-83721-0_9

under a certain blend concentration in common jet fuel. Thus, 50% vol. of FT-SPK can be blended with conventional jet fuel. Hydroprocessed esters and fatty acids (HEFA) were approved in 2011 to be blended up to 50% vol. Synthesized iso-paraffins (SIP) were approved in 2015 to be blended up to 10% vol. Alcohol-to-jet (ATJ) uses C2–C5 alcohols for the ATJ-SPK production, its allowed mixing limit is up to 50%vol. In this process, the use of intermediates such as isobutanol and ethanol was approved in 2016 and 2018, respectively. The main problem related to SIP (for a mixture greater than 10% vol.) is that it has a chemical composition composed of only 97% farnesane by weight. The main standard specifications for the approbation of a biojet fuel are summarized in Fig. 9.1.

The leading institutions responsible for biojet fuel certifications are found in USA and UK. Some specifications not considered in the ASTM D1655 standard can be found in the Aviation Fuel Properties Handbook. Due to the great complexity in the approval of new alternative fuels for aviation and the diversity of specifications, ASTM and DEF STAND 91-91 have elaborated a consistent international checklist, which combines the most rigorous requirements from jet fuel specifications. This checklist is known as Aviation Fuel Quality Requirements for Jointly Operated Systems (AFQRJOS). Additionally, ASTM (USA) and DEF STAND (UK) coordinate the creation of standard specification D7566 in 2009, which addressed new production routes for renewable jet fuels as shown in Fig. 9.1.

Brazil has a great infrastructure for the production of biomass and has demonstrated its capacity to transform it into biofuels. In addition, the country has areas of degraded pastures and livestock to be used for the raw materials crops, and this can be done without affecting the food supply [5]. Biojet fuel and renewable hydrocarbons have a competitive advantage for Brazil as they consist of interesting solutions

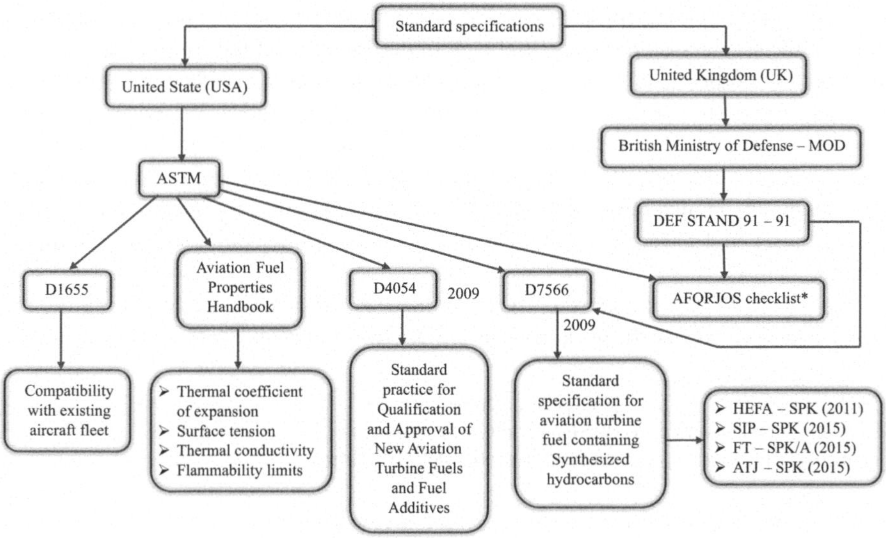

Fig. 9.1 Main international standard specifications for the approbation of a biojet fuel

for decentralized supply and the consequent reduction in logistics costs [4]. Currently, more than 80% of the national jet fuel production is concentrated in the Southeast region [6], while consumption is distributed through airfields scattered throughout the territory; in several cases in remote areas [5]. Transporting fossil jet fuel from those areas generates high operating costs for a reduced demand (~500 l day^{-1}). Therefore, those places tend to benefit more from decentralization initiatives [4].

Brazil's commitment to reducing GHG emissions caused by the aeronautical industry began with the creation of the "Brazilian Renewable Biokerosene Platform" in 2012, which aims at the sustainable production of biojet fuel from oilseeds and residues based on optimized regional logistics [7]. Due to the continental size of Brazil, each region needs to take advantage of its available land for the cultivation of raw materials, taking as reference the main airports and those that will be built in the future [5].

Since the Brazilian aviation sector is evolving considerably (see Fig. 9.2) the need to search for biofuels for this sector has increased as well. However, biojet fuels produced from biomass must meet rigorous international standards to maintain the performance and combustion characteristics of the reaction turbine.

Thus, this work can be divided into four main objectives, namely; (1) to analyze the conversion routes available internationally, their advances and limitations for the efficient conversion of potential raw materials (jatropha, sugarcane, eucalyptus, and agricultural residues) into biofuel for aviation, (2) to make a simplified techno-economic analysis of five conversion pathways currently available in the international market and estimate the selling price of a certain quantity of biojet fuel produced from available raw materials, (3) To show the potential and advantages of

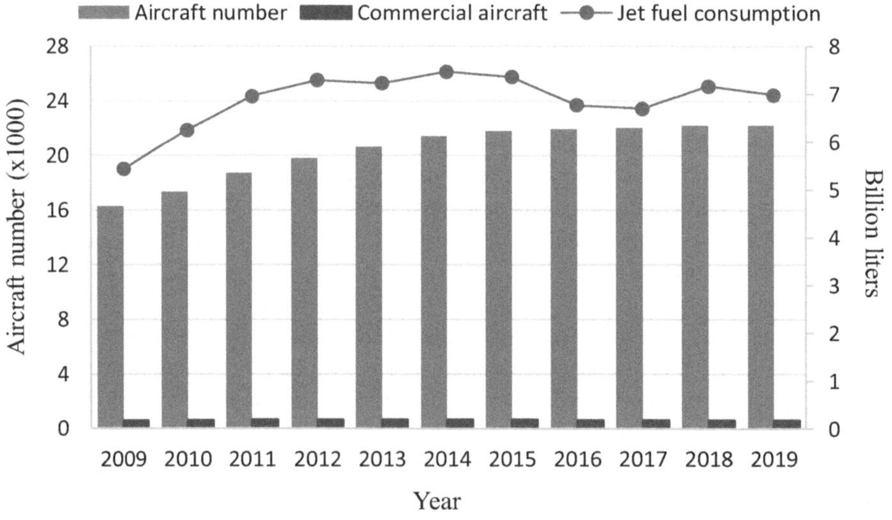

Fig. 9.2 Aeronautical activity and consumption of fossil jet fuel in Brazil in the last 10 years [8, 9]

biojet fuel production through an integrated system and (4) to estimate, through net CO_2, the environmental impacts generated by the biojet fuel production chain.

2 Potential Feedstock for Biojet Fuel Production

It is known worldwide that biomass is the main raw material for the production of biofuels used in road transport [10] due to its renewable nature [11]. In the same way, fuels obtained from biomass have the greatest potential to be used in air transport as this industry continues to develop [12]. From an environmental point of view, biofuels reduce CO_2 emissions into the atmosphere [13], but from a technical point of view, the combustion characteristics may change [14]. Currently, both internal combustion engines and gas turbines accept biofuels as long as they have similar or the same specifications as fossil jet fuel. This is the reason why biofuels have limited application in commercial aviation, especially because of their low aromatic content [15, 16]. Although aromatic compounds contribute to the formation of particulate matter, they are accepted for up to 25% vol. into fossil jet fuel [17].

The production of bioenergy and renewable fuels in Brazil is due to the great variety and available biomass in the country and especially to its great experience in the use of bioethanol and biodiesel in the ground transport sector. The development of large-scale crops and high social acceptance contributed to the gradual introduction into the Brazilian market [18, 19].

For ground transportation, renewable resources have been successfully applied to vehicles powered by diesel and gasoline, with biodiesel and ethanol, respectively, being the biofuels with the greatest impact in this sector. However, this energy source is inefficient in aeronautical applications due to the severe conditions (low temperature, payload, and range) to be overcome on a commercial flight [20, 21]. For this reason, it is necessary to find renewable alternative fuels that meet the demand while expecting the least environmental impact and a reduction in dependence on fossil fuels [22].

Ethanol was the first liquid biofuel to be blended with fossil fuels. This is because ethanol is produced from sugarcane, which is a renewable raw material. Subsequently, works such as Cheng et al. [23], Yun et al. [24], Mendez et al. [25] and Mendez et al. [26] showed that current combustion chambers accept mixtures of alcohols with fossil fuels, in specific proportions, maintaining a high performance without modifying the engine.

Various types of renewable raw materials can be used for the production of alcohols, such as cereals, algae, wood, and many crops. On the other hand, the fermentation of sugar and other carbohydrates such as cellulose, hemicellulose, lignin, xylan, and mannan can also be used [27].

Brazil is the world's second-largest producer of ethanol. Agricultural aircraft account for 75% of the sector sales. 25% of agricultural aircraft run on hydrated ethanol, which contributes both to reducing operating costs (up to 40%) and

improving the environment [28]. Furthermore, Brazil has the potential to increase biomass production mainly due to its available land and favorable climatic conditions [29].

Currently, Brazil has a land area of 850 Mha, of which 9% is used for seed crops in the country, including forestry, 13.2% is used for planted pastures, and 8% for natural pastures, totaling 30.2% of the Brazilian territory (approx. 257 Mha) [29]. In the period 2020/21, approximately 8.6 Mha was used for sugarcane cultivation, corresponding to 1.01% of the Brazilian territory or 3.7% of cultivated areas [30, 31]. It means that Brazil uses a small percentage of its land. Since the 2015/16 harvest, the sugarcane harvested area has oscillated around 8.7 Mha. This is because the leased areas were returned (and planted with other crops), a few projects focused on greenfields, and the stoppage of some existing units [31].

However, an increase of 11 Mha is expected for sugarcane cultivation for the 2029/2030 harvest period [32]. This increase in sugarcane cultivation is evidenced by the increase in the production of both anhydrous and hydrated ethanol in the last 10 years as shown in Fig. 9.3.

According to studies made by Cornelissen et al. [33], Prieler et al. [34], and Horta et al. [35], the expansion of biomass cultivation for renewable energy production would not affect food production simultaneously. This is because the use of sugarcane produces products and co-products that can be fully used for both energy generation and the food market as shown in Fig. 9.4. For this reason, sugarcane is a crop of great importance in Brazil, being part of its economy since the mid-sixteenth century.

The production of ethanol and n-butanol from sugarcane is well known and both are used in land transport, especially ethanol which has achieved a considerable

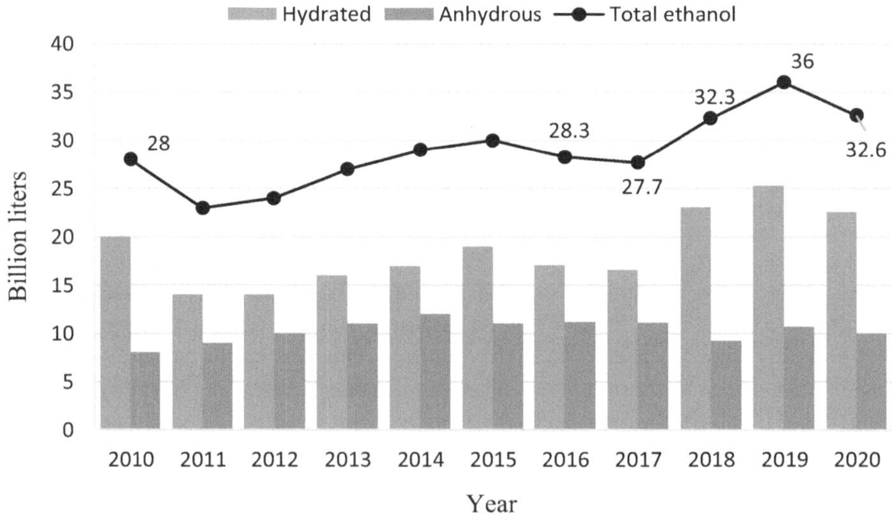

Fig. 9.3 Evolution of ethanol production in the 2010/2020 period [31]

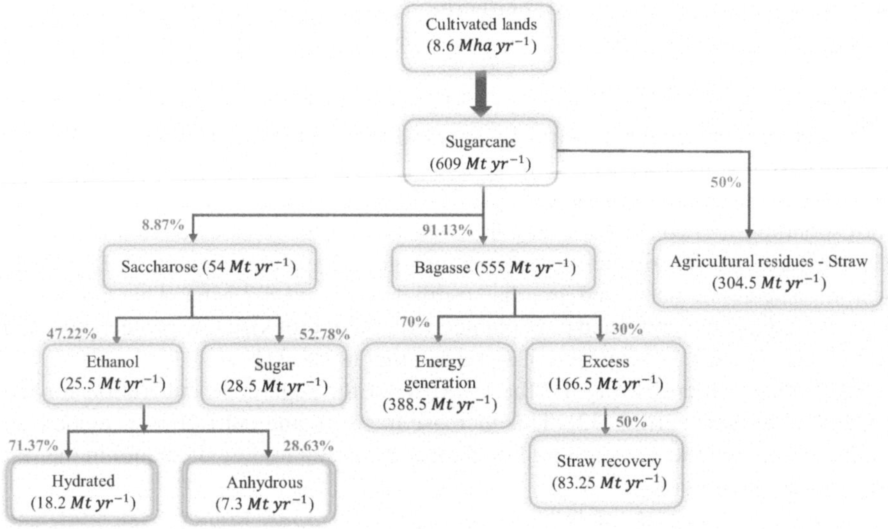

Fig. 9.4 Schematic representation of the main products and co-products obtained from sugarcane crops [30, 38]

Table 9.1 Comparison between the physical-chemical properties of ethanol, n-butanol, and fossil jet fuel [15]

Items	Ethanol	n—Butanol	Jet fuel
Molecular, weight (kg kmol^{-1})	46.07	74.12	≈185
Density (g ml^{-1} @ 20 °C)	0.789	0.810	0.775–0.84
Kinematic viscosity (mm^2 s^{-1} @ 40 °C)	1.08	2.23	8
Low heating value (MJ kg^{-1})	26.8	33.1	42.80–43.02
Freezing point (°C)	−114.3	−89.5	−47–−60
Flash point (°C)	8	35	38
Auto ignition temperature (°C)	434	385	210
Boiling point (°C)	78.4	117.7	150–170
Vapor pressure (kPa @ 38 °C)	13.8	2.27	14–21
Oxygen content (%)	34.8	21.6	–

level of development in the Brazilian market [13]. However, these fuels are not suitable for use on commercial flights. This is because they have unfavorable characteristics to be considered a "drop-in" biojet fuel such as low flash point, energy density, low-temperature properties, and high volatility. This last could cause evaporative loss or vapor lock in the fuel system, posing serious safety risks to crew and passengers [15]. Table 9.1 shows a comparison between the physical-chemical properties of ethanol, n-butanol, and fossil jet fuel.

On the other hand, in Brazil, eucalyptus and agricultural residues are considered potential feedstocks to be used in the F-T pathway for biojet fuel production. This is because Brazil has a high availability of eucalyptus wood with about 700 varieties

Table 9.2 Main characteristics of jatropha

Characteristics	Quantity	Units
Jatropha plant per hectare[a]	1100	Jatr ha^{-1}
Jatropha fruit per year	5194–6926	fruit $year^{-1}$
Fruit weight	1.53–2.85	g
Seeds weight	0.5–0.7	g
Seeds percent per fruit	64–75	%
Shell percent per fruit	15–36	%
Oil percent per seed	30–40	%
Seeds per hectare per year	3–5	ton ha^{-1} $year^{-1}$
Oil ton per year	1–2	ton ha^{-1} $year^{-1}$

[a]Considering the space among trees of 3 × 3 m

whose benefit is their adaptability to different climatic conditions, availability of land for cultivation, and high productivity [36].

Forestry represents a low-carbon economy, in addition to being used as the main feedstock in the paper industry (due to the cellulose content in its structure), the production of cleaner biofuels from wood (source of lignocellulosic biomass) such as ethanol and the manufacture of other products. These advantages have led private and government entities to strive to make Brazil a competitive market in the forestry sector [36]. According to Ibá (Industria Brasileira de Árvores), in 2015, the expansion of pulp mills helped maintain the Brazilian market as one of the largest producers of this input [36].

Another potential raw material considered in this study is jatropha fruit. Jatropha has several advantages to be considered as a raw material in biojet fuel production such as inedible (it is poisonous), pest resistance, a large amount of water is not necessary for its rapid growth for 40 years and each grain has a high oil content (30–40%) [15]. Since 2004, jatropha has been commercially produced in several Brazilian states, such as Minas Gerais (southeast), Bahia (northeast), Goiás (center-west), Mato Grosso (center-west), Maranhão (northeast), Rio de Janeiro (southeast) and Tocantins (southeast) [37]. The main characteristics of jatropha are shown in Table 9.2. Bergmann et al. [37], have reported that using 41,098.3 ha produced 116.3 ktons of jatropha in 2011. Thus, the amount of oil obtained was counted in a range from 1200 to 1500 kg ha^{-1} in 1 year.

3 Conversion Technologies for Biojet Fuel Production

3.1 Hydroprocessing Esters and Fatty Acids (HEFA) Process

The hydroprocessing esters and fatty acids (HEFA) process is a technology route used for biojet fuel production from triglycerides, whose long carbon chains ($C_{15} - C_{22}$) make them an excellent feedstock for this application [39].

Triglycerides can be found in vegetable oils, animal fats, used oils, and bio-oil. On the other hand, triglycerides are composed of other chemical substructures such as carbonyl group, glycerol compound, and fatty acids, which can be separated by catalytic hydrogenation using hydrogen [40]. The hydrogen demand in this process is determined by the degree of unsaturation and the chain length of the fatty acids [39]. Due to a possible high demand for biojet fuel production, it is necessary to look for alternatives for hydrogen production and increase the efficiency of existing technologies such as steam methane reforming, biogas production, and electrolysis [18].

Catalytic hydrogenation is carried out in a catalytic reactor, which is split into four stages:

1. Partial decarbonylation
2. Hydrodeoxygenation
3. Decarboxylation
4. Total decarbonylation

At the end of this process, long-chain paraffinic liquids ($C_{15} - C_{18}$, diesel range) are obtained. These chains are, then, reduced and isomerized in a hydrocracking reactor using hydrogen. The main product generated is biojet fuel, which has a long carbon chain of $C_8 - C_{12}$ and corresponds to 55%vol of the total of biofuels produced [41].

Biojet fuel, produced by the HEFA route, has properties compatible with fossil jet fuel such as cold flow, high thermal stability, high cetane number, low concentration of aromatic, etc.; therefore, it is considered a "drop-in" fuel [42]. It allows a maximum blend concentration of 50%vol of biojet fuel into fossil jet fuel, according to the international standard ASTM D7566 approved in September 2009 [42].

Other biofuels produced by the HEFA route are naphtha, propane, and diesel under concentrations of 8%, 11%, and 26%vol., respectively [41]. The HEFA route is the most attractive conversion technology due to its lower complexity compared to other conversion routes [12]. Wang [43] discussed the biojet fuel production from jatropha fruit. In this study, the biorefinery was divided into two stages: (1) from jatropha fruit to jatropha oil and (2) from jatropha oil to biojet fuel. In the first stage, 27.78 kg s^{-1} of jatropha fruit were processed to obtain 5.38 kg s^{-1} of jatropha oil, that is, the conversion efficiency was 19.38%. In the second stage, 5.38 kg s^{-1} of jatropha oil were processed to obtain 3 kg s^{-1} of biojet fuel, that is, the conversion efficiency was 55.75%.

In addition, it was reported that the operating costs of the first and second stages were 134 MUS$ year^{-1} and 116 MUS$ year^{-1}, respectively, which indicates a total investment of 150 MUS$ year^{-1} for every 85,147 tons of biojet fuel produced (assuming that the biorefinery works 90% of the days of the year). The sale price for jatropha oil and biojet fuel was estimated at 5.74 US$ gal^{-1} and 5.42 US$ gal^{-1}, respectively. The author concluded that the sale price of both products strongly depends on the cost of raw material, the cost of operation, the added value of the co-products, and the cost of energy.

3.2 Alcohol-to-Jet (ATJ) Process

Another conversion technology used for biojet fuel production is the alcohol-to-jet (ATJ) route. This pathway uses alcohols as a feedstock, which can be produced from sugarcane, agricultural residues (straw), sawdust, forest residues, and more [44, 45]. These alcohols go through an oligomerization process, which is divided into five stages [40], namely:

1. Dehydration
2. Oligomerization
3. Dimerization
4. Hydrogenation
5. Distillation

Ethanol, butanol, and hexanol are potential alcohols that can be used in this route. In general, any alcohol can be transformed into biojet fuel by the ATJ route, however, the type of alcohol to use will depend on the degree of complexity of obtaining it [46]. The lower the complexity, the more attractive it is as a feedstock.

Biojet fuel is the main product generated, representing 65% of the total biofuels produced. Biojet fuel produced by the ATJ route is also considered a "drop-in" fuel, which can be blended with fossil jet fuel in a maximum concentration of 50%vol., according to international standard ASTM D7566 [47]. Other biofuels produced by the ATJ route are naphtha and diesel in concentrations of 30% and 5% vol of the total biofuels, respectively [41]. In Brazil, sugarcane represents the main biomass used for the production of ethanol and butanol, which are the most attractive feedstocks for biojet fuel production via the ATJ route [12]. The use of ethanol as a feedstock for biojet fuel production was approved in April 2018, while butanol was already approved in April 2016 [42]. Butanol is preferred over ethanol because a butanol plant, using the ABE process, can be coupled to an existing ethanol plant, which would reduce the investment cost of production [12].

The analysis made by Klein et al. [48] reveals that 158 Mha of sugarcane, 162 Ml year^{-1} of biojet fuel can be produced with an internal rate of return (IRR) of 5.7% and a total investment of 303 MUS$ (or 303 MUS$ for 53,440 tons of biojet fuel). On the other hand, Pereira et al. [49]. have reported that from 1 ton of sugarcane processing 28 l of isobutanol can be produced. Therefore, from the data in Fig. 9.4, it is possible to obtain 1.84 l of isobutanol for each liter of ethanol anhydrous. The advantage of using isobutanol in the ATJ process is that it is not necessary to build new biorefineries for the production of isobutanol, but the ethanol production plants can be adapted for the co-production of ethanol and isobutanol, which would considerably reduce the investment. It is worth noting that atmospheric pressures are necessary for the process of dehydration and oligomerization of isobutanol, which also reduces the cost of the process. The amount of hydrogen required for this route was assumed to be 0.04 g per gram of butanol converted. The production cost of ATJ was assumed to be US$ 0.04 per gram of butanol converted [18].

3.3 Fischer–Tropsch (FT) Process

The Fischer-Tropsch process is also a technological route used for biojet fuel production from agricultural and forestry residues such as wood, bagasse, urban waste, and straw. This conversion route is formed by two joint processes [40], namely:

1. Gasification process
2. FT Synthesis process

The gasification process is a thermochemical process, that provides the feedstock for the FT synthesis process. The gasification process is divided into six stages [12], namely:

1. Biomass pretreatment
2. Gasification
3. Syngas cooling system
4. Acid gas removal system
5. Gas conditioning system
6. CO_2 removal system

At the end of this process, the most important parameter is H_2/CO ratio present in syngas. This ratio will determine the type of catalyst to be used in the FT synthesis to improve the process and increase its efficiency [40]. The common catalysts used are Fe ($H_2/CO \leq 0.6$) and Co ($H_2/CO \geq 2$) [50]. On the other hand, the FT synthesis process is made up of five stages [51], namely:

1. FT reactor
2. Hot separator
3. Cold separator
4. Hydrocracking
5. Distillation

At the end of this process, At the end of this process, long-chain hydrocarbons are obtained [12, 40, 42]. These hydrocarbons, then go through the hydrocracking process to produce diesel as the main product in a concentration of 58%vol of the total of biofuels generated, followed by naphtha and biojet fuel in a concentration of 23% and 19%vol., respectively [41, 48].

The biojet fuel produced by the FT route lacks aromatic, however, this problem can be overcome by blending biojet fuel with fossil jet fuel to a maximum concentration of 50%vol., according to ASTM D 7566 [42]. The FT process is the most attractive technological route with respect to raw materials since it does not compete with the food market.

In this context, eucalyptus is the most cultivated species in Brazil, occupying about 5.56 Mha of the country. There are 15 Brazilian states dedicated to the cultivation of eucalyptus, of which four are the largest producers: Minas Gerais (25.77%), São Paulo (17.97%), Mato Grosso do Sul (14.79%) and Bahia (11.61%). The annual yield of the fresh mass of eucalyptus is approximately 40 $m^3\ ha^{-1}$ and the density of the eucalyptus trunk is considered to be 560 $kg\ m^{-3}$ [52]. In addition, Brazil has 10

Mha of degradable land to promote forestry [36]. According to Klein et al. [48], this process has an investment cost of approximately 560 MUS$ per 0.5095 Mtons (or 631 $Ml_{bioject\ fuel}$ year^{-1}) of biojet fuel produced and a sale price of 0.36 US$ l^{-1}. It is worth noting that, the reduced sale price is due to Klein et al. [48] considering four plants operating through the F-T route for biojet fuel production.

3.4 Syngas Fermentation (SF) Process

The syngas fermentation process is a technology used for alcohol production, especially, ethanol and butanol, which uses CO-rich syngas as a feedstock [53]. Generally, this syngas is obtained from industrial processes that involve the combustion of fuels such as the steel industry, gasification of agro-industrial, forestry, and urban waste [54]. This syngas provides an adequate environment for bacterial growth, however, depending on the type of alcohol to be produced a certain bacteria must be selected.

The SF route is not a conventional route for biojet fuel production, however, it is possible to produce higher alcohols from it, which can be used as additives for fossil jet fuel, such as hexanol. The most promising bacteria for hexanol production, by the SF route, is *Clostridium carboxidivorans* P7, which also produces ethanol and butanol simultaneously [53]. However, the success of this technology depends on the optimal composition of $CO : H_2 : CO_2$ the syngas [55].

Hyun Ju Oh et al. [56], show that syngas with a composition of $CO : H_2 : CO_2 = 70 : 20 : 10$ was the most adequate concentration for the production of higher alcohols (butanol and hexanol) compared to $CO : H_2 : CO_2 = 40 : 30 : 30$ and $CO : H_2 : CO_2 = 60 : 20 : 20$. The SF route is divided into three stages:

1. Gasification of lignocellulosic biomass to produce CO-rich syngas
2. Conversion of syngas into crude alcohol
3. Distillation of alcohols to produce the desired purity

At the end of the process, ethanol, butanol, and hexanol are obtained in concentrations of 60%, 20%, and 20%vol of the total alcohols produced [54].

In the present study, points 2 and 3 were not considered, therefore, the quantities of ethanol, butanol, and hexanol assumed for the calculation were those reported by Phillips et al. [54], that is, 3.0 g $l_{syngas}{}^{-1}$, 1.0 g $l_{syngas}{}^{-1}$ and 1.0 g $l_{syngas}{}^{-1}$, respectively. Bagasse and straw can be used as input in a gasifier [54] for the production of 2 Nm^3 of syngas for each Kg of lignocellulosic biomass [55].

It should be noted that the SF route works under conditions inverse to those reported for the FT route, that is, the SF route needs CO-rich syngas, while the FT route needs H_2-rich syngas. The SF route has some advantages over the FT route such as [57]:

1. The fermentation process is carried out at low pressures and temperatures.
2. Less sensitive to changes in some parameters such as the presence of sulfur and flow rate variation.

3. CO is essential for the cellular growth of microorganisms resulting in the production of H_2-rich syngas at the exit of the process, which can be used as fuel for the generation of electrical energy.

The SF route turns out to be an attractive and promising technology for biofuel production since all the alcohols generated in the process have a wide application in the current chemical and automotive market.

Kick et al. [58], carried out a series of experiments with five mixtures of different types of biofuels to determine the most suitable mixture as an alternative to fossil jet fuel. The results showed that a mixture of synthetic kerosene, obtained through the Fischer-Tropsch pathway, and 20%vol. of hexanol meets the requirements to be a potential alternative fuel for aviation. However, Chuck and Donnelly [16] determined that the 10%vol. of hexanol does not affect the density and viscosity of the jet fuel, but its flash point is reduced below 38 °C, which represents a great risk for the crew and passengers. They concluded that hexanol should not be used as jet biofuel, however, it can be used as an additive up to 10% vol. Finally, Alves et al. [18] have reported that the investment cost of this technology is estimated at 259 MUS$ for 60,845.5 tons of hexanol produced (or 166.7 tons day^{-1}), and Phillips [54] estimated the sale price of hexanol at 1.53 US$ l^{-1}.

3.5 *Direct Sugar to Hydrocarbons (DSHC) Process*

The DSHC process is another technological route used for biojet fuel production from glucose ($C_6H_{12}O_6$), which is found in sugarcane, bagasse treatment, wood fermentation, etc. Glucose is transformed directly into biojet fuel without passing through intermediate alcohols as occurs in the ATJ route [59].

This technology depends on the development of genetic engineering, which modifies the way to metabolize sugar by microorganisms, so any raw material is considered fourth generation. Biojet fuel obtained through the DSHC route is a synthesized isoparaffinic (SIP or isoprenoid C15) also called farnesane ($C_{15}H_{32}$) [60]. In 2018, the international standard ASTM D7566 approved up to 30% vol. as the limit concentration of farnesane in petroleum-based jet fuel [20].

The DSHC route is the less attractive technology since produces a fuel made up of a single type of molecule, avoiding the generation of other fuels that could reduce its production cost.

Michailos [61] has reported that the efficiency of the biological process for the conversion of biojet fuel from sugarcane used as raw material was 12.1%, that is, from 1 kg of bagasse feedstock 0.121 kg of biojet fuel (farnesane) is produced. It is also reported that the cost of investment for this technology is estimated at 606 MUS$ for 80,000 tons of farnesane produced per year (this value was obtained from 10 ton h^{-1} and considering the operation time of the plant of 8000 h $year^{-1}$). The sale price of farnesane was estimated at 2.78 US$ l^{-1} and the CO_2 emissions generated in its process were estimated at 46.2 $gCO_{2-eq}MJ_{\text{biojet fuel}}^{-1}$, which can be reduced by up to 37 $gCO_{2-eq}MJ_{\text{biojet fuel}}^{-1}$ as long as the efficiency of the fermentation process is increased.

4 Greenhouse Gases (GHG) Emissions

Air commercial transport contributes to the global economy generating 3.5 trillion annually, however, this activity also produces up to 2.5% of global GHG emissions [41]. For this motive, the International Civil Aviation Organization (ICAO) aims to reduce 50% of the GHG emissions by 2050 about those generated in 2005 [41]. To achieve this goal, ICAO has created two entities: Carbon Offsetting and Reduction Scheme for International Aviation (CORSIA) and Committee on Aviation Environmental Protection (CAEP), which propose the use of biojet fuel in aircraft. It seems to be the most effective and fastest technical for reducing GHG emissions whose functional unit is represented as $gCO_{2-eq}/MJ_{biojet\ fuel}$ [62].

Biojet fuel is obtained from biomass, which will determine the environmental benefit of this biofuel. According to CORSIA, biojet fuel is environmentally friendly when its GHG emissions are lower than those generated by fossil jet fuel, which accounts for 89 $gCO_{2-eq}/MJ_{biojet\ fuel}$ [39]. Thus, the GHG emissions generated by a biojet fuel are evaluated using the Life Cycle Assessment (LCA), which is divided into two branches: Attributional Life Cycle Assessment (ALCA) and Consequential Life Cycle Assessment (CLCA) [42, 63].

ALCA considers the flow of energy and mass within a given evaluation boundary as shown in Fig. 9.5. Two techniques are used to evaluate the ALCA, that is, well-to-tank (WTT) and tank-to-wake (TTW). WTT involves four stages, namely [12]:

1. Raw material acquisition
2. Transportation of raw materials to biorefinery
3. Biojet fuel production (including H_2 production)
4. Transportation and distribution of biojet fuel

While TTW involves a single stage, that is, the use of biojet fuel in the aircraft turbine (combustion process). The sum of WTT and TTW analysis is known as well-to-wake (WTW) analysis [64].

On the other hand, CLCA is based on sensitivity analysis to estimate external effects on the production of biojet fuel such as [65, 66]:

- GHG emissions mitigation policy
- Increase in demand for some raw materials
- Fossil jet fuel prices
- Land use change (LUC)

LUC is a technique that evaluates the contribution of GHG emissions from the type of land to be used, which largely depends on the organic residues found inside and outside the soil. LUC changes depending on two factors, namely [67]:

1. Preparation of the soil intended for the production of certain raw materials.
2. Crop management after soil conversion

In general, the contribution of LUC GHG emissions can be assessed in two ways [67]:

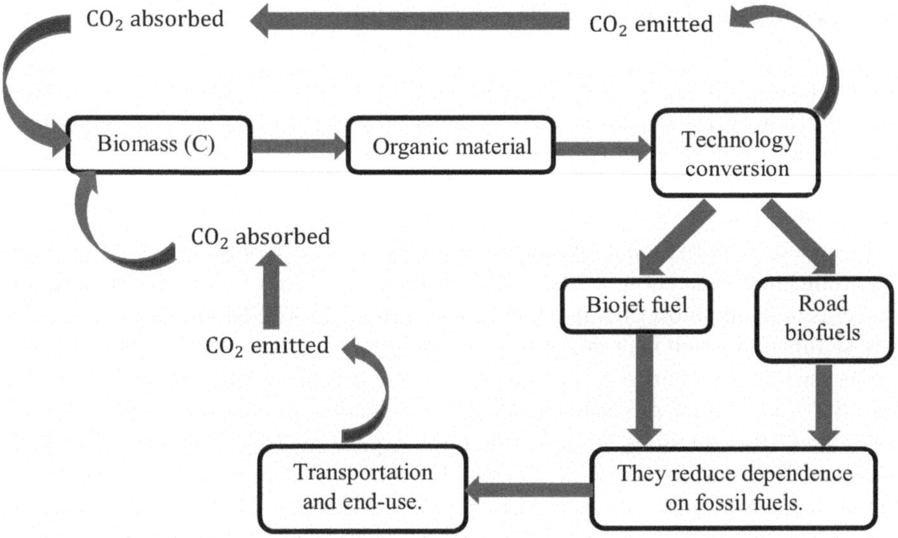

Fig. 9.5 Simplified scheme of net CO_2 emissions

1. Direct land use change (dLUC)
2. Indirect land use change (iLUC)

The dLUC assessment applies to land planned for the cultivation of certain raw materials. These types of lands can be those cultivated over a long time (established crops) or those cultivated annually [67]. Biomass grown in soil for a long time generates less impact on the dLUC, as is the case of jatropha, which does not require land work (tillage), in addition, has low input applications and whose benefit is related to the gain of soil organic carbon (SOC). Figure 9.6 shows the distribution of GHG emissions from jatropha in its production chain [68]. On the other hand, annual crops generate the greatest impact on dLUC, especially in carbon-rich ecosystems such as tropical forests and tropical shrublands [39]. The iLUC assessment applies to unplanned land, that is, those used to compensate for the deficit generated by the increase in demand for certain raw materials [39].

The land used for the cultivation of raw materials can be previously grown or abandoned lands. Previously grown lands are frequently used for the cultivation of raw materials for the production of vegetable oil and biodiesel. On the other hand, the use of abandoned land generates the greatest environmental impact due to its carbon sequestration capacity, which is estimated through net CO_2 [67].

The type of biomass to be converted will determine the environmental impact of biojet fuel production. Thus, the potential GHG emissions generated in the production of raw materials are evaluated through three important characteristics including:

1. CO_2 sequestration capacity from the environment, which offsets the CO_2 emissions generated in the combustion of fuels.
2. Carbon sequestration capacity in the soil

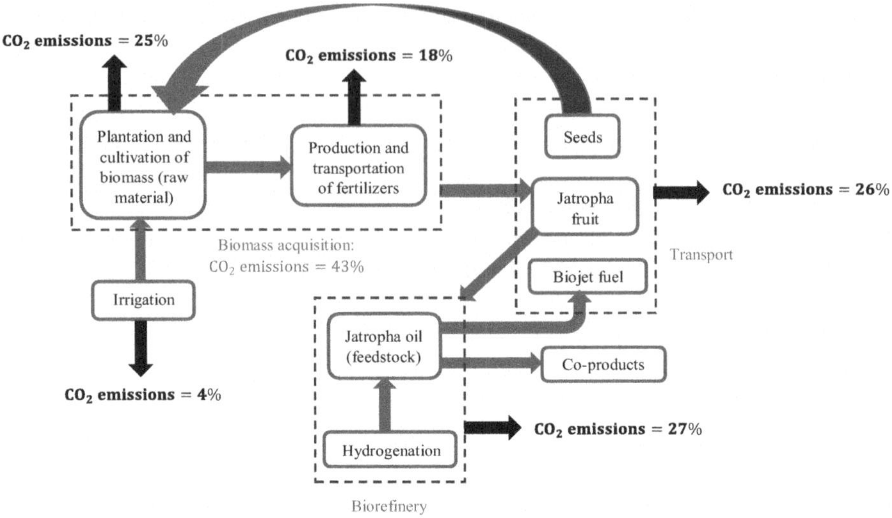

Fig. 9.6 Simplified diagram referring to the emissions generated in biojet fuel production from the jatropha fruit [68]

These first two characteristics represent the so-called "biomass credits"

3. Residual product generation
4. Management of crops, residues, and land conversion, which requires the use of energy and application of in situ chemicals of fossil origin such as fertilizers [69]. It represents a disadvantage as it would lead to a reduction in the environmentally positive aspects of biojet fuel compared to fossil jet fuel.

Water is an essential natural resource for agricultural production, which can be classified into green and blue water. Green water is produced naturally like rain, that is, it does not imply the generation of CO_2 emissions. On the other hand, blue water is chemically treated water, which implies the generation of CO_2 emissions [70, 71].

Water found in natural states such as lagoons, seas, and rivers are also used in industrial processes such as cooling and production systems [72]. However, excessive use of this resource can cause thermal pollution of the water [73]. In this context, production systems must be more efficient and responsibly use water. It requires large water-consuming facilities to implement water treatment units, which would lead to the reduction of freshwater demand while avoiding thermal pollution and other possible environmental impacts [74]. Water can also be used in the electrolysis process for the production of green hydrogen, which is a determinant factor in the production of biojet fuel and is also the most environmentally friendly technique [75].

Finally, GHG emissions estimates for each conversion route will depend on the approach used. Thus, we find four types [69]:

1. Mass allocation. This approach is applied when the GHG emissions generated in the production chain are distributed among the foods produced in a given conversion route.
2. Energy allocation. This approach is applied when the GHG emissions generated in the production chain are distributed among the products with the highest energy content produced in a given conversion route, that is, lower heating value (LHV) and high heating value (HHV).
3. Market value allocation. This approach is applied when the GHG emissions generated in the production chain are distributed among the products with lower sales prices in the market, which are generally food products.
4. Displacement. This approach is applied when the GHG emissions generated in the production chain are distributed among the total of products and co-products generated on a large scale to replace the current market fully.

5 Techno-Economic Analysis

ANAC [76] reports that the total operating costs in Brazilian airlines are divided into fuel cost (27.5%), lubricant cost (3.9%), aircraft maintenance (18.2%), workers (17.1%), operating expenses (12.7%), and others (20.6%). It is evident that the increase in aeronautical activity in the country (see Fig. 9.4) would force the players to reduce this dependence on fossil fuels as long as the cost of biojet fuel is attractive for its introduction in the market. Some enterprises linked to the aviation sector (EMBRAER, AIAB, TAM, GOL, Azul, and Trip) have worked in collaboration with biofuel production associations (União da Industria de Cana-de-açucae, Amyris Brasil, Algae Biotecnologia, and ABPM) testing biojet fuel in their commercial aircraft [77].

As mentioned above, biojet fuels are seen as an attractive alternative for reducing GHG emissions in air transport [78]. However, the large-scale application of biojet fuels remains a challenge due to uncertainty regarding land availability, variety of biomass crops, and variability in production. In addition, there are important techno-economic barriers to converting biomass to biojet fuel such as production costs, technological risks, and high costs for the development of new conversion technologies [18, 79].

The minimum sale prices of biojet fuel will depend on its production cost. At the same time, the cost of biojet fuel production (CBJP) depends on three factors, namely:

1. Costs at the biomass production plant or farms (CBP)
2. Costs at the greenfield biojet fuel plant or biorefinery (CGF)
3. Costs of transporting biojet fuel from the biorefinery to the airport (CT)
4. Recovery of investment in the sale of co-products (RISC)

Therefore, the final cost of biojet fuel production can be written as Eq. (9.1):

$$\text{CBJP} = \text{CBP} + \text{CGF} + \text{CT} - \text{RISC} \tag{9.1}$$

Biomass production cost (BPC) is the factor that most influences the cost of biojet fuel production. Thus, biomass production is influenced by four important parameters:

1. Land availability. It is related to the number of hectares for crops.
2. Agroecological sustainability. It is related to the type of land that is abandoned or displacement for other crops.
3. Specific biomass. It is related to the cultivation of potential raw materials based on their future application.
4. Production. It is related to the maximum biomass yield that can be achieved (ton year^{-1}).

The costs of transporting biojet fuel from the biorefinery to the airport (CT) depend on the type of route chosen, that is, it can be primary or secondary.

(a) Primary roads. These routes are normally used to transport fossil jet fuel to the current airports, that is, they correspond to the already established road networks.
(b) Secondary roads. They are the routes that do not intersect with the primary roads, that is, those that will be implemented for the transport of biojet fuel to the current and new airports. In addition, these routes will also be used to transport fossil jet fuel to the new airports. This is a perspective due to the increase in aviation activity.

Table 9.3 shows the values for primary and secondary roads. In this study, the cost of transporting biojet fuel through secondary roads will be considered.

Brazil as a strategic player presents advantages and disadvantages for biojet fuel production. Among the advantages we have:

(a) Experience in biofuels (ethanol and biodiesel), that is, the infrastructure could be refitted for future production of biojet fuel.
(b) Favorable climatic conditions for biomass production.

On the other hand, biojet fuels generated from different conversion pathways should have physical-chemical characteristics similar to conventional jet fuel. Table 9.4 shows the physical-chemistry characteristics of the biojet fuel produced from the potential feedstock proposed in this study. The biorefinery concept is based on a desirable model of integration of different conversion routes to produce a variety of products and take advantage of all the fractions present in the biomass, which would lead to the reduction of biojet fuel production costs due to the added value of side products [18], either by sale or implementation in the production chain.

The deductive method was used to show the content from the premises. Using the data collected in Sect. 3, seven main parameters were obtained for the

Table 9.3 Type of route and transportation costs of biojet fuel [80]

Road Type	Biojet fuel transportation cost (US$/t km)
Primary roads[a]	0.054
Secondary roads[b]	0.220

[a]Paved road on a regional or national scale
[b]Roads in poor conditions and segments with gravel road

Table 9.4 Main physical-chemical properties of biojet fuel obtained by different conversion routes

Items	Jet fuel	HEFA fuel	Farnesane	n—Hexanol	ATJ-SPK	ASTM D7566
Chemical formula	$C_{12}H_{23}$	$C_{18.68}H_{37.34}O_2$	$C_{15}H_{32}$	$C_6H_{13}-OH$	$C_{12.2}H_{26.4}$	–
MW[a] (kg kmol^{-1})	167.31	293.99	212.41	102.18	173.14	–
Density (kg m^{-3})	803.5	804	773.6	821.8	757	775–840
KV[b] (mm^2 s^{-1})	8	3.66	3.10	3.32	4.9	≤8 (@ −20 °C)
LHV (MJ kg^{-1})	43.28	43.4	44.06	39.10	44	≥42.8
Freezing point (°C)	−47	−39	< − 78	−46.7	< − 75	≤ − 47
Flash point (°C)	40	45	107.5	59	49	≥38
AT[c] (°C)	210	–	197	285	–	–
Boiling point (°C)	150–170	172–243	246.5	157	263	≤300
Vapor pressure (kPa)	14–21	–	0.0893	0.1333	–	–
% O_2 by mass	–	10.86	–	15.7	–	–
MAC[d] (%)	–	50	30	10	30	–
References	[15, 62, 81]	[62, 82, 83]	[81, 84]	[85, 86]	[87–89]	[83, 90, 91]

[a]Molecular Weight
[b]Kinematic Viscosity (@ −20 °C)
[c]Autoignition temperature
[d]Maximum Allowable Concentration in Fossil Jet Fuel

techno-economic evaluation of biojet fuel production. The summary of the results is shown in Table 9.5. In addition, Table 9.6 shows the assumed costs for the production of the biomass considered in this study. Thus, all this information was used to calculate the number of hectares necessary for raw material cultivation with which the objective of complete replacement can be achieved. The results are shown in Figs. 9.7 and 9.8. The main form of calculation was through conversion factors.

The techno-economic analysis revealed that biojet fuel production, through a single conversion route, is more attractive in the case of HEFA and FT processes, which present more competitive sales prices against fossil jet fuel compared to the other conversion routes as shown in Table 9.5. However, the range of sales prices found for the HEFA and FT routes corresponds to replacement percentages of 1.41% and 0.8%, respectively (see Table 9.8), of the 7500 Ml required to meet the demand for biojet fuel in Brazil. In addition, a considerable number of hectares are necessary for the production of low percentages (see Figs. 9.7 and 9.8), which

Table 9.5 Considered parameters for the techno-economic evaluation of biojet fuel production

Process	Raw material	Feedstock	Amount (Mt. year^{-1})	Biojet fuel (Mt. year^{-1})	C.E[a] (%)	Cost range (US\$. l^{-1}) [Refs.]	In this study (US\$. l^{-1})
HEFA	Jatropha fruit	Jatropha oil	0.0616	0.0344	55.84	1.52–1.78 [43, 48]	7.64
DSHC	Sugarcane	Bagasse	0.3197	0.0387	12.1	2.78–5.64 [48, 60, 61]	6.54
SF	Straw residues	Straw	1.00[b]	0.0448[b]	4.48	1.53–2.05 [48, 54]	5.30
F—T	Sugarcane Straw and Eucalyptus	Bagasse Straw and Eucalyptus	1.86[c]	0.0824	13.30	1.44–1.97 [48]	2.35
ATJ	Sugarcane	Isobutanol[d]	0.0207	0.0100	48.31	2.04–3.04 [48, 49]	10.67
Integrated system	Sugarcane Straw and Eucalyptus Jatropha fruit	Bagasse Straw and Eucalyptus Jatropha oil	0.2556	0.04827	18.88	–	1.09

HEFA Hydroprocessed esters and fatty acids (Jatropha oil), *DSHC* direct sugar hydrocarbons (farnesane), *SF* syngas fermentation (n-hexanol), *F—T* Fischer Tropsch, *ATJ* Alcohol To Jet

[a]CE: Conversion Efficiency

[b]If the 304.5 Mt of straw available (see Fig. 9.4) were used we would have 13.64 Mt of biojet fuel for every 304.5 Mt of straw. On the other hand, the produced hexanol is based on the literature (see Sect. 3.4)

[c]It is the total sum of three feedstock, i.e., Bagasse =1.09 Mtons year^{-1}; Eucalyptus wood =0.59 Mtons year^{-1}; Straw=0.18 Mtons year^{-1}

[d]According to the data in Sect. 3.2, it is possible to produce 122.63 Mt of biojet fuel for every 253.76 Mt of isobutanol

Table 9.6 Biomass production costs

Raw materials	US\$ ton^{-1}	Biomass (Mtons)	Total (MUS\$)	Potential regions[a]
Sugarcane	99.80	4.00	399.2	Southeast
Straw	73.12	1.18[b]	86.28	Southeast
Jatropha fruit	278.0	0.0616	17.12	Centre west, Northeast Southeast
Eucalyptus	57.10	0.59	33.69	South—Northeast
Bagasse	73.12	0.3197	23.38	Southeast
References	Alves et al. [18]	Table 9.6	Calculated	Cervi et al. [80]

[a]About Brazil

[b]This value indicates the sum of the straw used in the SF process accounts for 1.00 Mtons and the straw used in the F-T process accounts for 0.18 Mtons as shown in Table 9.5

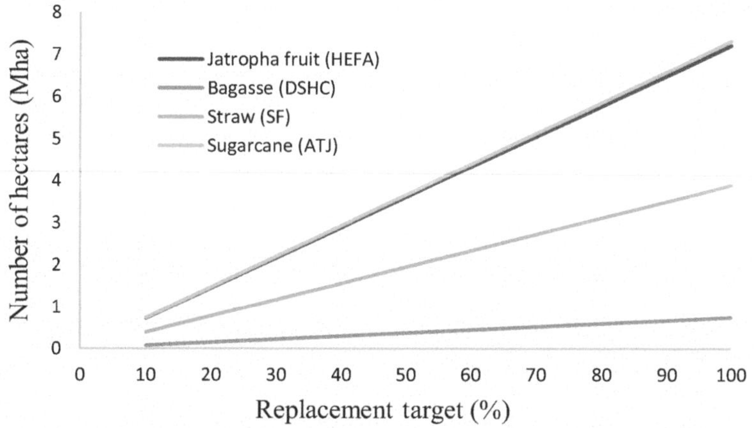

Fig. 9.7 Estimation of the number of hectares to be used for each raw material to meet a given replacement percentage

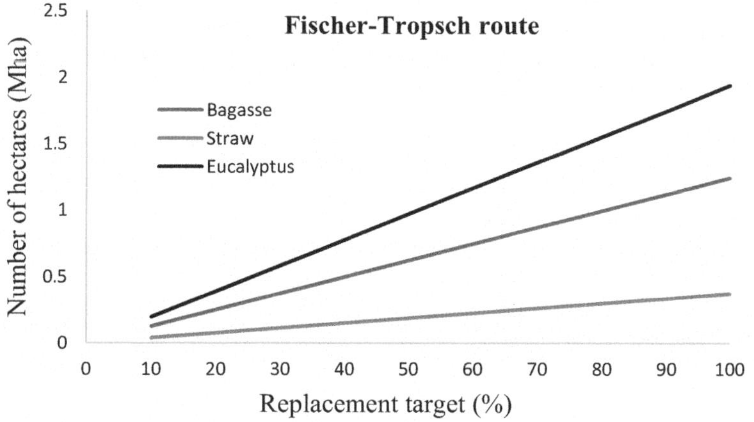

Fig. 9.8 Estimation of the number of hectares to be used by the FT route to meet a given replacement percentage

would result in an unviable application. This is because excessive use of land for biofuel production could originate a nonequilibrium in the food market, especially when the land is used for the cultivation of no-edible biomass as would be the case of jatropha fruit.

On the other hand, the FT route uses a smaller number of hectares for biojet fuel production compared to the other conversion routes, as shown in Fig. 9.8. However, this is because the replacement percentage for the FT route is lower than the HEFA route and, besides, the FT involves three different feedstocks, two of which (bagasse and straw) are co-products generated in sugarcane crop management. These residues positively impact the sales prices of biojet fuel, since bagasse and straw do not need special treatment for recuperation, but a logistic system is well established.

In this study, the sale price of biojet fuel produced by each route was also estimated. Table 9.5 shows that the values calculated are higher than the price range found in the literature, but it happened because the calculations did not consider the co-products generated in a specific conversion route, which would considerably reduce the sale price of biojet fuel.

In addition, it is observed that the higher the sale price, the greater the effects of co-products on the sales price of biojet fuel, especially in the case of raw materials with high acquisition costs. For example, the HEFA route generates co-products such as liquid and gaseous biofuels and biochemical in biojet fuel production, while the ATJ route, in addition to these, also generates electricity. Higher conversion efficiency (CE) indicates lower complexity in the technology for feedstock conversion, leading to higher biojet fuel production. On the other hand, lower conversion efficiency (CE) indicates higher technology complexity for feedstock conversion, leading to lower biojet fuel production and higher generation of co-products.

Now, we will choose certain concentrations to be produced, in an integrated system, to estimate the potential production of biojet fuel in Brazil and the number of hectares needed for the cultivation of these raw materials. In accordance with the substitution limits established by ASTM D7566, the replacement concentrations assumed to be blended in the jet fuel were as follows:

- HEFA: 30% $Vol_{\text{fossil jet fuel}}$
- DSHC: 10% $Vol_{\text{fossil jet fuel}}$
- SF: 10% $Vol_{\text{fossil jet fuel}}$
- FT: 30% $Vol_{\text{fossil jet fuel}}$
- ATJ: 20% $Vol_{\text{fossil jet fuel}}$

The biojet fuel production chosen was 7.5 billion l per year as it was the highest demand for fossil jet fuel in Brazil in 2014 (see Fig. 9.2).

The scheme of an integrated process is shown in Fig. 9.9. The values indicated (in blue) in each process are the conversion factors obtained from the literature (see Sect. 3) (Table 9.7).

Figure 9.10 shows that biojet fuel production in an integrated system required fewer hectares compared to individual conversion routes to achieve full replacement in the aviation fuel market. Table 9.5 shows the sales price of biojet fuel produced by a single integrated system, which was estimated at 4.37 US$ l^{-1}. Apparently, this process would not be competitive compared to other individual conversion routes. However, the replacement percentage of biojet fuel for a potential Brazilian demand is much lower compared to the integrated system, which would achieve a total replacement in the aviation fuel market as shown in Table 9.8.

On the other hand, each conversion route would require a considerable number of biorefineries to satisfy the demand for biojet fuel, meanwhile, an integrated system would require four biorefineries to generate more attractive prices (1.09 US$ l^{-1}) compared to other conversion routes (see Table 9.8) and a total of eight biorefineries to generate more competitive prices (0.55 US$ l^{-1}) against fossil jet fuel in the Brazilian market, as shown in Table 9.8.

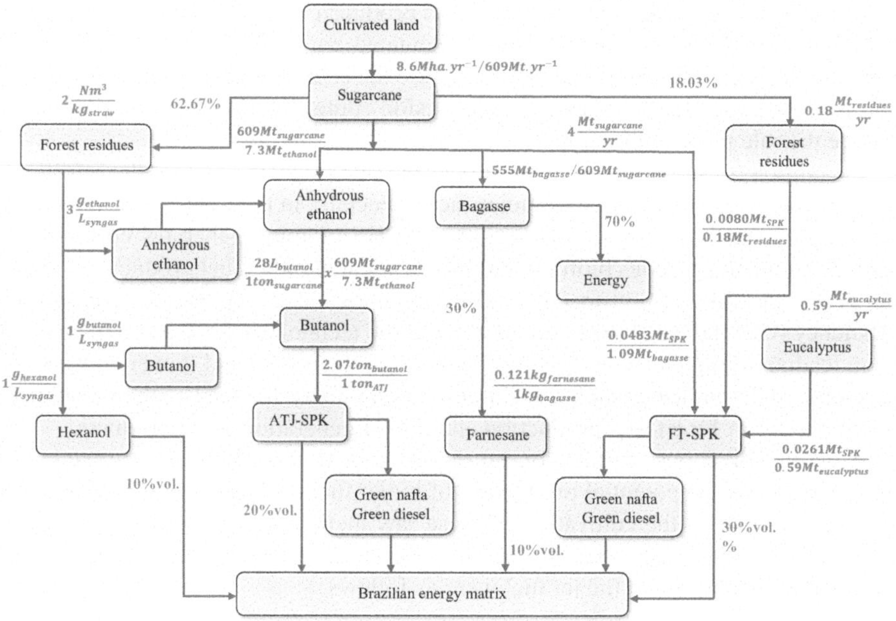

Fig. 9.9 Scheme of an integrated system for biojet fuel production from sugarcane as the main raw material

Table 9.7 Variation in the price of jet fuel in different years

Year	Sale price (US$ GJ⁻¹)	Sale price (US$ l⁻¹)	References
2015	23–33	0.80–1.15	[78]
2016	50–77	1.74–2.68	[92]
2017	39–81	1.36–2.82	[93]
2018	9–32	0.31–1.11	[48]
2019	20–97	0.70–3.37	[80]
2017	32–42	1.12–1.47[a]	[94]

[a]It indicates sales prices at Brazilian airports in 2017

In addition, if a surplus of jatropha oil is sold as a feedstock for the biodiesel market, the sale price of biojet fuel will be positively impacted since the acquisition price of the raw material (which represents around 50–70% of the production cost of biojet fuel) will also be reduced. It is evidenced in Table 9.8, where an integrated system, which has sugarcane as its only raw material, meets up to 70% of the replacement target, generating a sales price of 2.08 US$ l^{-1}, that is, jatropha oil is responsible for 52.4% of the biojet fuel sales price (52.4% of 4.37 US$ l^{-1}). This is because the production cost of jatropha fruit is much higher than that of other biomasses as shown in Table 9.6.

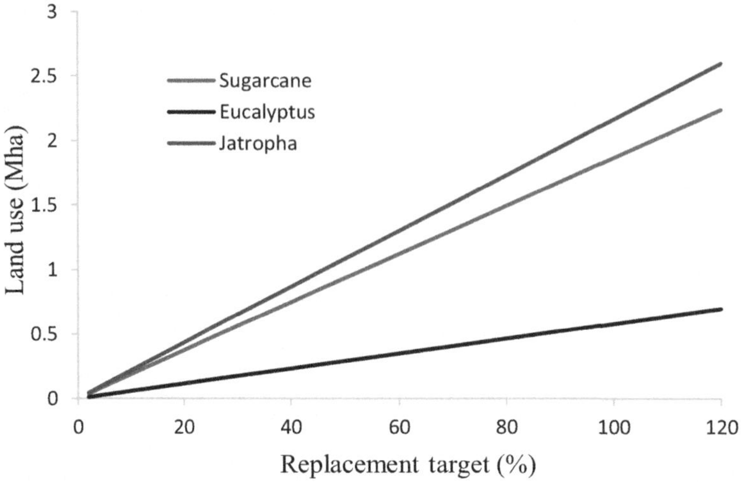

Fig. 9.10 Estimation of land use for biojet fuel production based on the replacement target

Table 9.8 Biojet fuel replacement percentages for each conversion route

Process	Feedstock	Biojet fuel (Ml year^{-1})	References	Replacement target (%) [a]	Number of biorefineries[b]
HEFA	Jatropha oil	105.90	[43]	1.41	71
ATJ	Butanol	70.59	[48]	0.94	106
DSHC	Bagasse	103.41	[61]	1.38	73
SF	Straw	74.06	[18]	0.99	101
FT	Bagasse, Straw and Eucalyptus	63.10	[48]	0.80	119
Integrated System	Butanol, Bagasse and straw	5250	–	70	2[a]
Integrated System	Butanol, Bagasse, Jatropha oil, and straw	7500	–	100	4[b]

[a]To obtain an approximate price of 1.09 US$ l^{-1} (that is, 1.04 US$ l^{-1}), the biojet fuel sales price without jatropha oil (2.08 US$ l^{-1}) is divided by two biorefineries
[b]The sales price of 1.09 US$ l^{-1} (see Table 9.5) is obtained by dividing 4.37 US$ l^{-1} (sales prices of each biorefinery) between 4 biorefineries
[c]The percentage is calculated based on the 7500 $Ml_{jet\ fuel}$ produced in Brazil in 2014 (see Fig. 9.2)
[d]The values were calculated by dividing. 7500 $Ml_{jet\ fuel}$ between $Ml_{bioject\ fuel}^{-1}$ of each route

On the other hand, Fig. 9.11 shows that the construction of biorefineries has a greater impact on the sales price of biojet fuel than the sales price of surplus jatropha oil in the HEFA process. Therefore, the use of jatropha fruit as the sole raw material for the total production of biojet fuel is unlikely, however, due to its large growing areas and growing conditions in Brazil, jatropha fruit becomes an interesting raw material for future small and medium-scale production of biojet fuel to supply some cities and states.

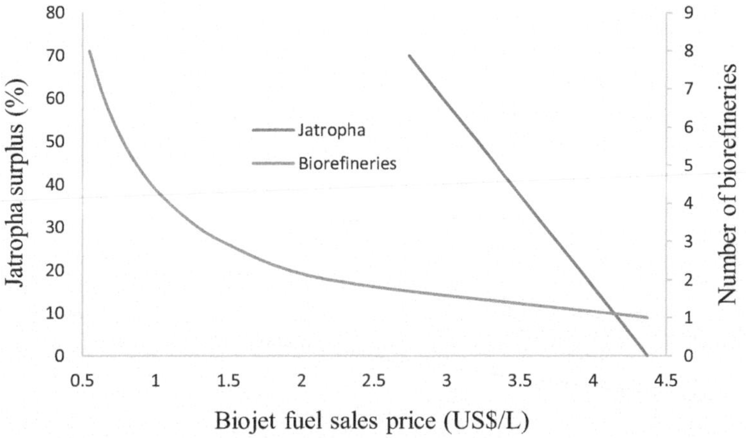

Fig. 9.11 Impact of the jatropha fruit surplus and the number of biorefineries on the sales price of biojet fuel

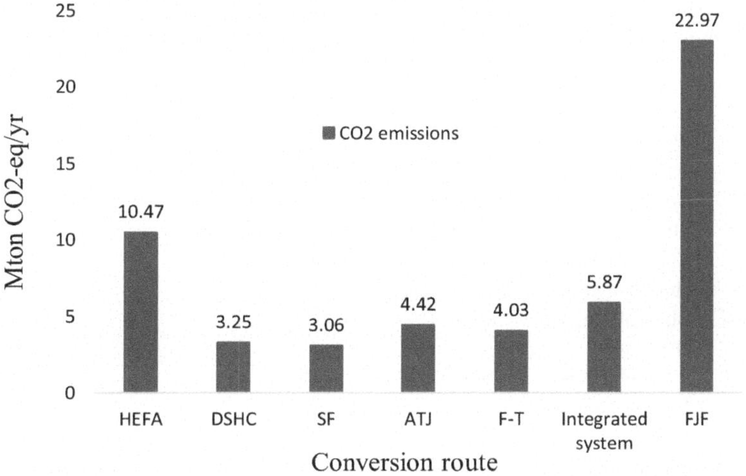

Fig. 9.12 Net CO_{2-eq} emissions involved in each conversion route for biojet fuel production (Mton year^{-1})

Figure 9.12 shows the net CO_2 emissions generated in each conversion route and the integrated system. The smallest contribution to net CO_2 emissions is given by the SF route; however, this route does not produce a biojet fuel itself, but rather an additive, i.e. hexanol, whose concentration in the biojet fuel mixture is limited to up to 10%vol.

On the other hand, the integrated system generates net CO_2 emissions much lower than those of fossil jet fuel even though the system unites 4 different technologies. It appears that multiconversion does not generate a negative impact on the

environment, considering the large volume of biojet fuel that will be produced to meet demand (7500 Ml $year^{-1}$).

6 Final Remarks

In this article, the potential raw materials for biojet fuel integrated production were evaluated. In addition, a techno-economic assessment was carried out to determine the sales price of biojet fuel produced and the number of hectares necessary for biomass cultivation to satisfy the demand of the Brazilian market.

Currently, the production of biojet fuel through a specific route depends on the availability of raw materials in a given region. However, its production capacity is limited and it requires the construction of a considerable number of biorefineries to generate competitive sales prices against fossil jet fuel (see Table 9.8). Therefore, the type of raw material used will determine the number of biorefineries to be built to achieve a given replacement target.

The integrated system has shown to be a promising alternative for biojet fuel production with the capacity to satisfy the demand of a continental-sized country like Brazil. In addition, the integration of different technologies leads to the full use of a certain raw material such as sugarcane. It also entails less sacrifice of land for the cultivation of biomass for biojet fuel production, which is beneficial to avoid competition with the food market.

The sales price of biojet fuel obtained in an integrated system turns out to be more attractive than the individual production of biojet fuel and more competitive compared to fossil jet fuel if an adequate number of biorefineries are built. Furthermore, integrated production generates a large amount of co-products, which can offset the initial investment resulting in a considerable reduction in the sales price of biojet fuel. Thus, the conversion efficiency of feedstock into biojet fuel depends on the complexity of the process, while the sales price of biojet fuel depends on the use of the co-products generated in the conversion route.

This study demonstrated that eight biorefineries are necessary to achieve the production of 7500 Ml $year^{-1}$ with a competitive selling price (0.55 US\$ $year^{-1}$) against fossil jet fuel. In addition, four biorefineries would be necessary to achieve the same production capacity (7500 Ml $year^{-1}$) with attractive selling price (1.09 US\$ $year^{-1}$) compared to individual conversion routes, which would have a production capacity lower than the demand, as shown in Table 9.8.

The environmental assessment demonstrated that despite the integration of four different technologies, the system emits a lower net amount of CO_2 than fossil jet fuel, indicating that an integrated system would be an alternative pathway toward environmental sustainability. However, the use of the integrating system leads to less land use for biomass cultivation, which is essential for the absorption of CO_2 from the environment. Therefore, a more detailed analysis is necessary to guarantee the balance between the CO_2 absorbed by the biomass and the CO_2 generated in the multiconversion.

Finally, the higher the replacement target for biojet fuel, the greater the number of conversion technologies that will be integrated to satisfy a given demand. However, the routes chosen must be appropriate, in the sense of using not only the final product, but also co-products of considerable economic value, reinforcing the concept of circular economy.

Acknowledgements The authors are grateful to Foundation for the Development of UNESP (FUNDUNESP), project number n[0] 3206/2021—CCP

References

1. Pavlenko, N., & Kharina, A. (2018). Policy and environmental implications of using HEFA+ for aviation. *International Council on Clean Transportation, 9*, 1–9.
2. Doliente, S. S., Narayan, A., Tapia, J. F. D., Samsatli, N. J., Zhao, Y., & Samsatli, S. (2020). Bio-aviation fuel: A comprehensive review and analysis of the supply chain components. *Frontiers in Energy Research, 8*, 1–38. https://doi.org/10.3389/fenrg.2020.00110
3. Gegg, P., Budd, L., & Ison, S. (2014). The market development of aviation biofuel: Drivers and constraints. *Journal of Air Transport Management, 39*, 34–40. https://doi.org/10.1016/j.jairtraman.2014.03.003
4. Calçado, D., Fleury, G., Henrique, T., Oliveira, J., Côrtes, R., Dupont, R., et al. (2019). *Brazil's action plan on CO2 emissions reduction from aviation*. Brasilia.
5. Ramos, C. L. (2022). Airline industry strives to eliminate its carbon footprint. *Revista Pesquisa Fapesp, 317*, 1–5. https://revistapesquisa.fapesp.br/en/airline-industry-strives-to-eliminate-its-carbon-footprint/. Accessed March 15, 2023.
6. RSB & A. (2021). *Feedstock availability for sustainable aviation fuel in Brazil: Challenges and opportunities*. Switzerland.
7. ICAO EP. *Brazilian Biojetfuel Platform (BBP). Back to Initiat Proj List 2012*, pp. 1–2. https://www.icao.int/environmental-protection/GFAAF/Pages/Project.aspx?ProjectID=31. Accessed April 1, 2023.
8. ANAC-Agência Nacional de Aviação Civil. *Aeronaves 2019*, p. 2. https://www.anac.gov.br/assuntos/dados-e-estatisticas/aeronaves
9. do Petróleo, A. N., & Tayane, F. (2021). Anuário estatístico brasileiro do petróleo, gás natural e biocumbustíveis: 2014. *Anp/Mme, 33*, 1–12.
10. Mahmudul, H. M., Hagos, F. Y., Mamat, R., Adam, A. A., Ishak, W. F. W., & Alenezi, R. (2017). Production, characterization and performance of biodiesel as an alternative fuel in diesel engines – A review. *Renewable and Sustainable Energy Reviews, 72*, 497–509. https://doi.org/10.1016/j.rser.2017.01.001
11. Valencia, M. J., & Cardona, C. A. (2014). The Colombian biofuel supply chains: The assessment of current and promising scenarios based on environmental goals. *Energy Policy, 67*, 232–242. https://doi.org/10.1016/j.enpol.2013.12.021
12. Escalante, E. S. R., Ramos, L. S., Rodriguez Coronado, C. J., & de Carvalho Júnior, J. A. (2022). Evaluation of the potential feedstock for biojet fuel production: Focus in the Brazilian context. *Renewable and Sustainable Energy Reviews, 153*, 111716. https://doi.org/10.1016/j.rser.2021.111716
13. Cremonez, A. P., Feroldi, M., Araújo, A. V., Negreiros Borges, M., Thompson, W. M., Feiden, A., et al. (2015). Biofuels in a Braziliam aviation: Current scenario and prospects. *Renewable and Sustainable Energy Reviews, 43*, 1063–1072.

14. Patra, J., Ghose, P., Datta, A., Das, M., Ganguly, R., Sen, S., et al. (2015). Studies of combustion characteristics of kerosene ethanol blends in an axi-symmetric combustor. *Fuel, 144*, 205–213. https://doi.org/10.1016/j.fuel.2014.12.036
15. Yilmaz, N., & Atmanli, A. (2017). Sustainable alternative fuels in aviation. *Energy, 140*, 1378–1386. https://doi.org/10.1016/j.energy.2017.07.077
16. Chuck, C. J., & Donnelly, J. (2014). The compatibility of potential bioderived fuels with Jet A-1 aviation kerosene. *Applied Energy, 118*, 83–91. https://doi.org/10.1016/j.apenergy.2013.12.019
17. Chen, K., Liu, H., & Xia, Z. (2013). The impacts of aromatic contents in aviation jet fuel on the volume swell of the aircraft fuel tank sealants. *SAE International Journal of Aerospace, 6*, 350–354. https://doi.org/10.4271/2013-01-9001
18. Alves, C. M., Valk, M., de Jong, S., Bonomi, A., van der Wielen, L. A. M., & Mussato, S. I. (2016). Techno-economic assessment of biorefi nery technologies for aviation biofuels supply chains in Brazil. *Biofuels, Bioproducts and Biorefining, 11*, 67–91. https://doi.org/10.1002/bbb.1711
19. Milanez, A. Y., Mancuso, R. V., Godinho, R. D., & Poppe, M. K. (2017). O Acordo de Paris e a transição para o setor de transportes de baixo carbono: o papel da Plataforma para o Biofuturo. *BNDES Setorial, 45*, 285–340.
20. de Souza, L. M., Mendes, P., & Aranda, D. (2018). Assessing the current scenario of the Brazilian biojet market. *Renewable and Sustainable Energy Reviews, 98*, 426–438. https://doi.org/10.1016/j.rser.2018.09.039
21. Lins, M. E., Oliveira, L. B., Da Silva, A. C. M., Rosa, L. P., & Pereira, A. O. (2012). Performance assessment of alternative energy resources in Brazilian power sector using data envelopment analysis. *Renewable and Sustainable Energy Reviews, 16*, 898–903. https://doi.org/10.1016/j.rser.2011.09.010
22. Wei, H., Liu, W., Chen, X., Yang, Q., Li, J., & Chen, H. (2019). Renewable bio-jet fuel production for aviation: A review. *Fuel, 254*. https://doi.org/10.1016/j.fuel.2019.06.007
23. Cheng, X., Chen, L., & Yan, F. (2014). Study of the characteristic of diesel spray combustion and soot formation using laser-induced incandescence (LII). *Journal of the Energy Institute, 87*, 383–392. https://doi.org/10.1016/j.joei.2013.07.001
24. Yun, H., Choi, K., & Lee, C. S. (2016). Effects of biobutanol and biobutanol-diesel blends on combustion and emission characteristics in a passenger car diesel engine with pilot injection strategies. *Energy Conversion and Management, 111*, 79–88. https://doi.org/10.1016/j.enconman.2015.12.017
25. Mendez, C. J., Parthasarathy, R. N., & Gollahalli, S. R. (2011). Performance and emission characteristics of a small scale gas turbine engine fueled with propanol/jet a blends. In *ASME 2011 international mechanical engineering congress and exposition IMECE 2011* (Vol. 4, pp. 1359–1370). ASME.
26. Mendez, C. J., Parthasarathy, R. N., & Gollahalli, S. R. (2014). Performance and emission characteristics of butanol/Jet A blends in a gas turbine engine. *Applied Energy, 118*, 135–140. https://doi.org/10.1016/j.apenergy.2013.12.011
27. Moliere, M. (2005). Expanding fuel flexibility of gas turbines. *Proceedings of the Institution of Mechanical Engineers, Part A: Journal of Power and Energy, 219*, 109–119. https://doi.org/10.1243/095765005X6818
28. Marsh, G. (2008). Biofuels: Aviation alternative? *Renewable Energy Focus, 9*, 48–51. https://doi.org/10.1016/S1471-0846(08)70138-0
29. Cantarella, H., Nassar, A. M., Cortez, L. A. B., & Baldassin, R. (2015). Potential feedstock for renewable aviation fuel in Brazil. *Environment and Development, 15*, 52–63. https://doi.org/10.1016/j.envdev.2015.05.004
30. EPE-Empresa de Pesquisa Energetica. (2019). Análise de Conjuntura dos Biocombustíveis Ano 2018. Minist Minas e Energ, p. 75. EPE-DPG-SDB-Bios-NT-01-2019-r0. https://www.epe.gov.br/sites-pt/publicacoes-dados-abertos/publicacoes/PublicacoesArquivos/publicacao-402/An%C3%A1lise_de_Conjuntura_Ano%202018.pdf. Accessed April 22, 2020.

31. EPE. (2021). Análise de Conjuntura dos Biocombustíveis – Ano 2020. EPE – Empres Pesqui Energética, p. 87.
32. CNA – Confederação da Agricultura e Pecuária do Brasil. (2018). O futuro é agro: 2018–2030, pp. 1–139. https://www.cnabrasil.org.br/assets/arquivos/plano_de_estado_completo_21x28cm_web.pdf. Accessed September 15, 2019.
33. Cornelissen, S., Koper, M., & Deng, Y. Y. (2012). The role of bioenergy in a fully sustainable global energy system. *Biomass and Bioenergy, 41*, 21–33. https://doi.org/10.1016/j.biombioe.2011.12.049
34. Prieler, S., Günther, F., & van Velthuizen, H. (2013). Land and the food – Fuel competition: Insights from modeling. *WIREs Energy and Environment, 2*, 199–217. https://doi.org/10.1002/wene.55
35. Horta Nogueira, L. A., & Silva, C. R. (2013). Biofuels in Brazil: Evolution, achievements and perspectives on food security. *Global Food Security, 2*, 117–125. https://doi.org/10.1016/j.gfs.2013.04.001
36. Mendes, L., Treichel, M., & Beling, R. R. (2016). *Anuario Brasileiro de Silvicultura (Brazilian forestry and timber yearbook)*. Gráfica Coan.
37. Bergmann, J. C., Tupinambá, D. D., Costa, O. Y. A., Almeida, J. R. M., Barreto, C. C., & Quirino, B. F. (2013). Biodiesel production in Brazil and alternative biomass feedstocks. *Renewable and Sustainable Energy Reviews, 21*, 411–420. https://doi.org/10.1016/j.rser.2012.12.058
38. Macedo, I. C. (2007). The current situation and prospects for ethanol. *Estudos Avançados, 21*, 157–165.
39. Seber, G., Escobar, N., Valin, H., & Malina, R. (2022). Uncertainty in life cycle greenhouse gas emissions of sustainable aviation fuels from vegetable oils. *Renewable and Sustainable Energy Reviews, 170*. https://doi.org/10.1016/j.rser.2022.112945
40. Okolie, J. A., Awotoye, D., Tabat, M. E., Okoye, P. U., Epelle, E. I., Ogbaga, C. C., et al. (2023). Multi-criteria decision analysis for the evaluation and screening of sustainable aviation fuel production pathways. *IScience, 26*. https://doi.org/10.1016/j.isci.2023.106944
41. Capaz, R. S., Guida, E., Seabra, J. E. A., Osseweijer, P., & Posada, J. A. (2021). Mitigating carbon emissions through sustainable aviation fuels: Costs and potential. *Biofuels, Bioproducts and Biorefining, 15*, 502–524. https://doi.org/10.1002/bbb.2168
42. Shahriar, M. F., & Khanal, A. (2022). The current techno-economic, environmental, policy status and perspectives of sustainable aviation fuel (SAF). *Fuel, 325*, 124905. https://doi.org/10.1016/j.fuel.2022.124905
43. Wang, W. C. (2016). Techno-economic analysis of a bio-refinery process for producing hydro-processed renewable jet fuel from jatropha. *Renewable Energy, 95*, 63–73. https://doi.org/10.1016/j.renene.2016.03.107
44. Kaltschmitt, M., & Neuling, U. (2017). *Biokerosene: Status and prospects*. https://doi.org/10.1007/978-3-662-53065-8
45. Yao, G., Staples, M. D., Malina, R., & Tyner, W. E. (2017). Stochastic techno-economic analysis of alcohol-to-jet fuel production. *Biotechnology for Biofuels, 10*(18), 1–13. https://doi.org/10.1186/s13068-017-0702-7
46. Zhang, M., & Yu, Y. (2013). Dehydration of ethanol to ethylene. *Industrial and Engineering Chemistry Research, 52*, 9505–9514.
47. Richter, S., Kukkadapu, G., Westbrook, C. K., Braun-Unkhoff, M., Naumann, C., Köhler, M., et al. (2022). A combined experimental and modeling study of combustion properties of an isoparaffinic alcohol-to-jet fuel. *Combustion and Flame, 240*. https://doi.org/10.1016/j.combustflame.2022.111994
48. Klein, B. C., Chagas, M. F., Junqueira, T. L., Rezende, M. C. A. F., de Fátima Cardoso, T., Cavalett, O., et al. (2018). Techno-economic and environmental assessment of renewable jet fuel production in integrated Brazilian sugarcane biorefineries. *Applied Energy, 209*, 290–305. https://doi.org/10.1016/j.apenergy.2017.10.079

49. Pereira, L. G., Dias, M. O. S., Junqueira, T. L., Pavanello, L. G., Chagas, M. F., Cavalett, O., et al. (2014). Butanol production in a sugarcane biorefinery using ethanol as feedstock. Part II: Integration to a second generation sugarcane distillery. *Chemical Engineering Research and Design, 92*, 1452–1462. https://doi.org/10.1016/j.cherd.2014.04.032
50. dos Santos, R. G., & Alencar, A. C. (2020). Biomass-derived syngas production via gasification process and its catalytic conversion into fuels by Fischer Tropsch synthesis: A review. *International Journal of Hydrogen Energy, 45*, 18114–18132. https://doi.org/10.1016/j.ijhydene.2019.07.133
51. de Klerk, A. (2016). *Aviation turbine fuels through the Fischer–Tropsch process*. Elsevier Inc. https://doi.org/10.1016/B978-0-12-804568-8.00010-X
52. Rosso, S., de Muniz, G. I. B., de Matos, J. L. M., Haselein, C. R., Hein, P. R. G., & de Carvalho Lopes, M. (2013). Estimate of the density of Eucalyptus grandis W. Hill ex Maiden using near infrared spectroscopy. *Cerne, 19*, 647–652. https://doi.org/10.1590/s0104-77602013000400015
53. Lee, S. H., Yun, E. J., Kim, J., Lee, S. J., Um, Y., & Kim, K. H. (2016). Biomass, strain engineering, and fermentation processes for butanol production by solventogenic clostridia. *Applied Microbiology and Biotechnology, 100*, 8255–8271. https://doi.org/10.1007/s00253-016-7760-9
54. Phillips, J. R., Atiyeh, H. K., Tanner, R. S., Torres, J. R., Saxena, J., Wilkins, M. R., et al. (2015). Butanol and hexanol production in Clostridium carboxidivorans syngas fermentation: Medium development and culture techniques. *Bioresource Technology, 190*, 114–121. https://doi.org/10.1016/j.biortech.2015.04.043
55. Lora, E. E. S., Andrade, R. V., Ángel, J. D. M., Leite, M. A. H., Rocha, M. H., de Sales, C. A. V. B., et al. (2012). Gaseificação e pirólise para conversão da biomassa em eletricidade e biocombustíveis. *Biocombustíveis, 1*, 411–498.
56. Oh, H. J., Ko, J. K., Gong, G., Lee, S. M., & Um, Y. (2022). Production of hexanol as the main product through syngas fermentation by Clostridium carboxidivorans P7. *Frontiers in Bioengineering and Biotechnology, 10*, 1–11. https://doi.org/10.3389/fbioe.2022.850370
57. Sun, X., Atiyeh, H. K., Huhnke, R. L., & Tanner, R. S. (2019). Syngas fermentation process development for production of biofuels and chemicals: A review. *Bioresource Technology Reports, 7*, 100279. https://doi.org/10.1016/j.biteb.2019.100279
58. Kick, T., Herbst, J., Kathrotia, T., Marquetand, J., Braun-Unkhoff, M., Naumann, C., et al. (2012). An experimental and modeling study of burning velocities of possible future synthetic jet fuels. *Energy, 43*, 111–123. https://doi.org/10.1016/j.energy.2012.01.035
59. Moraes, M. A. F. D., Nassar, A. M., Moura, P., Leal, R. L. V., & Cortez, L. A. B. (2014). Jet biofuels in Brazil: Sustainability challenges. *Renewable and Sustainable Energy Reviews, 40*, 716–726. https://doi.org/10.1016/j.rser.2014.07.210
60. Oßwald, P., Whitside, R., Schäffer, J., & Köhler, M. (2017). An experimental flow reactor study of the combustion kinetics of terpenoid jet fuel compounds: Farnesane, p-menthane and p-cymene. *Fuel, 187*, 43–50. https://doi.org/10.1016/j.fuel.2016.09.035
61. Michailos, S. (2018). Process design, economic evaluation and life cycle assessment of jet fuel production from sugar cane residue. *Environmental Progress & Sustainable Energy, 37*, 1227–1235. https://doi.org/10.1002/ep.12840
62. Gutiérrez-Antonio, C., Gómez-Castro, F. I., de Lira-Flores, J. A., & Hernández, S. (2017). A review on the production processes of renewable jet fuel. *Renewable and Sustainable Energy Reviews, 79*, 709–729. https://doi.org/10.1016/j.rser.2017.05.108
63. Blakey, S., Rye, L., & Wilson, C. W. (2011). Aviation gas turbine alternative fuels: A review. *Proceedings of the Combustion Institute, 33*, 2863–2885. https://doi.org/10.1016/j.proci.2010.09.011
64. Bauen, A., Howes, J., Bertuccioli, L. & Chudziak, C., 2009. Review of the potential for biofuels in aviation. Final Report Prepared for the Committee on, Climate Change. E4tech, Switzerland. https://doi.org/10.1007/978-981-99-8783-2_1.

65. Roos, A., & Ahlgren, S. (2018). Consequential life cycle assessment of bioenergy systems – A literature review. *Journal of Cleaner Production, 189*, 358–373. https://doi.org/10.1016/j.jclepro.2018.03.233
66. Earles, J. M., & Halog, A. (2011). Consequential life cycle assessment: A review. *International Journal of Life Cycle Assessment, 16*, 445–453. https://doi.org/10.1007/s11367-011-0275-9
67. Plevin, R. J., O'Hare, M., Jones, A. D., Torn, M. S., & Gibbs, H. K. (2010). Greenhouse gas emissions from biofuels' indirect land use change are uncertain but may be much greater than previously estimated. *Environmental Science & Technology, 44*, 8015–8021. https://doi.org/10.1021/es101946t
68. Bailis, R. E., & Baka, J. E. (2010). Greenhouse gas emissions and land use change from Jatropha curcas -based jet fuel in Brazil. *Environmental Science & Technology, 44*, 8684–8691. https://doi.org/10.1021/es1019178
69. Staples, M. D., Malina, R., Olcay, H., Pearlson, M. N., Hileman, J. I., Boies, A., et al. (2014). Lifecycle greenhouse gas footprint and minimum selling price of renewable diesel and jet fuel from fermentation and advanced fermentation production technologies. *Energy & Environmental Science, 7*, 1545–1554. https://doi.org/10.1039/c3ee43655a
70. Staples, M. D., Olcay, H., Malina, R., Trivedi, P., Pearlson, M. N., Strzepek, K., et al. (2013). Water consumption footprint and land requirements of large-scale alternative diesel and jet fuel production. *Environmental Science & Technology, 47*, 12557–12565. https://doi.org/10.1021/es4030782
71. Cox, K., Renouf, M., Dargan, A., Turner, C., & Klein-Marcuschamer, D. (2014). Environmental life cycle assessment (LCA) of aviation biofuel from microalgae, Pongamia pinnata, and sugarcane molasses. *Biofuels, Bioproducts and Biorefining, 8*, 579–593. https://doi.org/10.1002/bbb.1488
72. Atsonios, K., Kougioumtzis, M. A., Panopoulos, K. D., & Kakaras, E. (2015). Alternative thermochemical routes for aviation biofuels via alcohols synthesis: Process modeling, techno-economic assessment and comparison. *Applied Energy, 138*, 346–366. https://doi.org/10.1016/j.apenergy.2014.10.056
73. Rios Escalante, E. S., Perrella Balestieri, J. A., & Carvalho, J. A. (2022). The organic Rankine cycle: A promising technology for electricity generation and thermal pollution mitigation. *Energy, 247*. https://doi.org/10.1016/j.energy.2022.123405
74. Davis, R., Tao, L., Tan, E. C. D., Biddy, M. J., Beckham, G. T., Scarlata, C., et al. (2013). Process design and economics for the conversion of lignocellulosic biomass to hydrocarbons: Dilute-acid and enzymatic deconstruction of biomass to sugars and biological conversion of sugars to hydrocarbons. *National Renewable Energy Laboratory*, 147. https://www.nrel.gov/docs/fy15osti/62498.pdf. Accessed May 08, 2022.
75. Tao, L., Milbrandt, A., Zhang, Y., & Wang, W. C. (2017). Techno-economic and resource analysis of hydroprocessed renewable jet fuel. *Biotechnology for Biofuels, 10*, 1–16. https://doi.org/10.1186/s13068-017-0945-3
76. ANAC-Agência Nacional de Aviação Civil. (2019). *Anuário do Transporte Aéreo 2018 (in portuguese)*. ANAC-Agência Nacional de Aviação Civil.
77. Centro de Gestão e Estudos Estratégicos. (2010). Biocombustíveis aeronáuticos: Progressos e desafios. *Série Documentos Técnicos, 8*(5), 6.
78. de Jong, S., van Stralen, J., Londo, M., Hoefnagels, R., Faaij, A., & Junginger, M. (2018). Renewable jet fuel supply scenarios in the European Union in 2021–2030 in the context of proposed biofuel policy and competing biomass demand. *GCB Bioenergy, 10*, 661–682. https://doi.org/10.1111/gcbb.12525
79. Mawhood, R., Gazis, E., de Jong, S., Hoefnagels, R., & Slade, R. (2016). Production pathways for renewable jet fuel: A review of commercialization status and future prospects. *Biofuels, Bioproducts and Biorefining, 10*, 462–484. https://doi.org/10.1002/bbb
80. Cervi, W. R., Lamparelli, R. A. C., Seabra, J. E. A., Junginger, M., de Jong, S., & van der Hilst, F. (2019). Spatial modeling of techno-economic potential of biojet fuel production in Brazil. *GCB Bioenergy, 12*, 136–157. https://doi.org/10.1111/gcbb.12659

81. Richter, S., Kathrotia, T., Naumann, C., Kick, T., Slavinskaya, N., Braun-Unkhoff, M., et al. (2018). Experimental and modeling study of farnesane. *Fuel, 215*, 22–29. https://doi.org/10.1016/j.fuel.2017.10.117
82. Chaudhary, A., Gupta, A., Kumar, S., & Kumar, R. (2019). Pool fires of jatropha biodiesel and their blends with petroleum diesel. *Experimental Thermal and Fluid Science, 101*, 175–185. https://doi.org/10.1016/j.expthermflusci.2018.10.021
83. Wang, W. C., & Tao, L. (2016). Bio-jet fuel conversion technologies. *Renewable and Sustainable Energy Reviews, 53*, 801–822. https://doi.org/10.1016/j.rser.2015.09.016
84. Millo, F., Bensaid, S., Fino, D., Marcano, S. J. C., Vlachos, T., & Debnath, B. K. (2014). Influence on the performance and emissions of an automotive Euro 5 diesel engine fueled with F30 from Farnesane. *Fuel, 138*, 134–142. https://doi.org/10.1016/j.fuel.2014.07.060
85. Rajesh Kumar, B., & Saravanan, S. (2016). Use of higher alcohol biofuels in diesel engines: A review. *Renewable and Sustainable Energy Reviews, 60*, 84–115. https://doi.org/10.1016/j.rser.2016.01.085
86. De Poures, M. V., Sathiyagnanam, A. P., Rana, D., Rajesh Kumar, B., & Saravanan, S. (2017). 1-Hexanol as a sustainable biofuel in DI diesel engines and its effect on combustion and emissions under the influence of injection timing and exhaust gas recirculation (EGR). *Applied Thermal Engineering, 113*, 1505–1513. https://doi.org/10.1016/j.applthermaleng.2016.11.164
87. Kang, D., Kim, D., Kalaskar, V., Violi, A., & Boehman, A. L. (2019). Experimental characterization of jet fuels under engine relevant conditions – Part 1: Effect of chemical composition on autoignition of conventional and alternative jet fuels. *Fuel, 239*, 1388–1404. https://doi.org/10.1016/j.fuel.2018.10.005
88. Brooks, K. P., Snowden-Swan, L. J., Jones, S. B., Butcher, M. G., Lee, G.-S. J., Anderson, D. M., et al. (2016). *Low-carbon aviation fuel through the alcohol to jet pathway*. Elsevier Inc. https://doi.org/10.1016/b978-0-12-804568-8.00006-8
89. Prak, D. J. L., Jones, M. H., Trulove, P., Mcdaniel, A. M., Dickerson, T., & Cowart, J. S. (2015). Physical and chemical analysis of alcohol-to-jet (ATJ) fuel and development of surrogate fuel mixtures. *Energy and Fuels, 29*, 37660–33769. https://doi.org/10.1021/acs.energyfuels.5b00668
90. Hong, T. D., Soerawidjaja, T. H., Reksowardojo, I. K., Fujita, O., Duniani, Z., & Pham, M. X. (2013). A study on developing aviation biofuel for the tropics: Production process – Experimental and theoretical evaluation of their blends with fossil kerosene. *Chemical Engineering and Processing Process Intensification, 74*, 124–130. https://doi.org/10.1016/j.cep.2013.09.013
91. Dorrington, G. E. (2016). Certification and performance. In C. J. Chuck (Ed.), *Biofuels for aviation* (pp. 35–44). Elsevier Inc. https://doi.org/10.1016/b978-0-12-804568-8.00003-2
92. Diederichs, G. W., Ali Mandegari, M., Farzad, S., & Görgens, J. F. (2016). Techno-economic comparison of biojet fuel production from lignocellulose, vegetable oil and sugar cane juice. *Bioresource Technology, 216*, 331–339. https://doi.org/10.1016/j.biortech.2016.05.090
93. Santos, C. I., Silva, C. C., Mussatto, S. I., Osseweijer, P., van der Wielen, L. A. M., & Posada, J. A. (2018). Integrated 1st and 2nd generation sugarcane bio-refinery for jet fuel production in Brazil: Techno-economic and greenhouse gas emissions assessment. *Renewable Energy, 129*, 733–747. https://doi.org/10.1016/j.renene.2017.05.011
94. Roth, F. (2018). Costs analysis of aviation fuels in Brazil. https://ptx-hub.org/wp-content/uploads/2021/07/Cost-Analysis-of-Aviation-Fuels-in-Brazil-1.pdf. Accessed September 16, 2022.

Chapter 10
Advanced Biorefineries to Produce Sustainable Aviation Fuel

Araceli Guadalupe Romero-Izquierdo, Carlos Eduardo Guzmán-Martínez, Oscar Daniel Lara-Montaño, Salvador Hernández, and Claudia Gutiérrez-Antonio

1 Introduction

Over the past decade, the aviation sector has experienced significant growth driven by increasing global air traffic, advancements in aircraft technology, and expanding demand for air travel influenced by economic globalization. According to the Aerospace Forecast of the Federal Aviation Administration (FAA) for fiscal years 2024–2044, the industry is expected to continue expanding, with steady increases in passenger and cargo traffic projected through 2044, highlighting the aviation sector's critical role in the global economy [16]. This growth has brought substantial economic contributions and technological advancements, enhancing operational efficiencies and reducing environmental impacts. Integrating newer, more efficient aircraft models has been pivotal in meeting rising environmental and operational standards, significantly improving fuel efficiency, and aligning with global efforts to reduce the aviation sector's carbon footprint [26]. However, the rapid expansion of air travel has implied significant environmental challenges.

Despite improved fuel efficiency and adopting sustainable practices, the sector's overall emissions have continued to rise due to increased flight frequency and volume. This trend has crucial implications for global carbon budgets and the broader objectives of the Paris Agreement [18]. Economically, the aviation industry contributes to global GDP, supporting millions of jobs worldwide and fostering sectors

A. G. Romero-Izquierdo · C. E. Guzmán-Martínez · O. D. Lara-Montaño ·
C. Gutiérrez-Antonio (✉)
Facultad de Ingeniería, Universidad Autónoma de Querétaro, Campus Juriquilla,
Juriquilla, Querétaro, Mexico
e-mail: claudia.gutierrez@uaq.mx

S. Hernández
Departamento de Ingeniería Química, División de Ciencias Naturales y Exactas, Universidad de Guanajuato, Campus Guanajuato, Guanajuato, Mexico

M. Aslam et al. (eds.), *Sustainable Aviation Fuels*, Sustainable Aviation,
https://doi.org/10.1007/978-3-031-83721-0_10

such as tourism and international trade. The industry's resilience was particularly tested during the COVID-19 pandemic, which brought significant disruptions but also led to rapid adaptations. The pandemic underscored the aviation sector's importance to global mobility and economic continuity, highlighting its capacity for swift recovery and adaptation to global crises [2].

The environmental impact of aviation growth has become a central issue. Liu et al. [31] documented that the rebound in CO_2 emissions in 2021, following a temporary reduction during the pandemic, signals the ongoing challenge of aligning the sector with sustainable growth paths. With substantial emissions from domestic and international flights, the aviation industry plays a significant role in the global carbon budget, necessitating stringent measures to curb its environmental impact. The past decade has seen growth in the aviation sector and a significant shift towards sustainability, driven by innovations in fuel technology and biorefineries. As the industry moves forward, it must continue to balance growth with environmental stewardship to align with global climate goals and ensure sustainable development.

The aviation sector, a significant contributor to global CO_2 emissions, faces immense pressure to align with international climate goals. According to the International Energy Agency's "Net Zero by 2050: A Roadmap for the Global Energy Sector," the energy sector, including aviation, accounts for around three-quarters of global greenhouse gas emissions. Achieving net-zero emissions by 2050 is critical to limiting global temperature rise to 1.5 °C, necessitating a complete transformation of how energy is produced, transported, and consumed [7].

Despite technological advancements and improved fuel efficiency, the sheer increase in flight numbers has augmented overall emissions. In 2018 alone, global aviation emitted 918 million metric tons of CO_2, accounting for approximately 2.4% of global CO_2 emissions from fossil fuel consumption [1]. The European Aviation Environmental Report [15] underscores the environmental impacts of non-CO_2 emissions, such as nitrogen oxides (NOx) and particulate matter, significantly affecting climate change and local air quality. These pollutants contribute to the atmosphere's radiative forcing, exacerbating the greenhouse effect and impacting human health [15]. Addressing these non-CO_2 emissions is critical to mitigating the overall environmental impact of aviation.

Research provides a detailed analysis of the U.S. aviation sector's CO_2 emissions, indicating that if aircraft technology, operations, and fuels were frozen at 2019 levels, the sector would emit about 430 Mt CO_2 by 2050. Achieving net-zero emissions will require significant contributions from advanced aircraft technologies, operational improvements, and the increased use of sustainable aviation fuels (SAFs) [26]. To achieve substantial emissions reductions, various strategies, including the adoption of SAFs, are essential. There is a critical need for immediate and massive deployment of clean energy technologies and significant behavioral changes to meet these targets. However, even with the most optimistic scenarios, residual emissions from aviation are projected to remain a significant challenge [7].

The aviation sector's resilience during the COVID-19 pandemic demonstrated its critical role in global mobility and economic continuity. However, this period also highlighted the sector's environmental impact, with a temporary reduction in

emissions followed by a swift rebound as travel restrictions were lifted. This pattern underscores the ongoing challenge of decoupling emissions from traffic growth [2]. The introduction of sustainable aviation fuel (SAF) presents a promising solution. While technologies such as new engine architectures, the large-scale adoption of SAFs could drive significant emissions reductions. A significant increase in SAF production is required to meet IATA's goal of a 50% net reduction in CO_2 emissions by 2050 compared to 2005 levels, with an annual growth rate of 15% in SAF usage from 2030 to 2050 [4].

Various technologies have been explored to offer a solution for sustainable aviation achievement. Hydrogen as a fuel source offers considerable promise due to its potential for zero carbon emissions when used in fuel cells or as a direct combustion fuel. Studies have shown that hydrogen can be utilized effectively in aviation, especially with storage and fuel cell technology advancements. Airbus, for instance, has been actively working on the development of hydrogen-powered aircraft with the goal of launching such planes by 2035. However, the development and implementation of this technology face significant challenges, including the need for new hydrogen production, storage, and distribution infrastructure [4]. Electric aircraft development, particularly for short-haul flights, offers another avenue for reducing emissions. Battery technology has seen rapid advancements, and ongoing efforts exist to create electric planes that can serve regional routes. While current battery energy densities are insufficient for long-haul flights, continuous improvements in battery technology could eventually make electric aviation viable for a broader range of distances. The electrification of aviation will require substantial investments in new aircraft designs and infrastructure, but holds the potential for significant reductions in CO_2 emissions [4].

The aviation sector's contributions to CO_2 emissions and other pollutants have become a focal point in sustainability and climate change discussions. Initiatives like the Carbon Offsetting and Reduction Scheme for International Aviation (CORSIA) are critical in mitigating these impacts by setting ambitious carbon reduction targets and promoting the use of SAFs. Balancing the sector's growth with environmental stewardship is essential for aligning with global climate goals and ensuring sustainable development [10].

The aviation industry faces significant challenges in achieving sustainable growth, with reducing carbon emissions being a main concern. SAF emerges as one of the most promising alternatives to conventional jet fuel, offering the potential to significantly mitigate the sector's environmental impact. SAF is produced from renewable biomass sources and various waste streams, and can be blended with conventional jet fuel to power aircraft with minimal modifications to existing infrastructure [35, 58]. SAF offers substantial environmental benefits compared to conventional jet fuel. It can reduce life cycle greenhouse gas (GHG) emissions by up to 80%, depending on the feedstock and production processes used [14].

Using SAF also reduces non-CO_2 pollutants, such as particulate matter, which helps mitigate global warming and improve local air quality. Also, SAF reduces the nonvolatile particulate matter number emissions index by up to 70% compared to conventional fuels, directly influencing the microphysical properties of contrails

and potentially mitigating their climate impact [54]. This reduction in particulate emissions is crucial as it decreases contrail optical depth and possibly their lifetime, reducing the net radiative forcing associated with these cloud formations. These benefits collectively enhance SAF's role in addressing aviation's CO_2 and non-CO_2 impacts, aligning with broader environmental goals and contributing to the industry's efforts to mitigate climate change.

Several production pathways for SAF have been certified, including hydroprocessed esters and fatty acids (HEFA), Fischer-Tropsch (FT) synthetic paraffinic kerosene, and alcohol-to-jet (ATJ) fuel [57]. Each pathway utilizes different feedstocks, ranging from waste oils and fats to agricultural residues and municipal solid waste. The HEFA pathway is particularly notable for its maturity and scalability [28]. Despite its environmental advantages, the widespread adoption of SAF faces economic challenges. The production costs of SAF are currently higher than those of conventional jet fuel, largely due to the limited scale of production and the higher costs associated with feedstocks and conversion technologies. However, ongoing technological advancements and economies of scale are expected to reduce these costs over time, making SAF a more viable option for the aviation industry.

Developing SAF has been related to the ability to produce these fuels from various biomass through different processing routes. These pathways include HEFA, ATJ, and the Fischer-Tropsch process, each suited for different types of biomass and offering various yields and sustainability characteristics. This versatile approach allows for the adaptation of SAF production to locally available resources, enhancing the regional feasibility and adoption of these technologies. The flexibility in feedstock utilization is critical as it enables the integration of SAF production into existing agricultural and waste management systems, which can turn residual biomass into valuable aviation fuel, reducing waste and contributing to circular economy principles [49].

The ATJ and FT processes enable the use of solid biomass, such as forestry residues and agricultural by-products. The FT process involves gasifying biomass to produce a synthesis gas (syngas), which is then converted into liquid hydrocarbons through a series of chemical reactions. This pathway benefits regions with abundant biomass resources but low demand for food crops, avoiding the food-versus-fuel debate that often accompanies biofuel discussions [56]. Additionally, these technologies support the circular economy concept by optimizing resource use and reducing waste. By incorporating SAF into aviation, these methods diminish the reliance on fossil fuels and encourage economic development in rural areas where biomass resources are plentiful, providing a dual economic and environmental benefit.

SAF has proven its technical and environmental feasibility, achieving substantial reductions in greenhouse gas emissions for the aviation sector. Technologies like gasification-FT and HEFA have been shown to provide more than 80% direct GHG savings compared to conventional jet fuels when utilizing lignocellulosic biomass or used cooking oil, respectively [35]. These technologies are also compatible with existing fuel infrastructure, facilitating their integration into the current system

without significant new investments; thus promoting a smoother transition from fossil-based fuels to renewable alternatives.

Despite advancements in technical and environmental aspects, the economic feasibility of SAF remains a significant challenge. Production costs are higher than conventional aviation fuels due to the high cost of raw materials and the complexity of conversion processes. This financial aspect is crucial, especially given the competition with established road transport biofuels, which currently receive more substantial policy support in terms of subsidies and mandates. Additionally, the high capital expenditure (CAPEX) required for setting up production facilities and operational expenses (OPEX) for conversion processes like FT-SPK and HEFA contribute to the overall financial burden. Although HEFA has potentially lower OPEX due to simpler processing requirements, the absence of supportive policies for SAF highlights a critical need for policy intervention to enhance its economic attractiveness [35].

While the technical and environmental feasibilities of SAF production have been largely accomplished, economic feasibility remains an area where further efforts are necessary. Addressing this will require a multifaceted approach involving continued technological advancements, economic incentives, and regulatory support to ensure that SAF can be competitively priced and widely adopted in the aviation sector.

In this context, biorefineries emerge as a promising alternative that allows for the integral utilization of biomass. Biorefineries can process various types of biomass to produce biofuels, chemicals, and materials, maximizing the economic value of biomass and minimizing waste. The biorefinery concept integrates processes and technologies to convert biomass into fuels and valuable chemicals similar to petroleum refineries but focusing on renewable resources. Recent advancements have demonstrated the potential of biorefineries to significantly reduce greenhouse gas emissions and reliance on fossil fuels, supporting the transition toward a more sustainable economy [41, 42].

Furthermore, the diversification of biomass feedstocks, including agricultural residues, forestry by-products, and dedicated energy crops, enhances the sustainability and resilience of biorefinery operations. Therefore, developing and implementing biorefineries is crucial to achieving sustainable fuel production and reducing the aviation industry's environmental footprint.

2 Sustainable Aviation Fuel

Sustainable aviation fuel (SAF) is the non-conventional or non-fossil-derived aviation fuel, called by the International Association of Transport Aviation (IATA) as SAF to denote its sustainable, alternative, and renewable nature; it must fulfill certain sustainability criteria for demonstrating its reduction of CO_2 emissions in its life cycle assessment, along with ensuring that its raw material is not obtained from a crude oil source [25]. Also, its chemical nature must be similar to the conventional fuel, in order to be an accurate substitute to its fossil counterpart.

In order to analyze the SAF properties, it is important to know the characteristics of conventional jet fuel. This fuel has an average hydrocarbon distribution in the range of C8-C16, which comes from a cut of refining crude oil, approximately composed of 20 wt % paraffins, 40 wt % isoparaffins, and 20 wt % aromatic molecules. According to Amhamed et al. [5] there are several jet fuel types: Jet A for United States and Jet A-1 for Europe, Jet Propellant JP-4 for United States and F-40 for Europe, Jet Propellant JP-5 and F-44 for Europe, JP-8 and F-34 for United States and Europe, JP-7 and JP-8I100 for United States; specifically Jet A, Jet A-1, JP-5 and JP-8 contain C8–C18, whilst Jet B and JP-4 have C5 to C15, named naphtha-fuel type. It is important to mention that Jet A and Jet A-1 are used for commercial transport, whilst JP's for military aviation.

The physicochemical behavior of SAF is similar to fossil jet fuel; thus, both can be safely mixed at different degrees, without requiring the adaptation or modification of aircraft or engines, achieving the same combustion performance; this fuel is also called a "drop-in fuel", due to it can be directly incorporated into the conventional airport fueling schemes [9]. The SAF compounds include hydrocarbons in the same range as conventional jet fuel, lineal and branched, with an adequate freezing point, thermal stability, and volatility in order to address the flight extreme conditions. However, this fuel lacks aromatic compounds, which are responsible for ensuring elastomer seal swell in aircraft to prevent fuel leaks; this absence is neglected with the fossil blends, or with production processes capable of producing them [19]. Table 10.1 presents some physicochemical properties of conventional jet fuel and SAF, produced from Camelina and *Jatropha curcas*, two feedstocks of the second generation. As it can be observed, the boiling point and the freezing temperature are superior for SAF, whilst the energy content is roughly similar.

There are several processes for SAF production, outlined by the American Standard Testing Materials (ASTM), which provides the specifications for aviation turbine fuel containing synthesized hydrocarbons as SAF, involving the manufacture of SAF and its blends with conventional jet fuel (ASTM D7566). At this moment, there are seven certified production processes of SAF, which are provided next [20–22, 51].

Fischer-Tropsch Hydroprocessed Synthesized Paraffinic Kerosene (FT-SPK) Certified process in 2019, which can use natural gas, coal, and biomass as

Table 10.1 Main physicochemical properties of conventional jet fuel and SAF

	Jet A	Jet-A1	SAF from Camelina	SAF from *Jatropha curcas*
Boiling range (°C)	170–300	170–300	188–263	172–243
Freezing temperature (°C)	−40	−47	−63.5	−57
Flash temperatura (min, °C)	38	38	42	46.5
Density at 15 °C (kg/m^3)	775–840	775–840	751–840	751–840
Energy content (MJ/kg)	43.28	43.28	44	44.3

Extracted and modified from Gutiérrez-Antonio et al. [20]

feedstocks. The first stage of this process is the raw material gasification to produce syngas (a mixture of CO and H_2); then, the syngas is used as reactant in the Fischer-Tropsch reaction, next by hydroprocessing to obtain renewable hydrocarbons, including the sustainable aviation fuel. According to ASTM (D7566, Annex A1), the SAF produced is mixed with up to 50% volume of conventional jet fuel.

Synthesized Paraffinic Kerosene from Hydroprocessed Esters and Fatty Acids (HEFA-SPK) Certified process in 2011, using as raw materials vegetable oils, bio-oils from pyrolysis, waste oils, and animal fats. This process uses hydrodeoxygenation, hydroisomerization, and hydrocracking as reaction pathways, in order to produce light gases, naphtha, sustainable aviation fuel, and green diesel. The SAF produced must be used in mixtures up to 50% volume with jet fuel, according to the ASTM standard (Annex A2).

Synthesized Iso-Paraffins from Hydroprocessed Fermented Sugars (HFS-SIP) Certified process in 2014, which can use sugars and any biomass used for sugar production. The raw materials could be converted to farnesene and C15 hydrocarbons; then, these products are hydrotreated in order to generate SAF, which can be used in mixtures with 10% volume with conventional jet fuel, in accordance to the D7566 standard (Annex A3).

Synthesized Paraffinic Kerosene with Aromatics by Fischer-Tropsch Route (FT-SPK/A) Certified process in 2015, which can be used as raw materials natural gas, coal, and solid biomass, similar to the FT-SPK process. Its working principle points out that the feedstock is gasified in order to obtain syngas; this fuel is processed by the Fischer-Tropsch reaction followed by the alkylation of aromatics (mainly light aromatics) to produce a hydrocarbon blend with aromatic molecules. The produced SAF can be a mixture with up to 50% volume jet fuel, according to ASTM standards (Annex A4).

Alcohol-to-Jet Synthetic Paraffinic Kerosene (ATJ-SPK) Certified process in 2016, which uses as raw material the biomass with high content of starch and sugar, and the biomass capable of producing alcohol. This process involves three reactive zones: (1) alcohol dehydration, (2) oligomerization, and (3) alkene hydrogenation to produce linear hydrocarbons in the range of naphtha fuel, SAF, and green diesel. The SAF produced can be a mixture of up to 30% volume of conventional jet fuel, according to the ASTM standard (Annex A5).

Synthesized Kerosene from Hydrothermal Conversion of Fatty Acid Esters and Fatty Acids (CHJ) Certified process in 2018, which uses as feedstock vegetable oils, animal fats, and used cooking oils. Its working principle is a catalytic hydrothermolysis of raw material; wherein the SAF produced can be mixed with up to 50% of conventional jet fuel, according to the ASTM standard (Annex A6).

Synthesized Paraffinic Kerosene from Hydroprocessed Hydrocarbons, Esters and Fatty Acids (HC-HEFA-SPK) Certified process in 2020. This process is the same as HEFA-SPK, but as feedstock uses algae oil, in order to produce a broad range of hydrocarbons, including the SAF cut. According to the ASTM standard (Annex A7), the SAF produced is a mixture of up to 90% volume of conventional jet fuel.

Other routes are referrals of the mentioned processes [22, 37]. The synthetic paraffinic kerosene with aromatics using the ATJ route (ATJ-SKA), using C2-C5 alcohols from biomass, approved by ASTM in Annex A8. Then, in ASTM D1655 Annex A1 is located the co-hydroprocessing of esters and fatty acids in a conventional petroleum refinery, which uses vegetable oils, animal fats, used cooking oils with middle distillates petro-derived (HEFA-SKA); the SAF produced can be mixed with 50% volume of jet fuel, according to the standard. Another process in the same annex is the co-hydroprocessing of Fischer-Tropsch hydrocarbons in a conventional petroleum refinery; this process uses the produced Fisher-Tropsch hydrocarbons into a blend with petroleum, wherein the SAF produced is a mixture with up to 95% volume of conventional jet fuel.

There are other processes under evaluation by ASTM, which claim to increase the SAF fraction in the jet fuel mixture [22]. One of them is the synthesized aromatic kerosene (HDO-SAK), developed by Virent; this process uses plant sugars to produce the required aromatics for producing SAF with good properties, neglecting the conventional jet fuel without any aircraft modifications. The SAF produced was used in a commercial passenger flight last December 2021 by United Airlines using 100% SAF in one engine. Other interesting processing proposals are the integrated hydropyrolysis and hydroconversion (IH2) by Shell, Single reactor HEFA to produce drop-in liquid SAF and automotive fuel (DILSAAF) by Indian CSIR-IIP, pyrolysis of non-recyclable plastics (ReOIL) by OMV, co-processing of pyrolysis oil from used tires (TPO) by Philips 66, Methanol to jet (MTJ) by ExxonMobil, HEFA with higher cycloparaffins by Revo, Biomass pyrolysis by Alder, Biomass and/or waste pyrolysis by Green Lizard, Cycloalkanes from ethanol by Vertimass. Table 10.2 presents the current pathways for SAF and its mature level.

As mentioned before, ASTM has certified the production processes for SAF. This international organization provides the technical standards for certifying a wide range of materials, products, processes, systems, and services; specifically, the SAF produced must comply with the quality specifications to be used in the aircraft, regarding fuel composition and its properties in order to ensure a consistent fuel. ASTM has provided some standards related to the aviation sector, which are presented next [6, 21, 50]. It is important to mention that the process for achieving the new fuel certification is roughly 3–5 years, which implies several test stages, requiring up to 890,000 l of appropriate fuel blend.

- *D4054 standard* is a guide about the approval process of new fuels and additives for their use at commercial and military levels. This standard is named standard practice for evaluation of the new aviation turbine fuels and fuel additives. This standard includes the test for basic specification properties, expanded properties

Table 10.2 SAF pathways and their mature level

Processing route	Quality today	Blend limit (%)	Future 100% drop-in fuel
FT-SPK—Fischer-Tropsch Synthetic Paraffinic Kerosene	Yes	50	No[a]
HEFA-SPK—Hydroprocessed Fatty Esters and Fatty Acids Synthetic	Yes	50	No[a]
HFS-SIP—Hydroprocessed Fermented Sugars Synthesized iso-Paraffins	Yes	10	No[b]
FT-SKA—Fischer-Tropsch Synthetic Kerosene with Aromatics	Yes	50	Yes[c]
ATJ-SPK—Alcohol-to-jet Synthetic Kerosene with Aromatics	Yes	50	No[b]
CHJ—Catalytic Hydrothermolysis Jet	Yes	50	Yes[a]
HC-HEFA-SPK—Hydroprocessed Hydrocarbon Synthetic Paraffinic Kerosene	Yes	10	No[a]
ATJ-SKA—Alcohol-to-jet Synthetic Kerosene with Aromatics	No	50	Yes[a]
HEFA-SKA—Hydroprocessed Fatty Acids Synthetic Kerosene with Aromatics	No	50	Yes[a]
HDO-SAK—Hydrodeoxygenated Aromatic Kerosene	No	25	No[c]
CPK-0—Cycloparaffinic Kerosene	No	50	Yes[a]
HTL—Hydrothermal Liquefaction	No	50	Yes[a]

[a]Identical to Jet A/A-1—fleet-wide and infrastructure wide-compatible. Drop-in 100% SAF will need specification ASTM D7566 updated in the short-medium term

[b]Close to JetA/A-1 but not identical—not fleet-wide and infrastructure-wide compatible. Non-drop-in 100% SAF; it will have new specifications and separate infrastructure in the medium-long term

[c]Nothing like Jet A/A-1—not viable jet fuel

(fit-for-purpose properties), engine rig, component testing, and full-scale engine testing. This standard is the first stage for the approving route.

- *D7566 standard* is called the standard specification for aviation turbine fuel containing synthesized hydrocarbons. This standard contains the required properties for the certification of commercial aviation fuels, which include synthetic components; involving mixtures of Jet A and Jet A-1 with SAF produced from carbon, natural gas, biomass, and hydrogenated fats and oils by FT-SPK, HEFA-SPK, SIP, SPK/A and ATJ. Also, in this standard, the maximum blending ratio is specified for each processing route.
- *D1655 standard* is called standard specifications for aviation turbine fuels, and it is referred to as strong requirements for using SAF as fuel approved for flight. The D7566 standard includes all testing methods provided by ASTM D1655, thus, if the tested fuel fulfills the requirements of ASTM D7566, also fulfills the ASTM D1655 standard.
- *D7223 standard,* which focuses on the specification for aviation certification turbine fuel. This standard describes the accurate properties for the certification of

aviation fuels from conventional sources, excluding those presented by ASTM D1655.

The process for approving a new SAF, before its use, by ASTM is schematized in Fig. 10.1. As can be seen, each batch of jet fuel requires to be certified for previous use; thus, conventional jet fuel must be tested by the D1655 standard, whilst the pure SAF must be tested by the D7566 standard, according to its production route. Then, both certified fuels are mixed in the accurate proportion (allowed maximum mixture), in order to be tested by the D7566 standard, getting an automatic certification D1655, which specifies that this fuel can be used as drop-in fuel, without any modification of aircraft and engines. Other certifications of SAF to be approved by flight are Def Stan 91-091, CGSV 3.23-2020, DCSEA134 E, and MIL DTL 83133; verifying previously the aircraft documentation for any restrictions or limitations. Once the SAF certification, the fuel must be evaluated during aircraft and engine certification [6, 21].

In accordance to the SAF definition, the produced SAF must be certified for secure flights and must fulfill the environmental requirements to be a sustainable fuel. The SAF environmental regulations are included in the Carbon Offsetting and Reduction Scheme for International Aviation (CORSIA), which is the first global market-based measure for any sector, which offers a harmonized way to reduce emissions from international aviation, minimizing market distortion while respecting the capabilities of ICAO member; also, CORSIA offsetting the CO_2 emissions that cannot be reduced by the use of technological improvements, and sustainable aviation fuels with emissions units from the carbon market [23, 37]. It is important to mention that ICAO is the International Civil Aviation Organization, defined as a United Nations agency; ICAO was created to coordinate the principles and techniques of international air navigation, plan the development of international air transport, and ensure safety and structured order.

ICAO approved two fundamental documents in order to specify the requirements for eligible fuels of aviation, the first one called the CORSIA eligibility framework and requirements for the sustainability certification schemes, and

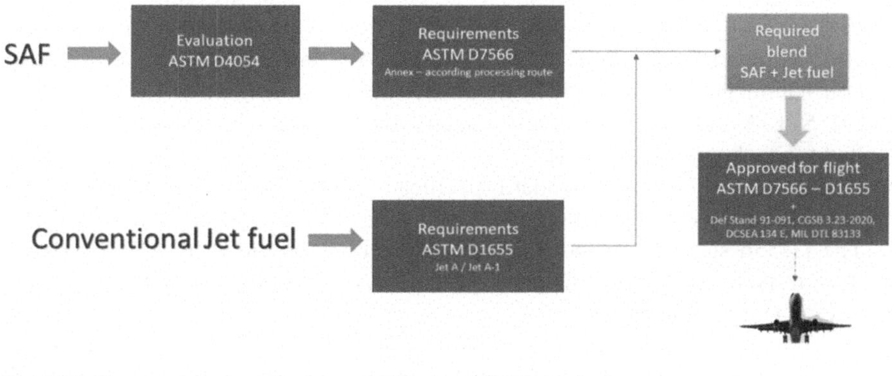

Fig. 10.1 Process for approving a new SAF using ASTM standards

CORSIA sustainability criteria for CORSIA eligible fuels; wherein are included the criteria for ensuring that the produced SAF fulfills the emission values to be considered as sustainable fuel, using certification schemes. The certification schemes provided by CORSIA, approved in 2020, are next: (1) International Sustainability and Carbon Certification (ISCC CORSIA), and (2) Roundtable on Sustainable Biomaterials (RSB).

ISCC has been a global certification system since 2010 with more than 180 members, driven by the ISCC association and a multi-stakeholder non-profit organization; this system is recognized in energy markets to demonstrate the sustainability requirements, including CORSIA. ISCC has broad experience in the transport sector, supporting the certification of advanced fuels, e-fuels, and recycled carbon fuels, such as SAF [24]. The sustainability certification provided by ISCC for SAF covers three fundamental points: (1) ensures the compliance of feedstocks with a defined set of sustainability requirements, (2) offers the traceability of sustainable material by SAF supply chains, and (3) provides the outline for verifying reductions in GHG emissions of the final SAF. Also, since November 2020 the ISCC CORSIA certification system has been available for economic operators and airlines in order to demonstrate their compliance with the CORSIA criteria for eligible fuels in the aviation sector [22].

On the other hand, RSB certification describes how to produce SAF from bio-based raw materials, enhancing long-term environmental and social outcomes; also, it specifies requirements for operators along the supply chain to produce adequate SAF for CORSIA. With this certification is possible to make powerful claims on greenhouse gas reduction, and other sustainability items, such as food security, environmental protection, and human rights. Specifically regarding greenhouse gas reduction, the certified SAF must demonstrate that its emissions reductions are at least 10% on a life cycle basis, compared to the baseline life cycle emissions value for aviation fuel, which is 89 g CO_{2eq}/MJ for jet fuel [43, 44].

3 Production of Sustainable Aviation Fuel Through Biorefineries Schemes

A biorefinery is a set of unit operations for the high-efficiency conversion of biomass feedstocks into biofuels, electrical or thermal energy, and multiple added-value products [46]. In other words, it is the facility where the processes and equipment are integrated for biomass conversion into biofuels, energy, and biobased chemicals. The biorefinery is like existing oil refineries, with various fuels and products generated from crude oil.

Biorefineries are not just theoretical concepts, but practical solutions that are essential to produce sustainable aviation fuel. To underscore the relevance of these schemes, let's consider some real-world applications.

Sadhukhan J. et al. [45] present a groundbreaking approach with comprehensive mathematical modeling equations for the design and sustainability evaluations of a

novel integrated biorefinery system. This system is a pioneering solution that produces renewable aviation fuel from biomass, demonstrating cutting-edge advancements in our field. It is a complex system, consisting of five intricate parts:

- The first one is biomass pyrolysis; the raw material is transformed into gas, bio-oil, and char.
- In the second stage, bio-oil is sent to hydrodeoxygenation and hydrocracking processing. The hydrodeoxygenation function decreases the oxygen presence in bio-oil molecules from 35% to 2%; this occurs by reacting it with hydrogen to generate water. As a secondary reaction, it has CO_2 and CO formation. The hydrocracking promotes the rupture of long carbon chains. Until now, aviation fuel has been obtained.
- Alkanes of short carbon chain are sent to the third stage: the alkane steam reforming and pressure swing adsorption. This step generates the hydrogen needed to carry on the second stage.
- The fourth stage employs a mixed ionic electronic conducting membrane for splitting high-pressure superheated steam into green hydrogen and oxygen.
- The last stage generates a combined heat and power generation using pyrolysis gas and carbon monoxide from the third stage as fuel with oxygen from the ionic electronic membrane [45].

The products from the biorefinery system are not just renewable jet fuel, char, and excess medium-pressure steam. They represent a significant step towards a greener future, reducing our reliance on fossil fuels and mitigating climate change. The authors proudly report that the overall biorefinery energy efficiency is 85.6%, an impressive testament to the sustainability of this innovative process.

Neves, R. C. et al. [34] presented the thermochemical routes to treat biomass to convert it into sustainable aviaition fuel production through fast pyrolysis, gasification, and Fischer-Tropsch synthesis, all of this in integrated sugarcane biorefineries. Neves, R. C. et al. [34] take the case of biorefineries in Brazil. These are first and second-generation biorefineries. The authors divide them into two groups: centralized and decentralized. In the first one, sugarcane and straw (from the field) are fed to a biorefinery to produce ethanol and sugar. All the bagasse generated is sent to pyrolysis, gasification, and Fischer-Tropsch synthesis. After these processes, the products obtained are sustainable aviation fuel, green diesel, green gasoline, and electricity. In decentralized biorefineries, straw passes directly to pyrolysis, gasification, and Fischer-Tropsch synthesis, and the sugarcane goes to produce sugar and ethanol; however, the sugarcane bagasse is sent to pyrolysis, gasification, and Fischer-Tropsch synthesis once the primary process (ethanol and sugar production) is left. An important thing that Neves R. C. et al. [34] highlighted is that sugarcane biorefinery pyrolysis must be self-sufficient in terms of heat and power.

Marangon B. B. et al. [32] propose a biorefinery using microalgae biomass as raw material to feed a conceptual biorefinery. One of the aims of this research is to treat wastewater by growing microalgae in open race ponds. The microalgae capture CO_2 gas coming from biorefinery. The wet microalgae are processed in a hydrothermal liquefaction reactor to obtain hydrochar, bio-oil, biogas, and an aqueous phase.

As mentioned, the gas is absorbed by microalgae; hydrochar is used as solid biofuel and catalyst support. The aqueous phase is employed in supercritical water reforming to produce hydrogen supplied to hydrotreatment. The bio-oil is sent to hydrotreatment for a rich hydrocarbon mixture, which must be separated in the distillation process. As a result, it is possible to obtain water-treated, light gases, gasoline, aviation fuel, and diesel. It must be mentioned that the work of Marangon B. B. et al. [32] is limited to a conceptual biorefinery.

Michaga, M. F. et al. [33] performed a simulation in Aspen Plus and Matlab about a biorefinery to produce sustainable jet fuel. It receives forest wastes that are size-reduced and dried. After that, biomass is sent to the gasification process to produce synthesis gas. From this gas, CO_2 is removed by using monoethanolamine (MEA). Finally, a Fischer-Tropsch reactor takes the clean syngas. The hydrocarbon mixture is separated by using a distillation column. The products from the biorefinery proposed by Michaga, M. F. et al. [33] are naphtha, sustainable aviation fuel, and green diesel. Besides, the CO_2 captured is compressed and sold as a product. The study of Michiga M. F. et al. [33] involves economic and environmental analysis. They conclude that this biorefinery would be economically feasible only if there is a reward for harmful emissions in the UK and the "renewable transport fuel obligation" policy was implemented.

Alves C. M. et al. [3] realize proposals for biorefineries in Brazil. Authors define it as raw material sugarcane for the Minas Gerais region and soybean for the Rio Grande do Sul region. For sustainable aviation fuel production, there are many processes that authors refer to, such as ATJ (alcohol to jet fuel), biomass gasification, Fischer-Tropsch Syngas to jet, biomass pyrolysis and hydrogenation, hydroprocessed esters and fatty acids. Alves C. M. et al. [3] consider the production of some chemicals, such as succinic acid, ethylene, lactic acid, butanol, isopropanol, 3-hydroxy propionic acid, 2,5-furan dicarboxylic acid, 1,3-propanediol, and 1,4-butanediol. They conclude that sustainable aviation fuel and biochemical co-production in a biorefinery context might be a promising alternative.

Puschnigg S. et al. [38] use residual softwood as biorefinery raw material. In this process, the raw material is pretreated to release simple sugars to be fermented. The pretreatment is based on an enzymatic process. In fermentation, the aim is to obtain isobutane; after its purification, the isobutane is transformed into sustainable aviation fuel by isobutane conversion to the SAF process. Some by-products obtained from remaining components after pretreatment are lignin-blend asphalt, phenol-formaldehyde resins, bioplastic filter, and ethanol. Products from fermentation residues are fertilizer sludge (with a high content of potassium and calcium) and biomass (which can be employed as raw material in other biorefinery) [38].

Romero A. et al. [42] developed a biorefinery to get sustainable aviation fuel using *Ricinus communis L.*, also known as Higuerilla. The design is modeled using Aspen Plus Software. This work employs all parts of the plant. The processes implemented are mechanical pressing, residue pyrolysis, shell pyrolysis, press cake pyrolysis, gasification, hydrodeoxygenation/hydrotreating, alcohol to jet fuel, and transesterification. As a product, Romero A. et al. [42] obtained 53.27% bio-oil, 9.41% naphtha, 9.19% biodiesel, 11.82% sustainable aviation fuel, 7.76% bio-char,

4.61% light gases, and 3.94% green diesel. In conclusion, according to the analysis the authors made, this biorefinery is self-sufficient in terms of electricity consumption.

Klein B. et al. [27] research is focused on designing a new pretreatment for use in biorefinery schemes. This process is called co-solvent enhanced lignocellulosic fractionation (CELF). The treatment consists of biomass exposure at temperatures of 140–160 °C with tetrahydrofuran as a co-solvent. This allows high yields in hemicellulose sugars, cellulose, and lignin intermediates. Once the CELF pretreatment is carried out in biorefinery, different products are obtained from pentoses and extractives, such as isobutyl acetate and adipic acid. From lignin, cycloalkanes, cellulose, ethanol, and isobutanol. Finally, Klein B. et al. [27] mentioned that ethanol from this biorefinery could be upgraded to sustainable aviation fuel.

Robertson, K. A. et al. [40] performed research testing different fertilizers to improve the growing rate of Switchgrass (*Panicum virgatum L.*). Unlike it seems, it does not have direct to sustainable aviation production, and the authors' primary objective is to increase the profitability of a biorefinery that uses it as raw material to transform from switch grass to sustainable aviation fuel through the pyrolysis process.

Deuber R. et al. [12] suggest incorporating hydrothermal liquefaction in Brazilian biorefineries to get sustainable aviation fuel. Thus, this research is focused on the techno-economic and life cycle assessment of sustainable aviation fuel production from sugarcane lignocellulosic residues, considering hydrothermal liquefaction integrated into first-generation ethanol distillery. Deuber R. et al.'s [12] research includes study cases of biorefineries that produce ethanol and sustainable aviation fuel (SAF) and are the only hydrothermal liquefaction of biomass that generates biojet fuel. The biomass involved are straw, sugarcane stalks, and bagasse. Their results point to a potential feasibility considering green house gas emission reduction and the minimum fuel selling price of SAF in Brazil.

Tanzil A. et al. [53] do not develop a biorefinery directly; however, their research tests five SAF technologies to identify SAF cost reduction opportunities and environmental benefits for being implemented in biorefinery schemes. The research was carried out using dry-grained corn. Tanzil A. et al. [52] implemented the processes of Virent's reforming, alcohol to jet fuel, direct sugar to hydrocarbon, fast pyrolysis, and gasification—Fischer-Tropsch. The best conditions (the minimum fuel selling price and greenhouse gas emissions) between these five processes were found in alcohol-to-jet fuel.

Trejo C. et al. [55] and English B. et al. [13] do not work directly with biorefinery to produce sustainable aviation fuel. However, these researchers consider the economic analysis of the supply chain to feed a biorefinery for SAF generation. Developing a supply chain is very important for biorefineries because if the raw material comes from places too far from the biorefinery, the transport cost could be so significant that the SAF price becomes inappropriate to sell.

Kumar D. et al. [29] simulate the biorefinery in SuperPro Designer Software. The biorefinery is fed by lipid-cane. Juice from this crop is extracted; its main components are oil and sugar. The lipid part is treated by hydrotreating to produce a

mixture of hydrocarbons; after that, these are separated, by distillation columns, into light gases, naphtha, and jet fuel. Sugars are fermented to produce ethanol. The bagasse is burnt as a solid fuel to generate steam and electricity. Although the research does not include the alcohol-to-jet fuel process, upgrading ethanol to sustainable aviation fuel is possible.

Santos C. et al. [47] present a techno-economic analysis and an environmental assessment of the whole production chain (biomass production, sugar extraction, biomass pretreatment, sugars fermentation, and products recovery and purification) of a fully autarkic sugarcane-based biorefinery for sustainable aviation fuel production. In Santos C. et al. [47] research, the conclusions are: (i) ethanol fermentation is considerably more economical than the farnesene route; (ii) upgrading lignin to SAF increases complexity without significant reductions on the minimum renewable jet fuel selling price, and (iii) bagasse should be upgraded to SAF via fast pyrolysis.

Lan K. et al. [30] converted waste paper sludge into sustainable aviation fuel. The biorefinery uses a catalytic sugar upgrading system that consists of ash removal, enzymatic hydrolysis, dehydration, aldol condensation, and hydroprocessing. Besides, the authors offer a life-cycle assessment of this biorefinery.

Batten R. et al. [8] investigated the potential for reducing greenhouse gas (GHG) emissions and displacing fossil fuel (petroleum-based jet kerosene) in the aviation fuel industry by using corn feedstocks in dry grind biorefineries to convert sugars to 1,4-dimethyl cyclo octane (DMCO), which qualifies as renewable jet-A.

Carrasco M. et al. [11] converted waste cooking oil through a biorefinery scheme. An intensification of the SAF process is made. The resulting biorefinery produces 306.4 kg/h of SAF and is self-sufficient in electricity requirements. The intensified biorefinery reduces energy consumption by 66.95% compared to conventional biorefinery. Other products obtained are biodiesel, glycerol, naphtha, and green diesel. In conclusion, converting waste cooking oil through an intensified biorefinery allows the generation of biofuels, bioenergy, and value-added products, simultaneously evaluating this hazardous residue.

Seufitelli G. et al. [48] investigated the economics of an integrated biorefinery converting hybrid poplar into jet fuel, xylitol, and formic acid. The process employs a combination of integrated biological, thermochemical, and electrochemical conversion pathways to convert the carbohydrates in poplar into jet fuel, xylitol, and formic acid production. In conclusion, the research presents that the minimum renewable jet fuel selling price for profitability decreases with increasing biorefinery scale. For all scales, the biorefinery presents favorable economics, leading to a minimum renewable jet fuel selling price lower than the current price for sustainable aviation fuel (SAF).

Rivas R. et al. [39] designed an intensified process to produce SAF from Mexican lignocellulosic biomass, with alcohols as intermediates. The process was modeled following a sequence of pretreatment/hydrolysis/fermentation/purification for the biomass-ethanol process and dehydration/oligomerization/hydrogenation/distillation for the ethanol-bio jet process under the concept of distributed configuration. To obtain a cleaner, greener, and cheaper process, Rivas R. et al. [39] intensified the purification zone of ethanol by employing a vapor-side stream distillation column

and a dividing wall column. Once designed, the entire process was optimized by using the stochastic method of differential evolution with a tabu list to minimize the total annual cost and with the Eco-indicator-99 to evaluate the sustainability of the process. The results show that savings of 5.56% and a reduction of 1.72% in Eco-indicator-99 were achieved with a vapor side stream column compared to conventional distillation.

Tanzil A. et al. [52] have several processes and raw materials to obtain SAF. About supplies, there are plant/animal oils and fats, wastewater (animal manure), forest residue, and sugar-rich feedstock. The processes concerned are hydrotreatment isomerization, pyrolysis, Fischer-Tropsch, gasification, fermentation, and alcohol oligomerization. Authors report jet fuel, diesel, naphtha, heat, and electricity as products.

García A. et al. [17] proposed a process that involves the conversion of biomass to butanol through the acetone-butanol-ethanol process and subsequent upgrading to SAF using the alcohol-to-jet (ATJ) pathway. Process intensification techniques have been integrated downstream with a simultaneous saccharification-fermentation (SSF) reactor and upstream with a Petlyuk distillation column, aiming to optimize the system configuration to minimize total annualized cost and environmental impact as measured by the Eco-Indicator 99 (EI99) method. Metaheuristic optimization techniques are applied. The findings of García A. et al. [17] suggest that the integrated biorefinery approach is a promising and efficient strategy for SAF production, contributing to expanding renewable energy sources in alignment with the United Nations' sustainable development goals.

Pipitone G. et al. [36] evaluated the technical, economic, and environmental feasibility of obtaining SAFs by hydrogenating vegetable oils thanks to in situ hydrogen production via aqueous phase reforming of glycerol by-products. The conceptual design of a conventional and advanced biorefinery is performed by Pipitone G. et al. [36]. As a result, it is essential to mention that only naphtha, sustainable aviation fuel, and green diesel have been obtained as products of this biorefinery. Techno-economic results showed that implementing the aqueous phase reforming led to a slight increase in fixed capital investment of 6.6% compared to the conventional one. In comparison, direct manufacturing costs decreased by 22%.

4 Perspectives

Sustainable aviation fuel represents the best strategy for the decarbonization of the aviation sector in the short term. This biofuel can be supplied to the aircraft, with no modification to the engines and the complete design of airplanes. Moreover, SAF can be produced from almost any type of raw material through several processing pathways. Depending on the processing route, SAF can be used up to 50% in volume with fossil jet fuel. SAF needs to fulfill technical and environmental norms, and also it must be economically competitive regarding its fossil counterpart. The technical aspects are covered by ASTM standards, regarding its main properties. On

the other hand, the RSB CORSIA standard establishes the reference value for the carbon footprint associated with this biofuel during the complete life cycle. Nowadays, both technical and environmental norms can be fulfilled. Nevertheless, the main challenge relies on the economic competitive regard of its fossil counterpart. In order to overcome this challenge several strategies have been proposed, beginning the use of biorefineries is one of them.

Biorefineries allow an increase in the financial profitability of biomass conversion. The main advantage is that at least one value-added product, one biofuel, and one bioenergy are produced. Since the profitability relies on several products, it could benefit to decrease the SAF price. In this research line, several works have been developed. From them, two important future trends are identified. The first one is the use of waste biomass since it represents a pollution problem. In this way, a problem can be solved along with the generation of renewable products. This line highlights the work of Carrasco et al. [11] for the conversion of waste cooking oil to SAF, other biofuels, and value-added products. One main advantage of these waste biorefineries for SAF production is that there is no competition for lands or resources for food production. The second one is the use of hydrothermal liquefaction as an alternative to the use of biomass with high moisture content, such as microalgae or organic waste. In this case, it is suggested to extract the value-added components first and then proceed to produce biocrude oil through this technology. It is well known that the drying of biomass in order to be processed represents the biggest part of the energy consumption of the conversion processes.

In all cases, the supply chain is a key piece in order to achieve the technical and environmental standards, as well as the economic viability. The production cost of a product depends on the processing capacity, which in this case also depends on the size of the supply chain. However, if the supply chain is bigger, more biomass can be recollected/harvested from different regions, but the environmental impact will increase. Moreover, the consideration of biorefineries will help to reduce the size of the supply chain, since the profitability of the scheme can be increased through the commercialization of value-added products, decreasing the need to process more biomass. Thus, the establishment of local biorefineries that supply the products close to where they can be used is the best alternative from economic, environmental, social, and technical points of view.

5 Concluding Remarks

In this chapter was presented a revision on the use of biorefineries for the production of sustainable aviation fuel. Sustainable aviation fuel represents the most promissory strategy in the short term for the decarbonization of aviation sector. This biofuel can be produced from several production routes, where just a fraction of biomass is considered. Another alternative is the production of SAF through biorefineries, with the main advantage of reaching economic viability.

The use of biorefineries to produce SAF is viable, and it was identified as the main future trend the use of waste biomass as feedstock for this processing scheme, as well as the hydrothermal liquefaction of wet biomass, residual or cultivated. These alternatives have been less explored in the literature, and they can be useful for SAF production considering environmental, technical, social, and economic aspects. This could boost the energy transition in the aviation sector, ensuring its sustainable development.

Acknowledgments The authors thank the Consejo Nacional de Humanidades, Ciencia y Tecnología (CONAHCYT) and the Universidad Autónoma de Querétaro (UAQ) for the support and scholarships granted of C.E. Guzmán Martínez and O.D. Lara-Montaño for the realization of this work. This research was also supported by the Bioproducts Sustainable Production Processes Research Group from the Universidad Autónoma de Querétaro (UAQ), Facultad de Ingeniería.

References

1. Abrantes, I., Ferreira, A. F., Silva, A., & Costa, M. (2021). Sustainable aviation fuels and imminent technologies—CO2 emissions evolution towards 2050. *Journal of Cleaner Production, 313*, 127937. https://doi.org/10.1016/j.jclepro.2021.127937
2. Abu-Rayash, A., & Dincer, I. (2020). Analysis of mobility trends during the COVID-19 coronavirus pandemic: Exploring the impacts on global aviation and travel in selected cities. *Energy Research & Social Science, 68*, 101693. https://doi.org/10.1016/j.erss.2020.101693
3. Alves, C. M., Valk, M., de Jong, S., Bonomi, A., van der Wielen, L. A., & Mussatto, S. I. (2017). Techno-economic assessment of biorefinery technologies for aviation biofuels supply chains in Brazil. *Biofuels, Bioproducts and Biorefining, 11*(1), 67–91.
4. Amankwah-Amoah, J., Debrah, Y., & Anang, S. (2023). Greening aviation in era of COVID-19: Towards conceptualizing and operationalizing decarbonization. *Journal of Environmental Management, 326*, 116649. https://doi.org/10.1016/j.jenvman.2022.116649
5. Amhamed, A., Al Assaf, A. H., Le Page, L., & Alrebei, O. (2024). Alternative sustainable aviation fuel and energy (SAFE)- A review with selected simulation cases of study. *Energy Reports, 11*(13). https://doi.org/10.1016/j.egyr.2024.03.002
6. ASTM. (2020). *ASTM D6751 – 20a standard specification for biodiesel fuel blend stock (B100) for middle distillate fuels* [WWW Document]. ASTM Int. URL https://www.astm.org/Standards/D6751.htm. Accessed June 28, 2021.
7. Bains, P., Kim, H., & Teter, J. (2022). *Decarbonising aviation in the IEA's net zero emissions by 2050 scenario*. International Energy Agency. https://www.icao.int/environmental-protection/Documents/EnvironmentalReports/2022/ENVReport2022_Art48.pdf
8. Batten, R., Galant, O., Karanjikar, M., & Spatari, S. (2023). Meeting sustainable aviation fuel policy targets through first-generation corn biorefineries. *Fuel, 333*, 126294.
9. Braun, M., Grimme, W., & Oesingmann, K. (2024). Pathway to net zero: Reviewing sustainable aviation fuels, environmental impacts and pricing. *Journal of Air Transport Management, 117*, 102580.
10. Capaz, R. S., Guida, E., Seabra, J. E. A., Osseweijer, P., & Posada, J. A. (2021). Mitigating carbon emissions through sustainable aviation fuels: Costs and potential. *Biofuels, Bioproducts and Biorefining, 15*(2), 502–524. https://doi.org/10.1002/bbb.2168
11. Carrasco-Suárez, M. T., Romero-Izquierdo, A. G., Gutiérrez-Antonio, C., Gómez-Castro, F. I., & Hernandez, S. (2022). Production of renewable aviation fuel by waste cooking oil processing in a biorefinery scheme: Intensification of the purification zone. *Chemical Engineering and Processing-Process Intensification, 181*, 109103.

12. Deuber, R. D. S., Bressanin, J. M., Fernandes, D. S., Guimarães, H. R., Chagas, M. F., Bonomi, A., & Watanabe, M. D. B. (2023). Production of sustainable aviation fuels from lignocellulosic residues in Brazil through hydrothermal liquefaction: Techno-economic and environmental assessments. *Energies, 16*(6), 2723.
13. English, B. C., Menard, R. J., & Wilson, B. (2022). The economic impact of a renewable biofuels/energy industry supply chain using the renewable energy economic analysis layers modeling system. *Frontiers in Energy Research, 10*, 780795.
14. European Aviation Safety Agency, & European Environment Agency. (2019). *European aviation environmental report 2019*. Publications Office. https://data.europa.eu/doi/10.2822/309946
15. European Union Aviation Safety Agency, & European Environment Agency. (2023). *European aviation environmental report 2022*. Publications Office. https://data.europa.eu/doi/10.2822/04357
16. FAA. (2024). *FAA aerospace forecast fiscal years 2024–2044*. Federal Aviation Administration. https://www.faa.gov/dataresearch/aviation/aerospaceforecasts/faa-aerospace-forecast-fy-2024-2044
17. García-Hernández, A. E., Segovia-Hernández, J. G., Sánchez-Ramírez, E., Zarazúa, G. C., Araujo, I. F. H., & Quiroz-Ramírez, J. J. (2024). Sustainable aviation fuel from Butanol: A study in optimizing economic and environmental impact through process intensification. *Chemical Engineering and Processing-Process Intensification, 200*, 109769.
18. Gössling, S., & Humpe, A. (2020). The global scale, distribution and growth of aviation: Implications for climate change. *Global Environmental Change, 65*, 102194. https://doi.org/10.1016/j.gloenvcha.2020.102194
19. Grote, M., Williams, I., & Preston, J. (2014). Direct carbon dioxide emissions from civil aircraft. *Atmospheric Environment, 95*, 214–224.
20. Gutiérrez-Antonio, C., Romero-Izquierdo, A. G., Gómez-Castro, F. I., & Hernández, S. (2021). *Production processes of renewable aviation fuel, production processes of renewable aviation fuel*. Elsevier.
21. ICAO. (2023). *SAF production technologies and certification* [WWW Document]. URL https://www.icao.int/environmental-protection/Documents/ACT-SAF/ACT SAF series 3 – SAF production technologies and certification.pdf. Accessed May 16, 2024.
22. ICAO. (2024a). *Conversion processes* [WWW Document]. ICAO Int. URL https://www.icao.int/environmental-protection/GFAAF/Pages/Conversion-processes.aspx. Accessed May 16, 2024.
23. ICAO. (2024b). *CORSIA News* [WWW Document]. URL https://www.icao.int/environmental-protection/CORSIA/Pages/default.aspx. Accessed May 18, 2024.
24. ISCC. (2021). *Sustainability certification of SAF* [WWW Document]. URL https://www.iscc-system.org/wp-content/uploads/2022/07/1.-Norbert-Schmitz_Sustainability-Certification-of-SAF.pdf. Accessed May 18, 2024.
25. ISCC. (2024). *Sustainable aviation fuel* [WWW Document]. URL https://www.iscc-system.org/markets/sustainable-aviation-fuels/. Accessed May 14, 2024.
26. Jensen, L. L., Bonnefoy, P. A., Hileman, J. I., & Fitzgerald, J. T. (2023). The carbon dioxide challenge facing U.S. aviation and paths to achieve net zero emissions by 2050. *Progress in Aerospace Sciences, 141*, 100921. https://doi.org/10.1016/j.paerosci.2023.100921
27. Klein, B. C., Scheidemantle, B., Hanes, R. J., Bartling, A. W., Grundl, N. J., Clark, R. J., & Cai, C. M. (2024). Economics and global warming potential of a commercial-scale delignifying biorefinery based on co-solvent enhanced lignocellulosic fractionation to produce alcohols, sustainable aviation fuels, and co-products from biomass. *Energy & Environmental Science, 17*(3), 1202–1215.
28. Kousoulidou, M., & Lonza, L. (2016). Biofuels in aviation: Fuel demand and CO2 emissions evolution in Europe toward 2030. *Transportation Research Part D: Transport and Environment, 46*, 166–181. https://doi.org/10.1016/j.trd.2016.03.018

29. Kumar, D., Long, S. P., & Singh, V. (2018). Biorefinery for combined jet fuel and ethanol production from lipid-producing sugarcane: A techno-economic evaluation. *GCB Bioenergy, 10*(2), 92–107.
30. Lan, K., Cruz, D., Li, J., Agyei Boakye, A. A., Park, H., Tiller, P., & Yao, Y. (2024). Life-cycle assessment of sustainable aviation fuel derived from paper sludge. *ACS Sustainable Chemistry & Engineering, 12*, 8379–8390.
31. Liu, Z., Deng, Z., Davis, S. J., Giron, C., & Ciais, P. (2022). Monitoring global carbon emissions in 2021. *Nature Reviews Earth & Environment, 3*(4), 217–219. https://doi.org/10.1038/s43017-022-00285-w
32. Marangon, B. B., de Siqueira Castro, J., & Calijuri, M. L. (2024). Aviation fuel based on wastewater-grown microalgae: Challenges and opportunities of hydrothermal liquefaction and hydrotreatment. *Journal of Environmental Management, 354*, 120418.
33. Michaga, M. F. R., Michailos, S., Akram, M., Cardozo, E., Hughes, K. J., Ingham, D., & Pourkashanian, M. (2022). Bioenergy with carbon capture and storage (BECCS) potential in jet fuel production from forestry residues: A combined Techno-Economic and Life Cycle Assessment approach. *Energy Conversion and Management, 255*, 115346.
34. Neves, R. C., Klein, B. C., da Silva, R. J., Rezende, M. C. A. F., Funke, A., Olivarez-Gómez, E., & Maciel-Filho, R. (2020). A vision on biomass-to-liquids (BTL) thermochemical routes in integrated sugarcane biorefineries for biojet fuel production. *Renewable and Sustainable Energy Reviews, 119*, 109607.
35. Ng, K. S., Farooq, D., & Yang, A. (2021). Global biorenewable development strategies for sustainable aviation fuel production. *Renewable and Sustainable Energy Reviews, 150*, 111502. https://doi.org/10.1016/j.rser.2021.111502
36. Pipitone, G., Zoppi, G., Pirone, R., & Bensaid, S. (2023). Sustainable aviation fuel production using in-situ hydrogen supply via aqueous phase reforming: A techno-economic and life-cycle greenhouse gas emissions assessment. *Journal of Cleaner Production, 418*, 138141.
37. Prussi, M., Lee, U., Wang, M., Malina, R., Valin, H., Taheripour, F., Velarde, C., Staples, M. D., Lonza, L., & Hileman, J. I. (2021). CORSIA: The first internationally adopted approach to calculate life-cycle GHG emissions for aviation fuels. *Renewable and Sustainable Energy Reviews, 150*, 111398.
38. Puschnigg, S., Fazeni-Fraisl, K., Lindorfer, J., & Kienberger, T. (2023). Biorefinery development for converting softwood residues into sustainable aviation fuel: Implications from life cycle assessment and energetic-exergetic analyses. *Journal of Cleaner Production, 386*, 135815.
39. Rivas-Interian, R. M., Sanchez-Ramirez, E., Quiroz-Ramírez, J. J., & Segovia-Hernandez, J. G. (2023). Feedstock planning and optimization of a sustainable distributed configuration biorefinery for biojet fuel production via ATJ process. *Biofuels, Bioproducts and Biorefining, 17*(1), 71–96.
40. Robertson, K. A., English, B. C., Clark, C. D., Thompson, J. M., Jensen, K. L., Menard, R. J., & Labbé, N. (2021). Optimal N application rates on Switchgrass for producers and a biorefinery. *Energies, 14*(23), 7912.
41. Romero-Izquierdo, A. G., Gómez-Castro, F. I., Hernández, S., & Gutiérrez-Antonio, C. (2022a). Computer aided-design of castor bean fruit-based biorefinery scheme to produce sustainable aviation fuel. *Chemical Engineering Research and Design, 188*, 746–763.
42. Romero-Izquierdo, A. G., Gutiérrez-Antonio, C., Gómez-Castro, F. I., & Hernández, S. (2022b). Advanced biorefineries for the production of renewable aviation fuel. In *Sustainable alternatives for aviation fuels* (pp. 103–124). Elsevier. https://doi.org/10.1016/B978-0-323-85715-4.00008-2
43. RSB. (2023). *RSB standard for ICAO CORSIA*. Switzerland.
44. RSB. (2024). *RSB CORSIA Certification* [WWW Document]. URL https://rsb.org/certification/certification-schemes/rsb-corsia-certification/. Accessed May 18, 2024.

45. Sadhukhan, J., & Sen, S. (2021). A novel mathematical modelling platform for evaluation of a novel biorefinery design with Green hydrogen recovery to produce renewable aviation fuel. *Chemical Engineering Research and Design, 175*, 358–379.
46. Sadhukhan, J., Ng, K. S., & Martinez-Hernandez, E. (2014). *Biorefineries and chemical processes. Design, integration, and sustainability analysis*. Wiley.
47. Santos, C. I., Silva, C. C., Mussatto, S. I., Osseweijer, P., van der Wielen, L. A., & Posada, J. A. (2018). Integrated 1st and 2nd generation sugarcane bio-refinery for jet fuel production in Brazil: Techno-economic and greenhouse gas emissions assessment. *Renewable Energy, 129*, 733–747.
48. Seufitelli, G. V., El-Husseini, H., Pascoli, D. U., Bura, R., & Gustafson, R. (2022). Techno-economic analysis of an integrated biorefinery to convert poplar into jet fuel, xylitol, and formic acid. *Biotechnology for Biofuels and Bioproducts, 15*(1), 143.
49. Shehab, M., Moshammer, K., Franke, M., & Zondervan, E. (2023). Analysis of the potential of meeting the EU's sustainable aviation fuel targets in 2030 and 2050. *Sustainability, 15*(12), 9266. https://doi.org/10.3390/su15129266
50. SkyNRG. (2023). *Sustainable aviation fuel certification and ASTM international: What is it and why does it matter?* [WWW Document]. URL https://skynrg.com/sustainable-aviation-fuel-certification-and-astm-international-what-is-it-why-does-it-matter/
51. Song, M., Zhang, X., Chen, Y., Zhang, Q., Chen, L., Liu, J., & Ma, L. (2023). Hydroprocessing of lipids: An effective production process for sustainable aviation fuel. *Energy, 283*, 129107.
52. Tanzil, A. H., Brandt, K., Wolcott, M., Zhang, X., & Garcia-Perez, M. (2021a). Strategic assessment of sustainable aviation fuel production technologies: Yield improvement and cost reduction opportunities. *Biomass and Bioenergy, 145*, 105942.
53. Tanzil, A. H., Zhang, X., Wolcott, M., Brandt, K., Stöckle, C., Murthy, G., & Garcia-Perez, M. (2021b). Evaluation of dry corn ethanol bio-refinery concepts for producing sustainable aviation fuel. *Biomass and Bioenergy, 146*, 105937.
54. Teoh, R., Schumann, U., Voigt, C., Schripp, T., Shapiro, M., Engberg, Z., Molloy, J., Koudis, G., & Stettler, M. E. J. (2022). Targeted use of sustainable aviation fuel to maximize climate benefits. *Environmental Science & Technology, 56*(23), 17246–17255. https://doi.org/10.1021/acs.est.2c05781
55. Trejo-Pech, C. O., Larson, J. A., English, B. C., & Yu, T. E. (2021). Biofuel discount rates and stochastic techno-economic analysis for a prospective pennycress (Thlaspi arvense L.) sustainable aviation fuel supply chain. *Frontiers in Energy Research, 9*, 770479.
56. Wang, W.-C., & Tao, L. (2016). Bio-jet fuel conversion technologies. *Renewable and Sustainable Energy Reviews, 53*, 801–822. https://doi.org/10.1016/j.rser.2015.09.016
57. Wang, M., Dewil, R., Maniatis, K., Wheeldon, J., Tan, T., Baeyens, J., & Fang, Y. (2019). Biomass-derived aviation fuels: Challenges and perspective. *Progress in Energy and Combustion Science, 74*, 31–49. https://doi.org/10.1016/j.pecs.2019.04.004
58. Xu, B., Ahmad, S., Charles, V., & Xuan, J. (2022). Sustainable commercial aviation: What determines air travellers' willingness to pay more for sustainable aviation fuel? *Journal of Cleaner Production, 374*, 133990. https://doi.org/10.1016/j.jclepro.2022.133990

Chapter 11
Performance, Combustion, and Emissions from Green Aviation Fuel

Jorge Aburto and Arick Castillo-Landero

1 Introduction

Nowadays, the aviation industry is considered to be the most popular means of long-distance transport for people and in recent years the use of this means of transport has been increasing, which is why this sector has set itself challenges where the decarbonization of the energy sector is taking place [1]. The aim is to have net-zero emissions by 2050 and since 2019 it has been committed to carbon-neutral growth. However, these goals are not easy to achieve, but scientists are currently working on the development of sustainable technologies that benefit climate change and the depletion of existing fossil fuel resources for the production of aviation fuels [2].

The global aviation sector accounts for 2–3% of all carbon dioxide (CO_2) emissions and 12% of all CO_2 emissions from the transport sector. This may be because the percentage of the population that uses this type of transport is small, but it tends to do so on a frequent basis nowadays. The use of biofuels is an alternative to reduce the climate impact of transport modes such as aviation, but compared to land transport, where ethanol or electricity can be used, aviation requires the use of energy-dense, non-oxygenated liquid hydrocarbon fuels. Therefore, the limitation between demand requirements and available biomass makes it difficult to produce using biomass conversion concepts. Despite this, work has been done on the production of biokerosene fuels from different types of biomass sources and on other occasions the use of blended fuels, i.e., a mixture of fossil fuels and biofuels to meet standards and demand but with contributions to decarbonization [3, 4].

Due to this concern about the increasing use of this mode of transport and the sustainable technologies required to be developed, the use of sustainable aviation

J. Aburto · A. Castillo-Landero (✉)
Energy Efficiency and Biofuels Research Division, Instituto Mexicano del Petróleo, Mexico City, Mexico
e-mail: acastillo@imp.mx

M. Aslam et al. (eds.), *Sustainable Aviation Fuels*, Sustainable Aviation,
https://doi.org/10.1007/978-3-031-83721-0_11

fuel (SAF) has been chosen as an alternative using renewable or CO_2-based fuel sources as raw material, which reduces emissions from air transport [1, 5].

To this end, the following are some of the actions that have been carried out or defined by institutions at the international level to seek strategies and comply with the established agreements.

The Air Transport Action Group (ATAG) set a net CO_2 emission reduction of 50% by 2050 compared to 2005, which amounts to around 325 $MtCO_2$. The Airports Council International (ACI) Europe has set a target of net-zero CO_2 emissions from airport operations within airport control. Furthermore, the Paris Agreement's Nationally Determined Contributions (NDCs) indicate that the European Union includes international aviation in its NDC, which means that the Paris Agreement does not explicitly exclude or include emissions from international aviation and shipping (IAS). Out of necessity due to the issue of fuel use and its impact, most governments now exempt taxation of jet fuel used on international routes, so carbon taxes and emissions trading schemes are becoming more widespread by directly or indirectly pricing CO_2 emissions from fuels and other sources of emissions through different mechanisms [6].

As a result, the U.S. Air Force has as a goal to start replacing 50% of the conventional fuel with SAF. Similarly, the European Union has begun to promote the use of renewable fuels by blending them with conventional fossil fuels to bring benefits to the environmental problems that exist today worldwide and thus achieve carbon neutral growth accompanied by institutions such as the Carbon Offsetting and Reduction Scheme for International Aviation (CORSIA) of the International Civil Aviation Organization (ICAO) [5]. The ICAO in Mexico has promoted the initiative to produce SAF as a need to drive towards independence from fossil fuels, have lower fuel consumption by having higher energy content, create opportunities for decent jobs in rural and indigenous communities, and an estimated 80% reduction of greenhouse gas emissions in the development of second-generation biofuels [7]. On the other hand, the International Air Transport Association (IATA) mentions that one of the promising ways to reduce greenhouse gas emissions is through the development of biofuels and synthetic fuels and defines SAF as a liquid aviation fuel that reduces CO_2 emissions by up to 80% [8, 9].

Today, the American Society for Testing and Materials (ASTM) has approved different routes for the production of SAF based on different technologies that nowadays have different levels of maturity to achieve the goals to get a green aviation industry. The SAF obtained, also called biojet fuel or synthetic paraffinic kerosene (SPK), is a mixture of renewable hydrocarbons with composition and properties similar to the fossil jet fuel [10, 11].

It exists three main factors that influence the use of biofuels or synthetic fuels for transportation in the whole world such as substituting fossil-based fuels, improving energetic security, and reducing greenhouse gas emissions (GHG) according to the IEA Bioenergy Report published in 2023. On the other hand, acidification, eutrophication, human toxicity, ecological toxicity, biodiversity, water, and terrestrial footprints are other indicators to consider in the life cycle assessment (LCA) [1].

Unlike fossil aviation fuels, SAF reduces CO_2 emissions throughout its life cycle. This is because the CO_2 absorbed by the plants during biomass growth comes from different sources such as CO_2 produced when the fuel is burned in the combustion engine and returned to the atmosphere. Figure 11.1a shows the life cycle of these

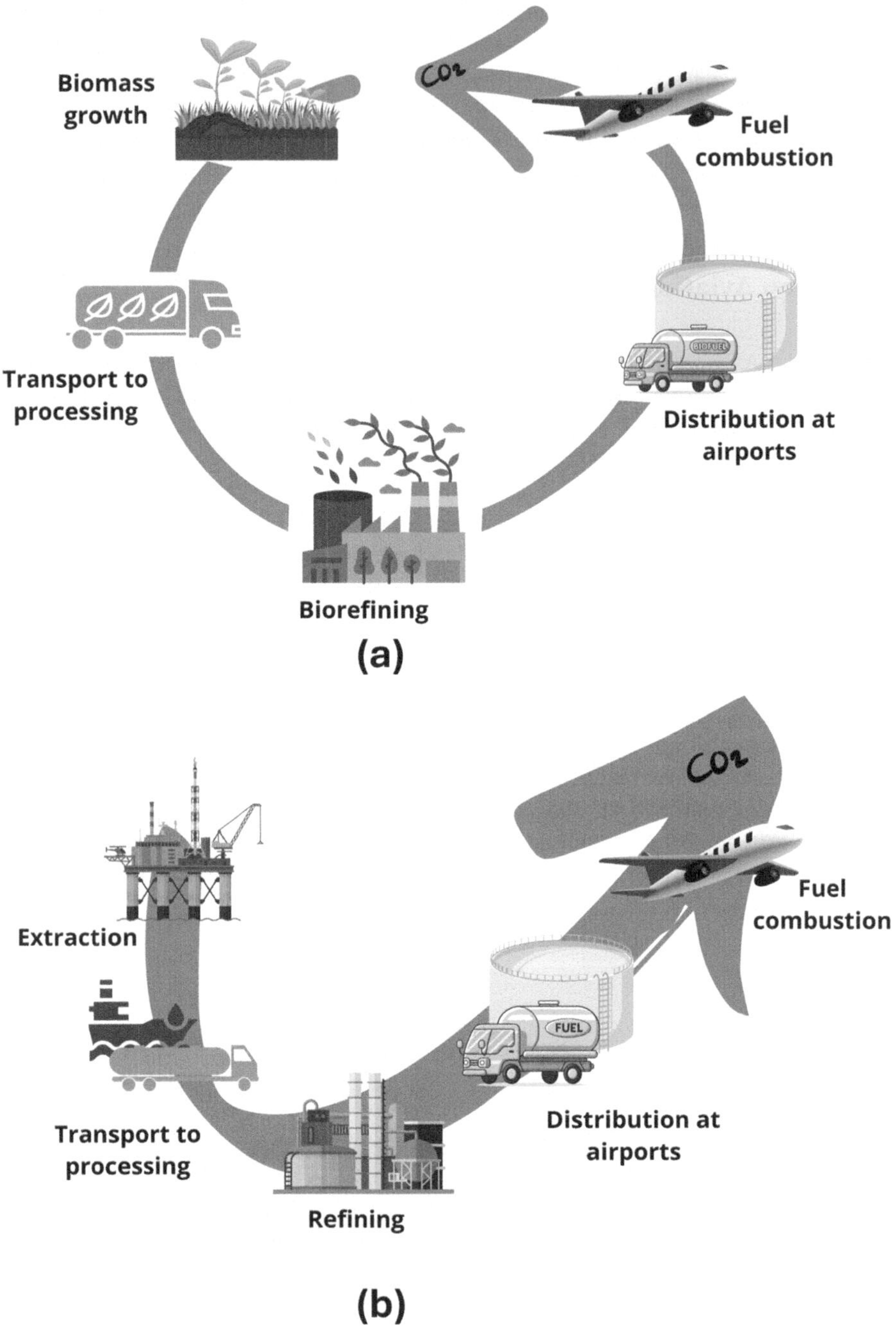

Fig. 11.1 Schematic of the carbon life cycle using aviation fuel from (**a**) renewable sources and (**b**) fossil sources

emissions from the growth of the biomass to the burning of the fuel, through the transport of the biomass to the production plant, the biorefinery, and on to its distribution to airports and loading onto aircraft. On the other hand, Fig. 11.1b shows the process using fossil fuel by extracting the crude oil, transporting it to the refinery for distribution and use in air transport, it can be seen that a cycle is not fulfilled due to the fact that the raw material does not use CO_2 as in the case of biomass [9].

2 Green Aviation Fuel

Nowadays SAF is considered one of the most promising medium-term solutions to reduce greenhouse gas emissions from the aviation sector in order to mitigate the effects of climate change. Currently, three key concepts can be defined to understand the different types of aviation fuels: conventional aviation fuels (CAF), SAF, and blended SAF. The concepts are shown in Fig. 11.2 [12, 13].

As shown in Fig. 11.2, the use of blended SAF comes from the D1655 and D7566 standards with a maximum allowable blend ratio of up to 50% in order to ensure compatibility with aircraft engines of all eras. The percentage of the blend will depend on the type of technology. For this purpose, Table 11.1 shows some ASTM-approved technologies to produce SAF and defines some of the possible raw materials based on the technology to be used. Figure 11.3 illustrates different pathways used to produce SAF based on different proposals using biomass feedstock.

The choice of technology depends on several factors such as the type of biomass, as biofuels can be classified as first, second, or third generation based on the type of biomass. Based on this, IATA has defined different classifications based on the type of generation feedstock, such as emission reduction potential, ability to meet broader sustainability criteria, and product costs. This description is defined in Table 11.2 [9].

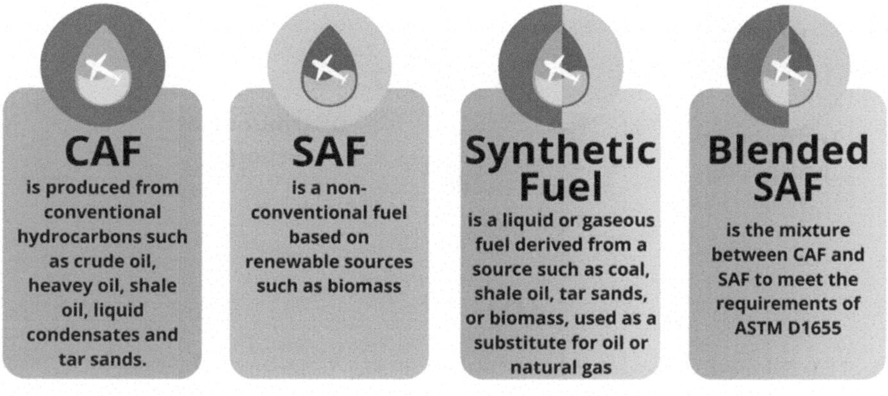

Fig. 11.2 Definition of CAF, SAF, synthetic fuel, and blended SAF

Table 11.1 Technology description to produce SAF approved by the ASTM

Technology	Description	Possible feedstocks	Blending limit (%)
Hydroprocessing of esters and fatty acids (HEFA-SPK)	The HEFA-SPK technology is the most consolidated pathway and certified process by the ASTM for producing SAF. It results lower technological complexity and minimum SAF selling price with higher energy efficiency [14]. HEFA process converts esters and fatty acids into hydrocarbon in the presence of H2, for removal of the oxygen atom from the oxygen. The process, as shown in Fig. 11.1a, consist in four steps: pretreatment, hydrodeoxygenation, isomerization, and product separation to produce biojet fuel, naphtha and green diesel [15].	Vegetable oils Animal fats Used cooking oils	50
Fischer-Tropsch (FT-SPK)	FT-SPK process implies converting the syngas to liquid hydrocarbons as shown in Fig. 11.1b. The process implies producing syngas via gasification of the biomass at high temperature. Before converting the syngas to liquid hydrocarbons, it exists some technology to adjusts the relation of H2 and CO such as autothermal reforming (ATR), partial oxidation (POX) and steam reforming (SMR) [16]. After that, hydrocracking, isomerization, and distillation technologies can be used to get high quality and lower aromatics fuels to meet the conventional jet fuel requirements [17–19].	Energy crops Lignocellulosic biomass Solid waste	50
Alcohol to jet (ATJ-SPK)	ATJ-SPK technology normally extracts polymer sugars from biomass feedstocks to decomposed to monomer sugar and fermented to get an alcohol platform molecule. The alcohol is dehydrated, oligomerized, and hydrogenated to a final fuel product [3].	Sugar Starch crops Lignocellulosic biomass	50
Isoparaffins from hydroprocessed fermented sugars (SIP)	SIP technology performs a biological conversion of biomass into synthetic isoparaffins. It involves pre-treatment to separate the sugars from the lignin. The sugars are then converted to farnesene by enzymatic hydrolysis and fermentation. A solid-liquid separation system is then applied to recover the farnesene and finally the farnesene is fed into a hydroprocessing step to obtain biofuel [20].	Biomass used for sugar production	10

(continued)

Table 11.1 (continued)

Technology	Description	Possible feedstocks	Blending limit (%)
Fischer-Tropsch plus aromatics (FT-SPK/A)	FT-SPK/A technology is closely related to FT-SPK. The difference lies in the presence of aromatics in the latter technology [21]. The aromatics used to produce FT-SPK/A can be obtained using a cut of benzene-rich light naphtha. It can then be oligomerised and hydrotreated to obtain the product and naphtha [22].	Coal Natural gas Biomass	50
Catalytic hydrothermolysis jet (CHJ-SPK)	It is also called hydrothermal liquefaction (HTL). The process involves converting biomass to biocrude oil by catalytic hydrothermolysis in which triglycerides are transformed to produce a crude hydrocarbon oil, and an aqueous phase containing glycerol and lower weight molecule is hydrotreated and separated [23].	Vegetable oils Animal fats Used cooking oils	50
Paraffinic kerosene produced from bio-derived hydroprocessed hydrocarbons (HC-HEFA-SPK)	HC-HEFA-SPK technology produces SAF though the catalytic pyrolysis of oil fats. The main characteristic is that the hydrocarbons are of biological origin such as algae [10, 24]	Algae	10

As mentioned in Table 11.2, the production costs and emission reduction potential are average, since on a commercial scale the process has implications based on biomass procurement, as it sometimes requires arable land use and in extreme cases deforestation. On the other hand, in the case of second-generation fuels, there are improvements compared to first-generation fuels, since biomass does not require land, it would be a second use that is given to it. On the other hand, the technology to carry out the production, due to the fact that they have already had a first use, requires greater neatness in its process and therefore has high production costs. Third-generation biofuels are abundant in nature and contribute positively to the reduction of GHG emissions compared to first- and second-generation biofuels, as they are made from by-products and residues and have a use for these wastes [9, 25].

Table 11.3 shows a comparison of the life cycle analysis of different published works comparing the global warming potential using different technologies and feedstocks. It should be noted that in the case of biofuel, the carbon intensity in the GWP category may vary depending on the biomass feedstock [26].

One of the main components to analyze in making the decision whether to use SAF or blending SAF is the similarities related to CAF under ASTM standards. The composition of CAF varies based on the technology used for its production and the fossil source used which include crude oil, natural gas, liquid condensates, heavy oil, shale oil, and tar sands. CAF has in its composition n-paraffins, iso-paraffins, and cycloparaffins between 70% and 85% of its total composition by mass, 25% or

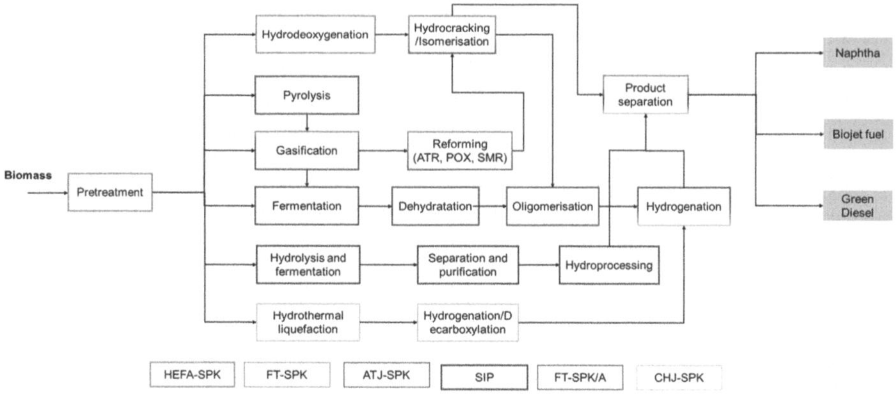

Fig. 11.3 Pathway technologies to produce SAF based on biomass transformation

Table 11.2 Generation feedstock and its characteristics

Generation feedstock	Biomass	Emissions reduction potential	Production Cost
First generation	Canola/Rapeseed oil Palm oil & derivatives Soybean oil Sunflower oil Corn	Mid	Mid
Second generation	Used cooking oil Inedible animal fats & Tallow Trap grease Fatty acid distillates	High	High
Third generation	Wet wastes Algae oils Forestry & Agri residues Food waste Municipal solid waste	High	Low

less of aromatics, and less than 1% olefins. Brooks et al. [30] describe the main characteristics of jet fuel such as clean combustion with high heat release per unit weight, fluidity at low fluid temperature, lower hydrogen/carbon ratio, lower heat release per unit weight, higher density, and lower freezing point than n-paraffins, prone to form soot during combustion, high energy density, good combustion characteristics, reactive and leads to gum formation [30]. Table 11.4 shows the comparison between the properties required to meet the ASTM D1655 standard of CAF and SAF which were collected by Chong and Ng [31].

Table 11.3 Comparison of the global warming potential (GWP) in different SAF or CAF production processes

Feedstock	Technology	Boundary system	SAF GWP kg CO_2 eq./GJ of fuel	Total energy input for product generation		References
				Nonrenewable fossil [MK]	Renewable [MJ]	
Camelina meal	HTL	Cradle to gate (final transportation is included)	47.0	–	–	[27]
Soybean oil	2019	Cradle to gate	40.1	0.4040	–	[28]
Palm oil	2019	Cradle to gate	14.2	0.1890	–	[28]
Palm oil	2021	Cradle to gate	4.8–14.2	–	–	[29]
Palm oil	2021	Cradle to gate	13.0	0.1437	1.962	[26]
Palm oil	2021	Cradle to gate	12.8	0.1395	1.962	[26]
Palm oil	2021	Cradle to gate	12.2	0.1160	1.962	[26]

Table 11.4 Comparison of some properties between CAF and SAF produce by FT-SPK, FT-SPK/A, ATJ-SPK, HEFA-SPK, and CHJ technology

Property	CAF	FT-SPK	FT-SPK/A	ATJ-SPK	HEFA-SPK	CHJ
Flash point, °C	≥38	≥38	≥38	≥38	≥38	≥38
Density at 15 °C, kg/m^3	773–830	730–770	755–800	730–770	730–773	775–840
Heat of combustion MJ/kg	≥42.80	44.10		43.20	44.15	
Freezing point, °C	≤−47	≤−40	≤−40	≤−40	≤−40	≤−40
Thermal stability						
Temperature, °C	≥260	≥325	≥325	≥325	≥325	≥325
Filter pressure drop, mmHg	≤25	≤25	≤25	≤25	≤25	≤25
Hydrocarbon composition						
Cycloparaffin, mass%	–	≤15	≤15	≤15	≤15	–
Aromatics, mass%	–	≤0.5	≤0.5	≤0.5	≤0.5	8.4–21.2
Aromatics, vol%	8.4–26.5	–	–	–	–	–

3 Combustion and Emissions of SAF

Once the impact of SAF production has been described and analyzed, another important point to consider is the combustion efficiency of this biofuel, which has a great impact on the environmental and economic viability of these fuels to understand the complete process from the LCA analysis. Combustion efficiency is a measure of how effectively a fuel converts its chemical energy into useful work during the combustion process. In terms of CAF, 1 kilogram of kerosene when burned

emits 1240 g of H_2O, 3150 g of CO_2, 6–20 g of NO_x, 1 g of SO_2, 0.7–2.5 g of CO, 0.1–0.7 g of unburned hydrocarbons (UHC), and 0.01–0.2 g of soot are produced as shown in Fig. 11.4 [32].

The values described above are checked with the proposed model for the combustion of a hydrocarbon fuel, sufficient air is required to convert the fuel completely into CO_2 and water vapor. Marszlek and Lis [24] defined a stoichiometric equation to calculate the air-fuel ratio by defining the molecular composition (C_aH_b) of a hydrocarbon fuel.

$$C_aH_b + \left(a + \frac{b}{4}\right)(O_2 + 3.773N_2) = aCO_2 + \frac{b}{2}H_2O + 3.773\left(a + \frac{b}{4}\right)N_2, \quad (11.1)$$

Within the combustion process, properties play a key role, as they are part of different processes that are fundamental to combustion performance. The combustion process consists mainly of three stages. The first stage is the spraying of the liquid fuel into small droplets; among the properties that are part of this stage is volatility. Once sprayed, the fuel is mixed with air and burned to near stoichiometric conditions. Finally, the burned gases are diluted with excess air [33]. There are three critical parts to perform the combustion: ignition and re-ignition altitude, lean blowing and exhaust emissions, and combustion efficiency. From this classification, Fig. 11.5 shows the fundamental processes and the impact of the different properties in carrying out the combustion process [34].

With respect to pollutants, the main characteristics of each can be described, for example, the formation of NO_x in aircraft turbines which is due to the thermal dissociation of N_2 and O_2 molecules present in the air and when exposed to high temperatures creates the Zeldovich mechanism (Eqs. 11.2 and 11.3) [35].

$$N_2 + O \rightarrow NO + N \quad (11.2)$$

$$N + O_2 \rightarrow NO + O \quad (11.3)$$

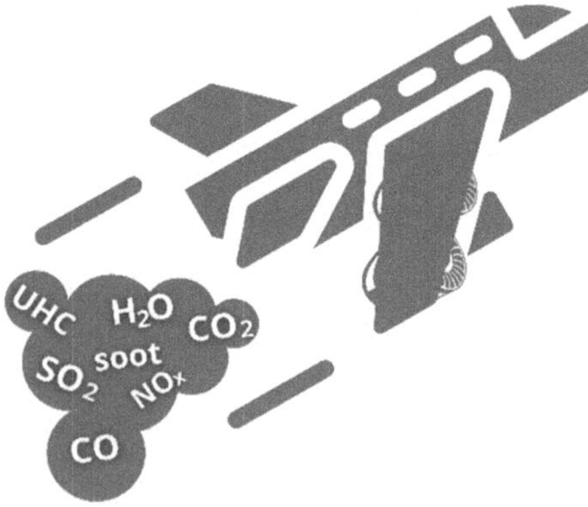

Fig. 11.4 Schematic representation of emissions in CAF combustion

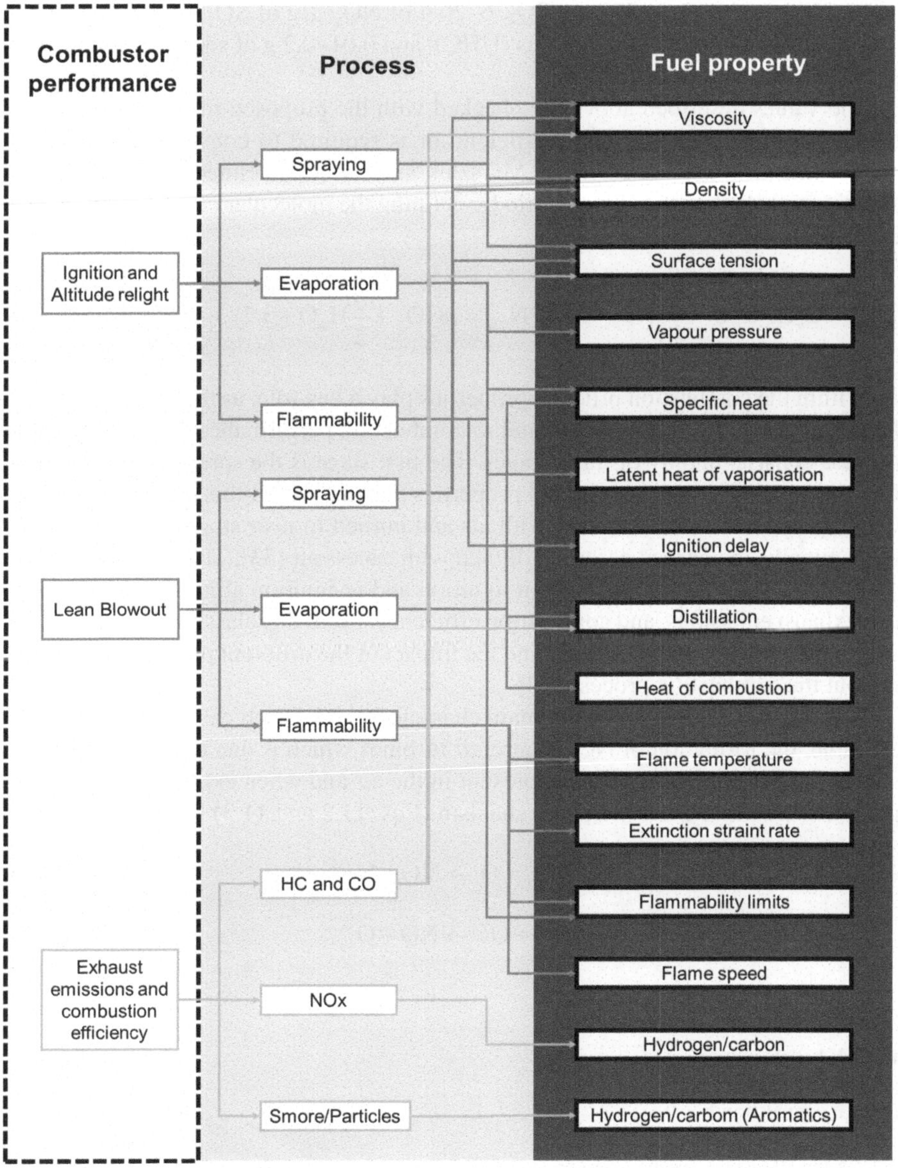

Fig. 11.5 Fuel properties impact in the combustion performance

On the other hand, there are the unburned hydrocarbons which are due to fuel that does not burn completely. UHC includes methane, formaldehyde, ethylene, acetaldehyde, acrolein, benzene, toluene, 1,3-butadiene, and higher aromatics species. The UHC emissions are higher when fuel consumption is the lowest (idle

condition) instead of when fuel consumption is the highest (takeoff conditions) [35].

SAF has demonstrated good efficiency during its combustion process, especially when mixed with 50% CAF as shown in Table 11.5 such as HEFA-SPK and FT-SPK fuels which achieve good combustion efficiency compared to CAF but it will depend on the biomass fed. ATJ-SPK shows a reduction in CO_2 and CO but its combustion increases NO_x and UHC emissions due to the higher average number of particles generated when the engine is fuelled with CAF and the lower average number of particles per kilogram of fuel used [36]. The cited works show either qualitatively or quantitatively the improvement over different SAF put into combustion.

From the results obtained when looking at the trend of emissions that sustainable fuels emit compared to fossil fuels, it can be seen that exhaust emissions are not necessarily lower, but in terms of life cycle analysis, they are considered less harmful to the environment.

In 2008 the first flight test was carried out using 20% SAF from coconut and babassu oil. From this first test, further studies were carried out by different commercial airlines. Table 11.6 shows the tests done by different airlines with SAF from different biomass sources. These tests have been part of the fuel certification process and have shown the acceptance of biofuels in the airline industry [41].

As the test flights are part of the rigorous process to obtain biojet fuel certification, it is expected that more flight tests will be conducted in the near future, as different alternative jet fuel methods are being developed. At the same time, the commitment shown by commercial airlines to the use of SAF indicates the growing acceptance of biofuels, which is important in the long term to achieve the goal of decarbonization in the aviation sector. After several studies it has been shown that SAFs have the potential to reduce GHG emissions compared to fossil fuels since biomass absorbs CO_2 from the atmosphere that is emitted during combustion, in terms of combustion there is not necessarily a better lower impact, but it does generate a cycle of CO_2 [42].

Table 11.5 Comparison of SAF emissions during combustion in aviation engines with respect to CAF

Biofuel	CO_2 emissions	CO emissions	NO_x emissions	UHC emissions	References
HEFA-SPK	20% reduction	20% reduction	No significance chance	Lower emissions due low aromatic content	[37]
HEFA-SPK	95% reduction	50% reduction	95% reduction	Not reported	[38]
FT-SPK	Lower emissions	Lower emissions	Lower emissions	Not reported	[39]
FT-SPK	Lower emissions	5–10% reduction	5% reduction	7% reduction at idle condition	[40]
ATJ-SPK	8 & reduction	23% reduction	50% increase	36% increase	[36]

Table 11.6 Flight tests recorded with different-biomass sustainable aviation fuel

Year	Biomass	Airline
2008	Babassu & coconut Jatropha	Virgin Atlantic Air New Zealand
2009	Algae & jatropha Camelina, jatropha & algae Camelina	Continental Airline Japan Airline KLM
2010	Jatropha	TAM Airlines
2011	Jatropha Camelina	Air China Interjet Boeing Honeywell ASA-Mexico
2012	Used cooking oil Sugarcane Brassica carinata	Etihad Airway All Nipon Airways Azul Airlines Air Canada
2013	Used cooking oil + palm	China Eastern Airlines
2014	Used cooking oil + animal fat	Boeing
2015	Used cooking oil	Hainan Airline
2016	Woody biomass	Alaska Airline
2017	Rapeseed	Saab
2018	Jatropha Brassica carinata	SpiceJet Qantas
2019	Salicornia	Etihad Airway

4 Conclusions

As can be seen in this chapter, the use of sustainable aviation fuels is an alternative to meet the objectives set by the aviation industry at the international level. At present, the implementation of these fuels from biomass sources is not intended to replace aviation fuels, but to accompany them in proportions that can range from 10% to 50% based on ASTM standards, which regulate the mixture of SAF and CAF with the aim of covering the properties required for aircraft engines to operate without any problems. Tests have been carried out currently with up to 50% blending and some test flights on 100% with SAF in different commercial flights and improvements have been seen in terms of environmental impact due to combustion.

On the other hand, it is important to mention that the biggest impact is in the production of SAF fuels. Since they come from non-fossil sources, this guarantees the use of clean raw material and that during the cycle of CO_2 emissions, the same that come out of the combustion of the biofuel, or their combinations, the ecosystem takes advantage of it for the growth of biomass.

Author Contributions All authors contributed to the study's conception and design. Material preparation, data collection, and analysis were performed by Jorge Aburto and Arick Castillo-Landero. The first draft of the manuscript was written by Arick Castillo-Landero and all authors

commented on previous versions of the manuscript. All authors read and approved the final manuscript.

Declarations The authors have no competing interests to declare that are relevant to the content of this article.

Jorge Aburto is an editorial board member for the book title "Sustainable Aviation Fuels: Recent Advances and Future Challenges" and was not involved in the editorial review, or the decision to publish, this article. All authors declare that there are no competing interests.

References

1. IEA Bioenergy. (2023). *How bioenergy contributes to a sustainable future*, International Energy Agency (www.iea.org).
2. Afonso, F., Sohst, M., Diogo, C. M. A., et al. (2023). Strategies towards a more sustainable aviation: A systematic review. *Progress in Aerospace Sciences, 137*, 100878.
3. Yao, G., Staples, M. D., Malina, R., & Tyner, W. E. (2017). Stochastic techno-economic analysis of alcohol-to-jet fuel production. *Biotechnology for Biofuels, 10*, 1–13.
4. Wang, M., Dewil, R., Maniatis, K., et al. (2019). Biomass-derived aviation fuels: Challenges and perspective. *Progress in Energy and Combustion Science, 74*, 31–49.
5. Bhatt, A. H., Zhang, Y., Milbrandt, A., et al. (2023). Evaluation of performance variables to accelerate the deployment of sustainable aviation fuels at a regional scale. *Energy Conversion and Management, 275*, 116441.
6. Cazzola, P., & Lassman, J. (2021). *Decarbonising air transport: Acting now for the future*. International Transport Forum (www.itf-oecd.org).
7. Buenrostro Domínguez, T. *Biocombustibles Sustentables de Aviación en México*.
8. Romero-Izquierdo, A. G., Gómez-Castro, F. I., Gutiérrez-Antonio, C., et al. (2021). Intensification of the alcohol-to-jet process to produce renewable aviation fuel. *Chemical Engineering and Processing: Process Intensification, 160*, 108270.
9. *IATA annual review*. (2021). International Air Transport Association (www.iata.org).
10. ASTM D7566-22 (2023). *Standard specification for aviation turbine fuel containing synthesized hydrocarbons*. American Society for Testing and Materials (www.astm.org).
11. ASTM D1655–09. (2009). *Standard specification for aviation turbine fuels*. American Society for Testing and Materials (www.astm.org).
12. IEA Bioenergy. (2021). *Progress in commercialization of biojet/sustainable aviation fuels (SAF): Technologies, potential and challenges*. IEA Bioenergy Task 39. Technology Collaboration Programme by IEA.
13. Ridjan, I., Mathiesen, B. V., & Connolly, D. (2016). Terminology used for renewable liquid and gaseous fuels based on the conversion of electricity: A review. *Journal of Cleaner Production, 112*, 3709–3720. https://doi.org/10.1016/j.jclepro.2015.05.117
14. Liu, S., Zhu, Q., Guan, Q., et al. (2015). Bio-aviation fuel production from hydroprocessing castor oil promoted by the nickel-based bifunctional catalysts. *Bioresource Technology, 183*, 93–100.
15. Martinez-Hernandez, E., Ramírez-Verduzco, L. F., Amezcua-Allieri, M. A., & Aburto, J. (2019). Process simulation and techno-economic analysis of bio-jet fuel and green diesel production—Minimum selling prices. *Chemical Engineering Research and Design, 146*, 60–70.
16. Julián-Durán, L. M., Ortiz-Espinoza, A. P., El-Halwagi, M. M., & Jiménez-Gutiérrez, A. (2014). Techno-economic assessment and environmental impact of shale gas alternatives to methanol. *ACS Sustainable Chemistry & Engineering, 2*, 2338–2344.

17. Lundberg, S., Fagerström, A., & Johansson, K. (2020). Life cycle assessment of gasification-based Fischer-Tropsch bio jet fuel production. Retrieved from IVL Svenska Miljöinstitutet website: https://urn.kb.se/resolve?urn=urn:nbn:se:ivl:diva-2792.
18. Wang, W.-C. (2016). Techno-economic analysis of a bio-refinery process for producing hydro-processed renewable jet fuel from jatropha. *Renewable Energy, 95*, 63–73.
19. Hsu, K.-H., Wang, W.-C., & Liu, Y.-C. (2018). Experimental studies and techno-economic analysis of hydro-processed renewable diesel production in Taiwan. *Energy, 164*, 99–111.
20. Serrano-Ruiz, J. C., & Dumesic, J. A. (2011). Catalytic routes for the conversion of biomass into liquid hydrocarbon transportation fuels. *Energy & Environmental Science, 4*, 83–99.
21. Tanzil, A. H., Brandt, K., Wolcott, M., et al. (2021). Strategic assessment of sustainable aviation fuel production technologies: Yield improvement and cost reduction opportunities. *Biomass and Bioenergy, 145*, 105942.
22. de Klerk, A., Chauhan, G., Halmenschlager, C., et al. (2024). Sustainable aviation fuel: Pathways to fully formulated synthetic jet fuel via Fischer–Tropsch synthesis. *Energy Science & Engineering, 12*, 394–409.
23. Yang, E., Chon, K., Kim, K.-Y., et al. (2023). Pretreatments of lignocellulosic and algal biomasses for sustainable biohydrogen production: Recent progress, carbon neutrality, and circular economy. *Bioresource Technology, 369*, 128380.
24. Marszałek, N., & Lis, T. (2022). The future of sustainable aviation fuels. *Combustion Engines, 61*(4), 29–40.
25. Bezergianni, S., & Dimitriadis, A. (2013). Comparison between different types of renewable diesel. *Renewable and Sustainable Energy Reviews, 21*, 110–116.
26. Castillo-Landero, A., Dominguillo-Ramirez, D., Aburto, J., et al. (2023). Improving the economic, environmental, and safety performance of bio-jet fuel production through process intensification and integration using a modularity approach. *ACS Sustainable Chemistry & Engineering, 11*, 660–669.
27. De Jong, S., Antonissen, K., Hoefnagels, R., et al. (2017). Life-cycle analysis of greenhouse gas emissions from renewable jet fuel production. *Biotechnology for Biofuels, 10*, 1–18.
28. Vasquez, M. C., Martinez, A., Castillo, E. F., & Silva, E. E. (2019). Holistic approach for sustainability enhancing of hydrotreated aviation biofuels, through life cycle assessment: A Brazilian case study. *Journal of Cleaner Production, 237*, 117796.
29. Julio, A. A. V., Batlle, E. A. O., Rodriguez, C. J. C., & Palacio, J. C. E. (2021). Exergoeconomic and environmental analysis of a palm oil biorefinery for the production of bio-jet fuel. *Waste and Biomass Valorization, 12*, 5611–5637.
30. Brooks, K. P., Snowden-Swan, L. J., Jones, S. B., et al. (2016). Low-carbon aviation fuel through the alcohol to jet pathway. In C. J. Chuck (Ed.), *Biofuels for aviation* (pp. 109–150). https://doi.org/10.1016/B978-0-12-804568-8.00006-8
31. Chong, C. T., & Ng, J.-H. (2021). Chapter 3 – Property specifications of alternative jet fuels. In C. T. Chong & J.-H. Ng (Eds.), *Biojet fuel in aviation applications* (pp. 143–174). Elsevier.
32. Braun-Unkhoff, M., Riedel, U., & Wahl, C. (2017). About the emissions of alternative jet fuels. *CEAS Aeronautical Journal, 8*, 167–180.
33. Chen, K., & Liu, H. (2013). The impacts of aromatic contents in aviation jet fuel on the volume swell of the aircraft fuel tank sealants. *SAE International Journal of Aerospace, 6*, 350–354.
34. Edwards, T., Moses, C., & Dryer, F. (2010). Evaluation of combustion performance of alternative aviation fuels. In *46th AIAA/ASME/SAE/ASEE joint propulsion conference & exhibit* (p. 7155). https://doi.org/10.2514/6.2010-7155
35. Wey, C., & Lee, C.-M. (2018). Clean combustion and emission control. In *Green aviation: Reduction of environmental impact through aircraft technology and alternative fuels* (pp. 137–163). CRC Press.
36. Kurzawska, P., & Jasiński, R. (2021). Overview of sustainable aviation fuels with emission characteristic and particles emission of the turbine engine fueled ATJ blends with different percentages of ATJ fuel. *Energies (Basel), 14*, 1858.

37. Klingshirn, C. D., DeWitt, M., Striebich, R., et al. (2012). Hydroprocessed renewable jet fuel evaluation, performance, and emissions in a T63 turbine engine. *Journal of Engineering for Gas Turbines and Power, 134*(5). https://doi.org/10.1115/GT2011-46572
38. Wang, Z., Feser, J. S., Lei, T., & Gupta, A. K. (2020). Performance and emissions of camelina oil derived jet fuel blends under distributed combustion condition. *Fuel, 271*, 117685.
39. Chan, T. W., Chishty, W. A., Canteenwalla, P., et al. (2016). Characterization of emissions from the use of alternative aviation fuels. *Journal of Engineering for Gas Turbines and Power, 138*, 011506.
40. Lobo, P., Hagen, D. E., & Whitefield, P. D. (2011). Comparison of PM emissions from a commercial jet engine burning conventional, biomass, and Fischer–Tropsch fuels. *Environmental Science & Technology, 45*, 10744–10749.
41. Zhao, X., Sun, X., Cui, X., & Liu, D. (2019). Chapter 5 – Production of biojet fuels from biomass. In M. Rai & A. P. Ingle (Eds.), *Sustainable bioenergy* (pp. 127–165). Elsevier.
42. Olcay, H., Malina, R., Lewis, K., et al. (2018). Environmental performance of alternative jet fuels. In *Green aviation: Reduction of environmental impact through aircraft technology and alternative fuels* (pp. 307–334). CRC Press.

Chapter 12
Challenges in the Environmentally Sustainable Aviation Ecosystem

Heini Noronen-Juhola, Risto Soukka, and Kaisa Grönman

1 Introduction

Air transport plays a major role in driving global economic growth through the creation of employment opportunities, facilitation of international trade, promotion of tourism, and support for sustainable development. It not only fosters global connectivity but also contributes significantly to social and economic prosperity [37].

In 2017, 1303 airlines transported four billion passengers to 3759 airports worldwide while carrying nearly 62 million tonnes of freight. The scale of this activity fuels the generation of 10.2 million direct jobs and contributes $704 billion to the global gross domestic product (GDP) [8]. To put this into perspective, this contribution equates to approximately 0.9% of the entire global GDP [8]. It is important to note that the impact of the aviation industry ripples extends beyond direct stakeholders, such as airlines, airports, and air navigation service providers, to indirect stakeholders, such as aircraft manufacturers and freight forwarders, among others.

Despite its global reach and impact, the aviation industry is not immune to global phenomena. Events like the 9/11 attacks in the USA in 2001, the 2008 financial crisis triggered by the Lehman Brothers bankruptcy, and the eruption of Iceland's Eyjafjallajökull volcano in 2010 have had significant downturn effects on aviation activities. The COVID-19 pandemic, which put the entire aviation industry on hold, served as a stark example of the industry's vulnerability [28, 29]. However, the aviation sector has demonstrated resilience in the face of such events, with remarkable recovery, evident in the overall reduction of airline seat offerings by only 15% in 2022 compared to 2020, despite a substantial 60% reduction during the peak of the pandemic in 2020 [36].

H. Noronen-Juhola (✉) · R. Soukka · K. Grönman
Sustainability Science, LUT School of Energy Systems, Lappeenranta-Lahti University of Technology, Lappeenranta, Finland
e-mail: Heini.Noronen-Juhola@student.lut.fi

M. Aslam et al. (eds.), *Sustainable Aviation Fuels*, Sustainable Aviation,
https://doi.org/10.1007/978-3-031-83721-0_12

Alongside its undeniable economic and social benefits, the aviation sector confronts substantial challenges. Foremost among these challenges is the issue of global carbon dioxide (CO_2) emissions from fossil jet fuel. Although commercial aviation is responsible for only 2–3% of global emissions, the industry's growth trajectory raises concerns, with emissions estimated to potentially reach as high as 22–30% by mid-century [54].

This chapter aims to comprehensively outline the challenges within the environmentally sustainable aviation ecosystem, with a particular emphasis on CO_2 emissions arising from flight operations. It is crucial to recognize that the aviation industry extends beyond the operations of airlines alone; it functions as an intricate ecosystem supported and influenced by various actors and stakeholders. In addition to airlines, airports, ground handling companies, air navigation service providers, and regulatory authorities play pivotal roles. Collaboration among these entities is essential to establish a successful and sustainable aviation ecosystem [9].

This article is grounded in a thorough literature review and is organized as follows: it begins with an introduction to the aviation industry, providing insights into its market dynamics and the evolving landscape of sustainability considerations. Next, it explores the aviation ecosystem, highlighting its key characteristics. This is followed by an exploration of the challenges posed by the aviation ecosystem, especially in terms of sustainability. The chapter culminates with a summary of the main findings and their significance within the context of the topic area.

2 The Aviation Industry

2.1 *Characteristics and the Market Situation*

Since the post-World War II formation of the aviation industry, it has experienced remarkable growth. Over the years, it has weathered several setbacks, including the energy crisis of the 1970s, the US banking crisis in 2008, and most recently, the COVID-19 pandemic [28, 31, 36]. The industry's character can be defined as volatile, often being one of the first to falter during crises but also displaying remarkable resilience in its swift recovery. Each recovery phase has typically ushered in a steeper upward trajectory in flight volume growth over subsequent years.

Projections indicate that the aviation industry will continue to grow significantly in the foreseeable future [36]. By 2050, it is anticipated that more than 10 billion passengers will travel by air, covering approximately 22 trillion kilometers each year. Although demographic shifts and population changes may temper growth compared to recent years, even before considering the impact of COVID-19, an annual growth rate of 3.1% from 2019 to 2050 is expected. This growth will be driven primarily by regions such as Asia-Pacific, the Middle East, Latin America, and Africa, although notable growth will persist in North America and Europe [9].

Examining growth from the perspective of aviation fuel consumption reveals another facet of this expansion. In 2018, international aviation consumed approximately 188 million tonnes of fuel, resulting in 593 million tonnes of CO_2 emissions. By 2050, fuel consumption is projected to surge to 1.9 to 2.6 times the 2018 figure, while revenue tonne kilometers (RTK) are forecasted to triple under the most recent estimates [36].

Several megatrends are influencing the industry's growth trajectories [29]. Digitalization stands out as a significant megatrend, offering tools to enhance operational efficiency and overall industry performance. Additionally, factors such as an aging population, urbanization, and rising prosperity contribute to the expansion of travel and the aviation sector. The future of air traffic, particularly in China, is poised for substantial growth [60]. However, the imperative of environmental sustainability may temper this growth if industry environmental initiatives do not lead to sustainable outcomes.

Current growth scenarios are contingent on how swiftly and robustly the industry rebounds from the COVID-19 pandemic and the potential emergence of other global economic downturns. Other substantial impacts may stem from the repercussions of the conflict in Ukraine, including high inflation and the closure of Russian airspace.

Traveling and the aviation industry hold deep emotional significance for people. Airlines and airports often evoke strong emotions, symbolizing the beginning of holidays and adventures. Consequently, interest in the industry and its services is expected to remain high in the future. Notably, in Europe, low-cost carriers, known for ferrying passengers to vacation destinations, experienced the most pronounced resurgence after the COVID-19 pandemic in 2022. Their commercial performance rebounded to 85% compared to 2019 levels, while traditional and regional airlines registered comparatively lower recoveries at 75% and 74%, respectively [23].

2.2 *The Environmental Goals of the Aviation Industry*

The International Civil Aviation Organization (ICAO) has laid out an ambitious objective for the aviation industry: to achieve net-zero carbon dioxide (CO_2) emissions by the year 2050 [37, 44]. This goal aligns with the broader global targets established by the United Nations [57]. In comparison to other modes of transportation, this endeavor is notably challenging. For instance, motorcycle transportation aims to reach net-zero emissions by 2040, rail transportation by 2050, and small truck transportation by 2060 [21].

Major industry players like the International Air Transport Association (IATA) have not only embraced but also actively implemented these ambitious targets [33]. This commitment necessitates that aviation-related companies align their environmental goals and work toward reducing their environmental footprint.

To expedite progress, several aviation organizations have published plans to reach interim targets. For example, Eurocontrol has introduced an initiative known as Skygreen, aiming to slash aviation industry CO_2 emissions by 55% by 2030

compared to 1990 levels [23, 24]. This substantial reduction is envisioned through a combination of modernizing and renewing aircraft fleets and increasing the utilization of sustainable aviation fuels. It's worth noting that achieving this goal may require new regulations to promote the blending of sustainable aviation fuels. In support of this, the European Commission approved the blending-enhancing regulation known as ReFuelEU in 2023 [27].

However, making these plans a reality hinges on unwavering industry commitment alongside substantial financial investments. According to Eurocontrol, the implementation of policy-driven decarbonization measures alone could increase airline costs by an estimated 54.8 to 62.0 billion euros by 2030 within the Eurocontrol member states [24].

Looking ahead to 2050, the potential for electric aviation and enhanced aviation technologies holds promise in supporting the use of sustainable aviation fuels (SAF) to achieve net-zero targets. Nonetheless, this optimism is counterbalanced by projections of a 44% increase in the number of flights compared to the 2019 baseline, amounting to a staggering 16 million flights per year [24].

2.3 *Environmental Impacts*

In the context of global warming potential, the aviation industry currently contributes approximately 2% of all human-induced CO_2 emissions, and within the broader transportation sector, it accounts for about 14% of emissions [24]. The primary source of these environmental impacts stems from the emissions generated during flight operations. However, it's important to recognize that a multitude of stakeholders within the aviation ecosystem influence emissions in the aviation industry.

Understanding and justifying these numbers, as well as exploring potential solutions, necessitates a comprehensive analysis of the calculation methods and system boundaries used. Typically, companies in the aviation sector adhere to reporting standards such as ISO 14040 and ISO 14067 [38]. In the European Union (EU), the Corporate Sustainability Reporting Directive (CSRD) came into effect in 2023 to enhance the quality of sustainability reporting [1].

Environmental reports are often included as part of a company's annual reports. Consequently, aviation companies invest significant effort into producing these reports, which are subject to review by environmental professionals. However, while reporting standards provide guidelines, there is a degree of flexibility that allows companies to make decisions about what data and information to include or exclude from their sustainability reports.

This emphasis on transparency and reporting standards plays a crucial role in helping stakeholders, including the public, investors, and regulators, understand the environmental impact of the aviation industry and evaluate the effectiveness of sustainability initiatives and mitigation strategies. As the aviation sector continues to navigate the challenges posed by climate change, these reporting practices will

likely play an increasingly vital role in driving progress toward environmental sustainability goals.

2.4 *Carbon Offsetting and Reduction Scheme for International Aviation (CORSIA)*

The International Civil Aviation Organization (ICAO) has introduced a carbon offsetting and reduction scheme for international aviation (CORSIA) as a comprehensive framework aimed at addressing the environmental impact of aviation [34]. CORSIA represents a critical initiative designed to guide the aviation industry in taking actions to reduce its environmental footprint [12, 49]. The overarching goal of CORSIA is to cap the industry's collective carbon dioxide (CO_2) emissions at the 2020 levels, even as the aviation sector continues to grow.

CORSIA provides airlines with clear guidelines and procedures for managing the calculation of their CO_2 emissions and implementing offsetting measures. This framework seeks to establish consistency across the aviation industry. For instance, it defines the types of eligible fuels that can be used in aircraft, enabling airlines to adopt environmentally friendly fuel options and thereby reduce their environmental impact. CORSIA also supplies airlines with a calculation tool that assists them in accurately assessing their CO_2 emissions. Moreover, it outlines the criteria for determining eligible emission units associated with various aviation fuels, which airlines can incorporate into their emissions calculations.

3 Aviation Ecosystem

When discussing aviation, the primary focus often centers around flying and the sustainability challenges associated with airlines. However, it is crucial to recognize that the environmental sustainability of flight operations is not solely the responsibility of airlines. In fact, there is a broader aviation ecosystem comprised of various stakeholders, each playing a significant role in addressing environmental concerns and promoting sustainable practices [36].

Beyond the airlines, some of the most operationally relevant stakeholders within the aviation ecosystem include airports, ground handling companies, air navigation services, and regulators.

3.1 Airlines

Airlines are companies that provide transportation services for both passengers and cargo [40]. Their primary tool for facilitating this transportation is the aircraft. Airlines establish transportation agreements with passengers through airline tickets and with cargo shippers via formal transportation agreements.

Airlines operate in various ways, including regular scheduled services and charter operations. Depending on market dynamics and their business strategies, airlines may enter into partnerships with other carriers. These partnerships can take the form of alliances or code-share agreements, enhancing their route networks and service offerings.

To operate legally, airlines must obtain an operating certificate from the aviation authority of the country in which they are based. Typically, this authority also serves as an aviation safety regulator, ensuring compliance with stringent safety standards.

The majority of environmental impacts within the aviation industry stem from aircraft operations, accounting for approximately 95% of aviation-related emissions [7]. These emissions primarily consist of carbon dioxide (CO_2) but also include other greenhouse gases like ozone (O_3), nitrous oxide (N_2O), methane (CH_4), sulfur oxides (SOX), and water vapor. Approximately 10% of emissions, excluding hydrocarbons and carbon monoxide, occur during the landing, departure, or taxiing phases. For carbon monoxide and hydrocarbons, this figure rises to about 30%. The remaining emissions are generated during the cruise phase of the flight.

In terms of environmental sustainability reporting, airlines typically adhere to ISO standards for greenhouse gas accounting. Targets and guidelines for emissions calculations often align with the Carbon Offsetting and Reduction Scheme for International Aviation (CORSIA), established by the International Civil Aviation Organization (ICAO). CORSIA has also gained approval from the International Air Transport Association (IATA), the industry's trade organization representing airlines worldwide.

3.2 Airports

Airports serve as aerodromes crucial for flight operations [48] and often act as home bases for airlines. Within the airport ecosystem, various stakeholders, including ground-handling companies and air navigation service providers, contribute to its smooth functioning.

Airports are either managed by airport operating companies or supervised by aviation authorities in specific countries. The primary clientele for airports consists of airlines, which pay fees for utilizing airport infrastructure in their operations. In addition to aircraft-related facilities such as parking positions, runways, and taxiways, airlines also pay for passenger-related amenities and services. Airports play a pivotal role in overseeing and coordinating many of these flight-related operations.

Other customers may include private aviation firms, helicopter service providers, or emergency aviation service providers, many of whom lease real estate from the airport.

Like airlines, airports must obtain operating certificates from the aviation authorities of their respective countries, subject to rigorous regulations governing airport infrastructure and operations. Environmental impacts at airports predominantly arise from airport infrastructure and the airport's own operations [2]. Airport-related CO_2 emissions account for 5% of aviation-related emissions [36]. These impacts can be mitigated through initiatives such as the adoption of renewable energy sources, improved insulation, energy-efficient heating and cooling systems, the transition to electric ground vehicles, and the use of LED-based traffic area lighting in ground operations. On the passenger side, promoting public transportation options for accessing and leaving the airport can reduce emissions. In flight-related operations, measures to reduce taxiing time can be implemented [5].

Sustainability targets at airports often align with the Airport Carbon Accreditation (ACA) program, established by the Airports Council International (ACI), the trade organization representing airports. Launched in 2009, the ACA program comprises six levels, each with specific criteria and objectives to guide airports in their pursuit of environmental sustainability. Further details can be observed in Table 12.1.

If the airport operator is a company, the environmental reporting requirements for the airports are typically based on the ISO standards. If the operator is an authority, an alternative reporting framework may apply.

3.3 *Air Navigation Services*

The management of almost all commercially used airspaces across the world falls under the responsibility of Air Navigation Service (ANS) operators [18]. The allocation of airspaces among different countries is determined under the auspices of the International Civil Aviation Organization (ICAO).

Airlines are the primary customers of ANS service providers, paying fees based on airspace usage. Flight routes are typically guided by air navigation services

Table 12.1 The Airport Carbon Accreditation (ACA) program levels

ACI ACA levels	Status	Actions
Level 1	Mapping	Footprint measurement
Level 2	Reduction	Carbon management towards a reduced carbon footprint
Level 3	Optimization	Third-party engagement in carbon footprint reduction
Level 4	Neutrality	Carbon neutrality for direct emissions by offsetting
Level 5	Transformation	Transforming airport operations and those of its business partners to achieve absolute emissions reductions
Level 6	Transition	Compensation for residual emissions with reliable offsets

according to the field flight plan, with some influence from the pilot. However, ultimate decisions are often made by air traffic controllers.

When it comes to environmental reporting, ANS companies typically adhere to ISO standards, assuming they operate as private companies. If the ANS operator functions as a government authority, different reporting frameworks may apply.

Environmental impacts for ANS service providers primarily stem from decisions related to flight routes. Airspaces may have restrictions or congestion, significantly affecting flight fuel consumption [18].

In 2022, the air navigation service trade organization, CANSO, introduced the CANSO GreenATM program to promote environmental sustainability among ANS providers [14]. CANSO GreenATM is an environmental accreditation initiative that evaluates organizations based on their efforts to reduce emissions within their airspace and minimize their direct carbon footprint.

The program comprises five levels of accreditation, each with specific criteria and goals aimed at recognizing and encouraging ANS providers' contributions to environmental sustainability. The levels of accreditation can be observed in Table 12.2.

The topic areas that are reviewed in the process can be found in Table 12.3.

The actions that focus specifically on the improved ATM directly affect flight time and fuel consumption. For example, the CDO operation can save up to 70% of the fuel during the arrival and landing phase based on the aircraft type [22].

3.4 Ground Handling Services

Ground handling services play a crucial role in the aviation industry, both upon arrival and before departure of an aircraft [55]. These services are provided by specialized companies operating at airports and establish a critical business relationship with airlines. Their activities can significantly impact the overall environmental footprint of a flight.

Ground handling services encompass a wide range of tasks, including fueling, aircraft maintenance, catering, ramp handling, check-in, gate services, arrival services, deicing, and towing. Environmental efforts within this sector often revolve around transitioning ground service equipment to electric vehicles and incorporating sustainable fuels into their operations. However, many ground handling

Table 12.2 The CANSO GreenATM Program levels of accreditation [14]

CANSO GreenATM levels	Status
Level 1	Informal arrangement
Level 2	Defined
Level 3	Managed
Level 4	Assured
Level 5	Optimized

Table 12.3 The topic areas for the CANSO GreenATM Program [14]

Governance	Infrastructure and utilities	Improved Air Traffic Management (ATM)	Other
Policy and plan	Energy management	Flexible Use of Airspace (FUA)	Sustainable procurement
Environmental management system	Power procurement and production	Meteorological information	Airport/community relations
Environmental culture	CNS flight rationalization	Improved surveillance coverage	Airspace change management
Environmental targets	CNS flight inspection	Airport Collaborative Decision Making (A-CDM)	Mobility management
		Surface management	
		Continuous Climb Operation (CCO)	
		Continuous Descent Operation (CDO)	
		Performance Based Navigation (PBN)	
		Wake turbulence optimization	
		Trajectory optimization	
		Air Traffic Flow Management (ATFM)	
		Research and development	

operations also have a direct influence on the environmental impact of flights. Examples include aircraft maintenance practices and the aircraft deicing process.

Environmental sustainability reporting within the ground handling industry typically adheres to ISO standards for environmental reporting. While there are no specific environmental programs tailored to this sector, the adoption of ISO standards ensures a standardized and systematic approach to measuring and reporting environmental performance. This consistency allows for the monitoring of sustainability efforts and the potential identification of areas where further environmental improvements can be made.

3.5 Aviation Authorities

The entire aviation ecosystem places paramount emphasis on safety and is subject to rigorous regulation [36]. Consequently, all aviation operations and infrastructure are under the purview of relevant authorities, primarily aviation safety authorities. Beyond ensuring safety and security, regulations can also influence flight sustainability measures and environmental permits required by airports.

Aviation-related regulations have a global scope and typically originate from ICAO. These regulations are then transposed into national aviation laws in various countries and enforced by national aviation safety authorities. In the European Union (EU), such regulations are adapted into EU-level directives and policies before being incorporated into the national legislation of EU member states. While aviation regulation has historically focused on safety and security, ICAO has long addressed environmental regulations, particularly regarding aircraft noise. However, the imperative of addressing climate change has elevated the prominence of environmental considerations within aviation regulation, leading to substantial growth in the environmental aspect of aviation regulation [36].

Other environmental regulations, such as corporate environmental restrictions, incentives, and emissions trading systems, also influence the aviation industry. In the EU, the ICAO net zero target aligns with initiatives like the European Green Deal scheme [26], the EU Emissions Trading Scheme (ETS) [20], and the Green Claims directive [25].

Tax policies are a valuable tool for promoting desired outcomes; however, their effectiveness in aviation can be challenging due to the international nature of the industry. Many aviation-related companies can easily register their businesses in countries offering favorable legislative or tax environments.

It is worth noting that there are no specific sustainability targets or dedicated environmental reporting requirements established for aviation authorities. Instead, their primary role is to enforce and oversee compliance with existing aviation regulations and safety standards.

3.6 Ecosystem Design

The most significant environmental impacts in the aviation industry stem from aircraft engines during flights, primarily attributable to the airlines. However, within the aviation ecosystem, several key stakeholders play essential roles in influencing aircraft engine-related CO_2 emissions in addition to their own emissions [16, 35, 47]. These stakeholders include airports, air navigation service providers, and ground-handling companies. Local aviation authorities and global regulators like ICAO also play pivotal roles in establishing safety standards and operational boundaries. Together, these stakeholders constitute the operational aviation ecosystem, as depicted in the inner circle presented in Fig. 12.1.

The stakeholders within the inner circle can influence each other in various ways. For instance, airports, ANS services, and GH services directly impact airline flights in multiple ways. Airlines, ANS services, and GH services are interdependent with airport infrastructure and services. ANS services play a role in determining flight routes but are also reliant on airports and the punctuality of GH services. The quality of GH services directly affects flight punctuality and, consequently, the efficiency of all airport-related operations. Aviation authorities oversee the regulatory aspects of operations for all stakeholders within the inner circle.

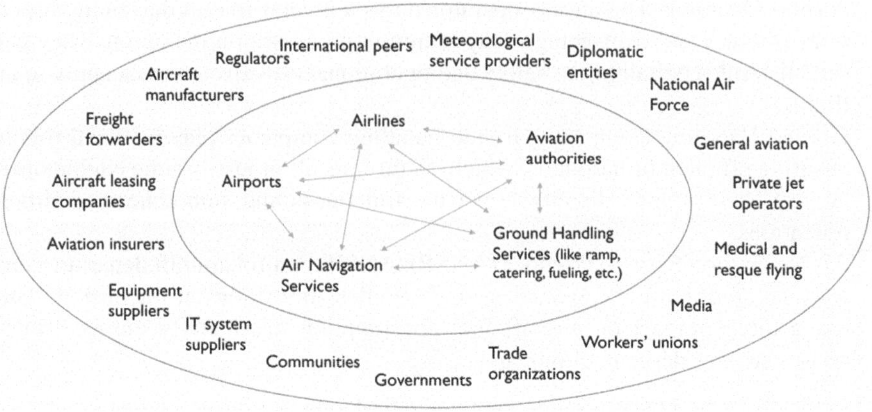

Fig. 12.1 Aviation ecosystem

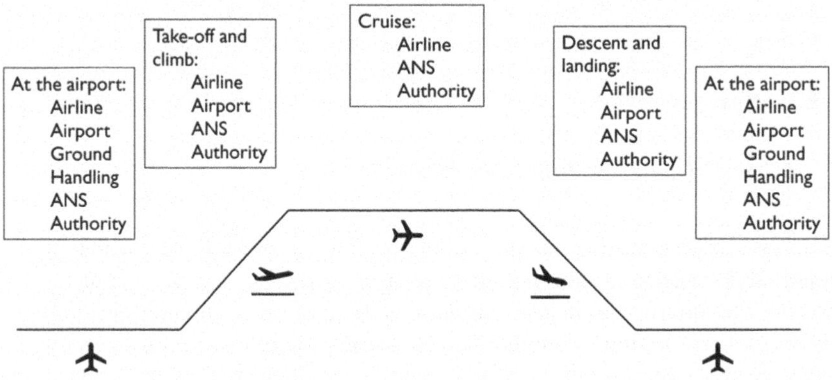

Fig. 12.2 Flight-related carbon footprint with the effects from the aviation ecosystem

In the outer circle, some stakeholders may consistently or occasionally exert an influence on those within the inner circle, albeit typically with an indirect operational impact. For example, business jet flights share the same airport infrastructure and airspace as commercial flights, potentially diverting operational capacity away from commercial operations due to their specific needs. Aircraft manufacturers design more energy-efficient aircraft, although their impact is not direct during flight operations.

When examining this ecosystem from the perspective of flight-time interactions, the visualization takes the form presented in Fig. 12.2.

All stakeholders within the inner circle of the aviation ecosystem exert significant influence on flight operations at the airport. Here's how each of them impacts these operations:

- Airport Operator: The airport operator plays a pivotal role in designing airport infrastructure and overseeing many common processes on the apron. They can establish rules pertaining to safety and operational effectiveness that apply to all flights.
- Ground Handling Company: Ground handling companies play a crucial role in ensuring efficient turnarounds, which, in turn, result in just-in-time engine operations. Their actions directly affect the timeliness and smoothness of airport operations.
- Air Navigation Service Provider (ANSP): ANSPs control aircraft departures and arrivals, contributing to queueless and seamless transitions at the airport. This has a direct impact on aircraft fuel consumption, particularly during critical phases such as descent, climb, and cruise.

In the air, during the cruise, descent, and climb phases, the air navigation service provider plays a significant role in selecting the flight path. Consequently, the ANSP, in conjunction with regulatory bodies, remains involved throughout the entirety of the journey, shaping the flight's trajectory and operational efficiency.

4 Challenges

4.1 Global Industry but Local Laws

Aviation is a global industry governed by regulations set by ICAO, which are then adopted and translated into national aviation laws by various countries [37]. However, the interpretation and implementation of these regulations may vary slightly from one region to another. Consequently, there can be variations in aviation laws from one country to another, including those related to environmental regulation. While ICAO establishes global regulations for environmental sustainability, the ultimate responsibility lies with individual countries to adapt these regulations to their local aviation laws [30, 42]. This diversity in the application of regulations means that achieving a globally unified approach to environmental sustainability in aviation is not guaranteed.

4.2 Reporting Inconsistencies in the Environmental Impact of an International Flight

For an airline, measuring and reporting flight time environmental impacts are relatively straightforward [7]. However, when a flight traverses multiple countries' airspaces during its route, the approach to reporting CO_2 emissions can vary based on the airline's reporting principles [38].

In Europe, the EU ETS legislation governs emissions from flights departing from, arriving in, or operating within the EU, as well as Iceland, Norway, and Liechtenstein (EEA area) [50]. This framework aligns with international laws.

However, for long-haul international flights that cross multiple airspaces, it becomes less clear how the countries overseeing those airspaces can report the associated environmental impacts [20]. International flights depart from one country and land in another, potentially passing through various countries' airspaces along the way. Additionally, the boundaries of airspaces controlled by different air navigation service providers may not always align with national borders. For instance, the United States' FAA controls a significant portion of the Atlantic Oceanic airspace, much of which is international airspace.

While very high altitudes lack airspace ownership, the highest on-route flight level (Class E airspace) currently reaches 18.3 km above sea level [45]. This restricts conventional combustion engine aircraft to controlled altitudes. Electric engines have the potential to operate at higher altitudes. Currently, upper airspace may be reserved for specific purposes like military or supersonic flights. The utilization of various airspace areas during a flight is illustrated in Fig. 12.3.

Reporting environmental impacts based on airspace borders is a feasible approach. The global division of airspaces was initially established and sanctioned by the ICAO Chicago Convention in 1944. While airspace borders often align with official country borders, there are instances, particularly over sea areas, where airspace control extends beyond national boundaries. Additionally, there are regions around the world where aircraft operate, but air traffic control infrastructure may be absent or limited.

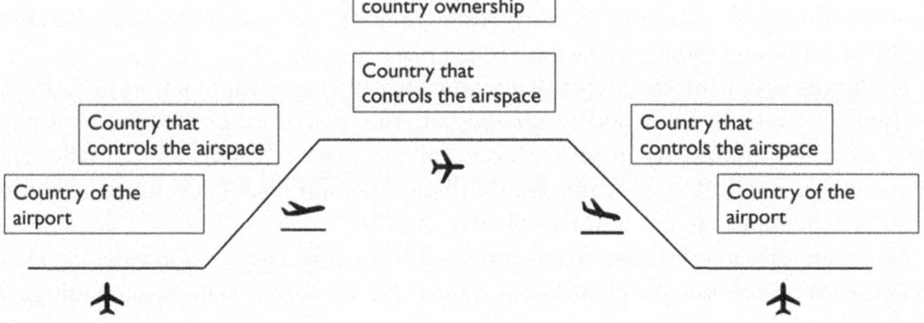

Fig. 12.3 Flight time carbon footprint in countries along the journey

4.3 Non-comparable System Boundaries in the Corporate Reporting

Companies have a degree of flexibility in how they report their environmental impacts. In the ISO standards for environmental reporting, such as ISO 14067 [38], system boundaries can be established around the legal entity of the company or the country of the company's origin. Many airlines, for example, utilize subcontractors, and their reporting may focus on actions related to the country of the company's origin. Depending on a company's policy, they have some discretion in defining their reporting boundaries, which may result in actions being excluded from the environmental reporting within the company's borders [61]. The ISO 14067 standard for carbon footprints allows the assessment conductor to choose the system boundaries for reporting [39]. This flexibility can lead to inconsistencies in environmental reports among different companies [20]. However, within the EU, the new directive for corporate sustainability reporting (CSRD) has the potential to address these issues and improve reporting consistency [1]).

4.4 Immaturity of Sustainable Solutions

There has been extensive discussion about how sustainable aviation fuels and electric aircraft will transform the nature of flying [10]. When the media discusses sustainable flying, it often revolves around one of these two options.

Sustainable aviation fuels come in various forms and are typically derived from organic materials, waste, or other renewable resources [15]. They are the subject of ongoing research, and new variations are continually entering the market [19]. One advantage is that sustainable aviation fuels can generally be used like traditional aviation fuel. However, there are challenges to consider [53]. Firstly, the cost of these fuels can be 2–4 times higher compared to traditional aviation fuels [3, 13]. Additionally, these fuels are less efficient, meaning more fuel is required for the same amount of flying. Secondly, the availability of sustainable aviation fuels has not yet stabilized, and at times, it can be challenging to obtain them at reasonable transportation costs. Thirdly, as production capacity increases, there may be challenges in sourcing sufficient raw materials [11].

Hydrogen is considered a promising new energy source for aviation [41]. It can be used in synthetic fuels and to charge batteries that store electricity. However, these solutions are mostly in pilot phases and are primarily suited for smaller and short-range aircraft. Battery technology still requires significant development before it can be applied to larger commercial aircraft [52].

Commercially viable solutions to support sustainable energy sources in the aviation ecosystem are not yet clear [4, 6]. Given the immature state of technological solutions, investing in infrastructure for electric aircraft charging, for example, presents challenges. There is a risk of investing in solutions that may ultimately prove

unsuitable. Determining the payback time and creating a viable business case for these emerging solutions poses a business risk that the ecosystem may be hesitant to undertake [52].

4.5 Narrow Focus from the Airline Perspective

The aviation industry's primary focus is on flight-related activities, and this also extends to discussions about environmentally sustainable aviation. It's commonly understood that a significant portion of environmental impacts in aviation arises from emissions during flight [46]. As a result, impacts from other aspects of the aviation ecosystem may not always be considered related, or various stakeholders may only discuss the environmental impacts that are directly associated with their operations. However, it's essential to recognize that the entire aviation ecosystem plays a role in flight-related sustainability, and these aspects should be included in discussions and calculations.

Research has been conducted extensively on flight time emissions and the airline's perspective [20]. However, there isn't a standardized approach for defining system boundaries to determine the extent of the value chain that should be considered beyond the flight itself.

Airlines typically report their carbon footprints in accordance with ISO standards. Since the standard allows reporting entities to choose which actions or operations to include in their calculations, airlines often focus on emissions that are directly attributable to the airline as a company. This typically encompasses flight-time CO_2 emissions and, in some cases, emissions related to real estate, such as building heating. However, actions performed by subcontractors or other stakeholders are often not extensively discussed or included in these calculations.

4.6 Incompleteness of CORSIA

CORSIA, developed by ICAO [34], aims to standardize the reporting and calculation of CO_2 emissions in the aviation industry. However, the scheme has several shortcomings that pose challenges [43, 51, 58]. These challenges are outlined in Table 12.4.

CORSIA has potential; however, it needs further development to achieve the targeted impact.

Table 12.4 CORSIA-related challenges and the explanations

Challenge	Issue
The program focuses only on the flight time CO_2 emissions.	The ecosystem's CO_2 emissions and the ecosystem effect are not considered.
Aviation ecosystem impacts are not covered.	The aviation ecosystem is creating emissions that would not be created without the flights.
The program is voluntary.	The best and most consistent results are gained if all the flights are covered.
The program omits the less developed countries.	Environmental impact is global; the involvement of all the countries is needed.
The program omits all the other emissions besides the CO_2.	Aircraft jet emissions also include other gases like ozone (O_3), nitrous oxide (N_2O), methane (CH_4), sulfur oxides (SOX), and water vapor.
The program has a strong focus on offsetting.	These are focusing mostly on carbon sink-related solutions, which are much less expensive that finding sustainable solutions from aviation.
The reduction targets are set for future growth; the targets are compared to a baseline that was set in the year 2019.	The industry has always exhibited a growth trend; using a baseline from earlier years would make the target more ambitious.
The program does not support the development of sustainable aviation solutions.	Sustainable aviation fuels and electric aviation are expected to have the biggest impact in achieving net-zero targets.

4.7 Non-aligned Formal Environmental Targets in the Aviation Ecosystem

All stakeholders within the aviation ecosystem have established their own environmental targets. In cases where applicable trade organizations are involved, they often define targets and guidelines [2, 14, 34]. While many of these targets overlap with those of other stakeholders, there is no coordination to establish a common set of goals. Consequently, some potential targets within the ecosystem may have been overlooked.

4.8 Independent Business Silos

In the aviation ecosystem, despite the interconnectedness of stakeholders, they remain separate entities, including companies, authorities, and organizations [56]. They often lack common strategies, targets, goals, or metrics, and interactions among them are limited. These entities conduct their operations independently, even though they may have business relationships with each other.

The aviation industry is known for its financial challenges, and companies within the ecosystem prioritize financial sustainability. Supporting each other with common financial goals is infrequent due to the competitive nature of the industry.

Moreover, these companies and organizations in the aviation ecosystem do not share common sustainability objectives. Each entity reports its environmental sustainability strategies and actions independently. CORSIA is the sole official CO_2 reduction scheme in the industry, but it does not account for ecosystem-wide effects, resulting in a lack of common environmental goals.

The only formal aspect that brings these entities together in terms of environmental sustainability is adherence to ISO standards for environmental reporting. However, because companies can independently define their reporting boundaries, there is no standardized method for environmental reporting in the aviation industry.

4.9 Limited Data Sharing

Legal company or organizational boundaries often hinder data sharing within the aviation ecosystem. Many airlines are publicly listed companies, which means sharing classified information is not a viable option. Additionally, the use of digital technologies and processes has introduced new challenges related to data confidentiality, integrity, and availability due to cyber vulnerabilities [17].

In the realm of passenger travel, passengers enter into agreements with airlines through tickets for their journeys. Airlines have numerous legal obligations to passengers, which contributes to their reluctance to share data. However, passenger data could be valuable for airports in terms of service and operations planning.

Due to various legal, regulatory, or security constraints, stakeholders within the aviation ecosystem have limited data-sharing practices. The data that is shared typically pertains directly to operational or safety-related matters and does not contain business or personal information.

This restricted data sharing can impede collaborative efforts and initiatives aimed at improving environmentally significant processes in environmental reporting.

4.10 Cost Intensiveness

Aviation is a capital-intensive industry with high operational costs and a susceptibility to rapid market changes. The COVID-19 pandemic had a significant impact on the aviation industry, leading to substantial financial losses for many companies [28, 31]. Additionally, fuel costs have risen sharply in recent times.

Given these factors, the aviation industry often sets ambitious economic targets, especially because many companies within the industry are large and publicly listed. During periods of volatility, such as the aftermath of the pandemic, the aviation sector tends to prioritize cost-saving measures, which can sometimes affect sustainability initiatives, particularly those related to expensive sustainable aviation fuels. It may be in the companies' interest to postpone or scale back environmental actions if it helps improve their financial stability.

4.11 Inadequate Toolset for Public Acceptance

Passengers and air cargo shippers are increasingly conscious of the environmental impacts of aviation [30]. In response, airlines are actively promoting their sustainability initiatives through traditional and social media to gain public acceptance.

Airlines have introduced programs and incentives that allow passengers and air cargo shippers to offset their flight-time CO_2 emissions [32, 59]. While this benefits airlines by showcasing their commitment to environmental sustainability, it also reduces the need for their direct investments in sustainability measures. However, there is currently no guarantee that the funds collected from passengers for these offset programs will be exclusively used for sustainable aviation fuels or other initiatives aimed at enhancing the aviation industry's environmental sustainability. Additionally, such programs are voluntary, and not all passengers participate, leading to unequal contributions.

The most efficient way for airlines to reduce flight-time CO_2 emissions would be to use sustainable aviation fuel on every flight, with any increased fuel costs being incorporated into ticket prices for passengers or air freight charges [59]. Whether passengers and shippers would continue to fly or use air cargo services as before or whether this approach would impact airlines' profitability remains to be seen.

5 Conclusions and Discussion

The challenges discussed in this article span across the aviation ecosystem and are interconnected among various stakeholders. Table 12.5 presents a summary of these challenges along with key characteristics.

The flight time CO_2 emissions constitute the most significant source of climate impact in aviation. As a result, airlines are the most significant stakeholders in the environmentally sustainable aviation ecosystem. Any changes airlines make to their fuel type, aircraft engine type, flight routings, or flight time operating patterns can have a significant impact on the environmental sustainability of the airline and, by extension, the entire aviation industry. However, airline decisions are not always made in isolation.

The aviation ecosystem is vast, and the importance of various stakeholders may vary depending on the situation. Nevertheless, the most crucial stakeholders regarding flight time operations and environmental impact are airlines, airports, ground handling companies, air navigation service providers, and aviation authorities.

To maximize environmental sustainability results concerning flight time CO_2 emissions, airlines and other key stakeholders in the aviation ecosystem need to establish a closer relationship with common targets, aligned action plans, and shared metrics. While each stakeholder currently develops its own environmental sustainability strategy and plan independently, achieving the maximum common benefits requires better alignment through regulation.

Table 12.5 Summary of the challenges and the key characteristics in the environmentally sustainable aviation ecosystem

Challenge	Key characteristics
Global industry but local laws	Aviation is a global industry, yet international aviation regulations are implemented as national laws. Interpretations and areas of interest might vary.
Reporting inconsistencies in the environmental impact of an international flight	A long-haul flight enters several air spaces during the span of the flight. Several stakeholders may report the environmental impacts, or some impacts may not be reported at all.
Non-comparable system boundaries in the corporate reporting	Companies have the freedom to define their own system boundaries for their environmental reporting. There is no guarantee that the selected system boundaries reflect the actual environmental impacts.
Immaturity of sustainable solutions	Sustainable aviation fuels are plagued by price and availability challenges. Many other sustainable energy sources for flying are still commercially immature. Electric-powered flights are also commercially in their infancy.
Narrow focus from the airline perspective	The aviation industry is mainly understood by the airlines. The holistic understanding of aviation ecosystem effects needs emphasizing.
Incompleteness of CORSIA	The carbon offsetting scheme for the aviation industry omits certain impacts and stakeholders. The implementation of the system is also voluntary.
Non-aligned formal environmental targets in the aviation ecosystem	Various stakeholders in the aviation ecosystem have their own official environmental targets, but these targets are not aligned. There are overlapping targets with no common reporting policies.
Independent business silos	Companies in the aviation ecosystem often engage in business relationships with each other. In some cases, the business goals can override the potential cooperation.
Limited data sharing	Regulatory and business reasons the companies in the aviation ecosystem limit their data sharing. This also limits the potential cooperation.
Cost intensiveness	Aviation is a very competitive industry, and companies are particularly cost-sensitive in the aftermath of the COVID-19 pandemic. If environmental actions cause costs, the willingness to take those actions might be challenged.
Inadequate toolset for public acceptance	The passengers and the cargo shippers are very aware of the environmental challenges of aviation. The increased costs will eventually be passed on to service prices, but the mechanism is still unsettled. The price sensitivity associated with the environmental costs remains unclear.

Another essential factor is expanding the scope of reporting on flight time environmental impacts in aviation. In addition to company-level environmental reporting, similar reporting should be conducted for the entire ecosystem. This could help reduce duplication in reporting efforts. However, achieving this would necessitate improved regulation to define and measure common environmental impacts, ultimately supporting the industry in analyzing and developing climate actions more comprehensively and realistically.

References

1. /43/EC and Directive 2013/34/EU, as regards corporate sustainability reporting (Text with EEA relevance), n.d.
2. ACI. (2021a). *Long-term carbon goal study for airports.*
3. ACI. (2021b). *Integration of Sustainable Aviation Fuels into the air transport system.*
4. ACI. (2021c). *Integration of hydrogen aircraft into the air transport system.*
5. ACI. (2019). *Guidance manual: Airport greenhouse gas emissions management.*
6. Ahmad, S., & Xu, B. (2021). A cognitive mapping approach to analyze stakeholders' perspectives on sustainable aviation fuels. *Transportation Research Part D: Transport and Environment, 100.* https://doi.org/10.1016/j.trd.2021.103076
7. Amicarelli, V., Lagioia, G., Patruno, A., Grosu, R. M., & Bux, C. (2021). Enhancing the sustainability of the aviation industry: Airlines' commitment to "green" practices. *Amfiteatru Economic, 23*, 934–947. https://doi.org/10.24818/EA/2021/S15/934
8. ATAG. (2018). *Aviation, benefits beyond borders.*
9. ATAG. (2021). *Waypoint 2050.*
10. Barke, A., Bley, T., Thies, C., Weckenborg, C., & Spengler, T. S. (2022). Are sustainable aviation fuels a viable option for decarbonizing air transport in Europe? An environmental and economic sustainability assessment. *Applied Sciences (Switzerland), 12.* https://doi.org/10.3390/app12020597
11. Bauen, A., Bitossi, N., German, L., Harris, A., & Leow, K. (2020). Sustainable aviation fuels status, challenges and prospects of drop-in liquid fuels, hydrogen and electrification in aviation. *Johnson Matthey Technology Review, 64*, 263–278. https://doi.org/10.1595/205651320x15816756012040
12. Bauen, A., Harris, A., Sim, C., Gudde, N., Prussi, M., & Scarlat, N. (2022). CORSIA lower carbon aviation fuels: An assessment of the greenhouse gas emission reduction potential. *Applied Sciences (Switzerland), 12.* https://doi.org/10.3390/app122211818
13. Brandt, K. L., Martinez-Valencia, L., & Wolcott, M. P. (2022). Cumulative impact of federal and state policy on minimum selling Price of sustainable aviation fuel. *Frontiers in Energy Research, 10.* https://doi.org/10.3389/fenrg.2022.828789
14. CANSO. (2022). *The guide to GreenATM accreditation a manual for the CANSO environmental accreditation programme.*
15. Chiaramonti, D., Talluri, G., Vourliotakis, G., Testa, L., Prussi, M., & Scarlat, N. (2021). Can lower carbon aviation fuels (LCAF) really complement sustainable aviation fuel (SAF) towards EU aviation decarbonization? *Energies (Basel), 14.* https://doi.org/10.3390/en14196430
16. Corrigan, S., Mårtensson, L., Kay, A., Okwir, S., Ulfvengren, P., & McDonald, N. (2015). Preparing for Airport Collaborative Decision Making (A-CDM) implementation: An evaluation and recommendations. *Cognition, Technology and Work, 17*, 207–218. https://doi.org/10.1007/s10111-014-0295-x
17. da Silva, S. J., & Silva, J. M. R. (2021). Cyber risks in the aviation ecosystem: An approach through a trust framework. In *Integrated Communications, Navigation and Surveillance Conference, ICNS.* Institute of Electrical and Electronics Engineers Inc. https://doi.org/10.1109/ICNS52807.2021.9441596
18. Dalmau, R. (2022). Predicting the likelihood of airspace user rerouting to mitigate air traffic flow management delay. *Transportation Research Part C: Emerging Technologies, 144.* https://doi.org/10.1016/j.trc.2022.103869
19. Dodd, T., & Yengin, D. (2021). Deadlock in sustainable aviation fuels: A multi-case analysis of agency. *Transportation Research Part D: Transport and Environment, 94*, 102799. https://doi.org/10.1016/j.trd.2021.102799
20. Efthymiou, M., & Papatheodorou, A. (2019). EU emissions trading scheme in aviation: Policy analysis and suggestions. *Journal of Cleaner Production, 237*, 117734. https://doi.org/10.1016/j.jclepro.2019.117734
21. Energy Agency, I. (2023). *Energy Technology Perspectives 2023.*

22. Errico, A., & Di Vito, V. (2019). Aircraft operating technique for efficient sequencing arrival enabling environmental benefits through CDO in TMA. In *AIAA Scitech 2019 Forum*. American Institute of Aeronautics and Astronautics Inc, AIAA. https://doi.org/10.2514/6.2019-1363
23. EUROCONTROL. (2022a). *Eurocontrol analysis paper: 2022 review.*
24. EUROCONTROL. (2022b). *Aviation Outlook 2050.*
25. European Commission. (2023). *Green claims directive.*
26. European Commission. (2019). *The European Green Deal.*
27. European Parliament. (2023). *ReFuel EU Aviation.*
28. Gole, I., Dobrea, R. C., & Gombos, C. C. (2021). Aviation industry – Challenges and uncertainties after the Covid-19 pandemic. *SHS Web of Conferences, 92*, 01010. https://doi.org/10.1051/shsconf/20219201010
29. Gössling, S., & Humpe, A. (2020). The global scale, distribution and growth of aviation: Implications for climate change. *Global Environmental Change, 65*, 102194. https://doi.org/10.1016/j.gloenvcha.2020.102194
30. Gössling, S., & Lyle, C. (2021). Transition policies for climatically sustainable aviation. *Transport Reviews, 41*, 643–658. https://doi.org/10.1080/01441647.2021.1938284
31. IATA. (2022a). *Global outlook for air transport times of turbulence.*
32. IATA. (2022b). *Aviation carbon offsetting guidelines for voluntary programs.*
33. IATA. (2021). *Net zero 2050 Progress tracking methodology.*
34. IATA. (2019). *An airline handbook on CORSIA.*
35. IATA. (2018). *Airport-collaborative decision making (A-CDM): IATA recommendations.*
36. ICAO. (2022). *Innovation for a green transition.*
37. ICAO. (2019). *Aviation benefits 2019.*
38. ISO. (2022). *Net zero guidelines.*
39. Kılıç, M., Uyar, A., & Karaman, A. S. (2019). What impacts sustainability reporting in the global aviation industry? An institutional perspective. *Transport Policy (Oxford), 79*, 54–65. https://doi.org/10.1016/j.tranpol.2019.04.017
40. Ko, Y. D. (2016). An airline's management strategies in a competitive air transport market. *Journal of Air Transport Management, 50*, 53–61. https://doi.org/10.1016/j.jairtraman.2015.10.005
41. Kramer, S., Andac, G., Heyne, J., Ellsworth, J., Herzig, P., & Lewis, K. C. (2022). Perspectives on fully synthesized sustainable aviation fuels: Direction and opportunities. *Frontiers in Energy Research, 9*. https://doi.org/10.3389/fenrg.2021.782823
42. Kulanovic, A., & Nordensvärd, J. (2021). Exploring the political discursive lock-ins on sustainable aviation in Sweden. *Energies (Basel), 14*. https://doi.org/10.3390/en14217401
43. Mai, D. T. T. (2021). Revising the EU ETS and CORSIA in times of the COVID-19 pandemic: Challenges for reducing global aviation emissions. *Climate Policy, 21*, 1357–1367. https://doi.org/10.1080/14693062.2021.1966358
44. McCausland, R. (2022). *Net zero resolution.*
45. NASA. (2021). *Cooperative upper class E airspace: Concept of operations and simulation development for operational feasibility assessment.*
46. Norton, T. M., & Norton, T. (2014). *Aircraft greenhouse gas emissions during the landing and takeoff cycle at Bay Area airports.*
47. Okwir, S., Ulfvengren, P., Angelis, J., Ruiz, F., & Núñez Guerrero, Y. M. (2017). Managing turnaround performance through collaborative decision making. *Journal of Air Transport Management, 58*, 183–196. https://doi.org/10.1016/J.JAIRTRAMAN.2016.10.008
48. Pacagnella, A. C., Hollaender, P. S., Mazzanati, G. V., & Bortolotto, W. W. (2021). Efficiency drivers of international airports: A worldwide benchmarking study. *Journal of Air Transport Management, 90*, 101960. https://doi.org/10.1016/j.jairtraman.2020.101960
49. Prussi, M., Lee, U., Wang, M., Malina, R., Valin, H., Taheripour, F., Velarde, C., Staples, M. D., Lonza, L., & Hileman, J. I. (2021). CORSIA: The first internationally adopted approach to calculate life-cycle GHG emissions for aviation fuels. *Renewable and Sustainable Energy Reviews., 150*, 111398. https://doi.org/10.1016/j.rser.2021.111398

50. Scheelhaase, J., Maertens, S., & Grimme, W. (2022). Options for improving the EU Emissions Trading Scheme (EU ETS) for aviation. In *Transportation research procedia* (pp. 193–202). Elsevier B.V. https://doi.org/10.1016/j.trpro.2021.11.111
51. Scheelhaase, J., Maertens, S., Grimme, W., & Jung, M. (2018). EU ETS versus CORSIA – A critical assessment of two approaches to limit air transport's CO2 emissions by market-based measures. *Journal of Air Transport Management, 67*, 55–62. https://doi.org/10.1016/j.jairtraman.2017.11.007
52. Scholz, A. E., Trifonov, D., & Hornung, M. (2022). Environmental life cycle assessment and operating cost analysis of a conceptual battery hybrid-electric transport aircraft. *CEAS Aeronautical Journal, 13*, 215–235. https://doi.org/10.1007/s13272-021-00556-0
53. Shahriar, M. F., & Khanal, A. (2022). The current techno-economic, environmental, policy status and perspectives of sustainable aviation fuel (SAF). *Fuel, 325*, 124905. https://doi.org/10.1016/j.fuel.2022.124905
54. Staples, M.D., Malina, R., Suresh, P., Hileman, J.I., Barrett, S.R.H., 2018. Aviation CO2 emissions reductions from the use of alternative jet fuels. *Energy Policy 114*, 342–354. https://doi.org/10.1016/j.enpol.2017.12.007
55. Studic, M., Majumdar, A., Schuster, W., & Ochieng, W. Y. (2017). A systemic modelling of ground handling services using the functional resonance analysis method. *Transportation Research Part C: Emerging Technologies, 74*, 245–260. https://doi.org/10.1016/j.trc.2016.11.004
56. The Business Times, S. (2017). *Singapore's aviation ecosystem needs new strategic thinking.*
57. United Nations. (2015). *2030 agenda for sustainable development.*
58. Wozny, F., Grimme, W., Maertens, S., & Scheelhaase, J. (2022). CORSIA—A feasible second best solution? *Applied Sciences (Switzerland), 12*. https://doi.org/10.3390/app12147054
59. Xu, B., Ahmad, S., Charles, V., & Xuan, J. (2022). Sustainable commercial aviation: What determines air travellers' willingness to pay more for sustainable aviation fuel? *Journal of Cleaner Production, 374*, 133990. https://doi.org/10.1016/j.jclepro.2022.133990
60. Zhou, W., Wang, T., Yu, Y., Chen, D., & Zhu, B. (2016). Scenario analysis of CO2 emissions from China's civil aviation industry through 2030. *Applied Energy, 175*, 100–108. https://doi.org/10.1016/j.apenergy.2016.05.004
61. Zieba, M., & Johansson, E. (2022). Sustainability reporting in the airline industry: Current literature and future research avenues. *Transportation Research Part D: Transport and Environment, 102*. https://doi.org/10.1016/j.trd.2021.103133

Chapter 13
Techno-economic Analysis and Life Cycle Assessment of Sustainable Aviation Fuel (SAF) Production

Luis Felipe Ramírez Verduzco, Judith Esperanza Cruz Ramírez, and Myriam Adela Amezcua-Allieri

Nomenclature

AD	Abiotic resource depletion
ATJ	Alcohol to jet
CO	Carbon monoxide
CO_2	Carbon dioxide
FT	Fischer-Tropsch
GHG	Greenhouse gases
GWP	Global warming potential
H_2	Hydrogen
HEFA	Hydroprocessed esters and fatty acids
HTL	Hydrothermal liquefaction
HTP	Human toxicity potential
IATA	International Air Transport Association
MVA	Mevalonate
OLD	Ozone layer depletion
POCP	Photochemical ozone creation potential
SAF	Sustainable aviation fuel
SIP	Synthesized isoparaffins
SIP-HFS	Synthetic isoparaffins produced from hydroprocessed fermented sugars
SPK	Synthetic paraffinic kerosene

L. F. R. Verduzco · J. E. C. Ramírez · M. A. Amezcua-Allieri (✉)
Mexican Institute of Petroleum, Energy Efficiency and Biofuels Management, Mexico City, Mexico
e-mail: mamezcua@imp.mx

M. Aslam et al. (eds.), *Sustainable Aviation Fuels*, Sustainable Aviation,
https://doi.org/10.1007/978-3-031-83721-0_13

1 Introduction

Renewable energy represents an option for clean energy generation; therefore, countries around the world are developing updated policies related to sustainable energy transition. The use of favorable environmental and techno-economical alternatives to produce fuels from biomass is not a new issue, but the need to transition to biofuels with low-carbon pool is growing due to the global efforts to replace stepwise and wisely the use of fossil fuels by other sources, like biofuels [1].

The production of aviation biofuels and derivatives is a highly energy-intensive industry that has partially replaced the use of fossil fuels, such as fuel oil, to diminish its operational costs and improve performance. Nevertheless, some contradictory data are analyzing the techno-economic performance resulting from the substitution of fossil fuels for biomass along with the environmental impacts derived from that substitution.

The production of sustainable aviation fuel (SAF) is of great importance due to several reasons. In the first place, it holds the potential to decrease greenhouse gas (GHG) emissions arising from the aviation sector. Given that the aviation industry is responsible for a significant amount of global carbon dioxide emissions, SAF can play a critical role in reducing the carbon footprint of air travel [2]. Second, the production of renewable jet fuel can broaden the feedstock portfolio of the biofuels industry, thereby creating new opportunities for the agricultural and forestry sectors. This has the potential to generate economic activity and employment in rural areas, particularly in the conversion of feedstocks. Third, the use of alternative jet fuel can improve energy security by reducing reliance on foreign oil and petroleum-based jet fuel. This can mitigate geopolitical risks associated with oil imports and enhance national energy independence. Fourth, SAF can contribute to sustainable development objectives by advocating for the use of renewable resources and reducing dependence on fossil fuels. This can help to mitigate the effects of climate change while promoting economic growth and social welfare.

The significance of SAF is undeniable, the chief aim of this chapter is to present an exhaustive outline of the TEA and LCA regarding the creation and utilization of SAF. It aims to provide readers with a detailed understanding of the major resources for green aviation fuel. Including feedstocks and conversion technologies. This chapter also discusses the challenges and opportunities in producing SAF, as well as the drivers and barriers to its adoption from the TEA and LCA point of view.

One of the primary and fundamental aims of this chapter is to emphasize and underscore the myriad and multifarious potential advantages, merits, and gains of employing sustainable aviation fuel meticulously and systematically. These manifold and diverse benefits encompass and encompass a wide range of factors, including but not limited to a marked and significant diminution and curtailment of greenhouse gas emissions, a notable and palpable amelioration, and enhancement of energy security, as well as a pronounced and conspicuous increase and augmentation in the utilization and deployment of renewable resources.

Another important goal of this chapter is to discuss current research efforts aimed at overcoming technical and economic challenges associated with SAF production. This includes information on methods to estimate investment and operation costs for SAF production. By highlighting these promising technologies, this chapter aims to inspire further research into SAF production. In this chapter, we compare the environmental impacts and techno-economic implementation of producing aviation biofuel from different feedstock and technologies.

2 Technical Description and Design: Feedstocks and Main Technologies Used to Produce SAF

In this section, we will delve into the technical aspects of producing SAF. We will discuss the various feedstocks and conversion technologies that are being used or developed for SAF production. The choice of feedstock and conversion technology can have a significant impact on the economic viability and environmental sustainability of SAF production. Therefore, it is important to understand the technical details of each option and their potential benefits and drawbacks. By the end of this section, you will have a better understanding of the technical aspects of SAF production and the factors that influence its feasibility and sustainability.

2.1 Feedstock

Biofuel can be produced from biomass, understood as organic matter of plant or animal origin, including residues that can be used as energy. The technologies used to produce SAF are varied and depend mostly on the type of feedstock because these contain unique properties that can affect the partitioning of greenhouse gas (GHG) emissions depending upon the allocation factor that is used.

Biofuel has been classified that often heard as "first generation" or "second generation" and sometimes even "third generation" [3]. First-generation biomass is represented by all crops that can be used for human consumption. The way to obtain them depends on the plant of origin: If it has a high sugar content, it is decided to convert it into alcohol by fermentation, as is the case with the production of ethanol from sugar cane. Instead, if it started from plants rich in fats or oils, a chemical reaction called transesterification is used, in which these oils are combined with alcohol to generate fatty esters, such as biofuel.

Second-generation biofuels are obtained from feedstock that cannot be used for human consumption, such as forestry and agricultural residues, which have a high content of cellulose and lignin, the main components of the plant cell wall. Recycled cooking oil can be considered a feedstock for second-generation biofuels since it is

no longer used for food. This is a good option as a feedstock because it is 2–3 times cheaper than virgin vegetable oil [4].

Another group of possible sources to produce biodiesel are fats of animal origin. The main animal fats used are tallow, white fat or lard, chicken fat, and yellow fat [5]. These fats, compared to plant cultures, offer a great economic advantage since they present a favorable conversion into biodiesel due to the high cetane number, no corrosion, and low concentrations of free fatty acids and water [6]. There are several ways to produce second-generation biofuels, with the biochemical and thermochemical pathways being the best known. The first uses microorganisms to reduce the complex chemical chains of cellulose molecules to simple sugars and then converts the sugars into biofuels. The thermochemical form uses high pressure and temperature to convert a wide variety of types of biomasses into fuels. One possible way is to make a synthesis gas, a mixture of carbon monoxide (CO), carbon dioxide (CO_2), and hydrogen (H_2). This gas is then used to produce energy in the form of heat. The other thermochemical form of conversion is pyrolysis, a chemical reaction that takes place in the absence of oxygen and by which organic matter is subjected to high temperatures, decomposes, and generates a mixture of carbon and tar.

Third-generation biofuels come from organisms that can make their food from solar energy and CO_2, including algae, which can be grown in photochemical reactors or outdoor swimming pools. After drying, the oil contained in its cells is extracted and transformed into biofuel by one of the above methods [7]. In this sense, lignocellulosic biomass is the most abundant renewable energy source in the world, composed mainly of aromatic biopolymers and polysaccharides. The agricultural, forestry, and industrial sectors depend on most of the lignocellulosic biomass. They are ideal according to the sugar content, the raw material that makes up this biomass to produce vegetable oils; coconut oil (3%), cottonseed (4%), olive (32%), palm kernel (4%), peanut (4%), rapeseed (14%), soybean (29%), and sunflower oil (8%) [8]. Within the range of biomass for the synthesis of biofuels include inedible crops, animal waste, and wood or vegetable residues, namely, microalgae, agro-industrial residues, and camelina.

Likewise, *Camelina satila* L. (Camelina) is a cold climate oilseed, which stands out for its high oil content (35–40% by weight). It has promising characteristics in terms of growth, making it ideal for meeting biofuel needs [9]. Another promising feedstock is poplar since it presents an attractive option to diversify and expand the available biomass, in the same way, that sugarcane is promising as a raw material since it presents density, viscosity, thermal conductivity, and thermal capacity. Like biofuel for kerosene-based aircraft [10]. Elgowainy et al. [11] carried out an analysis to generate biofuel FTJ biomass derived from cellulosic, corn straw, where the energy of the corn crop is not directly attributed to the corn straw since it is considered a by-product, therefore, it does not imply a change in use in the soil, being a good feedstock option for biofuels. Moretti et al. [12] reported potato by-products as feedstock for biofuels, these by-products being: potato peel (80%), gray gum (15%), and pressed pieces (5%).

The use of microalgae to produce biodiesel has emerged as a promising option since they have greater photosynthetic efficiency, are more effective in the

assimilation of CO_2 and other nutrients concerning plants, accumulate between 20% and 80% of triglycerides [13], does not require arable land, requires less renewable water consumption and can be grown in brackish water [14]. Regardless of the feedstock used, biofuel production processes must have a reduced associated environmental impact, which can help increase their sustainability. The use of renewable aviation fuel significantly reduces greenhouse gas emissions compared to fossil fuels Unlu [15].

Among all these three generations, the things that make it different are the final product of biofuel and the maturity of the science behind that fuel. First-generation biofuels are defined as those conventionally produced from food crops. Unfortunately, first-generation biofuels have withdrawn many disadvantages. They are being criticized for several reasons such as relatively low greenhouse gas (GHG) emissions avoidance, unsustainable production relating to deforestation, water use, and land management, competition for food crop feedstock's pushing up food commodity prices, and the need for generous government support schemes to remain competitive even after the technologies have become mature. In addition, some of the first-generation biofuels, such as biodiesel and ethanol (produced from corn) are not suitable fuels for powering commercial aircraft.

Many of these fuels do not meet the high performance or safety specifications for jet fuel. As a result, a lot of hope has been placed on second-generation biofuels. Second-generation biofuels produced via biomass and waste exploitation have significantly improved greenhouse gas emissions and the fuel's carbon footprint and have resulted in less environmental damage [16].

Biomass feedstocks do need to have the potential to displace large volumes of petroleum fuel to have considerable benefit. Any feedstock that is produced sustainably today could not only provide a benefit to those involved in the supply chain of fuel production and use, but it would also lead to valuable experience for the biofuels industry providing essential lessons in production and processing techniques. This experience should prove invaluable to the development of future sustainable feedstocks that could be used to create transportation fuels, such as jet fuel, with a minimum of arable land and freshwater (Prasad) [17].

2.2 *Main Technologies to Produce SAF*

2.2.1 Hydroprocessed Esters and Fatty Acids (HEFA)

The hydroprocessed esters and fatty acids (HEFA) process is currently the best known to produce SAF; it is like oil refining and a result is obtained with characteristics like conventional aviation fuel. This process consists of the reaction of vegetable oils in the presence of hydrogen and catalysts to produce aviation fuel, naphtha, diesel, and gasoline (Sonthalia) [18].

The process of producing biojet fuel through HEFA (Fig. 13.1) is usually divided into two steps: first, unsaturated fatty acids and triglycerides are converted to

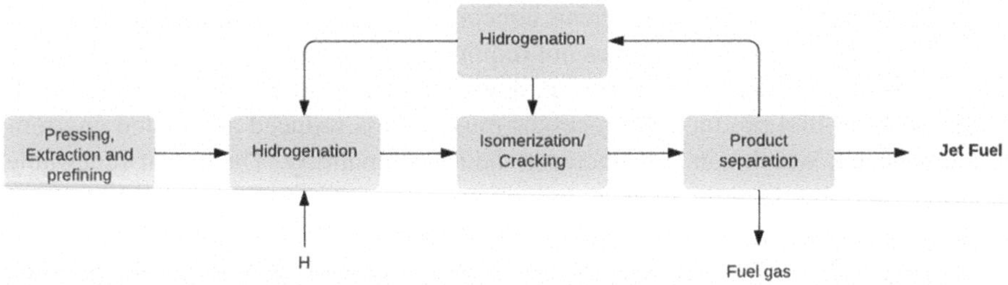

Fig. 13.1 HEFA/HJR technology

saturated fatty acids through catalytic hydrogenation, and then the saturated fatty acids are converted to straight alkane of C15-C18 through hydrotreating, and decarboxylation [19].

HEFA technology uses two basic chemical processes to convert vegetable oil into aviation biofuel. In the first step known as hydrodeoxygenation, hydrogen, and a catalyst are used to saturate the double bonds, extract all the oxygen from the oil, and split the main chain of propane, at the end of this process biofuel for aviation is obtained; however, parameters such as low-temperature performance or aromatic content are not sufficient to meet American Society for Testing and Materials (ASTM) specifications, which is why a second chemical process known as ısomerization and cracking is needed. In this second step, the length of the molecular chains is reduced to improve the performance of the biofuel at low temperatures.

In general, a biofuel production plant through the HEFA process follows the following steps: First, the seed oil is extracted from the storage tanks and is transported directly to a hydrotreater that contains hydrogen. The influent is cooled using steam generation and sent to the isomerization and cracking unit. Once isomerized, the resulting product is cooled with water at low temperatures before being sent to the separation tower where the gases, carbon dioxide, and excess hydrogen are separated from the liquid products. The hydrogen is separated from the carbon dioxide and again sent to the hydrotreater unit for recycling. Afterward, the liquid products are separated and sent to storage farms obtaining, depending on the concentration of hydrogen used in the hydrotreater, different concentrations of biofuel for aviation, glycerin, diesel, or naphtha (Starck) [20].

2.2.2 Fischer-Tropsch (FT)

The FT synthesis owes its name to two German chemists, doctors Franz Fischer (1877–1947) and Hans Tropsch, who discovered it and have already published their research results with a cobalt catalyst since 1923 (Casci) [21].

Fischer and Tropsch, both scientists at the Kaiser Wilhelm Institute in Berlin reported the reaction to produce synthol, a mixture of alcohols, aldehydes, and hydrocarbons, at 400–450 °C and 100–150 atm. By reducing the pressure and using Fe and Co bimetallic catalysts, they observed that a mixture of hydrocarbon chains

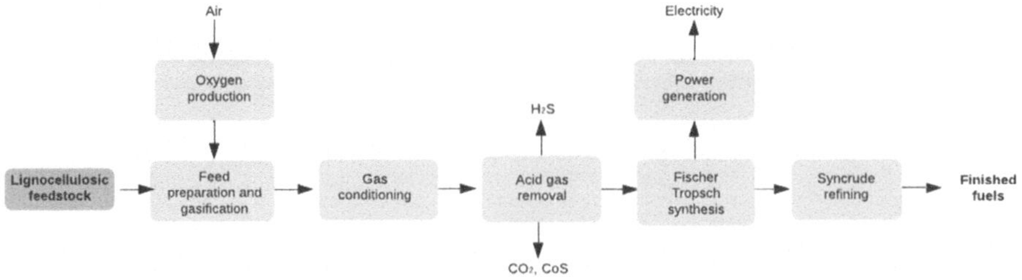

Fig. 13.2 Fischer-Tropsch technology

of different sizes was obtained through a polymerization reaction from CO and H_2; the reaction has since been called the Fischer-Tropsch Synthesis (Lapidus) [22].

The process involves the prior gasification of coal (or other sources such as biomass, charcoal, coke, and natural gas) by partial oxidation using oxygen as an oxidant and water vapor as a moderator at high temperatures, thus obtaining a mixture of H_2 and CO_2, synthesis gas, easy to separate from solid particles and acid gases (Fig. 13.2). In the second stage, the synthesis gas is transformed by a catalytic process, Co or Fe on silica at conditions between 450 and 650 K of temperature and 14–40 bar of pressure, almost exclusively in paraffin and alpha-olefins, through a mechanism in the chain, in which the probability of chain growth is practically independent of its size, obtaining a size distribution close to the Anderson-Schulz-Flory distribution. Two different FT processes have been certified by ASTM to date, one that produces a straight paraffinic jet fuel (SPK) and one that also produces additional aromatic compounds (SAK).

2.2.3 Alcohol to Jet (ATJ)

The biojet fuel obtained through the ATJ process is an aviation fuel produced from the conversion of short-chain alcohols, such as methanol, ethanol, and butanol, from various feedstocks (Romero-Izquierdo) [23]. The alcohols produced are subjected to a four-step upgrading process to obtain hydrocarbons in the jet fuel range (Fig. 13.3).

First, they are catalytically dehydrated to generate olefins, in this stage, the alcohol is reacted in the presence of a catalyst at temperatures ranging from 300 to 500 °C. The dehydration reaction also leads to the formation of ethers, which compete with the formation of olefins. To remove the water present in the reactor, the effluent stream is condensed by cooling the gas forming a dew, and then allowing unwanted products such as water, impurities, and alcohol to separate. Once the olefins are formed, these compounds are converted at conditions between 150 and 250 °C and 30 and 40 bar into a distillate containing unsaturated hydrocarbons in the range of diesel and kerosene through oligomerization with a range of C_6–C_{16}, typically of aviation fuels. Finally, the distillates are hydrogenated and distilled to obtain the range of kinds of paraffin that meet standard jet fuel specifications.

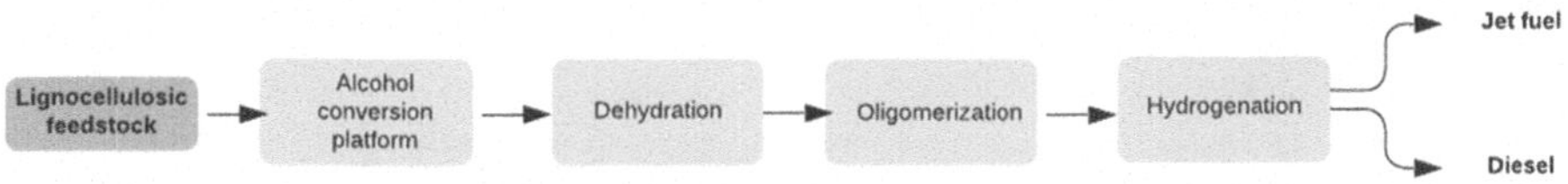

Fig. 13.3 Alcohol-to-jet technology

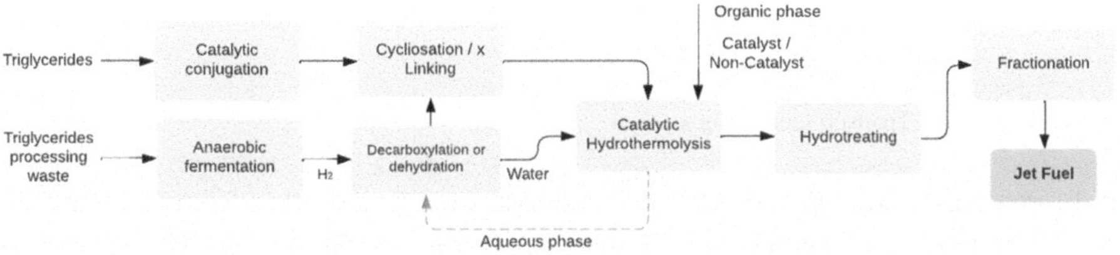

Fig. 13.4 Hydrothermal liquefaction technology

2.2.4 Hydrothermal Liquefaction (HTL)

One of the most promising thermochemical liquefaction processes is hydrothermal liquefaction (HTL). HTL is a technology that is capable of efficiently processing wet and dry biomass without restriction of lipid content, from lignocellulosic to organic waste (Fig. 13.4).

The product generated in this process is called bio-crude, which is the renewable equivalent of oil since it is an energy-dense intermediate that can be upgraded to a variety of liquid fuels. HTL generates bio-crude from organic matter thanks to specific characteristics, such as the presence of water in hydrothermal conditions, with temperatures ranging from 250 to 450 °C and pressures between 100 and 300 bar. Under these conditions, the water remains in a liquid state or a very dense supercritical state. This technique uses the specific characteristics of compressed hot water. Throughout the liquefaction process, the biomass undergoes a series of depolymerization reactions such as hydrolysis, dehydration, or decarboxylation, giving rise to insoluble products such as bio-crude oil or bio-carbon. In addition, other products are also generated, such as gases (CO_2, CO, H_2, or CH_4) or soluble organic substances (mainly acids or phenols) [24].

2.2.5 Synthesized Isoparaffins (SIP)

This method consists of the fermentation of sugars from the hydrolysis of biomass or direct sources, such as beets or sugar cane. First, the sugars are fermented by a genetically modified yeast (*Saccharomyces cerevisiae* PE-2) that transforms them into farnesene via the mevalonate (MVA) pathway.

The route starts from the acetyl-CoA produced from sugars and converts them through sequential steps into 3-hydroxy-3-methyl-glutaryl-CoA, which is transformed into mevalonate. Finally, after successive phosphorylations and

decarboxylation, isopentenyl diphosphate (IPP) and dimethylallyl diphosphate (DMAPP) are formed, the fundamental units whose condensation gives rise to farnesene. Farnesene has many unsaturations, which are associated with a low cetane number and low oxidative stability. Therefore, farnesene is subjected to hydrogenation giving rise to farnesane, which has a good cetane number and excellent cold flow properties. Farnesane is a 15-carbon molecule that is not found in petroleum but has a density like some of the hydrocarbons in petroleum that are used as jet fuels. The biojet fuel obtained by this process is called synthetic isoparaffins produced from hydroprocessed fermented sugars (SIP-HFS) (Wang) [25].

The synthetic isoparaffins produced from hydroprocessed fermented sugars (SIP-HFS) were approved by ASTM in 2014 for commercial use. But, unlike other SAFs, only up to 10% of this biofuel can be used blended with conventional Jet-A and Jet A-1 kerosene (International Civil Aviation Organization, 2020; Commercial Aviation Alternative Fuels Initiative, 2021) because the long carbon chains that compose it cause this fuel to have a high viscosity and a low combustion performance in aviation engines.

2.2.6 Catalytic Hydrothermolysis (CHJ)

Certified in 2020, this route has a 50% blend limit and uses vegetable or residual oils to produce biofuels. This is a route like HEFA; however, it presents significant differences in its processes. After cleaning and filtration, the oils and fats undergo hydrothermolysis, during which they are combined with water and brought to high temperatures and pressures. The resulting mixture is isomerized so that the generated molecules require less hydrotreating than in the HEFA route. It generates compounds such as paraffins, isoparaffins, cycloparaffins, and aromatic compounds.

3 Policies and Regulatory/Legal Requirements for the Adoption of SAF

One of the primary aims of this chapter is to underscore the prospective advantages of sustainable aviation fuel. These benefits encompass a decrease in greenhouse gas emissions, enhancement in energy security, and augmentation in the utilization of renewable resources [26].

To address this challenge, strong and supportive organizational initiatives, and policies are needed to promote SAF growth. Also, regulatory, and legal requirements for feedstocks, processes, and products are also required aimed at achieving SAF supply objectives. Figure 13.5 illustrates on a timeline diagram some initiatives taken by different organizations to promote the uptake of SAF.

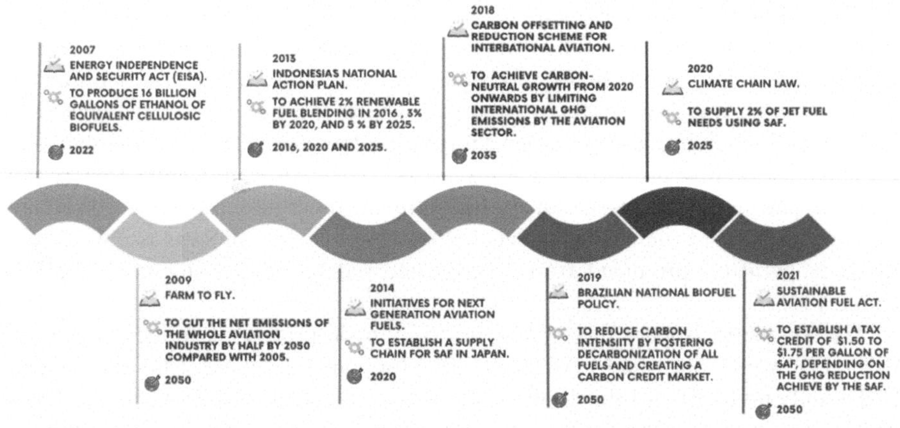

Fig. 13.5 Timeline of some initiatives implemented to promote the uptake of SAF. (Illustration adapted from Shariar et al. [27])

3.1 Initiatives and Partnerships

Various initiatives and partnerships have been launched to increase the adoption of SAF. To promote the use of renewable energy, some proposals have been put forward, including those established by the International Air Transport Association (IATA) and the European Union (EU). One of the key proposals, CORSIA, introduced by IATA, aims to offset and reduce carbon emissions from international aviation by encouraging airlines to compensate for any increase in emissions above their reference emissions from 2019 using SAF or by purchasing carbon credits. The second important initiative is RED II of the EU, which mandates member states to ensure that by 2030, a minimum of 14% of the energy utilized in transportation is derived from renewable sources [28–31].

Globally, there is a growing focus on eco-friendly aviation techniques as environmental concerns become more prominent. The United States Federal Aviation Administration (FAA) has taken an innovative approach to address this issue by launching its National Plan of Integrated Airport Systems (NPIAS) program. The program aims to promote the availability of SAF by investing in infrastructure developments that make it cost-effective and accessible for airlines. Additionally, the FAA envisions building strong partnerships between governments, airlines, and fuel suppliers worldwide to encourage the deployment of eco-friendly production units. KLM United Airlines and Qantas are examples of companies that have set a benchmark for the aviation industry through their successful collaborations, which were made possible through financial support from their respective governments [32, 33].

The Clean Skies for Tomorrow Coalition, established by the World Economic Forum, has become a critical undertaking in promoting the utilization of SAF. This coalition has effectively brought together a diverse group of key players in the

aviation industry, comprising airlines, airports, fuel suppliers, and government officials, to expedite innovation in SAF development and implementation. Intending to attain cost competitiveness on par with conventional jet fuels and significantly enhance access and availability, this partnership has made substantial progress since its inception.

However, there is still much work remaining to expand adoption rates while concurrently amending performance shortcomings. To bring these goals into fruition requires continued collaboration and investment from all stakeholders involved—airlines, government entities fuel providers and airports alike.

3.2 Policies

Lawmakers worldwide have undertaken an array of policies to stimulate the advancement of SAF production and consumption. Among these policies, the implementation of a low-carbon fuel standard (LCFS) has emerged as a pivotal one. The LCFS aims to incentivize fuel suppliers to gradually augment the proportion of LCFS in their offerings. California, being the pioneer state in the United States, has set a prime example in this regard. Other countries, such as Oregon and Canada, have emulated California's lead by enacting analogous policies [34].

One potential course of action is to enforce a regulation mandating the admixture of a specific proportion of SAF with traditional jet fuel. In 2019, Norway set a precedent by implementing such a mandate, which stipulates that 0.5% of the aggregate fuel supplied to airlines must consist of SAF. Other nations, such as France and Sweden, have disclosed their intentions to introduce analogous mandates [31, 35, 36].

Tax incentives represent a political instrument that can bolster the development of SAF production and consumption. A case in point is the United States' Renewable Fuel Standard (RFS), which offers tax credits to fuel blenders who use SAF. Several nations, such as the Netherlands, have taken the initiative to introduce a range of tax incentives that are aimed at incentivizing the enhancement and utilization of SAF [9, 37].

The utilization of governmental funds and subsidies can be strategically employed as a viable mechanism to bolster and foster the growth and maturation of SAF production facilities. A concrete illustration of this is evidenced in the United States where the Department of Energy has allocated and disbursed funding toward the pursuit and exploration of SAF development and generation with a particular focus on algae as a potential source. In a similar vein, the European Union has instituted and initiated a funding program with the express intent of buttressing and advancing the development and construction of SAF production facilities [38].

As the aviation sector perseveres in its endeavors toward mitigating its ecological imprint, it is anticipated that policies will expand in scope and extend globally in the forthcoming years. To incentivize the aviation industry to participate in the pressing

concern of climate change, lawmakers can facilitate the cultivation of SAF by proposing policies.

3.3 Standards

One organization that has supported the growth of SAF production and consumption is ASTM International, which is recognized worldwide for specifying different standards. By developing and publishing standards for SAF, ASTM helps ensure that the fuel is safe, reliable, and compatible with existing aircraft and infrastructure. For example, the specifications for synthetic aviation turbine fuels derived from hydroprocessed esters and fatty acids (HEFA) are established by ASTM D7566, which is considered a key ASTM standard for SAF. This standard provides detailed requirements for the composition, performance, and physical properties of HEFA-based SAF, which is currently the most widely used type of SAF. By providing clear and consistent specifications for HEFA-based SAF, ASTM D7566 helps to facilitate the production and use of the fuel by producers, airlines, and other stakeholders in the air travel industry [2, 39–41].

ASTM also develops standards for other types of SAF, including those derived from renewable diesel and alcohol-to-jet (ATJ) processes. ASTM D1655, which establishes the specifications for conventional aviation turbine fuel (jet fuel), has been modified to include a new annex for ATJ-based SAF. This annex provides detailed requirements for the composition, performance, and physical properties of ATJ-based SAF, enabling fuel producers and airlines to produce and use the fuel safely and reliably [25, 42, 43]. In addition to developing standards for SAF, ASTM also guides the testing and certification of SAF. ASTM D4054 provides a framework for the testing and evaluation of flight fuel additives, including those used in SAF. This standard helps to ensure that fuel additives used in SAF are safe, effective, and compatible with existing aircraft and infrastructure [44, 45]. ASTM standards for SAF are expected to play an increasing role in driving the growth of the market for biofuel in the future. Establishing a level of interaction between producers and suppliers allows healthy competition concerning conventional jet fuel. This is intended to encourage investment in SAF production facilities and support the growth of the industry.

The growth of SAF has been bolstered by the International Sustainability and Carbon Certification (ISCC), which has set a standard for sustainable biofuels, including SAF. The certification scheme provided by ISCC covers a range of sustainability aspects, such as greenhouse gas emissions, land use, and social and economic impacts. SAF producers can demonstrate their adherence to sustainability criteria and enter SAF markets by obtaining ISCC certification [46]. Moreover, the European Committee for Standardization (CEN) has formulated a series of guidelines for airplane fuel, which encompasses SAF. These standards put forth by CEN encompass different facets of aviation fuel, such as chemical structure, physical qualities, and efficacy. To guarantee the quality and suitability of their goods with

the present aviation infrastructure, SAF manufacturers can abide by CEN requirements [47].

Therefore, standards for SAF have been instrumental in supporting the growth of SAF production and consumption. By providing clear and consistent specifications for SAF, as well as guidance on testing and certification, standards have helped to ensure that the fuel is safe, reliable, and compatible with existing aircraft and infrastructure. As the aviation sector strives toward minimizing its ecological footprints, norms are expected to assume a progressively pivotal role in bolstering the expansion of SAF and mitigating the pressing issue of global warming.

4 Life Cycle Assessment: Main Results and Trends

A methodological tool that helps establish environmental indicators can be found in a life cycle assessment (LCA). According to Capaz et al. [48], over time requires standardizing the life cycle assessment process, creating the ISO 14014:2016, which defines it as a compilation and evaluation of inputs, outputs, and the potential environmental impact of a product throughout its life cycle, starting with natural resources at the beginning of the process and ending with the waste generated by the process, and the ISO 14044:2006, which defines four phases for this evaluation.

The CML Baseline 2000 Method Bueno [49] is used (unless clarified) to carry out the impact evaluation considering the following main categories: Global warming potential (GWP), Human toxicity potential (HTP), Photochemical Ozone Creation Potential (POCP), abiotic resource depletion (AD), acidification (AP), eutrophication (EP), ozone layer depletion (ODP), freshwater aquatic ecotoxicity (FAETP), seawater aquatic ecotoxicity (SAETP), and terrestrial ecotoxicity (TEP) potentials.

According to Agusdinata et al. [50], both economic and social factors are of equal importance to the environmental factor, and there were only two technologies for the production of biofuel for aviation which are the hydrotreating/hydrocracking process that uses vegetable oil and gasification followed by Fischer-Tropsch synthesis with lignocellulosic biomass, and after its analysis with different feedstocks, it was concluded that the cultivation stage is a factor with the greatest influence on the production of greenhouse gases is, followed by production.

Furthermore, as Elgowainy et al. [11] mention since that year fuel from the Fischer-Tropsch and HEFA processes already have regulations that allow its use with 50% conventional jet fuel, unlike that obtained from pyrolysis, which had not been certified for use in jet engines, the two technologies mentioned above are the ones that have had the greatest evolution and are the most widely used today.

To further support the use of this approach, global warming potential analysis is carried out by Zhu et al. [19] through biomass aqueous phase reforming (APR), hydrogenated esters and fatty acids (HEFA), and hydrothermal liquefaction (HTL) technologies, showing the production of biojet fuel through APR has the lowest

GPW, followed by HEFA and HTL; however, HEFA is the most mature technology up to now, having better opportunities and has the lowest cost.

As it was mentioned before, the economic, social, and environmental factors must be considered in aviation production fuel, and feedstock is one important factor that strongly impacts these aspects. Different authors have studied the impact of feedstock and the process used to produce aviation fuel on the environment, according to Urbanik et al. [51], GHG combustion emissions from the production of alternative fuels depend on each road, the specific case facility, feedstock source, and the supply route. In their study, they mentioned that GHG emissions from HEFA conversion come mainly from hydrogen production, and the path with the highest GHG production is with palm oil, also the production of fuel through ATJ with sugar cane has low emissions and high efficiency, concluding that fuels derived from sugar cane and sugar beet have higher GHG emissions savings due to their cultivation efficiencies.

According to Kolosz et al. [52], the best first-generation biomass to produce biojet fuel is sugarcane, the best second-generation biomass is jatropha as feedstock with ATJ process, and corn stover trough pyrolysis and for third-generation biomass is algae-derived for HEFA pathways. Resurreccion [9] mentions that camelina is an excellent rotation crop with wheat due to its ability to recycle soil nutrients, restore soil moisture, and break pest cycles, is highly competitive in that its wholesale price is on par with fossil fuel-based jet fuel. Not only the feedstock must be considered but also the path to obtaining it. Capaz et al. [48] concluded that the production of raw materials is usually the main contributor to greenhouse gas emissions and the environmental permanence depends largely on the conditions in which crops are grown.

Different authors have studied the impact of feedstock and the process used for the production of aviation fuel on the environment, Strappason [53] points out that, although in other countries there are several oil-producing crops, in Mexico only jatropha and castor beans were worked on, and the first one has mayor potential according to the hectares of land that can be used for its production, there is no competition with land for food production, that it promotes the use of land in debit and supports the rural and social development of peasants, and even though at that time many technologies were in their early stages of maturity, today this feedstock is still considered one of the most important.

Dangol et al. [54] carried out a study of biojet production using deoxygenation of camelina oil using the heterogeneous catalyst palladium on alumina and fractional distillation. The camelina grew in dry-land winter on a total of 60–120 ha of land as a rotational crop, the biojet fuel yield after distillation was 201 L/ha and only 38.4% of the vegetable oil was converted into jet fuel, using 2.9 MJ, and producing 420.1 gCO_2, with the cost of production of US$0.75 per liter and reducing the GHG emission 56.8% compared to Jet-A fuel.

Seber et al. [55] studied the life cycle assessment for different oils, obtaining that for jatropha, approximately 40 gCO_2eq/MJ is estimated for the production of biojet fuel, for pennycress 35–49 gCO_2eq/MJ, for tobacco 32.5–55.9 gCO_2eq/MJ, for castor between 41 and 78 gCO_2eq/MJ and Salicornia an approximate of 30.5–66.1

gCO_2eq/MJ, concluding that castor-HEFA fuel has the lowest value of emissions, the main contributors to GHG emissions are the farming and fuel production stages, being that natural gas and hydrogen consumption for HEFA conversion are key parameters.

Over time, and in search of stopping climate change from various areas, the production of aviation biofuel from waste feedstocks began to be studied and is currently one of the most used. Budsberg et al. [10] studied the production of biojet fuel using short-rotation woody crops as feedstock produced via the acetogen route with the help of the microorganism *Moorella thermoacetica*. Through two different simulations: natural gas steam reforming (NGSR) and lignin gasification (LG). The largest contributions to the GWP of poplar growth and harvesting are direct land use and nitrogen fertilizer, the largest source of biogenic CO_2 is biomass burning and the preferred energy from a life cycle impact standpoint is hog fuel with the NGSR simulation.

Lundberg et al. [56] carried out a study about the production of biojet fuel through the Fischer-Tropsch process using 66% root, 15% wood, and 14% bark as biomass, showing that transportation to and from the production site contributes the most to the GWP and 50% of the GWP contribution from biomass is derived from wood, concluding that for reduce the impact, the percentages of wood, root, and bark in the biomass can be changed to obtain the lower contribution of GWP.

Another example of an aviation biofuel production study from forest residue is Pierobon et al. [57], who studied the life cycle assessment of residual biomass, the production of biojet fuel consists of a pretreatment in which hemi cellulosic carbohydrates are converted to soluble sugars, enzyme production, and enzymatic hydrolysis are subsequently carried out, followed by fermentation, and upgrading, then an evaporation unit, a pyrolysis unit, wastewater treatment and finally a boiler, it was obtained that the main contributor to GWP is fermentation, carcinogens are dominated by the combustion of fuel in aircraft, the boiler is the main contributor to non-carcinogens, while the water treatment is the main one for ecotoxicity. They concluded that this path shows a reduction of more than 60% in GWP.

Moretti et al. [12] carried out a life cycle assessment using potato by-products, which are potato peel (80%), gray starch (15%), and pressed pieced (5%), obtaining that the main sources of the environmental impact of raw potatoes are the cultivation stage and its transportation to the processing industry, transportation process is responsible for 20% of the impact to climate change and 35% of the depletion of fossil fuels.

Vela-García et al. [8], who studied converting cellulose-derived alcohol intro biofuel for ATJ process and HEFA processing vegetable oil from oil crop residues, highlight that life cycle GHG emissions for the ATJ and HEFA routes were 75 gCO_2eq/MJ and 18 gCO_2eq/MJ, respectively, and that further development of microorganisms dedicated to simultaneous saccharification and co-fermentation concerning screening potential could improve the competitiveness of the concept.

Zhang et al. [58] analyzed the waste cooking oil as feedstock for producing biojet fuel, mentioning that has low energy consumption in the raw material stage and showed good prospects for its use as biofuel, being that this option can not only

recycle waste but also contribute to emission reduction for the aviation industry. They also mentioned that GHG emissions during transportation can be reduced with detailed raw material transportation route planning.

As previously mentioned, given the maturity of the HEFA technology, it is one of the most studied using first-generation biomass as feedstock (soybean and palm oil). According to the data collected, the biofuel that generates the least carbon footprint were those that were produced by the authors Wong [59] and Stratton et al. [60] with 29.9–50.8 gCO_2eq/MJ and 27.3 gCO_2eq/MJ for soybean and 21.9–31.5 gCO_2eq/MJ and 22.5–38.1 gCO_2eq/MJ for palm oil, respectively.

Bonomi et al. [61] present the LCA for FAME and HEFA technologies with different software, obtaining the lowest emission (16.90 gCO_2eq/MJ) of GHG for FAME with GHG Genius using soybean as a feedstock because the emissions associated with transport are low, and even if we could see minor values with used cooking oil, the reality is that the software does not consider some important emissions and stages in the process. Relating to HEFA technology, the less emission for soybean as feedstock was with VSB software (24.83 gCO_2eq/MJ) and for palm oil was with BioGrace (27.82 gCO_2eq/MJ) due to that CH_4 emissions were captured from the palm oil mill effluent pond. The technology that generates the least carbon footprint among those compared was Fischer-Tropsch, followed by ATJ.

Kolosz et al. [52] made the LCA showing that conventional crude oil has a GWP production of 85 gCO_2eq/MJ, natural gas produces approximately 100.4 gCO_2eq/MJ, and unconventional crude oil a result of 96.8–142.2 gCO_2eq/MJ; however, mitigation actions can be carried out through carbon capture technologies.

5 Techno-economic Analysis

The techno-economic analysis serves as a crucial tool in evaluating the feasibility and viability of SAF commercialization. It considers various aspects, including feedstock selection, conversion technologies, market dynamics, and environmental impacts. By providing a comprehensive understanding of the economic and technological aspects of SAF production, this analysis guides decision-making and helps accelerate the transition toward a more sustainable aviation industry.

Lim and colleagues [37] conducted a thorough examination of the economic prospects and difficulties linked to the production of biojet fuel, as seen through the eyes of biojet fuel producers. The opportunities identified in this study include the growing demand for biojet fuel, which has risen over time owing to increasing emissions and jet fuel demand (12–15%). Additionally, supportive government policies (12–15%) have gained some importance over time. On the other hand, the challenges faced by biojet fuel producers have shown varying trends. While the challenge of strict fuel standards has decreased in importance (18–10%), the challenges related to raw material availability (24–22%) and the lack of effective political incentives (9–11%) have remained relatively stable. Furthermore, the challenge of high production costs has increased significantly (22–29%).

5.1 Cost Breakdown, Evaluation of Capital, and Operating Expenses Over the SAF Production

In recent times, the concept of SAF has been garnering significant attention. It has emerged as a viable solution to mitigate the carbon footprint of the aviation sector. But one of the main challenges is determining the cost breakdown for SAF production, which depends on several factors such as the cost of raw materials, operating expenses, as well as localization, and size of industrial plants.

The cost of raw materials has a high impact on the final cost of SAF [62]. The cost of the feedstocks utilized in the production of various commodities may fluctuate following a wide array of factors that can exert a considerable influence on the pricing of these resources. Such factors could encompass the extent of accessibility of these feedstocks, the level of demand that exists for them, as well as the expenses incurred in transporting these materials from one location to another.

Since the production process for SAF involves converting these feedstocks into a usable form, any increase in the cost of raw materials will impact the final cost of SAF If a dearth in supply or a surge in demand for vegetable oils materializes, it follows that the outlay of manufacturing SAF through the utilization of vegetable oils will similarly experience an upswing.

To date, SAF is still a new industry [63], therefore is currently limited infrastructure for producing and transporting feedstocks. This can result in higher transportation costs for raw materials, which can also contribute to higher production costs and higher prices for SAF.

Thus, producers of SAF need to consider the cost and availability of raw materials when evaluating their production processes. Through the utilization of said approach, these individuals can guarantee that their ultimate commodity perseveres in its competitiveness with conventional jet fuels, all the while concurrently making a substantial contribution toward the cultivation of a more environmentally conscious and ecologically sustainable aviation industry.

The ensuing discourse expounds on several instances that underscore the inherent heterogeneity of expenses that may manifest in the process of producing SAF.

Klein-Marcuschamer et al. [64] provided a techno-economic analysis of the production of renewable aviation fuel from microalgae, *Pongamia pinnata* seeds, and sugarcane molasses. The study includes a detailed cost breakdown and evaluation of capital and operating expenses for each of the three routes. The authors found that depreciation, maintenance, insurance, and facility overhead were the most significant costs for all three processes. The paper provides valuable insights into the feasibility and cost-effectiveness of producing renewable aviation fuel from these sources.

Vela-García et al. [41] show that the cost of the raw material has a high impact on the final cost of the SAF, but the process can achieve a significant reduction of operating costs thanks to energetic optimizations.

Diederichs et al. in 2016 [62] compared the costs of producing jet fuel from lignocellulosic biomass (2G) versus two processes from first-generation (1G)

feedstocks, including vegetable oil and sugar cane juice. These authors also discuss the profitability of 1G jet fuel processes compared to 2G processes and identify the most promising 2G jet fuel options based on their economic viability.

Operating expenses are also a key factor to consider when evaluating the cost breakdown for SAF production because they represent a huge part of the total production costs. By analyzing and optimizing operating expenses, it is possible to reduce the cost of SAF production and make it more competitive with traditional fossil fuels. Additionally, understanding the breakdown of operating expenses can help identify areas where improvements can be made for energy efficiency and process optimization [25].

According to the study by Baral et al. published in 2019 [65], there are several steps involved in the production of SAF. These include feedstock supply, feedstock handling and preprocessing, pretreatment, conversion, recovery and separation, wastewater treatment, and on-site energy generation. Each of these steps requires resources such as energy, water, and labor.

One example of an operating expense in SAF production is feedstock supply planning. The study by Baral et al. [65] notes that different feedstocks have different costs associated with their transportation and storage. For example, if an SAF producer chooses to use corn stover as a feedstock, they will need to transport it from farms to their processing facility. This transportation cost can vary depending on factors such as distance traveled and mode of transportation. Another example of an operating expense is wastewater treatment. The study by Baral et al. [65] notes that during the production process, wastewater is generated that must be treated before it can be discharged into the environment. This treatment process requires energy and labor resources.

Lim et al. [37] shows the landscape of economic opportunities and challenges in SAF production from the point of view of the SAF producers, in terms of the opportunities, rising emission and jet fuel demand (12–15%) is an opportunity that has grown over time, while supportive government policies (12–15%) have slightly increased in importance, and in the challenges we see the challenge of strict fuel standards have decreased in importance (18–10%), the challenges of raw material availability (24–22%) and the lack of effective political incentives (9–11%) have remained fairly stable, and the challenge of high production costs has increased (22–29%).

Another crucial factor to consider when evaluating a capital investment is plant size [66]. Larger plants tend to have lower capital costs per unit of production due to economies of scale. Economies of scale enable larger plants to benefit from lower feedstock costs, improved operational efficiency, and better access to financing options. These factors contribute to the overall competitiveness of SAF in the market. Notably, companies such as Neste Corporation, Gevo Inc., and Alt Air Fuels, have emerged as key players in commercial SAF production, leveraging their expertise and infrastructure to provide sustainable alternatives to traditional jet fuels [38]. As the demand for SAF continues to grow, further advancements in plant size optimization and technological innovations are expected, leading to more affordable and accessible SAF in the future. Table 13.1 shows the main companies that produce SAF in the world, including their capacities and localization.

Table 13.1 Main companies that produce SAF in the world

Company	Country	Capacity (million Tons/year)	Localization
Neste Corp.	Finland	1	Singapur
Neste Corp.	Finland	0.8	Roterdam (the Netherlands)
Neste Corp.	Finland	0.8	Porvoo (Finland)
Gevo Inc.	EE.UU.	0.068	Luverne (Minnesota)
AltAir fuels	EE.UU.	0.115	Paramount (California)

Location is also a principal factor to consider when evaluating capital investment [50]. Location can impact both feedstock costs and infrastructure costs. For example, if a plant is located near a source of raw materials, feedstock costs may be lower than if it were located far away from such a source.

Gevo Inc., a United States-based biotechnology enterprise, is highly specialized in the sphere of renewable biofuel production, with a particular emphasis on the production of SAF. Through the utilization of a patented fermentation process, Gevo has been able to transform sugars into hydrocarbons, which can be used to generate renewable energy for the aerospace and aviation industries.

Neste Corporation (Finland) is a leading producer of renewable biofuels and has ventured into the production of SAF. They use a variety of sustainable raw materials such as used cooking oil and waste oil to produce advanced biofuels, including SAF.

AltAir Fuels, a pioneering United States-based organization, has successfully established a highly advanced biofuel production plant in the state of California. This company has been extensively engaged in the production of SAF through the utilization of cutting-edge hydroprocessing processes and revolutionary fermentation technology. Notably, AltAir Fuels has harnessed the potential of agricultural waste and vegetable oils as the primary raw materials to produce SAF, thereby emerging as a leader in the renewable energy sector.

Amyris, a biotechnology corporation situated in the United States of America, places its emphasis on the creation of renewable chemicals and biofuels. The corporation has dedicated itself to the development of fermentation technologies, which can be utilized to transmute sugars into hydrocarbons, thereby facilitating the production of SAF, which is an acronym for sustainable aviation fuel.

5.2 *Minimum Selling Price*

The minimum selling price (MSP) is important for SAF production as it enables cost recovery, promotes competitiveness, attracts investment, ensures long-term viability, and addresses environmental considerations. By setting an appropriate MSP, the SAF industry can grow, become economically sustainable, and contribute to reducing greenhouse gas emissions in the aviation sector.

The establishment of a competitive selling price for SAF can prove to be a daunting task owing to various obstacles that exist. These barriers are subject to variation based on factors such as the availability of feedstock, production technology,

market dynamics, and policy frameworks. Among the primary obstacles are the accessibility and cost of feedstock, the production scale, research and development expenses, policy support and market demand, and the infrastructure and distribution network. It is imperative to note that the SAF industry is rapidly developing, and the ongoing progress in technology, feedstock availability, policy support, and market demand can facilitate the resolution of these barriers over time. With the maturation of the industry, it is anticipated that economies of scale and improved efficiency will culminate in the reduction of production costs and the establishment of more competitive selling prices for SAF.

Various authors have researched the MSP issue for SAF production worldwide, and below are their findings.

The paper of Geeg et al. [67] stated that all respondents in their study acknowledged that the purchase price of aviation biofuels given current technology will be higher than the price of standard jet fuel. Biofuel producers stated "The price of bio-kerosene is at least twice the price of fossil jet fuel alone. And the price is a challenge. We are working hard to get the costs down, but this is a major issue for everybody." Estimates for price parity were given reluctantly, but most estimates were between 2015 and 2030.

McGarvey et al. [68] included a break-even price analysis for the SAF sold, which makes the net present value equal to zero. The break-even price was identified using the software @Risk, which records the distribution of break-even prices given the uncertainty in all input parameters. The study showed that break-even prices were found in each scenario, and that absent any policy, the project was not economically viable. However, once renewable identification number (RIN) and low-carbon fuel standard (LCFS) credits were applied, the project was viable in several cases. The credits assumed an SAF selling price of $2.66/gal when only RIN credit was applied and $3.36/gal when both RIN and LCFS credits were applied.

The work of Baral et al. [65] presented a detailed techno-economic and sensitivity analysis, including estimated MSP, and life cycle greenhouse gas (GHG) mitigation costs for five routes to four potential biojet fuel molecules. The study assesses the contributions from each stage of the entire supply chain (integrating feedstock supply planning and the downstream conversion processes) to the minimum selling price of the selected molecules, and their probabilistic distribution results are presented. These authors conclude that the optimal prices reported in the study can be considered as theoretical minima, with little room for further improvement without major changes in process configuration. The study also emphasizes that several factors contribute to fixing the MSP of SAF, including feedstock supply planning, biomass utilization, recovery and separation, and pretreatment. Therefore, it can be concluded that reducing these costs would require significant changes in process configuration or technological advancements.

Klein-Marcuschamer et al. [64] provided a techno-economic analysis of the production of renewable aviation fuel from microalgae, *Pongamia pinnata* seeds, and sugarcane molasses. The study included an estimation of the MSP for each of the three routes. The authors found that the MSP for microalgae-based SAF was higher than that for *Pongamia pinnata* and sugarcane molasses-based SAFs. The paper

provides valuable insights into the economic feasibility of producing renewable aviation fuel from these sources and can be used as a reference for future studies on SAF production.

Vela-García et al. [41] showed that the cost of the raw material has a high impact on the final cost of the SAF, but a significant reduction of operating costs can be achieved thanks to energetic optimizations of the process.

Shila & Johnson [69] calculated the gate MSP price for Camelina-derived hydro-processed renewable jet fuel. The gate MSP price is a function of capital costs, operating costs, and revenues. The economic analysis is based on the material and energy balances performed in similar studies from other literature sources. However, this price does not account for contractual agreements that may exist between fuel producers and aircraft operators to produce fuel for specified lengths of time, where the fuel price may not change with market price fluctuations.

Tanzil et al. [70] used the information obtained from the techno-economic analysis of six SAF production technologies to develop a simplified strategy for the estimation of MSP. The authors claim that the MSP is an important metric for determining the economic viability of SAF production and is used in strategic analysis to identify potential technologies that could compete with the fossil fuel industry.

Liu et al. [71] provided information about the MSP to produce SAF from jatropha and castor oilseeds in China. The study estimates the MSP by considering annual production costs, including capital and operating costs. Variable operating costs include feedstock planting costs, other raw materials, utility costs, and transport costs. Fixed operating costs include labor costs, sales costs, maintenance costs, property insurance costs, and taxes. The coproducts such as bio-oil, oily cake, propane, diesel, naphtha, and shell ash are assumed to be sold at current prices to offset the production cost. Prices were obtained from the WIND database.

Diederichs et al. [62] provided a techno-economic comparison of SAF production from different feedstocks and processes, including an evaluation of the MSP for the SAF produced. The authors compare the MSP for producing jet fuel from lignocellulosic biomass (2G) versus two processes from first-generation (1G) feedstocks, including vegetable oil and sugar cane juice. They evaluate the MSP for each process based on capital and operating expenses, as well as the cost of feedstock, labor, energy, and other inputs. The paper also discusses the profitability of 1G jet fuel processes compared to 2G processes and identifies the most promising 2G jet fuel options based on their economic viability.

Wang [72] reported information related to the MSP for SAF production. The study showed that various parameters such as feedstock cost, plant capacity, reactor construction, catalyst price/loading, and hydrogen cost cause high degrees of uncertainties in MSP. The authors found that the MSP value decreases by 16% if the selling price of bio-oil increases to $1.6/kg. It is also mentioned that doubling the profit needs a 21% higher fuel selling price.

The paper by Diniz et al. written in 2018 [73] provided information about the MSP for the SAF production process using the hydrodeoxygenation (HDO) process with camelina, carinata, and jatropha feedstocks. The authors present a stochastic

techno-economic analysis of the production of aviation biofuel from oilseeds, evaluating the economic viability of two product profiles: (i) Maximum diesel production and (ii) maximum jet fuel production (HRJ). The study considers uncertainties associated with feedstock prices, capital costs, operating costs, and product prices using Monte Carlo simulation.

The results show that the MSP for the HRJ product profile ranges from 2.5 to 3.5 USD/L depending on the feedstock price and capital cost assumptions. The authors also identify key factors that affect the MSP for SAF production, including feedstock prices, capital costs, operating costs, and product prices.

The article by Cervi et al. [74] provided information on the MSP for SAF production in Brazil. The authors calculate the MSP for different technological pathways, including sugar-to-jet (STJ), hydroprocessed esters and fatty acids (HEFA), and lignocellulosic-to-jet (LTJ) processes. They consider factors such as feedstock costs, labor costs, energy costs, and capital investment requirements in their calculations. The authors present their findings in the form of cost curves and sensitivity analyses, which allow for a comparison of the economic performance of different pathways under different scenarios.

In the work of Michailos & Bridgwate [75] the implementation of a discounted cash flow analysis was done to estimate the minimum selling price (MSP) of SAF, where it was obtained 1.98, 2.32, and 2.21 $/L for the hydroprocessing (HP), the gasification followed by Fischer-Tropsch synthesis (G + FT), and the zeolite cracking (ZC) routes, respectively. These estimates were based on simulations carried out in the Aspen Plus environment that aimed to establish and quantify mass and energy flows, which set the basis for a detailed cost breakdown of the examined SAF routes.

The paper of Tongpun et al. [76] provides information about the MSP for jatropha-derived hydroprocessed renewable jet fuel production. The study conducted a techno-economic analysis (TEA) of jatropha-derived HRJ production according to local situations, including farming processes for obtaining jatropha fruits and refinery processes for obtaining renewable aviation fuel. The results were compared to those from Thailand and Cambodia due to various local economies for discussing the differences in MSPs. Furthermore, based on the annual fuel consumption of one airline company in Taiwan, the MSPs for various blending proportions of renewable fuel into petroleum aviation fuel were listed and discussed to determine optimal blending ratios for various farming locations.

The hydroprocessed jet fuel provided an MSP of $2.22/kg jet fuel, lower than the ones of thermochemical ($2.44/kg) and hybrid ($2.50/kg) processes, according to Diederichs et al. [62]. Klein-Marcuschamer et al. [64] conducted economic evaluations of biojet fuels from microalgae and *Pongamia* oils and demonstrated that the prices per gallon of fuels were $31.98 and $8.9, respectively.

Figure 13.6 depicts the MSP as a function of the investment capital for various technologies and feedstocks utilized in the production of SAF. The technology employed in the production of SAF plays a critical role in determining the MSP. For instance, the use of alcohol-to-jet (ATJ) technology requires varying amounts of investment capital, leading to different MSPs. For instance, with an investment

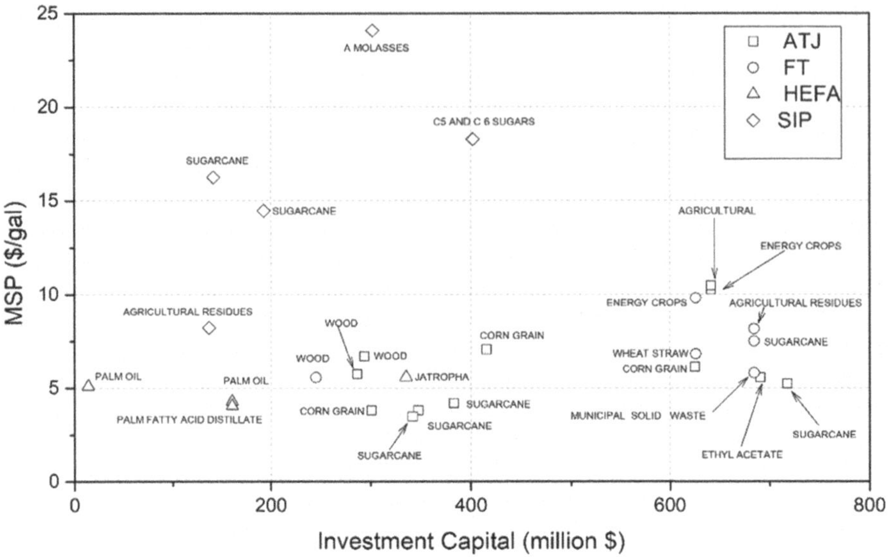

Fig. 13.6 Minimum selling price as a function of the investment capital for different technologies and feedstocks

capital of $700 million and sugarcane as the raw material, the MSP is $5/gal. However, if solid catalyst induced pyrolysis (SIP) technology is utilized with the same raw material, the required investment capital is lower, around $120 million. Nevertheless, opting for SIP technology would result in an increase in the MSP to $15/gal.

The hydroprocessed esters and fatty acids (HEFA) technology provide the lowest investment capital when palm oil is used as the raw material, resulting in an average MSP of $5/gal. Thus, based on the graph, HEFA technology is the most viable option.

On the other hand, Fischer-Tropsch (FT) technology requires the highest investment capital, ranging from $600 million to $700 million, resulting in an MSP of $5/gal to $10/gal. The technology utilizes residues as raw materials, which may be an excellent option for the future.

The implementation of SIP technology requires an investment capital between $150 million and $400 million, which is relatively low compared to ATJ and FT technologies. However, the MSPs are relatively high, ranging from $7.5/gal to $25/gal, which represents a disadvantage of using SIP technology.

Based on the observations presented in Fig. 13.7, it can be deduced that the MSP ranges from 4 ($/gal) to 10 ($/gal) while considering an operating cost of 100 (million $/year) to 200 (million $/year).

The implementation of HEFA technology proves to be beneficial in reducing the operating cost, ultimately leading to a lower MSP compared to the usage of SIP technology.

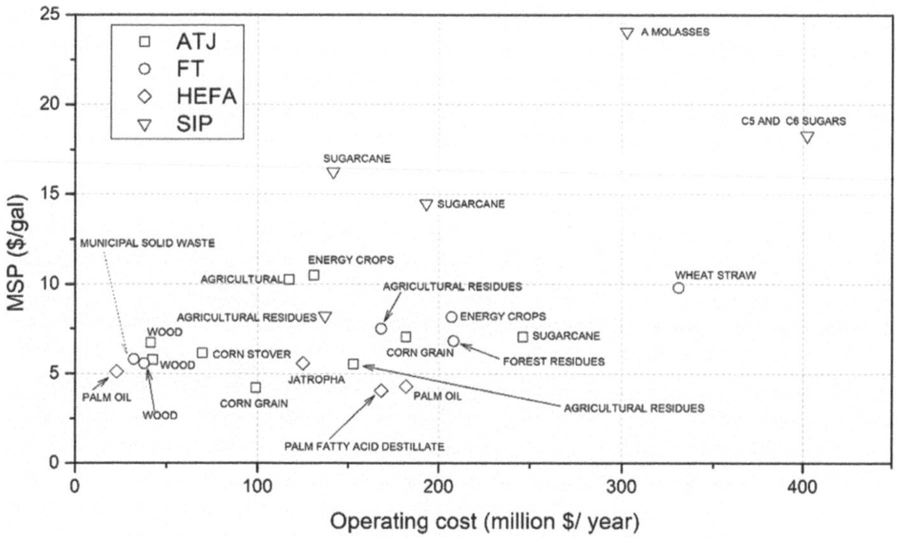

Fig. 13.7 Minimum selling price as a function of the operating cost for different technologies and feedstocks

In contrast, utilizing SIP technology incurs a higher operating cost, which increases the minimum selling price, ranging from 7 ($/gal) to 25 ($/gal), as depicted in the graph.

From the graph, it is evident that employing SIP technology using sugars as the feedstock incurs the highest operating cost of $402.23 million per year, resulting in an MSP of 18.28 ($/gal). Conversely, utilizing FT technology that utilizes wood as the feedstock incurs the lowest operating cost of $37.91 million, resulting in a minimum jet fuel selling price (MJSP) of 5.56 ($/gal).

The operating cost for ATJ technology ranges from $40 million to $250 million per year, leading to minimum selling prices ranging from $4/gal to $10/gal. The choice of raw material for SAF production significantly impacts the minimum selling price, with wood and corn proving to be the best options while sugarcane increases the MSP.

Figure 13.8 illustrates the varying production capacities of different technologies to produce SAF. To achieve an MSP range between 4 and 10 ($/gal), a production capacity from 60 to 62 (MMgal/year) using ATJ, HEFA, and FT technologies is necessary. These technologies utilize agricultural residues, municipal waste, and energy crops like corn, grain, palm oil, and hybrid poplar as feedstock.

The implementation of SIP technology increases the MSP, and the production capacity ranges from 15 to 20 (MMgal/year) when sugarcane is used as feedstock. However, if agricultural residues are used, the SAF production capacity would be 30 (MMgal/year), and the MSP would be 8 ($/gal). When wood is used as feedstock, an MSP range between 5 and 7 ($/gal) can be achieved, but the annual production would be less than 5 (MMgal/year), which is not beneficial. The use of

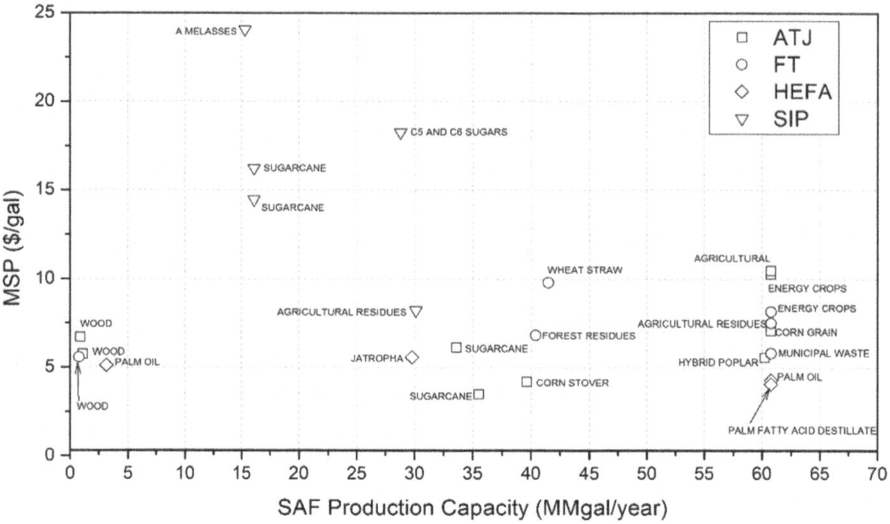

Fig. 13.8 Minimum selling price as a function of the production capacity for different technologies and feedstocks

jatropha as feedstock, particularly with HEFA technology, shows somewhat positive results, with an annual production capacity of 30 (MMgal/year) and an MSP close to 5 ($/gal), making it a viable option. Based on the observations from the graph, the best option is to employ HEFA technology with palm fatty acid distillate as a feedstock since it would result in an annual production capacity of 60 (MMgal/year), and the MSP would be less than 5 ($/gal).

5.3 *The Net Present Value (NPV) Associated with the SAF Production*

The significance of net present value (NPV) attributed to SAF production is noteworthy as it enables the evaluation of the economic feasibility and lucrativeness of investing in SAF production initiatives. NPV, a financial gauge that measures the worth of an investment, is ascertained by assessing the present value of the expected cash inflows in contrast to the present value of its cash outflows.

In the realm of SAF manufacturing, computation of the net present value (NPV) necessitates a meticulous assessment of forthcoming cash inflows linked with the production and trade of SAF, which incorporate proceeds from sales and expenses of production. These cash inflows are subsequently subjected to a discounting process to determine their present value via an appropriate discount rate. The discount rate accounts for the time value of money, manifesting the notion that future cash inflows are of lesser worth than contemporary cash inflows.

The NPV of a SAF production project provides insights into its profitability and the potential return on investment. A positive NPV indicates that the expected cash inflows exceed the initial investment and ongoing costs, suggesting that the project is financially viable and likely to generate a profit. From an alternative perspective, it can be argued that a negative net present value (NPV) serves as an indication that the proposed project lacks economic feasibility and has the potential to result in unfavorable financial outcomes.

The thorough evaluation and analysis of the net present value (NPV) of projects that are aimed at producing SAF is a critical and vital exercise aimed at providing insightful and informed guidance to various stakeholders. These stakeholders, which may include investors, project developers, and financial institutions, rely heavily on the results of such assessments to make well-informed and judicious investment and financing decisions regarding such ventures. Therefore, the assessment of the NPV of these SAF production projects is an indispensable tool that provides invaluable insight and guidance to the different stakeholders involved. It allows them to evaluate the economic feasibility and potential risks associated with SAF production, comparing it with alternative investment opportunities. Moreover, governments and policymakers may consider the NPV when formulating incentives, subsidies, or regulations to promote the development and adoption of SAF technologies.

The main findings found in the literature regarding the estimation of the NPV of different technological options used to produce SAF are shown below.

McGarvey et al. [68] included a stochastic techno-economic analysis of the catalytic hydrothermolysis aviation biofuel technology, which includes a financial analysis that calculates the NPV. The study found break-even prices for the SAF sold, which makes the net present value equal to zero. The break-even price was identified using the software @Risk, which records the distribution of break-even prices given the uncertainty in all input parameters. The study also found that absent any policy, the project was not economically viable, but once RIN and LCFS credits were applied, the project was viable for several cases.

Klein-Marcuschamer et al. in 2013 [64] provided a techno-economic analysis of the production of renewable aviation fuel from microalgae, Pongamia pinnata seeds, and sugarcane molasses. The study includes an economic and financial analysis that evaluates the NPV for each of the three routes. The authors found that all three routes had positive NPVs, with the microalgae-based SAF having the highest NPV. The payback period was estimated to be around 10 years for all three routes.

Zhao et al. [77] provided information about the NPV for eight different cellulosic biofuel production pathways, including SAF production. The study employs a techno-economic assessment-based financial analysis to derive NPVs for each pathway. The results of the analysis provide policymakers and private investors with information on how uncertainty affects the economic feasibility of each pathway. The results suggest that investment in none of the eight cases would be profitable at expected values and that most risk-averse investors would prefer the fast pyrolysis and hydroprocessing pathway.

The article by Budsberg et al. [10] provided information on the NPV and financial analysis of hydrocarbon biojet fuel production from poplar biomass. The authors estimated NPV values ranging from −1.2 to 3.5 billion USD, depending on facility capacity and discount rate. The payback period is estimated to range from 8 to 14 years, again depending on facility capacity and discount rate. These estimates consider several factors such as feedstock costs, capital and operating expenses, product yields, and revenue from the sale of biojet fuel. The article also contains the sensitivity of these economic and financial metrics to several factors such as feedstock costs and product yields.

The work by Diniz et al. written in 2018 [73] contains information about the NPV for SAF production using camelina, carinata, and jatropha feedstocks. The comparative variables of the hypothetical production facilities maximizing the production of renewable jet fuel were estimated. The camelina, carinata, and jatropha scenarios returned NPVs of −493, −717, and −1.627 US$MM, respectively. The conclusion about the NPV result for the SAF production is that the scenarios for camelina, carinata, and jatropha feedstocks returned negative NPVs. This suggests that the production of renewable jet fuel using these feedstocks may not be economically viable.

Michailos & Bridgwate [75] provided information about the use of NPV break-even assessment to estimate the minimum jet fuel price for the examined SAF routes. The simulations carried out in the Aspen Plus environment set the basis for a detailed cost breakdown of the SAF routes. The study concludes that the implementation of the Second-Generation Biofuel Producer Tax Credit (SGBPTC) generates positive NPVs for all cases when the respective MSPs are considered. However, without more specific information about the cases studied, it is difficult to provide a more detailed answer.

Drop-in fuels can be blended in high contents with fossil fuels, facilitating market penetration and social acceptance [78]. Also, drop-in fuels can be used in existing vehicles and distribution facilities without requiring new vehicles or infrastructure.

5.4 Internal Rate of Return Associated with the SAF Production

In the milieu of SAF production, the internal rate of return (IRR) serves as an instrumental metric for gauging the financial profitability and overall desirability of the project, as perceived by prospective investors. The IRR essentially embodies the annualized rate of return that an investment is anticipated to yield over the course of its entire duration, considering the magnitudes and temporal patterns of the associated cash flows.

The IRR helps stakeholders make informed financial decisions on SAF production. It enables them to test the financial viability of the project and compare it with

alternative investment opportunities. By considering the IRR, investors, and project developers can assess whether the potential returns meet their financial objectives and risk appetite.

The process of computing the IRR plays a key role in ascertaining the practicality of SAF production. When the IRR surpasses the predetermined minimum rate of return or hurdle rate, it signifies that the initiative is feasible and has the propensity to produce favorable cash flows. Such awareness is of utmost importance for obtaining the necessary funds and enticing potential investors to endorse the growth and expansion of SAF production facilities.

The analysis of IRR enables stakeholders to evaluate the susceptibility of the project toward fluctuations in vital variables like feedstock prices, production costs, or market demand. With the contemplation of divergent scenarios and customization of input assumptions, they can scrutinize the potential threats associated with the production of SAF. The assimilation of this data aids in the management of jeopardy and the process of making informed decisions.

The analysis of the IRR significantly impacts the determination of the MSP of SAF. The IRR serves as a crucial benchmark for establishing the selling price that effectively covers the production costs and generates the desired return. Profound insight into the IRR facilitates the development of a competitive pricing strategy for SAF, thereby rendering it more appealing to airlines and other key players in the aviation industry.

A high IRR indicates a lucrative investment opportunity. When SAF production projects demonstrate a favorable IRR, they attract investors, including financial institutions, venture capitalists, or private equity firms. The ability to showcase a competitive IRR increases the likelihood of securing funding and accelerating the development of SAF production infrastructure.

Understanding the IRR supports the growth of SAF production as a viable and sustainable alternative in the aviation industry. When stakeholders are confident in the financial feasibility and profitability of SAF projects, it encourages investment, innovation, and the scaling up of production capacity. The phenomenon, which occurs because of a particular action or event, serves to streamline, and expedite the progression of the aviation industry toward the adoption of increasingly eco-friendly practices that are conducive to the preservation and enhancement of the natural environment. Furthermore, this process also serves to mitigate the level of emissions of gases that are known to contribute significantly to the phenomenon of climate change and thereby promote a healthier, more sustainable planet. Below are the main findings obtained by different researchers related to the internal rate of return on this topic.

The Monte Carlo simulations made by Martínez-Hernández et al. [66] showed that there is a 50% risk of failure to achieve the 10% IRR target for both biojet fuel and green diesel-only cases. Additionally, there is more than a 20% probability of a negative IRR. The authors conclude that the green diesel option is a more attractive and less risky investment than biojet fuel. The biojet fuel production benefits from the credits from co-products and if larger plant sizes or lower vegetable oil prices can be obtained, the risks could be mitigated.

According to the work by Budsberg et al. [10], "selling the product for a higher price will increase the IRR while selling for a lower price will reduce the IRR." However, specific numerical results about the IRR are not provided by these authors. They also notice that "a high IRR would be required to attract investors" due to variability in historical petroleum prices and environmental subsidies.

The main results about the IRR obtained in the work by Diniz et al. [73] are as follows: The deterministic analysis of Camelina and Carinata diesel facilities returned IRRs of 25 and 18%, respectively. The stochastic analysis suggested that the probabilities of positive NPVs were 75, 59, and 15%, respectively, for camelina, carinata, and jatropha diesel plants. Jet fuel facilities presented probabilities of loss of 98, 99, and 100% for camelina, carinata, and jatropha scenarios, respectively. Only the Camelina and Carinata scenarios exceeded the minimum required IRR of 9.71% to have positive NPVs.

Neuling et al. [79] reported a financial analysis of SAF production technologies and found that the HEFA process was the economically most promising pathway based on their calculation of IRR for each process. However, it does not provide specific values for the IRRs of each process.

6 Market Analysis and Demand Forecasting

Market analysis and demand forecasting are fundamental aspects of the development of any industry, and SAF production is no exception. SAF is a sustainable biofuel that is used as an alternative to fossil fuels in the aviation industry. The aviation industry is increasingly interested in sustainable biofuels, and there are government policies and incentives that promote the development of the SAF industry. In addition, SAF production can be an interesting alternative to reduce greenhouse gas emissions and improve the sustainability of the aviation sector.

Biofuels are generated from sustainable origins, namely, crops and organic refuse, which ultimately incur expenses linked with the procurement of essential resources [80, 81]. In the case of SAF, like biodiesel and bioethanol, the costs of crops and vegetable oil raw materials can fluctuate due to factors such as food prices and weather conditions. In contrast, fossil fuels are procured via the extraction of petroleum and natural gas reserves, thereby giving rise to extraction and production expenditures, in addition to the geopolitical hazards that are associated with the process.

The processing of biofuels requires specific technologies. These processes can involve significant costs for equipment, energy, and specialized labor. Biofuels often require a shorter supply chain due to the local availability of raw materials, which can reduce transportation costs. The main sector where fuels are used is the transportation sector, being one of the largest energy consumers; the growing environmental awareness and stricter regulations on greenhouse gas emissions have driven the adoption of biofuels in the sector [82]. Liquid biofuels, such as SAF, bioethanol, and biodiesel, are used as blends in conventional fuels, thus reducing

dependence on fossil fuels and decreasing emissions of polluting gases; the agricultural sector also represents an important part of the demand for biofuels.

Energy crops, such as corn, sugarcane, palm oil, and sugar beet, are used for biofuel production. Finally, one of the most important sectors is energy generation which is adopting biofuels as an alternative to fossil fuels. Biomass, which includes agricultural, forestry, and animal-origin waste, is used for thermal and electrical energy production [83].

In Fig. 13.9, the process for obtaining biojet fuel is shown. It starts with biomass or renewable raw materials. When used as feedstock, it reduces the dependence on

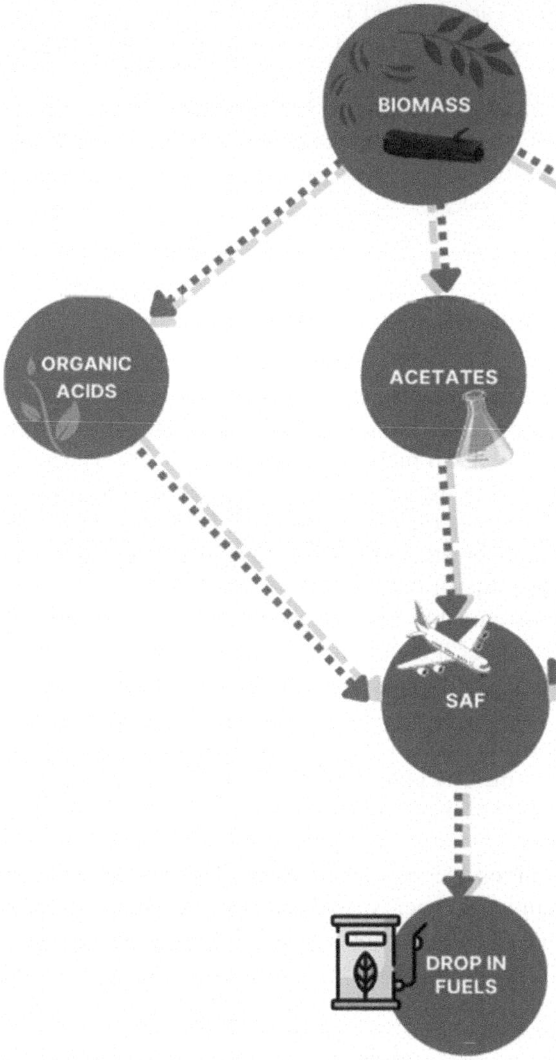

Fig. 13.9 Process for obtaining biojet fuel is shown

fossil fuels and decreases carbon emissions. Different technologies are employed for its production.

Current and projected demand for biofuels is high due to the need to reduce greenhouse gas emissions and energy security [30]. In addition, biofuels are presented as a viable alternative to traditional fossil fuels. Latin America is a region with great potential for the production and export of biofuels due to its favorable climate for the cultivation of raw materials such as sugar cane [84]. On the other hand, Europe has been a leader in the development of the biofuels market due to its commitment to reducing polluting emissions [85].

To implement the use of biofuels, favorable policies and regulations have been implemented. Emissions reduction targets, air quality standards, and renewable energy targets are driving demand for biofuels. Similarly, there is growing concern about the sustainability of biofuels, especially in terms of their environmental and social impact. Consumers and market players are demanding sustainably produced biofuels, driving the adoption of stricter certifications and standards, with these implementations the demand for biofuels is expected to continue to grow in the future due to several factors [86].

However, it is important to address the challenges associated with biofuel production, such as competition with food crops and the sustainable use of natural resources. Comprehensive strategies that foster the development of advanced and sustainable biofuels, such as production from algae and organic waste, will be required [87].

Distributors and retailers play a crucial role in the SAF supply chain, ensuring that products reach end consumers [88]. The sales prices of biofuels and fossil fuels are determined by factors such as production costs, supply and demand, taxes, and government subsidies [89]. It is important to assess its global environmental impact and its real contribution to the reduction of greenhouse gas emissions. Biofuels have a lower carbon footprint compared to fossil fuels, as the photosynthesis of the plants used to produce them absorbs carbon dioxide from the atmosphere. However, it is important to consider emissions associated with changes in land use and the production process, such as deforestation or intensive use of fertilizers [90].

One of the main environmental challenges associated with biofuels is the use of agricultural land to produce crops dedicated to their manufacture. The expansion of crops for biofuels can lead to the conversion of natural ecosystems, such as forests or grasslands, resulting in the loss of biodiversity and the release of large amounts of stored carbon [91]. The production of biofuels necessitates the presence of water for both the cultivation of the plants and industrial processing. The extent of water usage and its effects on the surrounding water resources are contingent on the specific biofuel type and the geographical location in question. Consequently, the employment of water resources can have a considerable influence on local water resources. A primary concern pertains to the competition that exists between crops utilized to produce biofuel and crops intended for food production. Given the rising demand for biofuels, there is a potential risk of food prices being impacted negatively, as well as agricultural land being allocated toward energy crop production as opposed to food production.

The production and consumption of biofuels in the region is primarily dominated by the United States and Canada, whose adoption has been bolstered by a combination of mixing mandate policies and tax incentives. The European Union has exhibited exemplary leadership in the biofuels market by demonstrating its commitment toward implementing progressive policies aimed at advocating its widespread consumption and mitigating carbon emissions. Brazil has emerged as a significant stakeholder in the production and export of biofuels, particularly ethanol derived from sugarcane. Meanwhile, Asia's main growth markets for biofuels are China and India, where blending policies and mandates are being implemented to address air pollution challenges and decrease dependence on fossil fuels.

Certain nations have instituted obstructions to trade, including levies and limitations, on the entrance of biofuels to safeguard their domestic sectors. These hindrances possess the potential to impede the attainment of global markets and restrict the commerce of biofuels.

Trade and environmental agreements, such as the Paris Agreement on climate change, possess the capability of exerting influence on the trade of biofuels through the promotion of favorable policies and regulations. The Paris Agreement, for instance, has resulted in heightened enthusiasm toward the reduction of greenhouse gas emissions and the augmentation of the proportion of biofuels. Collaboration among nations and the implementation of favorable policies are indispensable in cultivating a more unbarred and competitive global market for biofuels.

7 Sensitivity Analysis, Including Risk Management Strategies and Insurance Requirements

Sensitivity analysis is a useful tool for investigating uncertainties in selected parameters and identifying critical parameters that have the largest effects on the MSP. It is important for SAF production because it helps to investigate uncertainties in selected parameters such as capital costs, plant scale, and raw material costs, as well as uncertainties in reaction kinetics [92]. It helps to understand the impact of process design parameters on the economics of the plant and can guide decision-making to improve the economic viability of the SAF production process [62, 93].

According to Crawford et al. [94], sensitivity analysis is important for SAF production because the MSP to achieve reasonable returns on investment is sensitive to capital financing options due to high capital costs, which is important for attracting investors. Thus, sensitivity analysis is crucial to test the economic feasibility of SAF production and to identify potential risks and uncertainties.

Yang et al. for their part [95] argue that sensitivity analysis is important to show the impacts of variations in key cost variables on the economic feasibility of the project. Sensitivity analysis is used to determine the strength of the relation between uncertain inputs and the output, which is the MSP of jet fuel. By analyzing the sensitivity of the output to changes in the major cost variables, we can identify the most

critical factors affecting the economic feasibility of the project and provide insights into how to optimize the process.

Below are some results related to the sensitivity analysis reported by different authors. It will be evident that these results will vary depending on the type of process and biomass that is being transformed to obtain SAF.

The paper by Tao et al. [92] reports several results from the sensitivity analysis conducted on the Aspen Plus model for SAF production. The main results are as follows: The parameters of IRR, feedstock costs, total capital investment, process efficiency, plant capacity, and coproduct credit have the largest effects on the MSP. Variations in catalysts and associated reaction kinetics have effects on fuel yields and the MSP. Even when reactors and process conditions are cheaper and fewer, yield still plays a large role in the MSP. Pioneer plants have higher MSPs than mature plants. Sensitivity analysis provides important insights for the development of alcohol-to-jet (ATJ) technology by quantifying the cost impact of various process parameters on production cost and identifying critical parameters that need to be optimized to reduce MSP.

According to Shen et al. [93], the sensitivity analysis showed that the discount rate, total capital investment (TCI), and lignin feedstock prices have the largest impacts on the MSP. For instance, a variation of TCI in the range of −/+25% would result in a −/+$0.34/gal MSP. The magnitude of these impacts is a function of the range defined in the study.

The sensitivity analysis conducted in the work by Crawford et al. [94] shows that the MSP is sensitive to changes in feedstock cost, enzyme cost, and equity financing. The base case assumes a purchase delivered cost of feedstock at 77 USD t-1, enzymes at 5.22 USD kg-1 (0.13 USD L-1 fuel), and a 100% equity share. The analysis shows that the MSP per liter of jet fuel would increase if feedstock were purchased at 88, 99 USD t-1, or 110 USD t-1. Additionally, the price increase if hydrolysis enzymes are purchased for 10 USD kg-1 is investigated.

The paper by Yang et al. [95] provides several results about the sensitivity analysis conducted in the study. The most sensitive variables for jet fuel production in Scenario 1 were TCI, H2 market price, IRR, and feedstock price. Varying these variables resulted in changes in the MSP of jet fuel. The study found that H2 credit was the primary contributor to cost reduction, and H2 market price was negatively correlated with the variation of MSP of jet fuel. When H2 market price increased from $2.7/kg to $3.9/kg, the MSP of jet fuel decreased from $4.17 to $3.46/Gallon. The IRR indicated the opportunity cost of capital for the investor, and the woody biomass feedstock cost had an impact on the expense of the production process. Overall, the sensitivity analysis provided insights into the key cost drivers and helped identify opportunities for cost reduction and process optimization.

According to the paper by Diederichs et al. [62], the sensitivity analysis investigated the effects of economic parameter variation on the MSP of the SAF. The results showed that only a few parameters caused a change of more than 10% in the MSPs. The economic viability of the biochemical conversion to ethanol with an upgrading process with the highest base MSP did not increase significantly at the favorable parameters, and its lowest MSP of $3.02 per kg jet fuel is much higher

than the base MSPs of the other investigated processes. To increase the potential associated with the biochemical conversion to ethanol with upgrading processes, the enzyme cost must be reduced, while an increase in overall yield or ethanol broth concentration could also improve its prospect. Additionally, understanding the impact of process design parameters on MSP is critical to prioritize further research and determine how the coproduction of fuels and chemicals would affect the economics of the ethanol biorefinery.

Based on all the above, it is possible to establish the usefulness of carrying out sensitivity analysis studies as part of the techno-economic analysis of the processes. The sensitivity analysis shows that the MSP of jet fuel is sensitive to equity financing, which is important for attracting investors. Producing and selling intermediate products may be more profitable than converting all biomass into aviation fuel, and the biorefinery should be designed to produce multiple products to reduce risk and increase profitability. The sensitivity analysis highlights the importance of evaluating the economic feasibility of SAF production and identifying potential risks and uncertainties. The study's sensitivity analysis provided insights into the key cost drivers and helped identify opportunities for cost reduction and process optimization. The sensitivity analysis provides valuable insights into the economic viability of SAF production and can guide decision-making to improve the process.

8 Further Steps

It is well known the production of sustainable jet fuel is still in development; however, it is still important to think about the next steps, on the contrary, it is essential to continue investigating possible ways to optimize the process, reduce its cost, and protect the environment.

One of the best ways to reduce costs and mitigate the impact on the environment is using waste materials since some residues are considered renewable feedstocks with a high potential to produce energy through waste-to-energy technologies. With this, in recent years more research has been carried out at a laboratory scale and with simulators, and favorable results have been obtained toward the production of aviation biofuel.

Recently, Moreno Gómez et al. [96] proposed the conversion of waste chicken fat into aviation biofuel through an intensified hydrotreating process, since it is a low-cost and easily available feedstock, and results of a maximum yield of 47.46% achieved, with a 22% reduction in CO_2 emissions and a 33% cost reduction by applying thermally coupled distillation columns in direct sequence configuration.

Chen et al. [97] investigated the production of aviation biofuel through the hydroprocessing of waste cooking oil on different catalysts, the one that obtained the best results was a homemade NiAg/SAPO (silicoaluminophosphate)-11 catalyst using hydrotreated paraffins, it was found that the biofuel produced was very similar to conventional aviation fuel.

Shah et al. [98] studied the catalytic pyrolysis of agricultural waste (eucalyptus sawdust) and used frying oil and the implementation of hydrogenation on a Ni-Mo catalyst. Most of the oxygenated compounds were removed after hydrogenation, and the improved bio-oil showed properties like those of conventional aviation fuels. Furthermore, Why et al. [99] investigated a one-step catalytic deoxygenation process using palm-based feedstock to produce jet biofuel.

Sarker et al. (2012) studied the use of waste polystyrene cups as feedstock for biojet fuel production. The results showed a biofuel production yield of 23%. In addition, Natural State Research, Inc. (NSR) developed a simple and inexpensive procedure to convert hydrocarbon polymers from plastic waste to short- and medium-chain hydrocarbons from liquid fuels. Sarker (2011) also reported the conversion of plastic waste to aviation biofuel hydrocarbons by thermal liquefaction followed by partial cooling, condensation, and distillation of the plastic slurry that can produce aviation biofuel hydrocarbons meeting ASTM requirements [100].

Continuing with plastic waste as feedstock, Ali et al. [101] investigated the pyrolysis of these wastes including food bags, disposable cups, plates, and drinking water bottles where they were burned at 350–450 °C in the absence of oxygen in a graphite-catalyzed reactor. With this, the obtained pyrolysis oil was distilled and converted into jet fuel-type hydrocarbons, with a production of 80%. In addition, there have been other studies, Wang et al. [102] modeled the catalytic pyrolysis of low-density polyethylene on a synthesized nanocellulose-derived biochar (NCB) catalyst for biofuel production.

Another recently identified raw material was disposable COVID-19 protection masks, Park et al. (2021) investigated the pyrolysis and co-pyrolysis of waste masks mixed with food waste. There is potential to produce biofuels from the pyrolysis of these and, in the case of biofuel for jets, the yield can reach 18.4% [103].

Finally, another possible raw material is paper, since unlike other lignocellulosic materials, it contains a significant amount of cellulose, which is important since it reduces costs by requiring less pretreatment. Kubic et al. [104] presented a modeling investigation of two concepts of recycled paper as raw material, in which the potential feasibility of enzymatic and acid hydrolysis was estimated by performing a techno-economic analysis. However, the results revealed that the supply of paper cannot meet the total demand and therefore cannot compete with fossil fuels under current technological conditions, but it could be a scenario that can be studied further.

In addition to the raw material, the conditions in which the process is carried out can also significantly affect the cost and the impact on the environment, for this reason, catalysts are an important part of the production of aviation biofuel. Several studies have been carried out using different catalysts in different processes for different raw materials, and this step is extremely important as it allows us to determine which catalyst works best under certain conditions.

Zhang et al. [105] worked on the catalytic hydrogenation of waste cooking oil on the Ni-Mo/SBUY-MCM-41 hierarchical synthesized catalyst, which significantly improved the selectivity of cracking for jet fuel category hydrocarbons. Araby et al. [106] studied the hydrothermal catalytic cracking of waste cooking oil over a zinc

aluminate (ZnAl2O4) catalyst under different reaction conditions, the results showed that the aviation fraction was 49%, complying with ASTM standards.

We continue talking about studies that use waste cooking oil as feedstock, Wang et al. [107] studied the hydrodeoxygenation of this enzymatically pretreated feedstock with poor-grade lipids for aviation biofuel. For the Ni-Mo/γ-Al_2O_3 catalysts, obtaining optimal metal loading conditions, the yield was 97%. Long et al. [108] recently conducted computational and experimental research on the deoxygenation and upgrading of used oils to produce biofuels. The results indicate that pyrolysis is a suitable pretreatment method for used oil with impurities and pyrolysis oil can be efficiently upgraded using SAPO-11-Pd/C composite catalyst.

Liu [109] investigated the catalytic hydrocracking of polyethylene using a Pt/Al/MCM-48 bifunctional catalyst to produce jet biofuel in an autoclave reactor obtaining a high yield of 85.9%.

Until today, the production cost of "drop-in" fuels is approximately twice that of conventional fuel, for this reason, it is essential to search for the use of waste biomass to produce biofuel and the use of catalysts to optimize the process, reducing GHG emissions and lowering production costs. Within this search, the availability of feedstock and economic viability must also be considered, as well as the optimization of the location and the integration of processes to improve the profitability of production. However, techno-economic studies are significantly important and go hand in hand with this analysis, since without political support measures, none of the approved or studied pathways is economically viable.

The lack of information related to water footprint is a significant gap in the current state of the art in techno-economic analysis and life cycle assessment of sustainable aviation fuel production. As the aviation industry continues to transition toward low-carbon biofuels, it is important to address this gap and develop a better understanding of the water footprint of SAF production. Future steps could include conducting research to quantify the water footprint of different feedstocks and conversion technologies, as well as developing strategies to minimize water use and manage water resources sustainably. This information will be critical for decision-makers in the aviation industry to make informed choices about the most sustainable and economically viable options for SAF production.

9 Conclusions

The aviation industry has been under increasing pressure to reduce its environmental impact. One promising solution is the production of SAF from various feedstocks and process technologies. This chapter, it was reviewed the main aspects of the TEA and LCA of SAF to test its feasibility, profitability, and environmental impact.

The state of the art in TEA and LCA for SAF production is evolving as new feedstocks, conversion technologies, and production strategies are developed and optimized.

The production of SAF can be economically viable, especially if it is integrated with other industries such as waste management or agriculture. However, the profitability of SAF production is highly dependent on the price of crude oil and government policies such as tax incentives and mandates.

In terms of environmental impact, SAF production can significantly reduce greenhouse gas emissions compared to conventional jet fuel. But the environmental benefits of SAF production are highly dependent on the feedstock and production process used. For example, the use of waste oils and fats can result in significant greenhouse gas emissions reductions, while the use of lignocellulosic biomass can result in negative environmental impacts if not managed properly.

The use of waste material, including plastics, is one of the best ways to reduce costs and mitigate the impact on the environment, with a high potential to produce energy through waste-to-energy technologies. Another recently identified raw material was disposable COVID-19 protection masks using pyrolysis and co-pyrolysis of waste masks mixed with food waste. Finally, another possible raw material is paper, since unlike other lignocellulosic materials, it contains a significant amount of cellulose, which is important since it reduces costs by requiring less pretreatment.

The environmental impacts to produce aviation fuels depend on the technology as well as the feedstock. The technology that generates the least carbon footprint among those compared was Fischer-Tropsch, followed by ATJ. However, there is no preference for one of the technologies due to each one having advantages and disadvantages in addition the conditions in which the process is carried out significantly affect the cost and the impact on the environment. It was also identified that catalysts are an important part to produce aviation biofuel, particularly catalytic pyrolysis with Mo-Co/γ-Al_2O_3 catalyst produced the highest yield of renewable hydrocarbons with high CO_2 mitigation.

Until today, the production cost of "drop-in" fuels is approximately twice that of conventional fuel, for this reason, it is essential to search for the use of waste biomass to produce biofuel and the use of catalysts to optimize the process, reducing GHG emissions and lowering production costs.

Acknowledgments The authors thank the Mexican Institute of Petroleum for making this work possible.

Author Contributions All authors contributed to the study conception and design. Material preparation, data collection, and analysis for techno-economic analysis were performed by Luis F. Ramírez-Verduzco and Judith E. Cruz Ramírez. Material preparation, data collection, and analysis for life cycle assessment were performed by Myriam A. Amezcua-Allieri. The first draft of the manuscript was written by all the authors and all of them commented on previous versions of the manuscript. All authors read and approved the final manuscript.

Declarations The authors have no financial or proprietary interests in any material discussed in this article.

Funding No funding was received to assist with the preparation of this manuscript.Data AvailabilityNot applicable.

References

1. Khan, N., Sudhakar, K., & Mamat, R. (2021). Role of biofuels in energy transition, green economy and carbon neutrality. *Sustainability, 13*(22), 12374. https://doi.org/10.3390/su132212374
2. Adhikari, D. K. (2018). *Bio-jet Fuel* (pp. 187–201). https://doi.org/10.1007/978-3-319-67678-4_8
3. Singh, D., Sharma, D., Soni, S. L., Sharma, S., Kumar Sharma, P., & Jhalani, A. (2020). A review on feedstocks, production processes, and yield for different generations of biodiesel. *Fuel, 262*, 116553. https://doi.org/10.1016/j.fuel.2019.116553
4. Srivastava, A., & Prasad, R. (2000). Triglycerides-based diesel fuels. *Renewable and Sustainable Energy Reviews, 4*(2), 111–133. https://doi.org/10.1016/S1364-0321(99)00013-1
5. Balat, M., & Balat, H. (2010). Progress in biodiesel processing. *Applied Energy, 87*(6), 1815–1835. https://doi.org/10.1016/j.apenergy.2010.01.012
6. Robles-Medina, A., González-Moreno, P. A., Esteban-Cerdán, L., & Molina-Grima, E. (2009). Biocatalysis: Towards ever greener biodiesel production. *Biotechnology Advances, 27*(4), 398–408. https://doi.org/10.1016/j.biotechadv.2008.10.008
7. F. D. Asociación civil Ciencia Hoy, Díaz, M. S., & Villar, M. A., Ciencia Hoy. Asociación Civil Ciencia Hoy, 2017. Accessed: Jun. 16, 2023. [Online]. Available: https://ri.conicet.gov.ar/handle/11336/25791
8. Vela-García, N., Bolonio, D., García-Martínez, M.-J., Ortega, M. F., Almeida Streitwieser, D., & Canoira, L. (2021). Biojet fuel production from oleaginous crop residues: Thermoeconomic, life cycle and flight performance analysis. *Energy Conversion and Management, 244*, 114534. https://doi.org/10.1016/j.enconman.2021.114534
9. Resurreccion, E. P., Roostaei, J., Martin, M. J., Maglinao, R. L., Zhang, Y., & Kumar, S. (2021). The case for camelina-derived aviation biofuel: Sustainability underpinnings from a holistic assessment approach. *Industrial Crops and Products, 170*, 113777. https://doi.org/10.1016/j.indcrop.2021.113777
10. Budsberg, E., Crawford, J. T., Morgan, H., Chin, W. S., Bura, R., & Gustafson, R. (2016). Hydrocarbon bio-jet fuel from bioconversion of poplar biomass: Life cycle assessment. *Biotechnology for Biofuels, 9*(1), 170. https://doi.org/10.1186/s13068-016-0582-2
11. Han, J., Elgowainy, A., Cai, H., & Wang, M. Q. (2013). Life-cycle analysis of bio-based aviation fuels. *Bioresource Technology, 150*, 447–456. https://doi.org/10.1016/j.biortech.2013.07.153
12. Moretti, C., Vera, I., Junginger, M., López-Contreras, A., & Shen, L. (2022). Attributional and consequential LCAs of a novel bio-jet fuel from Dutch potato by-products. *Science of the Total Environment, 813*, 152505. https://doi.org/10.1016/j.scitotenv.2021.152505
13. Chisti, Y., & Yan, J. (2011). Energy from algae: Current status and future trends. *Applied Energy, 88*(10), 3277–3279. https://doi.org/10.1016/j.apenergy.2011.04.038
14. Fernandez-Linares, L. C., Montiel-Montoya, J., Millon-Oropeza, A., & Badillo-Corona, J. A. (2012). Produccion de biocombustibles a partir de microalgas. *Ra Ximhai: revista cientifica de sociedad, cultura y desarrollo sustenable, 8*, S101+. [Online]. Available: https://link.gale.com/apps/doc/A400038398/IFME?u=anon~fe3fa1ac&sid=googleScholar&xid=cf2d1249
15. Unlu, D., & Hilmioglu, N. D. (2018). Review of renewable biofuels in the aviation sector. In *Advances in sustainable aviation* (pp. 25–39). Springer International Publishing. https://doi.org/10.1007/978-3-319-67134-5_3
16. Mohd Noh, H., & Md Zulkifly, N. A. (2012). Short-term and Long-term feedstock bio jet fuel for green environment of air transport in climate change awareness. *Applied Mechanics and Materials, 225*, 572–577. https://doi.org/10.4028/www.scientific.net/AMM.225.572
17. Prasad, M. N. V. (2020). Biomass energy sources and conversion technologies for production of biofuels. In P. Pathak & R. R. Srivastava (Eds.), *Alternative energy resources. The handbook of environmental chemistry* (pp. 115–132). Springer. https://doi.org/10.1007/698_2020_613

18. Sonthalia, A., & Kumar, N. (2019). Hydroprocessed vegetable oil as a fuel for transportation sector: A review. *Journal of the Energy Institute, 92*(1), 1–17. https://doi.org/10.1016/j.joei.2017.10.008
19. Zhu, X., Xiao, J., Wang, C., Zhu, L., & Wang, S. (2022). Global warming potential analysis of bio-jet fuel based on life cycle assessment. *Carbon Neutrality, 1*(1), 25. https://doi.org/10.1007/s43979-022-00026-4
20. Starck, L., Pidol, L., Jeuland, N., Chapus, T., Bogers, P., & Bauldreay, J. (2016). Production of Hydroprocessed Esters and Fatty Acids (HEFA) – Optimisation of process yield. *Oil & Gas Science and Technology – Revue d'IFP Energies nouvelles, 71*(1), 10. https://doi.org/10.2516/ogst/2014007
21. Casci, J. L., Lok, C. M., & Shannon, M. D. (2009). Fischer–Tropsch catalysis: The basis for an emerging industry with origins in the early 20th century. *Catalysis Today, 145*(1–2), 38–44. https://doi.org/10.1016/j.cattod.2008.08.032
22. Lapidus, A. L. (2005). Fundamentals of the Fischer-Tropsch synthesis. In *Sustainable strategies for the upgrading of natural gas: Fundamentals, challenges, and opportunities* (pp. 173–183). Springer. https://doi.org/10.1007/1-4020-3310-9_9
23. Romero-Izquierdo, A. G., Gómez-Castro, F. I., Gutiérrez-Antonio, C., Hernández, S., & Errico, M. (2021). Intensification of the alcohol-to-jet process to produce renewable aviation fuel. *Chemical Engineering and Processing – Process Intensification, 160*, 108270. https://doi.org/10.1016/j.cep.2020.108270
24. Grande, L., Pedroarena, I., Korili, S. A., & Gil, A. (2021). Hydrothermal liquefaction of biomass as one of the most promising alternatives for the synthesis of advanced liquid biofuels: A review. *Materials, 14*(18), 5286. https://doi.org/10.3390/ma14185286
25. Wang, W.-C., & Tao, L. (2016). Bio-jet fuel conversion technologies. *Renewable and Sustainable Energy Reviews, 53*, 801–822. https://doi.org/10.1016/j.rser.2015.09.016
26. Lee, D. S., et al. (2009). Aviation and global climate change in the 21st century. *Atmospheric Environment, 43*(22–23), 3520–3537. https://doi.org/10.1016/j.atmosenv.2009.04.024
27. Shahriar, M. F., & Khanal, A. (2022). The current techno-economic, environmental, policy status and perspectives of sustainable aviation fuel (SAF). *Fuel, 325*, 124905. https://doi.org/10.1016/j.fuel.2022.124905
28. Prussi, M., et al. (2021). CORSIA: The first internationally adopted approach to calculate life-cycle GHG emissions for aviation fuels. *Renewable and Sustainable Energy Reviews, 150*, 111398. https://doi.org/10.1016/j.rser.2021.111398
29. Martinez-Valencia, L., Garcia-Perez, M., & Wolcott, M. P. (2021). Supply chain configuration of sustainable aviation fuel: Review, challenges, and pathways for including environmental and social benefits. *Renewable and Sustainable Energy Reviews, 152*, 111680. https://doi.org/10.1016/j.rser.2021.111680
30. Kandaramath Hari, T., Yaakob, Z., & Binitha, N. N. (2015). Aviation biofuel from renewable resources: Routes, opportunities and challenges. *Renewable and Sustainable Energy Reviews, 42*, 1234–1244. https://doi.org/10.1016/j.rser.2014.10.095
31. Ng, K. S., Farooq, D., & Yang, A. (2021). Global biorenewable development strategies for sustainable aviation fuel production. *Renewable and Sustainable Energy Reviews, 150*, 111502. https://doi.org/10.1016/j.rser.2021.111502
32. Mueller, D., Hoard, S., Smith, P., Sanders, C., & Gaffney, M. J. (2019). Airport management perspectives on aviation biofuels: Drivers, barriers, and policy requirements in the US Pacific Northwest. *International Journal of Aviation Management, 4*(3), 163. https://doi.org/10.1504/IJAM.2019.098380
33. Smith, P. M., Gaffney, M. J., Shi, W., Hoard, S., Armendariz, I. I., & Mueller, D. W. (2017). Drivers and barriers to the adoption and diffusion of Sustainable Jet Fuel (SJF) in the U.S. Pacific Northwest. *Journal of Air Transport Management, 58*, 113–124. https://doi.org/10.1016/j.jairtraman.2016.10.004

34. Ringsred, A., van Dyk, S., & (Jack) Saddler, J. (2021). Life-cycle analysis of drop-in biojet fuel produced from British Columbia forest residues and wood pellets via fast-pyrolysis. *Applied Energy, 287*, 116587. https://doi.org/10.1016/j.apenergy.2021.116587
35. Atmowidjojo, A., et al. (2021). Supporting clean energy in the ASEAN: Policy opportunities from sustainable aviation fuels initiatives in Indonesia and Malaysia. *IOP Conference Series: Earth and Environmental Science, 940*(1), 012031. https://doi.org/10.1088/1755-1315/940/1/012031
36. Santos, K., & Delina, L. (2021). Soaring sustainably: Promoting the uptake of sustainable aviation fuels during and post-pandemic. *Energy Research and Social Science, 77*, 102074. https://doi.org/10.1016/j.erss.2021.102074
37. Lim, M., (Marty) Luckert, M. K, & Qiu, F. (2023). Economic opportunities and challenges in biojet production: A literature review and analysis. *Biomass and Bioenergy, 170*, 106727. https://doi.org/10.1016/j.biombioe.2023.106727
38. Wang, W.-C., et al. (2016, July). *Review of Biojet fuel conversion technologies*. Golden, CO (United States). https://doi.org/10.2172/1278318.
39. Yang, J., Xin, Z., (Sophia) He, Q, Corscadden, K., & Niu, H. (2019). An overview on performance characteristics of bio-jet fuels. *Fuel, 237*, 916–936. https://doi.org/10.1016/j.fuel.2018.10.079
40. van Dyk, S., Su, J., Ebadian, M., O'Connor, D., Lakeman, M., & (Sophia) Saddler, J. (2019). Potential yields and emission reductions of biojet fuels produced via hydrotreatment of biocrudes produced through direct thermochemical liquefaction. *Biotechnology for Biofuels, 12*(1), 281. https://doi.org/10.1186/s13068-019-1625-2
41. Vela-García, N., Bolonio, D., Mosquera, A. M., Ortega, M. F., García-Martínez, M.-J., & Canoira, L. (2020). Techno-economic and life cycle assessment of triisobutane production and its suitability as biojet fuel. *Applied Energy, 268*, 114897. https://doi.org/10.1016/j.apenergy.2020.114897
42. Tian, B., et al. (2021). Measurement and simulation of sooting characteristics by an ATJ-SKA biojet fuel and blends with Jet A-1 fuel in laminar non-premixed flames. *Combustion and Flame, 233*, 111582. https://doi.org/10.1016/j.combustflame.2021.111582
43. Prussi, M., O'Connell, A., & Lonza, L. (2019). Analysis of current aviation biofuel technical production potential in EU28. *Biomass and Bioenergy, 130*, 105371. https://doi.org/10.1016/j.biombioe.2019.105371
44. Heyne, J., Rauch, B., Le Clercq, P., & Colket, M. (2021). Sustainable aviation fuel prescreening tools and procedures. *Fuel, 290*, 120004. https://doi.org/10.1016/j.fuel.2020.120004
45. Chiaramonti, D. (2019). Sustainable aviation fuels: The challenge of decarbonization. *Energy Procedia, 158*, 1202–1207. https://doi.org/10.1016/j.egypro.2019.01.308
46. Feige, A., & Pforte, L. (2018). Sustainability certification in the aviation industry. In *Biokerosene* (pp. 403–432). Springer. https://doi.org/10.1007/978-3-662-53065-8_17
47. Thrän, D., & Ponitka, J. (2016). Government policy on delivering biofuels for the aviation sector. In *Biofuels for aviation* (pp. 295–313). Elsevier. https://doi.org/10.1016/B978-0-12-804568-8.00013-5
48. Capaz, R. S., & Seabra, J. E. A. (2016). Life cycle assessment of Biojet fuels. In *Biofuels for aviation* (pp. 279–294). Elsevier. https://doi.org/10.1016/B978-0-12-804568-8.00012-3
49. Bueno, C., Hauschild, M. Z., Rossignolo, J. A., Ometto, A. R., & Mendes, N. C. (2016). Sensitivity analysis of the use of life cycle impact assessment methods: A case study on building materials. *Journal of Cleaner Production, 112*, 2208–2220. https://doi.org/10.1016/j.jclepro.2015.10.006
50. Agusdinata, D. B., Zhao, F., Ileleji, K., & DeLaurentis, D. (2011). Life cycle assessment of potential biojet fuel production in the United States. *Environmental Science & Technology, 45*(21), 9133–9143. https://doi.org/10.1021/es202148g
51. Urbanik, M., & Tchórzewska-Cieślak, B. (2015). Ecological aspects of the natural gas use. *Journal of Civil Engineering, Environment and Architecture, XXXII*(1), 409–417. https://doi.org/10.7862/rb.2015.29

52. Kolosz, B. W., Luo, Y., Xu, B., Maroto-Valer, M. M., & Andresen, J. M. (2020). Life cycle environmental analysis of 'drop in' alternative aviation fuels: A review. *Sustain Energy Fuels, 4*(7), 3229–3263. https://doi.org/10.1039/C9SE00788A
53. Strapasson, A., Falcão, J., Rossberg, T., Buss, G., Woods, J., & Peterson, S. (2019). Land use change and the European biofuels policy: The expansion of oilseed feedstocks on lands with high carbon stocks. *OCL, 26*, 39. https://doi.org/10.1051/ocl/2019034
54. Dangol, N., Shrestha, D. S., & Duffield, J. A. (2020). Life-cycle energy, GHG and cost comparison of camelina-based biodiesel and biojet fuel. *Biofuels, 11*(4), 399–407. https://doi.org/10.1080/17597269.2017.1369632
55. Seber, G., Escobar, N., Valin, H., & Malina, R. (2022). Uncertainty in life cycle greenhouse gas emissions of sustainable aviation fuels from vegetable oils. *Renewable and Sustainable Energy Reviews, 170*, 112945. https://doi.org/10.1016/j.rser.2022.112945
56. Lundberg, S., Fagerström, A., & Johansson, K. (2020). *Life cycle assessment of gasification-based Fischer-Tropsch bio jet fuel production*.
57. Pierobon, F., Eastin, I. L., & Ganguly, I. (2018). Life cycle assessment of residual lignocellulosic biomass-based jet fuel with activated carbon and lignosulfonate as co-products. *Biotechnology for Biofuels, 11*(1), 139. https://doi.org/10.1186/s13068-018-1141-9
58. Zhang, Z., Wei, K., Li, J., & Wang, Z. (2022). Life-cycle assessment of bio-jet fuel production from waste cooking oil via Hydroconversion. *Energies (Basel), 15*(18), 6612. https://doi.org/10.3390/en15186612
59. Wong, H. M. (2008). *Life-cycle assessment of greenhouse gas emissions from alternative jet fuels*. Thesis submitted to the Engineering Systems Division in Partial Fulfillment of the Requirements for the Degree of Master of Science in Technology and Policy. Accessed: Jul. 22, 2023. [Online]. Available: http://lae.mit.edu/uploads/LAE_report_series/2008/LAE-2008-005-T.pdf
60. Stratton, R. W., Wong, H. M., & Hileman, J. I. (2010). *Life cycle greenhouse gas emissions from alternative jet fuels: Partner Project 28 report*. The Partnership for Air Transportation Noise and Emission Reduction. Accessed: Jul. 22, 2023. [Online]. Available: http://web.mit.edu/aeroastro/partner/reports/proj28/partnerproj28-2010-001.pdf
61. Bonomi, A., Klein, B. C., Chagas, M. F., & Souza, N. R. D. (2018). *Technical report: Comparison of Biojet life cycle analysis tools*. Phase 2, Part 1: FAME and HVO/HEFA. https://www.researchgate.net/publication/349173815_Comparison_of_Biofuel_Life_Cycle_Analysis_Tools_Phase_2_Part_1_FAME_and_HVOHEFA. Accessed 22 Jul 2023.
62. Diederichs, G. W., Ali Mandegari, M., Farzad, S., & Görgens, J. F. (2016). Techno-economic comparison of biojet fuel production from lignocellulose, vegetable oil and sugar cane juice. *Bioresource Technology, 216*, 331–339. https://doi.org/10.1016/j.biortech.2016.05.090
63. Zhang, L., Butler, T. L., & Yang*, B. (2020). Recent trends, opportunities and challenges of sustainable aviation fuel. In *Green energy to sustainability* (pp. 85–110). Wiley. https://doi.org/10.1002/9781119152057.ch5.
64. Klein-Marcuschamer, D., et al. (2013). Technoeconomic analysis of renewable aviation fuel from microalgae, *Pongamia pinnata*, and sugarcane. *Biofuels, Bioproducts and Biorefining, 7*(4), 416–428. https://doi.org/10.1002/bbb.1404
65. Baral, N. R., et al. (2019). Techno-economic analysis and life-cycle greenhouse gas mitigation cost of five routes to bio-jet fuel blendstocks. *Energy & Environmental Science, 12*(3), 807–824. https://doi.org/10.1039/C8EE03266A
66. Martinez-Hernandez, E., Martinez-Hernandez, E., Ramírez-Verduzco, L. F., Amezcua-Allieri, M. A., & Aburto, J. (2019). Process simulation and techno-economic analysis of bio-jet fuel and green diesel production — Minimum selling prices. *Chemical Engineering Research & Design, 146*, 60. https://doi.org/10.1016/j.cherd.2019.03.042
67. Gegg, P., Budd, L., & Ison, S. (2014). The market development of aviation biofuel: Drivers and constraints. *Journal of Air Transport Management, 39*, 34–40. https://doi.org/10.1016/j.jairtraman.2014.03.003

68. McGarvey, E., & Tyner, W. E. (2018). A stochastic techno-economic analysis of the catalytic hydrothermolysis aviation biofuel technology. *Biofuels, Bioproducts and Biorefining, 12*(3), 474–484. https://doi.org/10.1002/bbb.1863
69. Shila, J., & Johnson, M. E. (2021). Techno-economic analysis of Camelina-derived hydro-processed renewable jet fuel within the US context. *Applied Energy, 287*, 116525. https://doi.org/10.1016/j.apenergy.2021.116525
70. Tanzil, A. H., Brandt, K., Wolcott, M., Zhang, X., & Garcia-Perez, M. (2021). Strategic assessment of sustainable aviation fuel production technologies: Yield improvement and cost reduction opportunities. *Biomass and Bioenergy, 145*, 105942. https://doi.org/10.1016/j.biombioe.2020.105942
71. Liu, H., Zhang, C., Tian, H., Li, L., Wang, X., & Qiu, T. (2021). Environmental and techno-economic analyses of bio-jet fuel produced from jatropha and castor oilseeds in China. *International Journal of Life Cycle Assessment, 26*(6), 1071–1084. https://doi.org/10.1007/s11367-021-01914-0
72. Wang, W.-C. (2016). Techno-economic analysis of a bio-refinery process for producing hydro-processed renewable jet fuel from Jatropha. *Renewable Energy, 95*, 63–73. https://doi.org/10.1016/j.renene.2016.03.107
73. Diniz, A. P. M. M., Sargeant, R., & Millar, G. J. (2018). Stochastic techno-economic analysis of the production of aviation biofuel from oilseeds. *Biotechnology for Biofuels, 11*(1), 161. https://doi.org/10.1186/s13068-018-1158-0
74. Cervi, W. R., Lamparelli, R. A. C., Seabra, J. E. A., Junginger, M., Jong, S., & Hilst, F. (2020). Spatial modeling of techno-economic potential of biojet fuel production in Brazil. *GCB Bioenergy, 12*(2), 136–157. https://doi.org/10.1111/gcbb.12659
75. Michailos, S., & Bridgwater, A. (2019). A comparative techno-economic assessment of three bio-oil upgrading routes for aviation biofuel production. *International Journal of Energy Research*, er.4745. https://doi.org/10.1002/er.4745
76. Tongpun, P., Wang, W.-C., & Srinophakun, P. (2019). Techno-economic analysis of renewable aviation fuel production: From farming to refinery processes. *Journal of Cleaner Production, 226*, 6–17. https://doi.org/10.1016/j.jclepro.2019.04.014
77. Zhao, X., Brown, T. R., & Tyner, W. E. (2015). Stochastic techno-economic evaluation of cellulosic biofuel pathways. *Bioresource Technology, 198*, 755–763. https://doi.org/10.1016/j.biortech.2015.09.056
78. Schemme, S., Samsun, R. C., Peters, R., & Stolten, D. (2017). Power-to-fuel as a key to sustainable transport systems—An analysis of diesel fuels produced from CO 2 and renewable electricity. *Fuel, 205*, 198–221. https://doi.org/10.1016/j.fuel.2017.05.061
79. Neuling, U., & Kaltschmitt, M. (2018). Techno-economic and environmental analysis of aviation biofuels. *Fuel Processing Technology, 171*, 54. https://doi.org/10.1016/j.fuproc.2017.09.022
80. Bullerdiek, N., Neuling, U., & Kaltschmitt, M. (2021). A GHG reduction obligation for sustainable aviation fuels (SAF) in the EU and in Germany. *Journal of Air Transport Management, 92*, 102020. https://doi.org/10.1016/j.jairtraman.2021.102020
81. Panoutsou, C., et al. (2021). Advanced biofuels to decarbonise European transport by 2030: Markets, challenges, and policies that impact their successful market uptake. *Energy Strategy Reviews, 34*, 100633. https://doi.org/10.1016/j.esr.2021.100633
82. Richard, T. L. (2010). Challenges in scaling up biofuels infrastructure. *Science (1979), 329*(5993), 793–796. https://doi.org/10.1126/science.1189139
83. Charles, M. B., Ryan, R., Ryan, N., & Oloruntoba, R. (2007). Public policy and biofuels: The way forward? *Energy Policy, 35*(11), 5737–5746. https://doi.org/10.1016/j.enpol.2007.06.008
84. Janssen, R., & Rutz, D. D. (2011). Sustainability of biofuels in Latin America: Risks and opportunities. *Energy Policy, 39*(10), 5717–5725. https://doi.org/10.1016/j.enpol.2011.01.047
85. Tiwari, R., et al. (2023). Environmental and economic issues for renewable production of bio-jet fuel: A global prospective. *Fuel, 332*, 125978. https://doi.org/10.1016/j.fuel.2022.125978

86. Nair, S., & Paulose, H. (2014). Emergence of green business models: The case of algae biofuel for aviation. *Energy Policy, 65*, 175–184. https://doi.org/10.1016/j.enpol.2013.10.034
87. Sadatshojaei, E., Wood, D. A., & Mowla, D. (2020). Third generation of biofuels exploiting microalgae. In Inamuddin & A. Asiri (Eds.), *Sustainable green chemical processes and their allied applications. Nanotechnology in the life sciences* (pp. 575–588). Springer. https://doi.org/10.1007/978-3-030-42284-4_21
88. Ebrahimi, S., Haji Esmaeili, S. A., Sobhani, A., & Szmerekovsky, J. (2022). Renewable jet fuel supply chain network design: Application of direct monetary incentives. *Applied Energy, 310*, 118569. https://doi.org/10.1016/j.apenergy.2022.118569
89. Timilsina, G. R., Csordás, S., & Mevel, S. (2011). When does a carbon tax on fossil fuels stimulate biofuels? *Ecological Economics, 70*(12), 2400–2415. https://doi.org/10.1016/j.ecolecon.2011.07.022
90. Havlík, P., et al. (2011). Global land-use implications of first and second generation biofuel targets. *Energy Policy, 39*(10), 5690–5702. https://doi.org/10.1016/j.enpol.2010.03.030
91. Demirbas, A. (2008). Biofuels sources, biofuel policy, biofuel economy and global biofuel projections. *Energy Conversion and Management, 49*(8), 2106–2116. https://doi.org/10.1016/j.enconman.2008.02.020
92. Tao, L., Markham, J. N., Haq, Z., & Biddy, M. J. (2017). Techno-economic analysis for upgrading the biomass-derived ethanol-to-jet blendstocks. *Green Chemistry, 19*(4), 1082–1101. https://doi.org/10.1039/C6GC02800D
93. Shen, R., Tao, L., & Yang, B. (2019). Techno-economic analysis of jet-fuel production from biorefinery waste lignin. *Biofuels, Bioproducts and Biorefining, 13*(3), 486–501. https://doi.org/10.1002/bbb.1952
94. Crawford, J. T., Shan, C. W., Budsberg, E., Morgan, H., Bura, R., & Gustafson, R. (2016). Hydrocarbon bio-jet fuel from bioconversion of poplar biomass: Techno-economic assessment. *Biotechnology for Biofuels, 9*(1), 141. https://doi.org/10.1186/s13068-016-0545-7
95. Yang, Z., et al. (2018). Process design and economics for the conversion of lignocellulosic biomass into jet fuel range cycloalkanes. *Energy, 154*, 289–297. https://doi.org/10.1016/j.energy.2018.04.126
96. Moreno-Gómez, A. L., Gutiérrez-Antonio, C., Gómez-Castro, F. I., & Hernández, S. (2021). Modelling, simulation and intensification of the hydroprocessing of chicken fat to produce renewable aviation fuel. *Chemical Engineering and Processing – Process Intensification, 159*, 108250. https://doi.org/10.1016/j.cep.2020.108250
97. Badoga, S., Alvarez-Majmutov, A., Xing, T., Gieleciak, R., & Chen, J. (2020). Co-processing of hydrothermal liquefaction biocrude with vacuum gas oil through Hydrotreating and hydrocracking to produce low-carbon fuels. *Energy & Fuels, 34*, 7160–7169. https://doi.org/10.1021/acs.energyfuels.0c00937
98. Shah, Z., Veses, R. C., Vaghetti, J. C. P., Amorim, V. D. A., & da Silva, R. (2019). Preparation of jet engine range fuel from biomass pyrolysis oil through hydrogenation and its comparison with aviation kerosene. *International Journal of Green Energy, 16*(4), 350–360. https://doi.org/10.1080/15435075.2019.1566730
99. Why, E. S. K., et al. (2022). Single-step catalytic deoxygenation of palm feedstocks for the production of sustainable bio-jet fuel. *Energy, 239*, 122017. https://doi.org/10.1016/j.energy.2021.122017
100. Sarker, M., Rashid, M. M., Molla, M., & M. S. [Department of R. and D. Rahman Natural State Research Inc 37 Brown House Road (2nd Floor) Stamford CT-06902 (United States)]. Polystyrene (PS) waste plastic conversion into aviation /kerosene category of fuel by using fractional column distillation process," Vol. 3, 2012.
101. Ali, Z., Rathnakumar, P., Ashfaq Hussain, M., Roma, E., Nagaral, M., & Umar, M. (2022). Jet fuel produced from waste plastic with graphite as a catalyst. *Materials Today Proceedings, 52*, 716–723. https://doi.org/10.1016/j.matpr.2021.10.131

102. Wang, C., et al. (2021). Catalytic upcycling of waste plastics over nanocellulose derived biochar catalyst for the coupling harvest of hydrogen and liquid fuels. *Science of the Total Environment, 779*, 146463. https://doi.org/10.1016/j.scitotenv.2021.146463
103. Park, C., Choi, H., Andrew Lin, K.-Y., Kwon, E. E., & Lee, J. (2021). COVID-19 mask waste to energy via thermochemical pathway: Effect of co-feeding food waste. *Energy, 230*, 120876. https://doi.org/10.1016/j.energy.2021.120876
104. Kubic, W. L., Moore, C. M., Semelsberger, T. A., & Sutton, A. D. (2021). Recycled paper as a source of renewable jet fuel in the United States. *Frontiers in Energy Research, 9*. https://doi.org/10.3389/fenrg.2021.728682
105. Zhang, Z., Wang, Q., & Zhang, X. (2019). Hydroconversion of waste cooking oil into bio-jet fuel over NiMo/SBUY-MCM-41. *Catalysts, 9*(5), 466. https://doi.org/10.3390/catal9050466
106. El-Araby, R., Abdelkader, E., El Diwani, G., & Hawash, S. I. (2020). Bio-aviation fuel via catalytic hydrocracking of waste cooking oils. *Bulletin of the National Research Centre, 44*(1), 177. https://doi.org/10.1186/s42269-020-00425-6
107. Wang, M., He, M., Fang, Y., Baeyens, J., & Tan, T. (2017). The Ni-Mo/γ-Al2O3 catalyzed hydrodeoxygenation of FAME to aviation fuel. *Catalysis Communications, 100*, 237–241. https://doi.org/10.1016/j.catcom.2017.07.009
108. Long, F., et al. (2022). An effective strategy for waste oil deoxygenation and upgrading for hydrocarbon biofuels production: A computational and experimental investigation. *Journal of the Energy Institute, 100*, 109–119. https://doi.org/10.1016/j.joei.2021.11.003
109. Liu, Y. (2020). Hydrocracking of polyethylene to jet fuel range hydrocarbons over Bifunctional catalysts containing Pt- and Al-modified MCM-48. *Reactions, 1*(2), 195–209. https://doi.org/10.3390/reactions1020014

Chapter 14
Sustainable Aviation Biofuels: Progress and Challenges

Viviane de Castro Bizerra, Kaiany Moreira dos Santos, Paulo Gonçalves de Sousa Júnior, José Roberto de Matos Filho, Antônio Luthierre Gama Cavalcante, Francisco Simão Neto, Valdilane Santos Alexandre, Érico Carlos de Castro, Maria Cristiane Martins de Souza, and José C. S. dos Santos

1 Introduction

As climate change threatens life on Earth, the need for measures to reduce future emissions of greenhouse gases (GHG) increases [1]. According to the latest Intergovernmental Panel on Climate Change (IPCC) report, before 2040, the limit of 1.5 °C will be reached, even for scenarios with very low GHG emissions [2]. The transport sector is responsible for most GHG emissions, of which aviation is part [3]. Although the aviation sector has changed how we connect with the world, the high dependence of the aviation sector on fossil fuels represents more than 99% of the global aviation kerosene supply [4]. The international aviation industry uses approximately 349 million tons of jet fuel annually, corresponding to ~114 billion gallons. Some projections suggest a doubling in the next two decades [5]. Aviation contributes about 3.5–4% of global anthropogenic climate, mainly CO_2 [4].

V. de Castro Bizerra · K. M. dos Santos · P. G. de Sousa Júnior · J. R. de Matos Filho · V. S. Alexandre · M. C. M. de Souza · J. C. S. dos Santos (✉)
Instituto de Engenharias e Desenvolvimento Sustentável, Universidade da Integração Internacional da Lusofonia Afro-Brasileira, Campus das Auroras, Redenção, Brazil
e-mail: jcs@unilab.edu.br

A. L. G. Cavalcante
Departamento de Química Analítica e Físico-Química, Universidade Federal do Ceará, Campus do Pici, Fortaleza, Brazil

F. S. Neto
Departamento de Engenharia Química, Universidade Federal do Ceará, Campus do Pici, Fortaleza, Brazil

É. C. de Castro
Grupo de Química de Materiais Avançados (GQMat), Departamento de Química Analítica e Físico-Química, Universidade Federal do Ceará – UFC, Campus do Pici, Fortaleza, Brazil

M. Aslam et al. (eds.), *Sustainable Aviation Fuels*, Sustainable Aviation,
https://doi.org/10.1007/978-3-031-83721-0_14

Following the COVID-19 pandemic, aviation CO_2 emissions increased rapidly in 2022, reaching almost 800 Mt. CO_2, about 80% of the pre-pandemic level [6].

The International Air Transport Association (IATA) has set a cap on net CO_2 emissions from aviation from 2020 and a 50% reduction in CO_2 emissions by 2050 compared to 2005 levels. So, try to urgently reduce GHG emissions [7]. The International Civil Aviation Organization (ICAO) also joined this long-term for net-zero CO_2 emissions from international aviation by 2050, ensuring the maximum effectiveness of the Carbon Offsetting and Reduction Scheme for International Aviation (CORSIA) [8, 9]. Since fuel is the lifeblood of aviation, a viable fuel alternative that reduces CO_2 emissions is needed [10, 11]. Using low-emission fuels is one of several ways to halt emissions growth and get on track to net-zero emissions by 2050 [6].

The Third Conference on Aviation and Alternative Fuels (CAAF/3) in Dubai in November 2023 was a critical strategic event focused on sustainable aviation fuels (SAFs) [12]. To foster and make SAF production economically competitive, CORSIA was globally adopted [13]. A CORSIA-eligible fuel is one of the methods an airline can use to claim emissions reduction [3, 14]. SAFs are a viable alternative to reduced environmental pollution [11, 13, 15, 16]. They are considered the short-term solution to reduce fossil CO_2 emissions from aviation [13, 17]. SAFs are produced from biological or waste hydrocarbons instead of from fossil sources [5]. ICAO has certified SAFs production for different conversion technologies and biomass types with specific benefits and shortcomings [3, 5].

This review aims to analyse the recent advances and future challenges in SAFs, providing new insights into them. A bibliometric analysis was performed from 2014 to 2024, evaluating studies from several parameters from academic papers published via the creation of a database. This approach was used to help researchers understand the changing characteristics of SAFs and to identify fundamental influences.

2 Aviation Fuels

2.1 Aviation Fuels Production Processes

Aviation fuels are primarily produced from refining crude oil or, in some cases, from alternative sources such as biofuels [18]. The main production processes include fractional distillation, hydrotreating, catalytic cracking, isomerization and adding additives [19]. Fractional distillation is the fundamental refining process that separates the components of crude oil into fractions based on their boiling points [20]. Crude oil is heated in a distillation column, where rising vapours condense at different levels of the column, allowing the extraction of specific fractions. In aviation fuels, fractional distillation isolates the range of hydrocarbons required to produce aviation kerosene, also known as Jet Fuel [21].

Aviation kerosene is a fraction of diesel and gasoline in terms of molecular weight and volatility [22]. The fractional distillation process must be carefully controlled to ensure that the jet fuel has the proper density, freezing point and stability for use in aircraft [23]. Any variation in these parameters may negatively affect engine performance or flight safety. Hydrotreatment is a methodology that involves the reaction of hydrocarbons with hydrogen at high temperatures and pressures using specific catalysts [24]. In aviation fuels, hydrotreatment removes impurities such as sulphur, nitrogen and heavy metals that can compromise fuel quality and cause engine corrosion or damage [25].

By reducing sulphur concentrations, hydrotreatment also contributes to reducing pollutant emissions during the combustion of aviation fuel. Furthermore, the process helps improve the fuel's thermal and oxidative stability, essential for maintaining engine performance and safety during long-distance flights [26]. Catalytic cracking is breaking down heavy hydrocarbons into smaller, lighter molecules, improving the production efficiency of aviation fuels [27]. Cracking is carried out in a fluid catalytic cracking unit (FCC), where crude oil is heated and passed through a catalyst to trigger chemical reactions that transform heavy hydrocarbons into lighter fractions [28]. Aviation fuel [29] can also increase the production of aromatic compounds, which contribute to the energy density of the fuel. Still, it must be controlled to avoid problems such as forming residues in the engine.

Isomerization is defined as a methodology that aims to alter the molecular structure of hydrocarbons, converting linear chains into branched structures. This improves fuel's performance properties, such as stability and resistance to oxidation [30]. In the production of aviation fuels, isomerization is used to adjust the structure of hydrocarbons to meet desired performance specifications. The isomerization process helps optimize the combustion of aviation fuel, providing more efficient burning and less deposit formation. Furthermore, it contributes to a more adequate cetane number, improving the ignition capacity of the fuel in jet engines [29]. Isomerization is essential in ensuring that aviation fuels have the properties necessary for safe and efficient performance.

Adding additives is fundamental to fuel production processes, as they aim to improve their performance and safety properties [31]. Additives may include antioxidants to prevent fuel degradation during storage, antifreeze agents to reduce the freezing point and ensure fluidity at high altitudes and corrosion inhibitors to protect the internal parts of engines and fuel tanks [32]. Other additives include anti-electrostatic agents to minimize the formation of static charges during fuelling, detergents to keep the fuel system clean and improve engine operability and deposit control agents to prevent the build-up of residues that could harm engine performance [22]. The appropriate choice and dosage of additives are essential to ensure that aviation fuel meets the quality and safety standards required by the aeronautical industry.

2.2 Type of Aviation Fuels Used

Aviation fuels are designed to meet specific aircraft performance requirements. Jet A and Jet A-1, AvGas (Aviation Gasoline) and aviation biofuels are the main types. Jet A and Jet A-1 are the two main types of jet fuel used in jet aircraft [33]. They have similar chemical compositions, both being petroleum-derived hydrocarbons. Their main difference is their freezing point: Jet A has a freezing point of −40 °C, while Jet A-1 has a lower freezing point of −47 °C, making it more suitable for international flights involving extremely low temperatures at high altitudes [34].

Both fuels are formulated to meet performance, thermal and oxidative stability requirements, and they have low sulphur content to minimize corrosion and pollutant emissions [35]. Additionally, they may contain additives such as anti-icing agents, antioxidants and corrosion inhibitors to ensure safety and efficiency during flight [36]. Jet A-1 is the most widely used fuel type due to its greater versatility in flight conditions.

AvGas, or aviation gasoline, is used in aircraft equipped with piston engines, such as small planes and training planes. Unlike aviation kerosene, AvGas is a highly refined gasoline with specific additives to ensure efficient and safe burning. The chemical composition of AvGas is carefully controlled to maintain a high octane rating, which helps prevent premature detonation or "knocking" in engines [37].

In addition to high octane, AvGas may contain additives such as tetraethylene to increase stability and reduce the risk of corrosion or water contamination. AvGas is characterized by its controlled volatility, allowing piston engines to operate efficiently over various temperatures and altitudes [38]. Due to the widespread use of piston engines in general aviation, AvGas is a prominent fuel in the industry [39].

Next, aviation biofuels are presented as alternative fuels from renewable sources, such as residual biomass and agro-industry waste [24, 40]. They are developed to reduce dependence on fossil fuels and greenhouse gas emissions associated with air transport [41, 42]. Aviation biofuels can be blended with Jet A or Jet A-1 in specific proportions, usually around 50%. Using these mixtures does not require changes to aircraft engines [43].

Additionally, aviation biofuels offer significant environmental benefits, such as reducing carbon dioxide emissions throughout their life cycle [44]. Its production can contribute to the diversification of the energy matrix and generate economic opportunities in regions that produce the biomass used [32]. However, there are still challenges related to the cost of production and the availability of raw materials for commercial-scale biofuel production [31]. Still, the gradual adoption of aviation biofuels is a growing trend in the global aviation industry.

2.3 Properties and Composition of Aviation Fuels

Aviation fuels have specific characteristics to ensure safe and efficient aircraft performance. Global specifications and standards, such as ASTM D1655 for Jet A and Jet A-1, establish quality requirements for aviation fuels. Among these, density stands out, which is a relevant property, as it influences engine performance and burning efficiency. Generally, the density of Jet A and Jet A-1 fuels is standardized at around 0.775 to 0.840 kg/l at 15 °C [19]. A controlled density is essential to ensure the accuracy of calculating the weight of the fuel on board [40].

Another characteristic concerns the flash point, the minimum temperature at which the fuel releases flammable vapours when exposed to a flame. In the case of aviation fuels, it is essential to have a high flash point to ensure safety during storage, handling and operation [33]. The flash point of Jet A and Jet A-1 fuels must be at least 38 °C [45].

Next, the freezing point is highlighted, which refers to the minimum temperature at which the fuel begins to solidify or form crystals, which can block filters and fuel lines [18]. For Jet A and Jet A-1 fuels, the freezing point must be below −40 °C and −47 °C ensure fuel fluidity at high altitudes and low temperatures [46].

In addition to the points already exposed, it is possible to highlight the thermal and oxidative stability of fuels. This is the ability of fuels to resist degradation and deposit formation under high-temperature conditions and oxygen exposure [47]. This property is essential to maintain engine efficiency and prevent the formation of residues that can cause problems such as clogging of filters or damage to turbines [48].

Some fuel components, such as cetane, also need to be controlled. Therefore, the cetane number is relevant information about the fuel used. This index is a measure of the fuel's ignitability. Although most appropriate for piston engine fuels such as AvGas, it is also an essential parameter for jet aviation fuels [44]. Fuels with adequate cetane numbers guarantee more uniform combustion [49].

Finally, aviation fuels generally comprise hydrocarbons, especially alkanes, cycloalkanes and aromatics. The composition is carefully controlled to achieve desired properties such as density, flash point and thermal and oxidative stability [24]. Specific additives, such as antioxidants, antifreeze agents and corrosion inhibitors, are added to improve performance [50].

These characteristics and compositions are controlled, as they influence the aircraft's performance and flight range. Furthermore, they provide safety against fires or unintentional explosions and even prevent engine malfunctions.

3 Aviation Fuels and Sustainable

3.1 *Production Methods*

The use of sustainable aviation fuels (SAFs) has been growing with the intention of minimizing the impacts of fossil fuel combustion. Consequently, new methods have been developed, and some already known routes have been enhanced, focusing on reducing carbon emissions, minimizing environmental impacts generated by conventional fuel combustion by-products and optimizing stages for greater economic viability [51]. Methods may include the production of biofuels, synthetic fuels from renewable sources and methods utilizing clean energy sources for fuel production. Figure 14.1 presents some of these methods for sustainable fuel production used in the aviation industry.

3.1.1 Biofuels

Biofuels represent a viable alternative to conventional fuels used in the aviation industry. The advantages of this alternative include the wide availability of raw materials, such as urban waste and biomass from algae, wood and oils, as well as the low carbon dioxide emission rate emitted during combustion [52]. The method chosen to produce these biofuels depends on the origin of the biomass and its

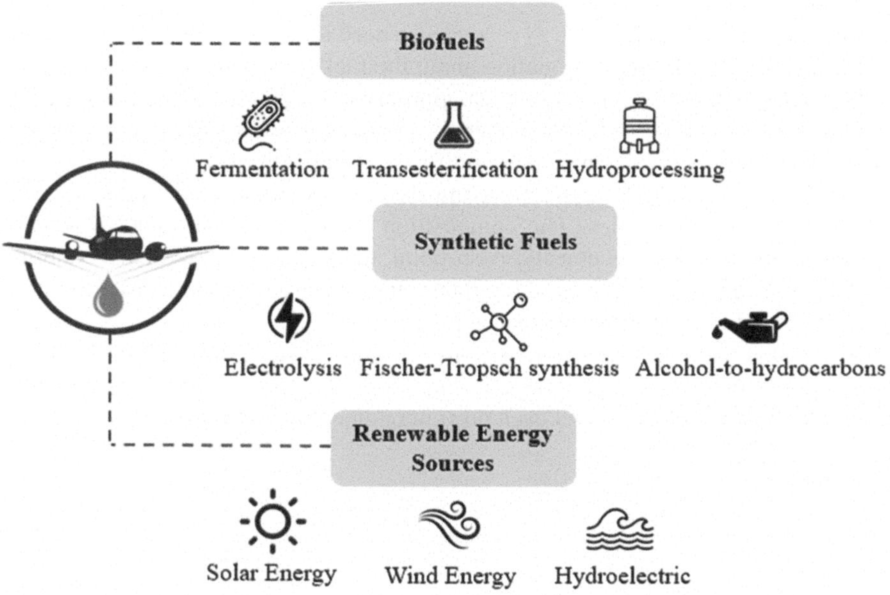

Fig. 14.1 Production methods and resources used to obtain sustainable fuels for the aviation industry

composition. The main methods include fermentation techniques, which involve the transformation of biomass sugar into ethanol using microorganisms. The raw materials used in this method are biomass from corn plantations, sugarcane and cellulosic biomass [53].

Transesterification is another widely used method for obtaining SAFs. In this process, vegetable oils, used cooking oil or animal fats are converted into esters that compose biodiesel. This reaction involves an alcohol reacting with fatty acids, leading to the formation of long-chain esters. A biocatalyst can be added to the medium, enabling the biodiesel production [54]. Pyrolysis is also an alternative method for obtaining biofuels, where biomass is heated without oxygen. Bio-oil, syngas and biochar are produced. Bio-oil can be refined and transformed into liquid fuels used by the aviation sector [55].

Gasification represents a thermochemical procedure transforming biomass into synthesis gas (syngas) comprising carbon monoxide, hydrogen and methane [22]. It serves as a versatile fuel source for power generation or can undergo further conversion into liquid fuels or chemicals [56]. Conversely, hydrothermal liquefaction entails the conversion of moist biomass into bio-oil through elevated temperature and pressure in the presence of water. This method accommodates various feedstocks, including algae, wastewater sludge and agricultural residues [57].

Anaerobic digestion, a biological process, decomposes organic matter devoid of oxygen, yielding biogas. This biogas finds utility in heating, electricity generation, or as a vehicular fuel. Additionally, algae serve as viable biofuel sources [58]. Algae biomass cultivation transpires in ponds, bioreactors, or photobioreactors, subsequently undergoing conversion into biofuels such as biodiesel, bioethanol or biogas via amalgamated processes like extraction, fermentation or anaerobic digestion [59].

3.1.2 Synthetic Fuels

Synthetic fuels, called e-fuels or power-to-liquid fuels, are generated by converting renewable electricity and carbon dioxide into liquid hydrocarbon fuels. These fuels can be synthesized using diverse processes, including electrolysis, Fischer-Tropsch synthesis and methanol-to-hydrocarbons conversion [60]. Synthetic fuels present several benefits, such as compatibility with current infrastructure, high energy density and diminished emissions compared to conventional fuels. Nevertheless, the substantial production costs and energy-intensive processes pose notable hurdles to the extensive adoption of synthetic fuels in aviation [61].

3.1.3 Renewable Energy

Renewable energy sources such as solar, wind and hydroelectric power can be used to produce SAFs through electrolysis and biomass conversion processes. Electrolysis entails splitting water molecules into hydrogen and oxygen using renewable electricity, facilitating the production of hydrogen-based aviation fuels [62]. Biomass

conversion technologies, including gasification and pyrolysis, can also be integrated with renewable energy sources to generate syngas or bio-oils for aviation fuel synthesis. Despite offering clean and abundant energy options, the intermittent nature and infrastructure requirements of renewable energy sources necessitate addressing to ensure the reliable and scalable production of aviation fuels [63].

The imperative transition towards sustainable production of SAFs is crucial for mitigating climate change and advancing environmental stewardship within the aviation industry. While biofuels, synthetic fuels and renewable energy sources present promising alternatives to conventional aviation fuels, their widespread adoption hinges upon sustained research, technological innovation and policy backing [64].

3.2 Properties and Composition

SAFs have diverse properties and compositions, depending on the production method and feedstock used. SAFs are derived from renewable feedstocks such as biomass, algae or renewable electricity. Unlike conventional fossil fuels, which are finite and contribute to carbon emissions, SAFs are produced from sustainable sources that can be replenished over time [65].

One of the primary goals of SAFs is to reduce greenhouse gas emissions compared to conventional aviation fuels. SAFs typically have a lower carbon footprint because they are produced from biomass or synthesized using renewable energy sources, reducing life-cycle emissions [66]. These sustainable fuels are designed to be compatible with existing aircraft and infrastructure without requiring modifications or significant changes to engine technology. This ensures seamless integration into the aviation industry and enables airlines to use SAFs as drop-in replacements for conventional jet fuels [67].

The chemical composition of SAFs can vary depending on the production method and feedstock. Typical components of SAFs include hydrocarbons such as alkanes, aromatics and oxygenates [68]. Synthetic fuels may contain carbon monoxide, hydrogen and other intermediate products of synthesis processes. SAFs derived from renewable biomass sources are typically biodegradable, meaning natural processes can break them down over time. This property reduces environmental impact in case of spills or leaks during handling and transportation [69].

These sustainable fuels are expected to perform comparably to conventional jet fuels regarding energy density, combustion efficiency and aircraft performance. They should meet stringent aviation fuel specifications and regulatory requirements to ensure safety and reliability during flight operations [70].

Therefore, sustainable aviation fuels offer a promising pathway to decarbonize the aviation sector and mitigate climate change while ensuring continued access to air transportation services [71]. Continued research and development efforts are essential to optimize SAF production methods, improve performance characteristics and scale up production to meet growing demand from the aviation industry.

3.2.1 Potential Applications

SAFs have many potential applications across the aviation sector and in related industries. Some of their critical applications include commercial aviation demand to replace conventional aviation fuels in commercial airline operations [72]. Military aviation can be another way to use SAFs, reducing its reliance on fossil fuels and improving operational sustainability. SAFs can be used in military aircraft for training exercises, reconnaissance missions and operational deployments, enhancing energy security and resilience [73]. In general aviation, SAFs can be used in various applications, including flight training, recreational flying and aerial work operations. General aviation companies can incorporate SAFs into their operations to support sustainability goals and reduce emissions from aviation activities [74].

SAFs are crucial in driving research and development efforts to advance alternative aviation fuels and technologies. Research institutions, government agencies and industry partners collaborate to develop new production methods, improve fuel performance and address technical challenges associated with SAF deployment [75]. Overall, while sustainable aviation fuels offer numerous advantages regarding environmental, economic and social benefits, addressing their associated challenges is essential to realizing their full potential and accelerating the transition to a sustainable aviation industry.

4 Materials and Methods

Bibliometric analysis is used as a fundamental data collection tool that helps to determine specific studies in the desired area, such as the number of scientific productions, citations, relevance, authors, institutions, keywords and other aspects in a detailed manner [76]. For this work, an advanced bibliometric analysis was carried out, with the central objective being scientific productions involving aviation fuels. The data analysed were obtained through the collection of the Web of Science (WoS) platform, currently owned by Clarivate Analytics, which has a wide range of scientific productions, such as review articles, conference articles and others [77].

The final data obtained was obtained from the search of a central database, from the query of the terms "aviation fuels" in the search field that contained "all fields" corresponding to all segments and areas of scientific publications that included the search words. Next, the topic referring to the "year of publication" was researched in the following line, selecting the 10 years, from 2014 to 2024, filtering publications from 2014 to April 2024 [78]. The central database included some research refinements, such as selecting only articles and reference articles and publishing only in English [79].

The software used to construct all the figures presented in the work was the VOSviewer software (version 1.6.18), available for specific bibliometric studies. VOSviewer imports the database you want to analyse, which includes sources from periodicals, newspapers, keywords, countries, institutions and authors, to build

bibliometric visualization network maps. Other software was also used to catalogue the data obtained from the WoS central database, such as Word, Excel and Google Spreadsheets, for the construction of tables, graphs and spreadsheets, filtered and entered manually, which helped identify essential information throughout the process study.

All information was considered for the construction of the work. The analyses included the most relevant countries in the area of aviation fuels, the authors who publish the most on the topic, the most appropriate scientific publications, citation frequencies, pioneering institutions in the region, the number of publications based on the delimitations of the years, the project financing agencies in the area, the most mentioned keywords that are emerging in the clusters presented in the visualization network maps, these being the quantitative data presented throughout the figures and tables.

Citation analyses are standard in studies based on bibliometrics, as they analyse the count of times other articles cite an article, presenting the relevance and impact of scientific publications. These analyses are carried out in co-citations and co-occurrences, constructed via VOSviewer software, which was used throughout the work for graphical representations from the database collected from WoS, which comprehensively identifies the data of each article inserted into the platform, which also helps in categorizing data and discussing the results obtained. Throughout the discussions of this work, the results obtained from the bibliometric tools used are presented, highlighting that all analyses provide unique research data according to the insertion of study objects.

5 Results and Discussion

5.1 Bibliometric Analysis

5.1.1 Results of Publication: Overall

A search was initially carried out in the Web of Science (WoS) database between 2014 and April 2024, which resulted in 541 academic productions. 2022 is the year with the most publications, totalling 109 works. The year 2023 stands out with the highest number of citations, with 2357, as shown in Fig. 14.2. It is possible to highlight the most cited work, with 300 citations, Production of Renewable Jet Fuel Range Alkanes and Commodity Chemicals from Integrated Catalytic Processing of Biomass" [80]. The article presents the results of experimental studies and technical-economic analysis of a catalytic process for converting integral biomass into aviation fuels with maximum carbon yields12 [81].

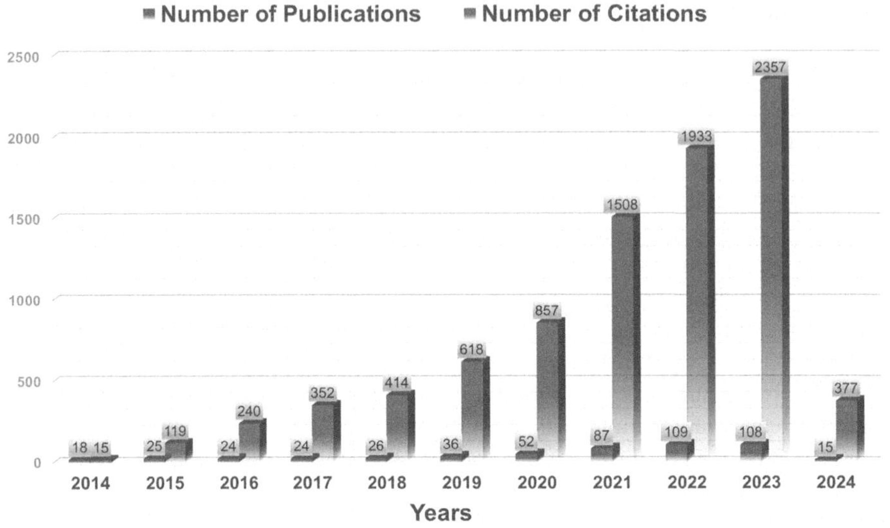

Fig. 14.2 Number of publications and citations from 2014 to April 2024 (10 years). (Figure built by Vosviewer)

Although the article above is the most cited, it is worth highlighting that the most relevant work, according to WoS, is: "A Recent Review of Aviation Fuels and Sustainable Aviation Fuels" [82]. This work discusses the importance of aviation fuels and the current challenges to mitigate global warming. The article begins by providing a detailed overview of the current status of aviation fuels, including their growth, types and emissions trends [82]. Despite a slight drop in work publications in 2015 and stagnation in 2016 and 2017, the high exponential growth of the area in the last 10 years is visible. One of the main reasons for this growth is the expansion of the use of biofuels as a measure of the reduction of greenhouse gas (GHG) emissions emitted by aircraft engines and reactions [83].

Several factors drive the increase in studies and the search for aviation fuels, including the growing flight demand. According to the International Civil Aviation Organization (ICAO), in 2022, there was a significant increase in the number of air passengers, with an increase of around 47% compared to the previous year. At the same time, airlines experienced substantial revenue growth of 70%, driven predominantly by international routes. Concerning annual revenue from airline passengers, ICAO observed an estimated increase of 50% from 2021 to 2022. This scenario reflects not only the increase in the movement of people travelling by plane but also the expansion of operations and air services globally. However, it is essential to highlight that this growing demand for aviation fuels raises concerns about current efforts to mitigate global warming [82].

5.1.2 Contributions of Countries/Regions

The analysis revealed the impact of the 10 most influential countries in research on aviation fuels, highlighting the increase in articles over time. This was done considering the period from 2014 to April 2024. As shown in Fig. 14.3a, the United States leads in terms of scientific production in the area, contributing with 166 publications (30.93%), followed by China with 139 articles (25.87%) and by Germany with 67 (12.48%) of the 537 articles collected in the database [84].

Figure 14.3b represents the number of citations by country, resulting in 11,422 citations. Notably, the United States was cited 3792 times, corresponding to 33.2% of the total citations. In second place, we have China, with 3092 citations, representing 27.05% of the total, followed by Germany, with 1111 citations, corresponding to 9.73%. These numbers highlight the outstanding role of the United States in this field of research on aviation fuels, both in terms of the number of publications and impact, as indicated by the citations received. The United States leadership in scientific production and the repercussions of the topic are relevant points to be considered in academic analyses and evaluations [85].

Figure 14.3c presents a network of connections between countries, representing scientific collaboration based on academic work. Each point on the map represents a country, and we see how the United States, China and Germany stand out, with the lines indicating research collaboration. The more lines, the greater the collaboration between countries. The colours represent groupings of nations with a high degree of cooperation. It is interesting to note the strong connection between the United States

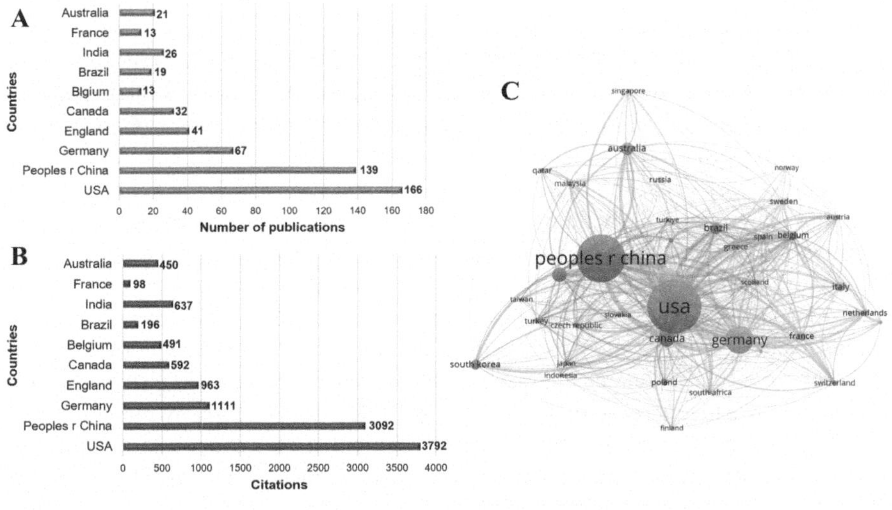

Fig. 14.3 (a) Table shows the distribution of publications by country. (b) The table illustrates the number of citations by country. (c) A network map representing a bibliographic collaboration between countries using two works as a starting point, in which the lines and colours indicate the different degrees of cooperation, was generated using Vosviewer

and China in all map colours, revealing significant patterns, such as regional groupings or global research centres [86].

5.1.3 Contributions of Institutions

Figure 14.4 presents the visualization map of collaboration between organizations from 2014 to April 2024, which reveals a complex and dynamic network, where the intensity of partnerships is indicated by the thickness of the lines and cooperation groups are highlighted by different colours [87]. The three predominant institutions identified as leaders in collaboration are Purdue University, Washington State University and the Chinese Academy of Sciences. These institutions stand out for the density of connections and centrality in cooperation clusters, suggesting a significant role in conducting joint research and advancing knowledge in their areas of specialization. Therefore, it is clear that bibliometric analysis, represented by the map, is a valuable tool for understanding trends and opportunities in the academic and industrial scenario, emphasizing the importance of interinstitutional collaboration for innovation and the development of strategic policies. Therefore, it is clear that there is a need to improve the interaction of studies carried out for greater cooperation between the specialists involved in conducting bibliometric analyses, considering the quality of the studies carried out [88].

Table 14.1 lists the ten institutions with the highest academic output on aviation fuels from 2014 to April 2024. The institutions are ranked in order, with Purdue University in India taking the top with 22 publications. Tianjin University and the Chinese Academy of Sciences, both in China, followed in second and third place,

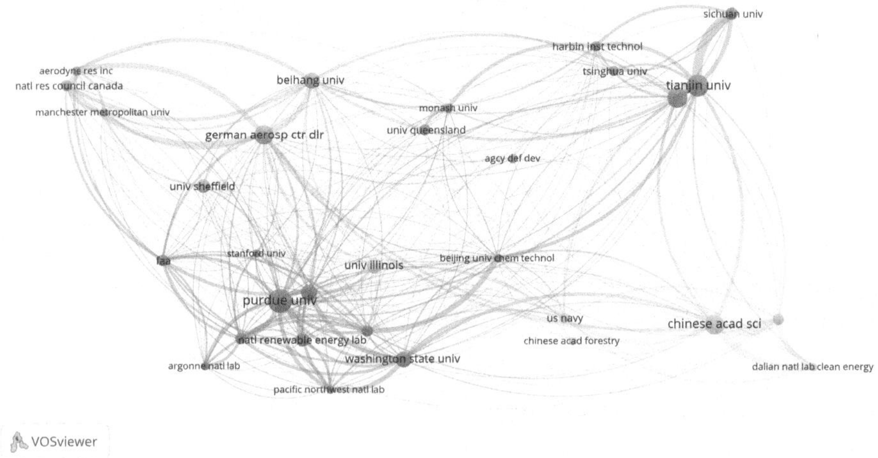

Fig. 14.4 Network visualization map from 2014 to April 2024 based on collaboration between organizations (institutions) of two or more registries. The thickness of the line connecting the two institutions indicates the degree of collaboration, and the colours indicate groups of organizations with a high level of cooperation. (Figure built by Vosviewer)

Table 14.1 The ten institutions with the highest academic productions on aviation fuels from 2014 to April 2024

Rank	Institutions	Countries/regions	Number of publications
1	Purdue University	India	22
2	Tianjin University	China	20
3	Chinese Academy Science	China	18
4	German Aerospace Center DLR	Germany	17
5	Beihang University	China	13
6	University Dayton	USA	13
7	Washington State University	USA	13
8	Federal Aviation Administration	USA	8
9	Massachusetts Institute of Technology	USA	8
10	University Chinese Academy of Science	China	8

Table constructed by Vosviewer

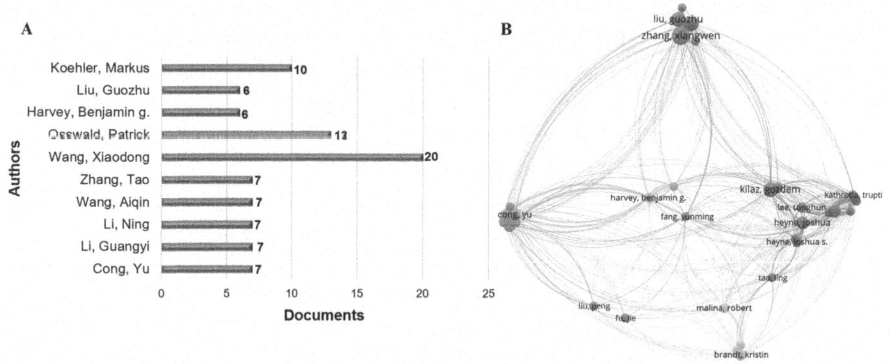

Fig. 14.5 (**a**) The table represents the number of documents published by authors. (**b**) Network map of authors who published at least two articles. The colours indicate collaborations between authors who have a high degree of cooperation. (Generated via VOSviewer)

with 20 and 18 publications, respectively. Observing the table in general, we notice that China stands out the most, with 59 publications, right after the United States with 42 works, followed by India with 22 academic productions. It is worth noting that the data analysed was finalized in April 2024 but may change as the platform is updated over the years, so the perspective is that the available data will be intensified over time [89].

5.1.4 Contributions of Authors

The visualization in a bar graph clearly and objectively presents the number of documents produced by different authors, as shown in Fig. 14.5a [85]. The author Wang, Xiaodong, stands out, leading the ranking with a total of 20 documents,

while Osswald, Patrick, reaches several 13 publications, followed by Koehler, Markus with ten published works. Analysis of this data reveals essential insights into authors' productivity and impact on the academic landscape. Despite Wang, Xiaodong standing out as the author of the most significant number of publications, he is only found as a co-author of more relevant publications, such as the work: "Production of Aviation Fuel Range Quadricyclane with 2,5-Hexanedione" [90]. The article addresses the production of high-density quadricyclanes in the C12 aviation fuel range with 2,5-hexanedione (HD), a hydrogenolysis product of cellulose.

Figure 14.5b shows a complex and colourful network visualization, representing relationships and connections between prominent authors in the field. The names are linked by coloured lines that indicate information cooperation. The variety of colours represents different types or strengths of connections. The circular layout with the names arranged around the circle and lines connecting them in the middle indicates network analysis between researchers, such as a bibliometric or co-authorship analysis, where the proximity and strength of the lines between names indicate the frequency and the strength of collaboration. Notably, there is a more significant partnership and work relationship between the authors in blue, red and green.

5.1.5 Journal Analysis

Table 14.2 presents a detailed view of academic production in several scientific journals, especially in aviation fuels. Notably, the newspaper *Fuel*, based in the Netherlands, leads the list with 69 documents. In second place is *Energy & Fuels* magazine from the United States, with 32 publications, followed by *Energies* magazine from Switzerland, with 23 publications. These numbers reflect aviation fuel research's diversity and international reach, with significant contributions from different countries and regions.

Table 14.2 Relevance of aviation fuel journals, considering the number of articles published and their Impact Factor (IF)

Rank	Journal title	Country	Documents	IF
1	*Fuel*	Netherlands	69	7.4
2	*Energy & Fuels*	United States	32	5.3
3	*Energies*	Switzerland	23	3.2
4	*Renewable and Sustainable Energy Reviews*	United Kingdom	16	16.7
5	*Frontiers in Energy Research*	Switzerland	14	3.4
6	*Combustion and Flame*	United States	10	4.4
7	*Journal of Analytical and Applied Pyrolysis*	Netherlands	10	6.0
8	*Energy*	United Kingdom	9	9.0
9	*Proceedings of the Combustion Institute*	United Kingdom	9	3.4
10	*Journal of Cleaner Production*	United Kingdom	9	11.1

These are generated via VOSviewer

When considering the Impact Factor (IF), which quantifies the frequency of citations of average articles in a journal during a specific period and indicates its importance and influence in its field, the UK journal *Renewable and Sustainable Energy Reviews* has the highest IF, with 16.7. Next, we have *Fuel* magazine, with an IF of 7.4, and *Energy* magazine, with an IF of 9.0. These numbers indicate the recognition and relevance of journals in the academic scenario, demonstrating the quality and impact of published research.

Analysing the table by country, we note that the Netherlands has 79 documents and an IF of 13.4. The United States, in turn, contributes 42 papers and an IF of 9.7, highlighting the significant scientific production in this area. Switzerland presents 37 papers and an IF of 6.6, while the United Kingdom stands out with 43 documents and a total IF of 40.2. These data reveal these countries' strong presence and impact on the academic production of aviation fuels.

5.1.6 Most Articles Cited

Table 14.3 presents the main articles cited in the last 10 years and contrasts between the year of publication and the volume of citations derived from that article. All articles deal with the bioethanol fermentation industry, which is the largest fermentation industry in the world [22, 61, 80, 91–97]. Notably, the conversion of bioethanol into liquid fuels for transportation, namely gasoline, jet and diesel hydrocarbons, has received increasing attention. In this sense, the article entitled "Production Of Renewable Jet Fuel Range Alkanes and Commodity Chemicals from Integrated Catalytic Processing of Biomass" is the most cited in this field of publication due to its primary, being an article from 2014 with an innovative character and full of possibilities in the era [96–100]. The article presents the results of experimental studies and technical-economic analyses of a catalytic system for converting total biomass into drop-in aviation fuels with maximum carbon yields. Collectively, the authors of this work have published more than 100 articles relating to the area to date (2024) [22, 94–97].

From this perspective, articles 2 and 3 in this table were published in the same period and had similar relevance. However, both were cited in experimental articles confirming the dialogic of some process optimization experiments, thus enabling the development of various bioethanol production techniques for air transport [72, 74, 101]. In contrast and complementarity, the two articles presented the challenges and future directions of the bioethanol route for aviation fuel in the last decade (2011–2020) [72, 74, 80, 91, 101].

Therefore, articles 4, 5, 6, and 7 presented the situation of bioethanol approximately 5 years after the articles above, presenting advances in the area and the possibilities found by large research laboratories to solve the problems exposed by previous research [102–106]. In this sense, the authors highlight the importance of the Guerbet reaction and ethylene oligomerization, which were and are two essential routes for upgrading ethanol to produce jet precursors [107–111]. Furthermore, essential catalysts, bifunctional acid-base catalysts and transition metal catalysts for

Table 14.3 The most cited articles on aviation fuels

Ranking	Title	Journal	Year published	Citations	References
1	Production of Renewable Jet Fuel Range Alkanes and Commodity Chemicals from Integrated Catalytic Processing of Biomass	*Energy & Environmental Science*	2014	300	[70]
2	Aviation biofuel from Renewable Resources: Routes, Opportunities and Challenges	*Renewable & Sustainable Energy Reviews*	2015	236	[71]
3	The Compatibility of Potential Bioderived fuels with Jet A-1 aviation kerosene	*Applied Energy*	2014	176	[72]
4	Renewable Bio-jet Fuel Production for Aviation: A Review	*Fuel*	2019	164	[73]
5	Transforming Carbon Dioxide into Jet Fuel Using an Organic Combustion-synthesized Fe-mn-k Catalyst	*Nature Communications*	2021	145	[74]
6	Biomass-derived Aviation Fuels: Challenges and Perspective	*Progress in Energy and Combustion Science*	2019	145	[75]
7	A Review on the Production of Renewable aviation Fuels from the Gasification of Biomass and Residual Wastes	*Bioresource Technology*	2020	143	[76]
8	Recent Development in Studies of Alternative Jet Fuel Combustion: Progress, Challenges, and Opportunities	*Renewable & Sustainable Energy Reviews*	2016	138	[77]
9	Sustainable Alternative Fuels in Aviation	*Energy*	2017	135	[78]
10	An Overview on Performance Characteristics of Bio-jet Fuels	*Fuel*	2019	129	[79]

alcohol production in the distillate range via the Guerbet reaction were highlighted as promising alternatives for overcoming hitherto insoluble and borderline problems for the bioenergy market. Finally, the articles present similar citations, with over 100 citations over 4 years. Being considered temporal transition articles by some researchers in the field of bioethanol [72, 74, 80, 101, 112]. In summary, all the articles presented in the table are cited more than 100 times, demonstrating their

notoriety and relevance to the research and development scenario of biofuels for air transport [22, 61, 80, 91–94].

5.1.7 Research Area

As the world strives to efficiently reduce greenhouse gas emissions from the production and consumption of fossil fuels to mitigate the impacts of climate change, communities that depend on coal, oil, and natural gas as economic drivers are likely to face significant challenges. Figure 14.6 illustrates the role of fossil and plant-based sources [22, 61, 91–94]. This research was due to the energy crisis and environmental pollution, forcing industries to develop efficient and robust energy storage (or conversion) systems. However, it is still a significant challenge [95–99]. Even though there is a concentration of research in two areas, the others shown in the graph present an absolute interdependence in the market, as the improvement of processes is based on the ideas of thermodynamics and physical chemistry that are supported by the development of new materials to enable the optimized and sustainable production desired by the market.

5.1.8 Keyword Analysis

Table 14.4 and Figs. 14.7, 14.8, and 14.9 present the relationship of connectivity and interdependency between the main words in the context of biofuels and sustainability, which are Jet fuel, Aviation fuels, combustion Pyrolysis, Biomass Oxidation Aviation Kerosene Biofuels, Emissions [72, 74, 101, 110–112]. An overconcentration of research in Jet fuel is noticeable, as the area needs innovative solutions to deal with the demand for air traffic, which is immense in a globalized world, especially in the economic north, where developed countries are directly connected and need to align their airlines. To boost the space-time relationship, which is already very small on this side of the planet [105–110].

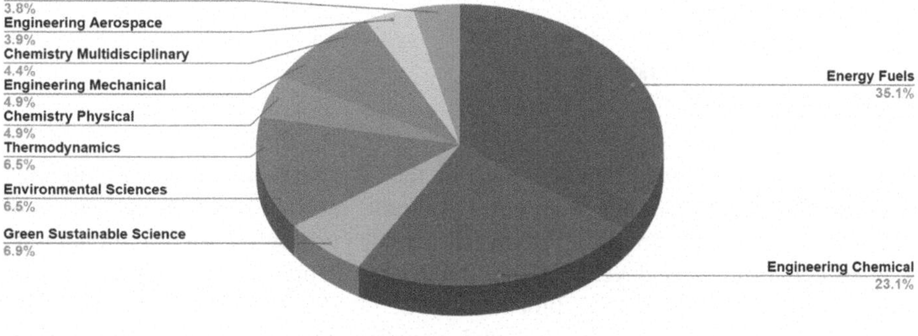

Fig. 14.6 Distribution of aviation fuel research areas according to the Web of Science database

Table 14.4 Ranking of the 20 most prominent keywords cited in the articles analysed

Ranking	Keywords	F	Total link strength	Ranking	Keyword	F	Total link strength
1	Jet fuel	78	368	11	Performance	37	188
2	Aviation fuels	73	360	12	Diesel	28	178
3	Combustion	66	326	13	Conversion	31	163
4	Pyrolysis	45	281	14	Kinetics	28	155
5	Biomass	47	251	15	Sustainable aviation fuels	43	145
6	Oxidation	39	219	16	Sustainable aviation fuel	41	143
7	Aviation	48	218	17	Oil	27	140
8	Kerosene	36	214	18	Hydrodeoxygenation	24	134
9	Biofuels	45	204	19	Hydrogen	25	134
10	Emissions	38	188	20	Life-cycle assessment	23	134

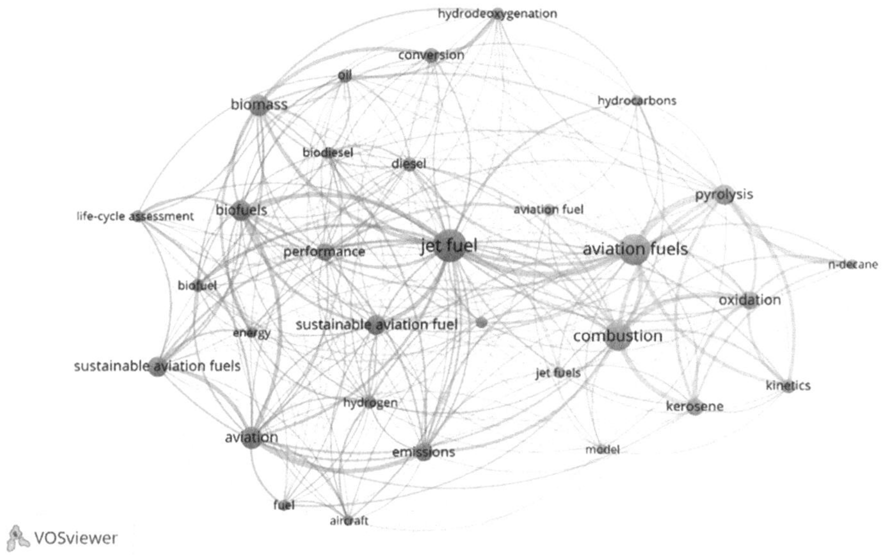

Fig. 14.7 Visualization of keywords related to the co-occurrence search of all search keywords relevant to aviation fuels. (Figure built by Vosviewer)

In this sense, connections between large companies and industries become critical in globalization and informational traffic. This theme is very innovative, and its development is still very early and requires several optimizations, whether in production, storage or transport [103–107]. It is expected that the demand for aviation fuels will continue to grow exponentially in the coming years and that it will continue to depend on kerosene as a primary source of energy [102, 103]. Even though efficiency and operational improvements are plausible ways to minimize

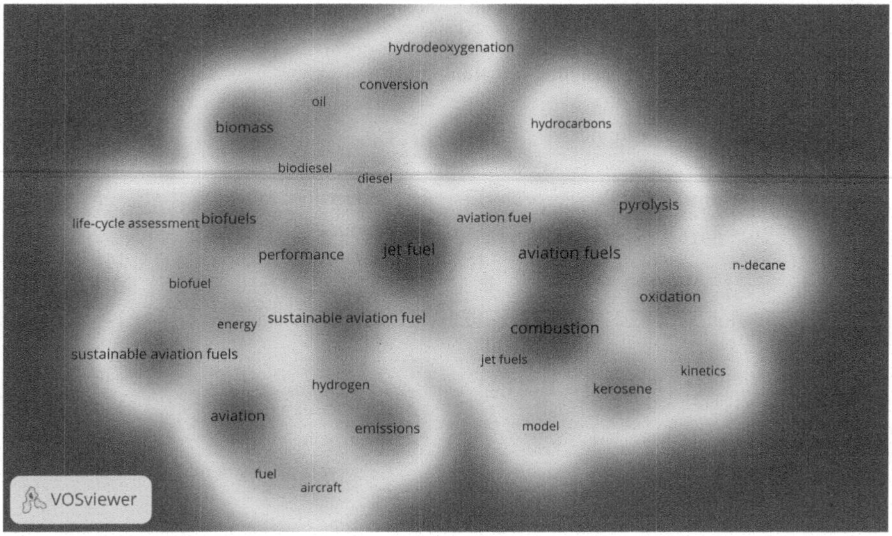

Fig. 14.8 A keyword density map generated by VOSviewer shows the hotspot in aviation fuel research

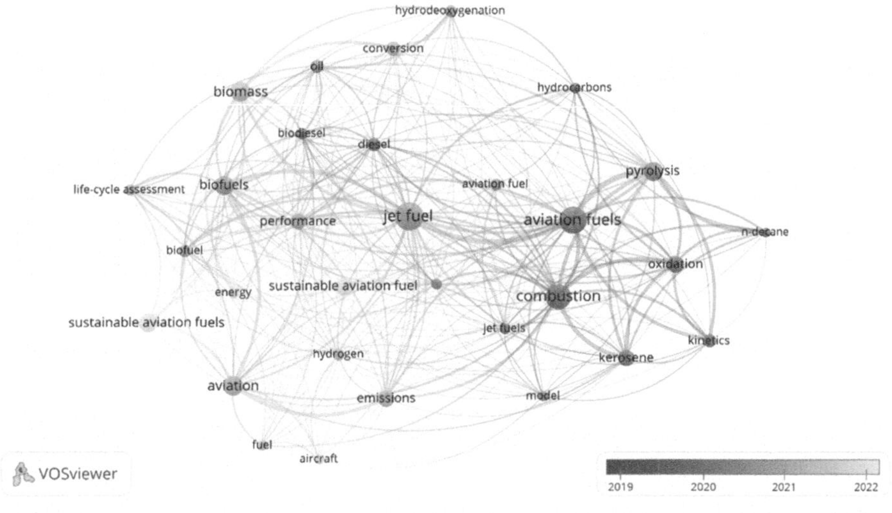

Fig. 14.9 Visualization of the overlay map of the co-occurrence analysis of aviation fuel keywords

greenhouse gas emissions, decarbonization will rely heavily on alternatives to low-carbon kerosene consumption.

Therefore, co-citation analyses are based on anticipating a shared theme between the co-cited studies. Two different studies are co-cited when they are cited by at least one study [72, 74, 80, 91, 101]. Using the ten most cited studies as we do, four research clusters are revealed [91, 92, 96–99]. The co-citation network analysis

creates four clusters. Furthermore, a co-occurrence analysis was performed to clarify potential trends and the intellectual structure of the literature on the most frequently cited keywords in the current literature related to biofuels for air transport.

5.1.9 Contribution of Funding Agencies

The co-authors' network of countries reflects the most influential countries in this field of research. It shows the level of communication between countries as the globalization process deepens through the globalized and economic north, where the world's greatest globalized potentials are geographically positioned [105, 107, 108]. The unequal distribution of power in the international investigation into aviation carbon emissions is mainly reflected in the contribution and cooperation of different countries in this northern domain of the globalized world [22, 94–97]. Table 14.5 shows the distribution of the leading national powers in aviation carbon emissions research. According to the table, the United States, the United Kingdom and China are the most active countries. Furthermore, they are home to the most advanced research centres in the contemporary world regarding research into aviation fuels [72, 101].

Research results from these countries jointly promote the development of aviation carbon emissions theory and practice [72, 74, 101, 111, 112]. Research in China and the United States is relatively interconnected and has formed synergy with other countries. Subsequently, the statistics of the proportion of collaborative articles, the United States, the United Kingdom and China, as the world's leading economies and aviation powers, are essentially concerned about aviation carbon emissions and have the most significant number of research results in recent years, far ahead of other countries. All in all, it appears to be a promising power in the

Table 14.5 Top ten related funding agencies

Funding agencies	Countries/ regions	Records	Percentage (%)
National Natural Science Foundation of China (NSFC)	China	104	44.06
European Union (EU)	–	25	10.59
United States Department of Energy (DOE)	United States	24	10.16
US Federal Aviation Administration Office of Environment and Energy Through Ascent	United States	16	6.77
United States Department of Defense	United States	14	5.93
UK Research Innovation (UKRI)	United Kingdom	12	5.08
FAA Center Of Excellence For Alternative Jet Fuels and the Environment	United States	12	5.08
Engineering Physical Sciences Research Council (EPSRC)	United Kingdom	11	4.66
National Science and Technology Major Project	China	09	3.81
National Science Foundation (NSF)	United States	09	3.81

development of new technological modes with high energy efficiency and respect for sustainability preached by global organizations [105–107, 109, 110].

Furthermore, the United States played a central role in bioenergy studies, collaborating with all members of the globalized network to produce and develop new bioenergy technologies. The cooperation between the United States and China was the most noticeable among the scientific population and bore fruit. Accordingly, China has also played a key role in bioenergy research and cooperated closely with countries such as Canada, Japan and England [106–112].

6 Strengths and Limitations

Analyses conducted on sustainable aviation fuel trends have revealed studies that can lead to new promising technologies related to the topic. Some of these works include "The Impact of Sustainable Aviation Fuels on Aircraft Fuel Line Ice Formation and Pump Performance" [113], which investigated pressure drop through a fuel filter made by Airbus for different types of SAFs, along with visual analyses of ice build-up on the filter mesh. Results showed that conventional fuel blocked the filter more than the tested SAF. However, SAF blends behaved differently, with the fuel-to-jet A-1 ratio affecting pressure differentials at different temperatures, attributed to increased aromatic content in the SPK blend. According to the authors, the experimental results of this work will help the aerospace industry predict the ice temperature that could lead to the blockage of a specific type of fuel and understand the ice accumulation process in fuel systems containing SAF. Recommendations for future work include investigating the role of surfaces and flow rate in ice accumulation and release in different SAF blends [113].

The second work, titled "Sustainable aviation fuel (SAF) production through power-to-liquid (PtL): A combined techno-economic and life cycle assessment" [114], critically evaluates a power-to-liquid (PtL) system for sustainable aviation fuel (SAF) production, addressing technical, economic and environmental aspects. This system involves several components, including a direct air capture (DAC) unit, offshore wind farm, alkaline electrolyzer and a refinery plant with Fischer-Tropsch reactors, evaluating conversion efficiency, integration and optimization, economic evaluation, life-cycle assessment, water footprint, including certification and mandates. According to the authors, the assumptions made in this study demonstrated that Power to Liquids has the potential to significantly reduce greenhouse gas emissions and thus aid in decarbonizing the aviation industry. However, supportive policies are needed for further development and, eventually, large-scale deployment. More research and data collection for pilot/demo plants will support future investigations and applications [114].

A third and final study, "Synergy of blending HEFA with alternative petroleum fractions" [115], investigates the use of alternative petroleum fractions such as hydrocracked kerosene (Jet HC) and hydro-treated heavy FCC naphtha (FCC HN), in combination with hydrogenated esters and fatty acids (HEFAs), which are the

main components of current SAFs. It analyses the properties of these components, their miscibility and their limitations and potential for expanding use in aviation fuel. The authors state that future research will focus on the hydrogenation of aromatics to improve all problematic parameters found in the study [115].

It is stated that there are several benefits of the possible use of SAFs in aviation transport, such as more excellent compatibility between engines, greater flexibility and fewer emissions. Depending on the raw material used and the development of technologies, it is possible to use 100% of the SAF and reduce greenhouse gas emissions by up to 94% [116]. In this context, a substantial increase in demand for studies in this area is foreseen. Moreover, collaboration and partnership levels between institutions and countries are expected to experience significant growth driven by these factors. These trends, in turn, are likely to result in substantial advancements and the introduction of new technologies in the analysed domain. Despite the numerous advantages highlighted in this context, it is crucial to recognize that multiple databases, such as Scopus, SPELL and SciELO, host various publications covering various topics, including the proposed theme.

7 Future Challenges of Aviation Fuels

7.1 Technical, Economical, and Sustainable Feasibility Aspects

To begin with, the aviation industry has achieved several milestones in reducing emissions, including improving aircraft fuel efficiency and better air traffic control to promote air travel with more excellent safety, stability, cost and sustainability [3, 54, 117–120]. However, an optimized, sustainable aviation fuel technology has not yet been developed. This situation encourages continued research and development of new innovative aviation biofuel technologies [16, 56, 121, 122].

From this perspective, to significantly minimize the environmental impacts of aviation and achieve the CO_2 targets proposed in the Paris Agreement, the efficiency improvements of conventional aircraft technologies are not satisfactory [11, 123–127]. Meanwhile, new technologies, such as electric or hybrid electric propulsion systems, have been analysed as promising proposals with enormous potential to enable more sustainable and energy-efficient aviation [11, 128–130].

Therefore, it is highlighted that conventional kerosene-based jet engines must be replaced by battery-powered propulsion engines based on fuel cells or hybrid electrics, drastically minimizing or eliminating greenhouse gas emissions. (GHG) during the flight [66, 131–133]. Furthermore, alternative aviation fuels, which are produced from bio-raw materials, as well as so-called electro fuels (e-fuels), which are synthesized from renewable energy sources (RES), are analysed by several centres of research as promising options to replace kerosene of fossil origin as an energy carrier [120, 134–137]. Therefore, lightweight strategies are being explored

and developed to reduce aircraft weight and fuel consumption, making flights more efficient [138].

Therefore, many research centres are investing in methodologies for assessing sustainability and the engineering of complex systems, such as aircraft, which require a substantial and solid amount of data [120, 137, 139–141]. Due to rapid globalization, the volume of information is significant and its substantial value for engineering purposes is available. Therefore, the data needs to be available for the foreground and background systems to analyse in detail every nuance of the flight plans executed by the aircraft.

Therefore, foreground data models refer to the processes related to the system under consideration throughout its life cycle, repositories of experimental results from measurements, simulations, calculations or reports about materials, components or processes, data industry-specific or internal company data [120, 136, 137, 139].

In line with data analysis, there is a need to decarbonize the aviation sector, and research into new propulsion concepts has intensified in recent years [66, 133–136]. The research centres emphasize that generally, this research analyses a conventional jet engine, which generates fewer climate-damaging emissions through improved combustion processes and greater efficiency [11, 129–132]. However, the fuel is used as an energy source, so efforts are made to reduce dependence on fossil kerosene. In particular, biofuels are based on raw materials of plant origin and electronic fuels, such as hydrogen and electric fuels, which are potential alternatives to fossil kerosene [126–129].

In this sense, biofuels have many advantages over fossil fuels, notably lower sulphur content, greater sustainability and lower carbon emissions (80–90% lower than fossil fuels), which are significant benefits to aircraft engines [126]. In contrast, the high production cost of biofuels is currently much higher than that of fossil fuels [16, 122–124]. In this sense, the primary sources spent on producing biofuels are raw material, transport of raw material, densification and storage and pre-treatment processes that make the process poorly scalable. The economic part of the aviation sector will be covered in greater detail in the next section.

7.2 Economic Analysis

As for economic evaluation, there are several methodological approaches for calculating costs and performance to evaluate the scalability of new technologies for the production, storage and distribution of new sources for aircraft [16, 54, 56, 120, 121]. We can mention Life Cycle Assessment (LCA), which includes Life-Cycle Costing (LCC) as a methodology to quantify the economic aspects analyzed throughout the process, incorporating all costs and benefits associated with the product under analysis. It is generally the most used, and by doing so, potential economic critical points that are little noticed in the industry or sector can be identified. It is noteworthy that for the assessment based on LCC, the Society of Toxicology

and Environmental Chemistry (SETAC)—European Working Group on Life-Cycle Costing defined three types of LCC that differ in terms of the external costs to be considered: LCC conventional, environmental LCC and social LCC [3, 117–119].

In the conventional CCV methodology, all costs associated with the product's life cycle are assessed and are directly covered by the producer or user. The approach is focused on internal production and use costs [3, 54, 56, 119, 120]. It should be noted that these are all costs on which a company bases its price. They include equipment, energy, labour, accommodation and general expenses. Using conventional LCC, a product is analysed from the perspective of a market actor (manufacturer or customer) and is generally not accompanied by separate LCA results [121]. In the environmental CCV, all life-cycle costs of a product that are covered by one or more market players (supplier, manufacturer, consumer or end-of-life player) are assessed [11, 125–127].

Therefore, it should be noted that the societal CCV expands the environmental CCV by assessing additional external costs, generally in monetary terms (based on willingness-to-pay methods). The perspective of the analysis is that of society in general, nationally and internationally, including governments. Other approaches, such as eco-efficiency, technical-economic assessment, as well as input-output analysis, allow the quantification of environmental impacts with economic parameters [3, 54, 56, 119–121].

It is essential to understand the technical, environmental and economic performance of alternative aviation fuels and future engine and propulsion technologies for future aircraft. The sector's economy is promising due to the high demand for biofuels, the high flight traffic and the volume of people travelling in a globalized world [16, 122]. It is a sector that has been growing exponentially for many decades; however, recent environmental issues have the development of more sustainable technologies from an ecological point of view and balanced concerning the industrial scalability of processes [123, 124].

7.3 Opportunities and Future Applications

The rise of sustainability is a latent and essential issue facing the current airline sector. In general, sustainability requires industries to introduce viable solutions that can perpetually support the environmental, economic and social results necessary for current and future generations of human beings. It is noteworthy that the future perspectives regarding biofuels for aircraft seem promising and hopeful, with a significant increase in research, analysis and development of new tools [16, 54, 56, 121–123].

Hydroisomerization is perhaps the best technological alternative to improve the chemical and physical properties of cold flow of aviation biofuels [117, 119]. Hydroisomerization is one of the most critical reactions in petroleum transformation processes [124]. During the Second World War, it increased the octane number of gasoline by producing isobutene [11, 125]. Currently, hydroisomerization is

widely applied to hydrocarbons with chains of different lengths. It improves the properties of gasoline, diesel and lubricating oils, increasing octane and cetane numbers and improving viscosity, fluidity and freezing point [130].

It is worth mentioning that there are challenging aspects associated with hydroisomerization catalysts [131–133]. The use of hydrogen and catalysts containing noble metals greatly increases operational and procedural costs [66, 134, 135]. In this sense, searching for catalysts from non-noble metals is one of the market's primary needs. Notably, catalysts developed must have balanced hydrogenation functionality to reduce hydrogen consumption [120, 136].

Biohydrogen has emerged as a formidable alternative as an aviation fuel, highlighting that the energy density of hydrogen, in terms of energy-to-weight ratio, is three times greater than that of conventional kerosene, making it more efficient and viable for longer flights In large aircraft that support high payload capacity [137, 139, 140]. However, in its liquid form, biohydrogen occupies a much larger volume than conventional kerosene, making its storage and storage unfeasible [138, 141, 142]. Still, the main challenge of hydrogen as an aviation fuel is related to the mass penalty induced by the onboard storage system [138, 143–147].

Furthermore, it should be noted that three broad categories of biofuels are rapidly expanding, which are distinguished according to the raw material used as a base: Carbohydrate rich biomass, lignocellulosic biomass and petroleum-based biomass [145, 146]. All three have relevant characteristics. However, petroleum derivatives have the prerogative of non-renewability [16, 121]. On the other hand, biomass is highly preeminent, gaining prominence on the world stage due to its high efficiency, scalability, environmental character of constant reuse, cyclical renewal and keeping the chain self-supplied [123, 124].

In summary, the opportunities are diverse, and the possibilities of success are encouraging for developers of solutions for aviation fuels. However, some challenges need to be overcome. For this, more significant investment in research, storage structure and market education is essential to a sustainable and innovative segment [11, 125].

8 Conclusions

The study highlighted a dynamic panorama related to sustainable aviation fuels (SAFs), especially regarding current research that addresses the topic and the challenges faced in the growth of production in the area. Productivity in this field is still low compared to other research on energy fuels, as the work presented articles from the last 10 years in aviation fuels. Current trends indicate that it is possible to reduce the emission of carbon dioxide (CO_2) and the impacts caused by the greenhouse effect by adopting SAFs in aviation transport, in addition to future trends, which foresee an increase in the consumption of more sustainable and accessible fuels for the aviation sector. However, the costs for developing scientific research in the area are still high, as well as for applications, as it is still economically limited, in

addition to the limited availability of raw materials to be inserted on a large scale and research financing, significantly affecting production. It was concluded that the United States is the country that produces the most research in the area of sustainable fuels for aviation transport, with 166 articles, which is directly related to the significant use of SAFs in commercial flights in the country, and in second place, China published 139. Therefore, growth prospects are favourable for the area. New research projects addressing the topic are promising, innovative and capable of applications, such as the growth of low-carbon aviation fuels (LCAF) and other cleaner energies for aviation. There are many interests in reducing environmental impacts in the aviation sector, such as the International Civil Aviation Organization (ICAO), which aims to reduce CO_2 emissions in international aviation by up to 5% over the next 6 years [147]. However, cooperation between institutions, countries, authors and funding agencies should be encouraged to increase future research, giving more incredible notoriety in the area, as well as providing the development of new technologies to be applied in different types of aviation transport, with proposals that address cost-benefits, sustainability, assistance in technological advancement and the integration of researchers who contribute to the growth of work aimed at SAFs.

Acknowledgements The authors express their gratitude for the support received from the Instituto de Engenharias e Desenvolvimento Sustentável (IEDS) da Universidade da Integração Internacional da Lusofonia Afro-Brasileira (Unilab). They would also like to thank the following Brazilian Scientific and Technological Development Agencies for their financial support: Fundação Cearense de Apoio ao Desenvolvimento Científico e Tecnológico (FUNCAP) (PS1-0186-00216.01.00/21; 31052.000417/2023; UNI-0210-00537.01.00/23), Conselho Nacional de Desenvolvimento Científico e Tecnológico (CNPq) (311062/2019-9; 440891/2020-5; 307454/2022-3) and Coordenação de Aperfeiçoamento de Ensino Superior (CAPES) (finance code 001).

References

1. Engelmann, L., Arning, K., & Ziefle, M. (2024). One step closer: Laypeople's perception of production steps for manufacturing CO2-based jet fuel. *Energy, Sustainability and Society, 14*, 9. https://doi.org/10.1186/s13705-024-00441-8
2. Cambridge University Press. (2023). (IPCC) Intergovernmental Panel on Climate Change (2023) climate change 2022 – Impacts, adaptation and vulnerability.
3. Ershov, M. A., Savelenko, V. D., Burov, N. O., et al. (2023). An incorporating innovation and new interactive technology into obtaining sustainable aviation fuels. *Energy, 280*, 128156. https://doi.org/10.1016/j.energy.2023.128156
4. Quante, G., Voß, S., Bullerdiek, N., et al. (2024). Hydroprocessing of fossil fuel-based aviation kerosene – Technology options and climate impact mitigation potentials. *Atmospheric Environment: X, 22*, 100259. https://doi.org/10.1016/j.aeaoa.2024.100259
5. Report. Livermore C (United S (2024) Myers CJA-A (2024) Sustainable aviation fuels from 30, 000 feet – Scoping.
6. (2024) (2023) (IEA) Transport – Energy system. Aviation.
7. International Air Transport Association (2021) IATA (2021) Resolution on the Industry's commitment to reach net zero carbon emissions by 2050.

8. (2024) (ICAO) International Civil Aviation Organization (2022) Long term global aspirational goal (LTAG) for international aviation. https://www.icao.int/environmental-protection/Pages/LTAG.aspx. Accessed 16 May 2024.
9. Organization (ICAO) International Civil Aviation. (2023). Third conference on aviation and alternative fuels (CAAF/3).
10. Wang, F., & Rijal, D. (2024). Sustainable aviation fuels for clean skies: Exploring the potential and perspectives of strained hydrocarbons. *Energy & Fuels, 38*, 4904–4920. https://doi.org/10.1021/acs.energyfuels.3c04935
11. Watson, M. J., Machado, P. G., da Silva, A. V., et al. (2024). Sustainable aviation fuel technologies, costs, emissions, policies, and markets: A critical review. *Journal of Cleaner Production, 449*, 141472. https://doi.org/10.1016/j.jclepro.2024.141472
12. (2023) Industries (ICCAIA) International Coordinating Council of Aerospace (2024) Annual report 2023.
13. Almena, A., Siu, R., Chong, K., et al. (2024). Reducing the environmental impact of international aviation through sustainable aviation fuel with integrated carbon capture and storage. *Energy Conversion and Management, 303*, 118186. https://doi.org/10.1016/j.enconman.2024.118186
14. Davis, C. D., Sreekumar, S., Altman, R., et al. (2024). Geospatially explicit technoeconomic assessment of sustainable aviation fuel production: A regional case study in Virginia. *Fuel Communications, 19*, 100114. https://doi.org/10.1016/j.jfueco.2024.100114
15. Hebish, A. A., Zulkifli, N. W. M., Julkapli, N. M., & Patah, M. F. A. (2024). The impact of operating conditions on H2-free catalytic deoxygenation of triglycerides for bio-jet fuel synthesis: A step towards sustainable aviation. *Applied Catalysis O: Open, 190*, 206926. https://doi.org/10.1016/j.apcato.2024.206926
16. Główka, M., Wójcik, J., Boberski, P., et al. (2024). Sustainable aviation fuel – Comprehensive study on highly selective isomerization route towards HEFA based bioadditives. *Renewable Energy, 220*, 119696. https://doi.org/10.1016/j.renene.2023.119696
17. Kurzawska-Pietrowicz, P. (2023). Life cycle emission of selected sustainable aviation fuels – A review. *Transportation Research Procedia, 75*, 77–85. https://doi.org/10.1016/j.trpro.2023.12.010
18. Zhang, H., Fang, Y., Wang, M., et al. (2020). Prospects and perspectives foster enhanced research on bio-aviation fuels. *Journal of Environmental Management, 274*, 111214. https://doi.org/10.1016/j.jenvman.2020.111214
19. Liu, Y., Chen, L., Chen, Y., et al. (2023). Pilot study on production of aviation fuel from catalytic conversion of corn stover. *Bioresource Technology, 372*, 128653. https://doi.org/10.1016/j.biortech.2023.128653
20. Lovestead, T. M., Burger, J. L., Schneider, N., & Bruno, T. J. (2016). Comprehensive assessment of composition and thermochemical variability by high resolution GC/QToF-MS and the advanced distillation-curve method as a basis of comparison for reference fuel development. *Energy & Fuels, 30*, 10029–10044. https://doi.org/10.1021/acs.energyfuels.6b01837
21. Harries, M. E., Huber, M. L., & Bruno, T. J. (2019). A distillation approach to phase equilibrium measurements of multicomponent fluid mixtures. *Energy & Fuels, 33*, 7908–7915. https://doi.org/10.1021/acs.energyfuels.9b01366
22. Shahabuddin, M., Alam, M. T., Krishna, B. B., et al. (2020). A review on the production of renewable aviation fuels from the gasification of biomass and residual wastes. *Bioresource Technology, 312*, 123596. https://doi.org/10.1016/j.biortech.2020.123596
23. Müller-Casseres, E., Szklo, A., Fonte, C., et al. (2022). Are there synergies in the decarbonization of aviation and shipping? An integrated perspective for the case of Brazil. *iScience, 25*, 105248. https://doi.org/10.1016/j.isci.2022.105248
24. Marangon, B. B., de Siqueira Castro, J., & Calijuri, M. L. (2024). Aviation fuel based on wastewater-grown microalgae: Challenges and opportunities of hydrothermal liquefaction and hydrotreatment. *Journal of Environmental Management, 354*, 120418. https://doi.org/10.1016/j.jenvman.2024.120418

25. Wang, M., Chen, M., Fang, Y., & Tan, T. (2018). Highly efficient conversion of plant oil to bio-aviation fuel and valuable chemicals by combination of enzymatic transesterification, olefin cross-metathesis, and hydrotreating. *Biotechnology for Biofuels, 11*, 30. https://doi.org/10.1186/s13068-018-1020-4
26. Kim, T.-W., Kim, E. S., Kim, J.-R., et al. (2019). Effects of zeolite supports on the production of fuel-range hydrocarbons in the hydrotreatment of various vegetable oils with platinum-based catalysts. *Journal of Nanoscience and Nanotechnology, 19*, 2443–2448. https://doi.org/10.1166/jnn.2019.16111
27. Rodriguez, Y., Guerra, R., Vizuete, K., et al. (2023). Kinetic study of the catalytic cracking of waste motor oil using biomass-derived heterogeneous catalysts. *Waste Management, 167*, 46–54. https://doi.org/10.1016/j.wasman.2023.05.027
28. Meirer, F., Kalirai, S., Weker, J. N., et al. (2015). Agglutination of single catalyst particles during fluid catalytic cracking as observed by X-ray nanotomography. *Chemical Communications, 51*, 8097–8100. https://doi.org/10.1039/C5CC00401B
29. Sun, A., Ye, C., Yao, C., et al. (2023). Morphological and compositional analysis on thermal deposition of supercritical aviation kerosene in micro channels. *Molecules, 28*, 4508. https://doi.org/10.3390/molecules28114508
30. Deng, Y., Cao, K.-K., Wetzel, M., et al. (2023). Carbon-neutral power system enabled e-kerosene production in Brazil in 2050. *Scientific Reports, 13*, 21348. https://doi.org/10.1038/s41598-023-48559-7
31. Sheng, X., Li, N., Li, G., et al. (2015). Synthesis of high density aviation fuel with cyclopentanol derived from lignocellulose. *Scientific Reports, 5*, 9565. https://doi.org/10.1038/srep09565
32. Cremonez, P. A., Feroldi, M., de Oliveira, C. D. J., et al. (2015). Environmental, economic and social impact of aviation biofuel production in Brazil. *New Biotechnology, 32*, 263–271. https://doi.org/10.1016/j.nbt.2015.01.001
33. Jing, L., El-Houjeiri, H. M., Monfort, J.-C., et al. (2022). Understanding variability in petroleum jet fuel life cycle greenhouse gas emissions to inform aviation decarbonization. *Nature Communications, 13*, 7853. https://doi.org/10.1038/s41467-022-35392-1
34. Wang, B.-Y., Zeng, P., He, R., et al. (2021). Single-pulse shock tube experimental and kinetic modeling study on pyrolysis of a direct coal liquefaction-derived jet fuel and its blends with the traditional RP-3 jet fuel. *ACS Omega, 6*, 18442–18450. https://doi.org/10.1021/acsomega.1c02530
35. Zhao, X., Taheripour, F., Malina, R., et al. (2021). Estimating induced land use change emissions for sustainable aviation biofuel pathways. *Science of the Total Environment, 779*, 146238. https://doi.org/10.1016/j.scitotenv.2021.146238
36. Cho, C. H., Han, H. S., Sohn, C. H., & Han, J. S. (2021). Ignition delay time and oxidation of a kerosene aviation fuel and a blended jet50-bio50 fuel. *ACS Omega, 6*, 26646–26658. https://doi.org/10.1021/acsomega.1c04002
37. Gilbertson, R., Friedman, E., Nguyen, V. T., et al. (2023). The role of physician advocacy in supporting policy change that reduces leaded aviation gasoline emissions. *Journal of Community Health, 48*, 446–449. https://doi.org/10.1007/s10900-023-01186-9
38. Frank, D., Neubauer, G., Bauer, C., et al. (2022). Exhaust emission measurements from a spark-ignition engine using fuels with different ethanol content for aircraft applications. *ACS Omega, 7*, 29923–29933. https://doi.org/10.1021/acsomega.2c02907
39. Hasanudin, H., Asri, W. R., Said, M., et al. (2022). Hydrocracking optimization of palm oil to bio-gasoline and bio-aviation fuels using molybdenum nitride-bentonite catalyst. *RSC Advances, 12*, 16431–16443. https://doi.org/10.1039/D2RA02438A
40. Mofijur, M., Ashrafur Rahman, S. M., Nguyen, L. N., et al. (2022). Selection of microalgae strains for sustainable production of aviation biofuel. *Bioresource Technology, 345*, 126408. https://doi.org/10.1016/j.biortech.2021.126408

41. Kim, S., Im, H., Yu, J., et al. (2023). Biofuel production from Euglena: Current status and techno-economic perspectives. *Bioresource Technology, 371*, 128582. https://doi.org/10.1016/j.biortech.2023.128582
42. Chang, D., Li, J., Yang, Y., & Gan, Z. (2022). Soot morphology and nanostructure differences between chinese aviation kerosene and algae-based aviation biofuel in free jet laminar diffusion flames. *ACS Omega, 7*, 11560–11569. https://doi.org/10.1021/acsomega.1c05125
43. Moore, R. H., Thornhill, K. L., Weinzierl, B., et al. (2017). Biofuel blending reduces particle emissions from aircraft engines at cruise conditions. *Nature, 543*, 411–415. https://doi.org/10.1038/nature21420
44. Yang, L., Hu, Y.-J., Wang, H., et al. (2023). Uncertainty quantification of CO2 emissions from China's civil aviation industry to 2050. *Journal of Environmental Management, 336*, 117624. https://doi.org/10.1016/j.jenvman.2023.117624
45. Liu, S., Zhu, Q., Guan, Q., et al. (2015). Bio-aviation fuel production from hydroprocessing castor oil promoted by the nickel-based bifunctional catalysts. *Bioresource Technology, 183*, 93–100. https://doi.org/10.1016/j.biortech.2015.02.056
46. Chaoyue, L., Shiyu, F., Lei, X., et al. (2021). Gas–liquid mass transfer characteristics of aviation fuel scrubbing in an aircraft fuel tank. *Scientific Reports, 11*, 15426. https://doi.org/10.1038/s41598-021-94786-1
47. Hu, D., Lin, W., Zeng, J., et al. (2019). Profiling the microbial contamination in aviation fuel from an airport. *Biofouling, 35*, 856–869. https://doi.org/10.1080/08927014.2019.1671977
48. Skowron, A., Lee, D. S., De León, R. R., et al. (2021). Greater fuel efficiency is potentially preferable to reducing NOx emissions for aviation's climate impacts. *Nature Communications, 12*, 564. https://doi.org/10.1038/s41467-020-20771-3
49. Bojanowski, C. L., Crookes Goodson, W. J., & Robinson, J. B. (2016). Use of bacteriophage to prevent Pseudomonas aeruginosa contamination and fouling in Jet A aviation fuel. *Biofouling, 32*, 1163–1170. https://doi.org/10.1080/08927014.2016.1239084
50. Adekitan, A. I., Shomefun, T., John, T. M., et al. (2018). Dataset on statistical analysis of jet A-1 fuel laboratory properties for on-spec into-plane operations. *Data in Brief, 19*, 826–834. https://doi.org/10.1016/j.dib.2018.05.083
51. Mawhood, R., Gazis, E., de Jong, S., et al. (2016). Production pathways for renewable jet fuel: A review of commercialization status and future prospects. *Biofuels, Bioproducts and Biorefining, 10*, 462–484. https://doi.org/10.1002/bbb.1644
52. Heyne, J., Rauch, B., Le Clercq, P., & Colket, M. (2021). Sustainable aviation fuel prescreening tools and procedures. *Fuel, 290*, 120004. https://doi.org/10.1016/j.fuel.2020.120004
53. Chiaramonti, D. (2019). Sustainable aviation fuels: The challenge of decarbonization. *Energy Procedia, 158*, 1202–1207. https://doi.org/10.1016/j.egypro.2019.01.308
54. Undavalli, V., Gbadamosi Olatunde, O. B., Boylu, R., et al. (2023). Recent advancements in sustainable aviation fuels. *Progress in Aerospace Sciences, 136*, 100876. https://doi.org/10.1016/j.paerosci.2022.100876
55. Wang, Z. J., Staples, M. D., Tyner, W. E., et al. (2021). Quantitative policy analysis for sustainable aviation fuel production technologies. *Frontiers in Energy Research, 9*. https://doi.org/10.3389/fenrg.2021.751722
56. Rogachuk, B. E., & Okolie, J. A. (2024). Comparative assessment of pyrolysis and Gasification-Fischer Tropsch for sustainable aviation fuel production from waste tires. *Energy Conversion and Management, 302*, 118110. https://doi.org/10.1016/j.enconman.2024.118110
57. Bashir, M. A., Lima, S., Jahangiri, H., et al. (2022). A step change towards sustainable aviation fuel from sewage sludge. *Journal of Analytical and Applied Pyrolysis, 163*, 105498. https://doi.org/10.1016/j.jaap.2022.105498
58. Bauen, A., Bitossi, N., German, L., et al. (2020). Sustainable aviation fuels. *Johnson Matthey Technology Review, 64*, 263–278. https://doi.org/10.1595/205651320X15816756012040
59. Said, Z., Nguyen, T. H., Sharma, P., et al. (2022). Multi-attribute optimization of sustainable aviation fuel production-process from microalgae source. *Fuel, 324*, 124759. https://doi.org/10.1016/j.fuel.2022.124759

60. Tanzil, A. H., Brandt, K., Wolcott, M., et al. (2021). Strategic assessment of sustainable aviation fuel production technologies: Yield improvement and cost reduction opportunities. *Biomass and Bioenergy, 145*, 105942. https://doi.org/10.1016/j.biombioe.2020.105942
61. Wei, H., Liu, W., Chen, X., et al. (2019). Renewable bio-jet fuel production for aviation: A review. *Fuel, 254*, 115599. https://doi.org/10.1016/j.fuel.2019.06.007
62. Wang, W.-C., & Tao, L. (2016). Bio-jet fuel conversion technologies. *Renewable and Sustainable Energy Reviews, 53*, 801–822. https://doi.org/10.1016/j.rser.2015.09.016
63. Gutiérrez-Antonio, C., Gómez-Castro, F. I., de Lira-Flores, J. A., & Hernández, S. (2017). A review on the production processes of renewable jet fuel. *Renewable and Sustainable Energy Reviews, 79*, 709–729. https://doi.org/10.1016/j.rser.2017.05.108
64. Ng, K. S., Farooq, D., & Yang, A. (2021). Global biorenewable development strategies for sustainable aviation fuel production. *Renewable and Sustainable Energy Reviews, 150*, 111502. https://doi.org/10.1016/j.rser.2021.111502
65. Ahmad, S., Ouenniche, J., Kolosz, B. W., et al. (2021). A stakeholders' participatory approach to multi-criteria assessment of sustainable aviation fuels production pathways. *International Journal of Production Economics, 238*, 108156. https://doi.org/10.1016/j.ijpe.2021.108156
66. Emmanouilidou, E., Mitkidou, S., Agapiou, A., & Kokkinos, N. C. (2023). Solid waste biomass as a potential feedstock for producing sustainable aviation fuel: A systematic review. *Renewable Energy, 206*, 897–907. https://doi.org/10.1016/j.renene.2023.02.113
67. Galadima, A., & Muraza, O. (2015). Catalytic upgrading of vegetable oils into jet fuels range hydrocarbons using heterogeneous catalysts: A review. *Journal of Industrial and Engineering Chemistry, 29*, 12–23. https://doi.org/10.1016/j.jiec.2015.03.030
68. Kosir, S., Heyne, J., & Graham, J. (2020). A machine learning framework for drop-in volume swell characteristics of sustainable aviation fuel. *Fuel, 274*, 117832. https://doi.org/10.1016/j.fuel.2020.117832
69. Wu, X., Jiang, P., Jin, F., et al. (2017). Production of jet fuel range biofuels by catalytic transformation of triglycerides based oils. *Fuel, 188*, 205–211. https://doi.org/10.1016/j.fuel.2016.10.030
70. Diederichs, G. W., Ali Mandegari, M., Farzad, S., & Görgens, J. F. (2016). Techno-economic comparison of biojet fuel production from lignocellulose, vegetable oil and sugar cane juice. *Bioresource Technology, 216*, 331–339. https://doi.org/10.1016/j.biortech.2016.05.090
71. Lin, J.-K., Nurazaq, W. A., & Wang, W.-C. (2023). The properties of sustainable aviation fuel I: Spray characteristics. *Energy, 283*, 129125. https://doi.org/10.1016/j.energy.2023.129125
72. Kramer, S., Andac, G., Heyne, J., et al. (2022). Perspectives on fully synthesized sustainable aviation fuels: Direction and opportunities. *Frontiers in Energy Research, 9*. https://doi.org/10.3389/fenrg.2021.782823
73. Platzer, M. F. (2023). The prospects for green aviation by 2050. *Progress in Aerospace Sciences, 141*, 100926. https://doi.org/10.1016/j.paerosci.2023.100926
74. Gonzalez-Garay, A., Heuberger-Austin, C., Fu, X., et al. (2022). Unravelling the potential of sustainable aviation fuels to decarbonise the aviation sector. *Energy & Environmental Science, 15*, 3291–3309. https://doi.org/10.1039/D1EE03437E
75. Zhang, L., Butler, T. L., & Yang*, B. (2020). Recent trends, opportunities and challenges of sustainable aviation fuel. In *Green energy to sustainability* (pp. 85–110). Wiley
76. Balz, T. (2022). Scientometric full-text analysis of papers published in remote sensing between 2009 and 2021. *Remote Sensing, 14*, 4285. https://doi.org/10.3390/rs14174285
77. Singh, V. K., Singh, P., Karmakar, M., et al. (2021). The journal coverage of web of science, scopus and dimensions: A comparative analysis. *Scientometrics, 126*, 5113–5142. https://doi.org/10.1007/s11192-021-03948-5
78. Aldridge, J. L., & Diekmann, F. (2019). Bibliometric profile of an agbioscience research enhancement grant program. *Journal of Agricultural & Food Information, 20*, 98–128. https://doi.org/10.1080/10496505.2018.1546184

79. Sikandar, H., Vaicondam, Y., Khan, N., et al. (2021). Scientific mapping of industry 4.0 research: A bibliometric analysis. *International Journal of Interactive Mobile Technologies (iJIM), 15*, 129. https://doi.org/10.3991/ijim.v15i18.25535
80. Bond, J. Q., Upadhye, A. A., Olcay, H., et al. (2014). Production of renewable jet fuel range alkanes and commodity chemicals from integrated catalytic processing of biomass. *Energy & Environmental Science, 7*, 1500–1523. https://doi.org/10.1039/C3EE43846E
81. da Silva, J. L., Sales, M. B., de Castro, B. V., et al. (2023). Lipase from Yarrowia lipolytica: Prospects as an industrial biocatalyst for biotechnological applications. *Fermentation, 9*, 581. https://doi.org/10.3390/fermentation9070581
82. Qasem, N. A. A., Mourad, A., Abderrahmane, A., et al. (2024). A recent review of aviation fuels and sustainable aviation fuels. *Journal of Thermal Analysis and Calorimetry, 149*, 4287. https://doi.org/10.1007/s10973-024-13027-5
83. de Rocco, G. K., & Henkes, J. A. (2020). Biocombustíveis sustentáveis para a aviação no Brasil. *Revista Gestão & Sustentabilidade Ambiental, 9*, 191. https://doi.org/10.19177/rgsa.v9e42020191-226
84. Guimarães, C. E. C., Neto, F. S., de Castro, B. V., et al. (2023). Sustainable bioethanol production from first- and second-generation sugar-based feedstocks: Advanced bibliometric analysis. *Bioresource Technology Reports, 23*, 101543. https://doi.org/10.1016/j.biteb.2023.101543
85. Borges, P. T., Sales, M. B., César Guimarães, C. E., et al. (2024). Photosynthetic green hydrogen: Advances, challenges, opportunities, and prospects. *International Journal of Hydrogen Energy, 49*, 433–458. https://doi.org/10.1016/j.ijhydene.2023.09.075
86. Melo, R. L. F., Sales, M. B., de Castro, B. V., et al. (2023). Recent applications and future prospects of magnetic biocatalysts. *International Journal of Biological Macromolecules, 253*, 126709. https://doi.org/10.1016/j.ijbiomac.2023.126709
87. Cavalcante, I. O., Simão Neto, F., da Silva Sousa, P., et al. (2024). Evolving sustainable energy technologies and assessments through global research networks: Advancing the role of blue hydrogen for a cleaner future. *RSC Sustainability, 2*, 348–368. https://doi.org/10.1039/D3SU00309D
88. da Silva, M. R., Hayashi, C. R. M., & Hayashi, M. C. P. I. (2011). Análise bibliométrica e cientométrica: desafios para especialistas que atuam no campo. *InCID: Revista de Ciência da Informação e Documentação, 2*, 110–129. https://doi.org/10.11606/issn.2178-2075.v2i1p110-129
89. de Castro, B. V., Leandro Fernandes Melo, R., Simão Neto, F., et al. (2024). Exploring the synergy of ionic liquids and lipase in sustainable enzymatic engineering. *Journal of Molecular Liquids, 399*, 124373. https://doi.org/10.1016/j.molliq.2024.124373
90. Liu, C., Hu, Y.-C., Yu, Z., et al. (2022). Production of aviation fuel range quadricyclane with 2,5-hexanedione. *ACS Sustainable Chemistry & Engineering, 10*, 17221–17229. https://doi.org/10.1021/acssuschemeng.2c05128
91. Kandaramath Hari, T., Yaakob, Z., & Binitha, N. N. (2015). Aviation biofuel from renewable resources: Routes, opportunities and challenges. *Renewable and Sustainable Energy Reviews, 42*, 1234–1244. https://doi.org/10.1016/j.rser.2014.10.095
92. Chuck, C. J., & Donnelly, J. (2014). The compatibility of potential bioderived fuels with Jet A-1 aviation kerosene. *Applied Energy, 118*, 83–91. https://doi.org/10.1016/j.apenergy.2013.12.019
93. Yao, B., Xiao, T., Makgae, O. A., et al. (2020). Transforming carbon dioxide into jet fuel using an organic combustion-synthesized Fe-Mn-K catalyst. *Nature Communications, 11*, 6395. https://doi.org/10.1038/s41467-020-20214-z
94. Wang, M., Dewil, R., Maniatis, K., et al. (2019). Biomass-derived aviation fuels: Challenges and perspective. *Progress in Energy and Combustion Science, 74*, 31–49. https://doi.org/10.1016/j.pecs.2019.04.004

95. Zhang, C., Hui, X., Lin, Y., & Sung, C.-J. (2016). Recent development in studies of alternative jet fuel combustion: Progress, challenges, and opportunities. *Renewable and Sustainable Energy Reviews, 54*, 120–138. https://doi.org/10.1016/j.rser.2015.09.056
96. Yilmaz, N., & Atmanli, A. (2017). Sustainable alternative fuels in aviation. *Energy, 140*, 1378–1386. https://doi.org/10.1016/j.energy.2017.07.077
97. Yang, J., Xin, Z., Corscadden, K., & Niu, H. (2019). An overview on performance characteristics of bio-jet fuels. *Fuel, 237*, 916–936. https://doi.org/10.1016/j.fuel.2018.10.079
98. Yuan, R., Pu, J., Wu, D., et al. (2022). Research priorities and trends on bioenergy: Insights from bibliometric analysis. *International Journal of Environmental Research and Public Health, 19*, 15881. https://doi.org/10.3390/ijerph192315881
99. Li, X., Tang, J., Li, W., et al. (2023). A bibliometric analysis and visualization of aviation carbon emissions studies. *Sustainability, 15*, 4644. https://doi.org/10.3390/su15054644
100. Narayanan, M. (2024). Promising biorefinery products from marine macro and microalgal biomass: A review. *Renewable and Sustainable Energy Reviews, 190*, 114081. https://doi.org/10.1016/j.rser.2023.114081
101. Kumar, T. P., & Dulloo, R. (2024). *Sustainability in aviation industry- A bibliometric analysis* (pp. 848–857). Atlantis Press.
102. Dalmiya, A., Sheyyab, M., Mehta, J. M., et al. (2023). Derived cetane number prediction of jet fuels and their functional group surrogates using liquid phase infrared absorption. *Proceedings of the Combustion Institute, 39*, 1495–1504. https://doi.org/10.1016/j.proci.2022.08.104
103. Ahmed, M., Alam, M. N., Abdullah, A., & Ahmad, Z. (2023). *Bio-jet fuel: An overview of various feedstock and production routes* (p. 030007). AIP Publishing.
104. Zhang, X., Song, M., Liu, J., et al. (2023). Synthesis of high density and low freezing point jet fuels range cycloalkanes with cyclopentanone and lignin-derived vanillins. *Journal of Energy Chemistry, 79*, 22–30. https://doi.org/10.1016/j.jechem.2022.12.017
105. Romanczyk, M. (2024). Chemical compositional analysis of jet fuels: Contributions of mass spectrometry in the 21st century. *Mass Spectrometry Reviews, 43*, 345–368. https://doi.org/10.1002/mas.21825
106. Dong, S., Liu, Z., & Yang, X. (2024). Hydrothermal liquefaction of biomass for jet fuel precursors: A review. *Chinese Chemical Letters, 35*, 109142. https://doi.org/10.1016/j.cclet.2023.109142
107. Tiwari, R., Mishra, R., Choubey, A., et al. (2023). Environmental and economic issues for renewable production of bio-jet fuel: A global prospective. *Fuel, 332*, 125978. https://doi.org/10.1016/j.fuel.2022.125978
108. Kathrotia, T., Oßwald, P., Naumann, C., et al. (2021). Combustion kinetics of alternative jet fuels, Part-II: Reaction model for fuel surrogate. *Fuel, 302*, 120736. https://doi.org/10.1016/j.fuel.2021.120736
109. Oßwald, P., Zinsmeister, J., Kathrotia, T., et al. (2021). Combustion kinetics of alternative jet fuels, Part-I: Experimental flow reactor study. *Fuel, 302*, 120735. https://doi.org/10.1016/j.fuel.2021.120735
110. Goh, B. H. H., Chong, C. T., Ong, H. C., et al. (2022). Recent advancements in catalytic conversion pathways for synthetic jet fuel produced from bioresources. *Energy Conversion and Management, 251*, 114974. https://doi.org/10.1016/j.enconman.2021.114974
111. Muldoon, J. A., & Harvey, B. G. (2020). Bio-based cycloalkanes: The missing link to high-performance sustainable jet fuels. *ChemSusChem, 13*, 5777–5807. https://doi.org/10.1002/cssc.202001641
112. Jia, T., Zhang, X., Liu, Y., et al. (2021). A comprehensive review of the thermal oxidation stability of jet fuels. *Chemical Engineering Science, 229*, 116157. https://doi.org/10.1016/j.ces.2020.116157
113. Ugbeh-Johnson, J., & Carpenter, M. (2023). The impact of sustainable aviation fuels on aircraft fuel line ice formation and pump performance. *The Aeronautical Journal, 127*, 1287–1307. https://doi.org/10.1017/aer.2023.6

114. Rojas-Michaga, M. F., Michailos, S., Cardozo, E., et al. (2023). Sustainable aviation fuel (SAF) production through power-to-liquid (PtL): A combined techno-economic and life cycle assessment. *Energy Conversion and Management, 292*, 117427. https://doi.org/10.1016/j.enconman.2023.117427
115. Kittel, H., Horský, J., & Šimáček, P. (2024). Synergy of blending HEFA with alternative petroleum fractions. *Fuel, 359*, 130390. https://doi.org/10.1016/j.fuel.2023.130390
116. Department of Energy U.S. Alternative fuels data center – Sustainable aviation fuel. https://afdc.energy.gov/fuels/sustainable-aviation-fuel#:~:text=Sustainable%20aviation%20fuel%20(SAF)%20is,how%20the%20fuel%20is%20produced. Accessed 16 May 2024.
117. Bhatt, A. H., Zhang, Y., Milbrandt, A., et al. (2023). Evaluation of performance variables to accelerate the deployment of sustainable aviation fuels at a regional scale. *Energy Conversion and Management, 275*, 116441. https://doi.org/10.1016/j.enconman.2022.116441
118. Colelli, L., Segneri, V., Bassano, C., & Vilardi, G. (2023). E-fuels, technical and economic analysis of the production of synthetic kerosene precursor as sustainable aviation fuel. *Energy Conversion and Management, 288*, 117165. https://doi.org/10.1016/j.enconman.2023.117165
119. Burov, N. O., Savelenko, V. D., Ershov, M. A., et al. (2023). Knowledge contribution from science to technology in the conceptualization model to produce sustainable aviation fuels from lignocellulosic biomass. *Renewable Energy, 215*, 118898. https://doi.org/10.1016/j.renene.2023.06.019
120. Misra, P., Alvarez-Majmutov, A., & Chen, J. (2023). Isomerization catalysts and technologies for biorefining: Opportunities for producing sustainable aviation fuels. *Fuel, 351*, 128994. https://doi.org/10.1016/j.fuel.2023.128994
121. Lan, K., Cruz, D., Li, J., et al. (2024). Life-cycle assessment of sustainable aviation fuel derived from paper sludge. *ACS Sustainable Chemistry & Engineering, 12*, 8379. https://doi.org/10.1021/acssuschemeng.4c00795
122. Märkl, R. S., Voigt, C., Sauer, D., et al. (2024). Powering aircraft with 100% sustainable aviation fuel reduces ice crystals in contrails. *Atmospheric Chemistry and Physics, 24*, 3813–3837. https://doi.org/10.5194/acp-24-3813-2024
123. Cui, Q., & Chen, B. (2024). Cost-benefit analysis of using sustainable aviation fuels in South America. *Journal of Cleaner Production, 435*, 140556. https://doi.org/10.1016/j.jclepro.2024.140556
124. Aghamiri, A. R., & Lahijani, P. (2024). Catalytic conversion of biomass and plastic waste to alternative aviation fuels: A review. *Biomass and Bioenergy, 183*, 107120. https://doi.org/10.1016/j.biombioe.2024.107120
125. Morganti, K., Moljord, K., Pearson, R., et al. (2024). Low carbon transportation fuels: Deployment pathways, opportunities and challenges. *Energy & Environmental Science, 17*, 531–568. https://doi.org/10.1039/D3EE01328F
126. Poblador-Ibanez, J., & Nocivelli, L. (2024). Toward a real-fluid modeling framework for sustainable aviation fuels. *Fuel Communications, 18*, 100100. https://doi.org/10.1016/j.jfueco.2023.100100
127. Sapra, H., Hessel, R., Miganakallu, N., et al. (2024). Computational fluid dynamics and machine learning-based piston-bowl optimization for energy-assisted compression ignition of low cetane number sustainable aviation fuel blends. *Energy Conversion and Management, 300*, 117929. https://doi.org/10.1016/j.enconman.2023.117929
128. Nasrullah, M. N. C. H., Kustanto, M. N., Darsin, M., et al. (2024). *Advancements and challenges in the search for lead-free aviation fuel: A review.* Proceeding of International Conference on Artificial Intelligence, Navigation, Engineering, and Aviation Technology (ICANEAT), vol. 1 (pp. 123–128). https://doi.org/10.61306/icaneat.v1i1.216
129. Moodley, D., Botha, T., Crous, R., et al. (2024). Catalysis for sustainable aviation fuels: Focus on Fischer-Tropsch catalysis. In *Catalysis for a sustainable environment* (pp. 73–116). Wiley.
130. Ding, Y., Zheng, P., Yang, L., et al. (2024). The present and future of sustainable aviation fuels in China. In *Annual report on China's petroleum, gas and new energy industry (2022–2023)* (pp. 333–356). https://doi.org/10.1007/978-981-99-7289-0_19

131. Arias, A., Nika, C.-E., Vasilaki, V., et al. (2024). Assessing the future prospects of emerging technologies for shipping and aviation biofuels: A critical review. *Renewable and Sustainable Energy Reviews, 197*, 114427. https://doi.org/10.1016/j.rser.2024.114427
132. Degirmenci, H., Uludag, A., Ekici, S., & Karakoc, T. H. (2023). Challenges, prospects and potential future orientation of hydrogen aviation and the airport hydrogen supply network: A state-of-art review. *Progress in Aerospace Sciences, 141*, 100923. https://doi.org/10.1016/j.paerosci.2023.100923
133. Grimme, W. (2023). The introduction of sustainable aviation fuels—A discussion of challenges, options and alternatives. *Aerospace, 10*, 218. https://doi.org/10.3390/aerospace10030218
134. Shehab, M., Moshammer, K., Franke, M., & Zondervan, E. (2023). Analysis of the potential of meeting the EU's sustainable aviation fuel targets in 2030 and 2050. *Sustainability, 15*, 9266. https://doi.org/10.3390/su15129266
135. Ballal, V., Cavalett, O., Cherubini, F., & Watanabe, M. D. B. (2023). Climate change impacts of e-fuels for aviation in Europe under present-day conditions and future policy scenarios. *Fuel, 338*, 127316. https://doi.org/10.1016/j.fuel.2022.127316
136. Su-ungkavatin, P., Tiruta-Barna, L., & Hamelin, L. (2023). Biofuels, electrofuels, electric or hydrogen?: A review of current and emerging sustainable aviation systems. *Progress in Energy and Combustion Science, 96*, 101073. https://doi.org/10.1016/j.pecs.2023.101073
137. Ansell, P. J. (2023). Review of sustainable energy carriers for aviation: Benefits, challenges, and future viability. *Progress in Aerospace Sciences, 141*, 100919. https://doi.org/10.1016/j.paerosci.2023.100919
138. Omidkar, A., Xu, H., Li, Z., et al. (2023). Techno-economic and life cycle assessment of renewable diesel production via methane-assisted catalytic waste cooking oil upgrading. *Journal of Cleaner Production, 414*, 137512. https://doi.org/10.1016/j.jclepro.2023.137512
139. Detsios, N., Theodoraki, S., Maragoudaki, L., et al. (2023). Recent advances on alternative aviation fuels/pathways: A critical review. *Energies (Basel), 16*, 1904. https://doi.org/10.3390/en16041904
140. Manigandan, S., Ryu, J. I., Praveen Kumar, T. R., & Elgendi, M. (2023). Hydrogen and ammonia as a primary fuel – A critical review of production technologies, diesel engine applications, and challenges. *Fuel, 352*, 129100. https://doi.org/10.1016/j.fuel.2023.129100
141. Hussain, M. H., & Biradar, C. H. (2023). Production of hydroprocessed renewable diesel from Jatropha oil and evaluation of its properties. *Materials Today Proceedings, 72*, 1420–1425. https://doi.org/10.1016/j.matpr.2022.09.341
142. Drexler, M., Haltenort, P., Arnold, U., et al. (2023). Progress in the anhydrous production of oxymethylene ethers (OME) as a renewable diesel fuel in a liquid phase process. *Catalysis Today, 424*, 113847. https://doi.org/10.1016/j.cattod.2022.07.015
143. Yan, P., Wang, H., Liao, Y., & Wang, C. (2023). Synthesis of renewable diesel and jet fuels from bio-based furanics via hydroxyalkylation/alkylation (HAA) over SO42-/TiO2 and hydrodeoxygenation (HDO) reactions. *Fuel, 342*, 127685. https://doi.org/10.1016/j.fuel.2023.127685
144. Cung, K., Buffaloe, G., Briggs, T., et al. (2023). *Engine and emissions performance of renewable diesel in a heavy-duty diesel engine: A single-cylinder engine experiment.*
145. Fani, K., Lycourghiotis, S., Bourikas, K., & Kordouli, E. (2023). Influence of natural mordenite activation mode on its efficiency as support of nickel catalysts for biodiesel upgrading to renewable diesel. *Nanomaterials, 13*, 1603. https://doi.org/10.3390/nano13101603
146. Petropoulos, G., Zafeiropoulos, J., Kordouli, E., et al. (2023). Influence of nickel loading and the synthesis method on the efficiency of Ni/TiO2 catalysts for renewable diesel production. *Energies (Basel), 16*, 4333. https://doi.org/10.3390/en16114333
147. International Civil Aviation Organization (ICAO) Sustainable Aviation Fuel (SAF). https://www.icao.int/environmental-protection/pages/SAF.aspx. Accessed 15 May 2024.

Chapter 15
Sustainable Aviation Fuel: A Future Outlook

Claudia Gutiérrez-Antonio

1 Introduction

In 2023, global carbon dioxide emissions increased by 410 million tonnes, amount that represents a growth of 1.1% respect 2022 [20]. It is important to highlight that this growth is lower than the value of 1.3% reported in the previous year [20]. This decrease in the growth of global carbon dioxide emissions is due to the rise in the use of renewable sources for energy production. Indeed, the global rise in carbon dioxide emissions was due to the expansion of carbon used to generate electricity, which was the consequence of the fall of hydropower generation because of drought.

In particular, the transport sector contributes to more than a third of carbon dioxide emissions from end-use sectors [21]. The operation of road vehicles contributes with 74.5% of the total emissions of the sector, followed by aviation with 11.6% and shipping with 10.6% [21]. In spite of the increment in electric vehicles, the transport sector still relies mainly on fossil fuels. For this reason, the energy transition in this sector must be driven by the development of biofuels, which allows its decarbonization without a change in infrastructure. In the transport sector, aviation accounts for a relatively small share of global emissions, but is one of the most challenging sectors to decarbonize [21]. According to the International Energy Agency, in 2022 aviation accounted for 2% of global energy-related CO_2 emissions, having grown faster in recent decades than rail, road, or shipping. This fast growth, in combination with being the most challenging sector to decarbonize, was the main driver of the proposal of a strategy to achieve sustainable development of the aviation industry.

C. Gutiérrez-Antonio (✉)
Facultad de Ingeniería, Universidad Autónoma de Querétaro, Campus Juriquilla, Juriquilla, Querétaro, Mexico
e-mail: claudia.gutierrez@uaq.mx

M. Aslam et al. (eds.), *Sustainable Aviation Fuels*, Sustainable Aviation,
https://doi.org/10.1007/978-3-031-83721-0_15

Hence, in 2009 the International Air Transport Association (IATA) along with the International Civil Aviation Organization (ICAO) established objectives to guarantee the sustainable development of the aviation sector [14]:

- An annual average fuel efficiency improvement of 1.5% from 2009 to 2020
- Stabilize net CO_2 emissions at 2020 levels with carbon-neutral growth
- Reduce aviation's net CO_2 emissions to half of what they were in 2005, by 2050

The aforementioned objectives were really ambitious, and the aviation industry was the first one to propose them within the transport sector. In order to achieve these objectives, the four-pillar strategy was proposed, which consists of [14]:

- The development of new, more efficient aircraft and engines can substantially decrease CO_2 emissions
- Operational measures include identifying weight savings in the current fleet, allowing the aircraft to burn less fuel
- Navigational improvements, making better use of airspace and streamlining the routes taken by aircraft to cut down on flight time
- Development of alternative fuels along with a Global Market-Based Measure

Thus, the implementation of these strategies will allow to reach the sustainability of the aviation sector, Fig. 15.1. The analysis of these strategies leads to identify that the first three could contribute to a lesser extent, due to the technological advances reached. In counterpart, the development of alternative aviation fuels was identified as the most promissory to reduce the carbon footprint; moreover, the use of alternative aviation fuels can be implemented in the short term, with the main objective of avoiding modifications or replacement of the existing aircraft. This constraint was

2009
Objectives establishment

Implementation of four
pillar strategy

2050
Sustainable aviation sector

Fig. 15.1 Flight path to reach sustainability in aviation sector. (Own elaboration with images taken from Pixabay [34])

established since the modifications in existing aircraft will lead to the re-certification of all fleets, which will decrease the profitability of this industry.

Over time, alternative aviation fuels have been called with different names, among which can be listed synthetic paraffinic kerosene, biokerosene, aviation biofuel, biojet fuel, renewable aviation fuel, and recently the most common term is sustainable aviation fuel (SAF). In the last decade and a half, this biofuel has attracted the attention of the scientific community, as can be observed in the number of scientific documents as well as patents; for instance, in 2009, 187 documents were reported in a search conducted at ScienceDirect [10] with the keywords *sustainable aviation fuel*; in the same search, 2203 documents were reported in 2024. Therefore, in this chapter, future outlook regarding what has been done and where the efforts need to be focused on sustainable aviation fuel are discussed.

2 Sustainable Aviation Fuel

Nowadays, aircraft for passenger and freight transportation use fossil jet fuels, which are commercially known as Jet A and Jet A-1; the main difference between them is the sulfur content which is minor in Jet A-1. Fossil jet fuel is produced from oil, and it consists of paraffinic, naphthenic, and aromatics compounds whose boiling point range corresponds to C8–C16. This composition confers physicochemical properties of the aviation turbine fuel for its use in commercial and military aircraft. All the properties that need to be fulfilled by fossil jet fuel are contained in the ASTM D1655 standard [1]. The properties include composition, volatility, fluidity, combustion, corrosion, thermal stability, contaminants, and additives [12].

On the other hand, the renewable counterpart of fossil jet fuel is sustainable aviation fuel (SAF). This biofuel is composed of paraffinic and naphthenic compounds, while the aromatic compounds content varies depending on the raw material and processing route. The difference in the contents of aromatic compounds does not affect the physicochemical properties of SAF; nevertheless, the aromatic compounds interact with the elastomers presented in the seal tanks, so its absence could produce leaks. Due to this, the ASTM D7566 standard requires a minimum aromatic composition of 8.4% vol in the blends of renewable and fossil jet fuels [2, 32]; thus, the amount of aromatic compounds present in SAF determines the maximum biofuel blend ratios, and this percentage varies for each conversion pathway being 50% vol the maximum mixing ratio allowed [32]. It is important to mention that the physicochemical properties of SAF are equal, and in some cases better, than those reported for fossil jet fuel, Table 15.1.

SAF can be produced from almost any type of biomass through different production processes. According to Singh [39], biomass is simply recycled CO_2 with solar energy and can be found in various versatile forms such as agriculture, forest products, animal and fish products and residues, aquatic biomass, algae and microorganisms, manure, as well as household, community, and industrial organic waste and effluent. There are several reported classifications for biomass, considering its

Table 15.1 Some physicochemical properties of sustainable aviation fuel and fossil jet fuel

Property	Jet A	SAF (produced from *Jatropha curcas*)
Boiling range (°C)	170–300	172–243
Freezing temperature (°C)	−40	−57
Density at 15 °C (kg/m³)	775–840	751–840
Energy content (MJ/kg)	43.28	44.3

Own elaboration with data from Gutiérrez-Antonio et al. [12]

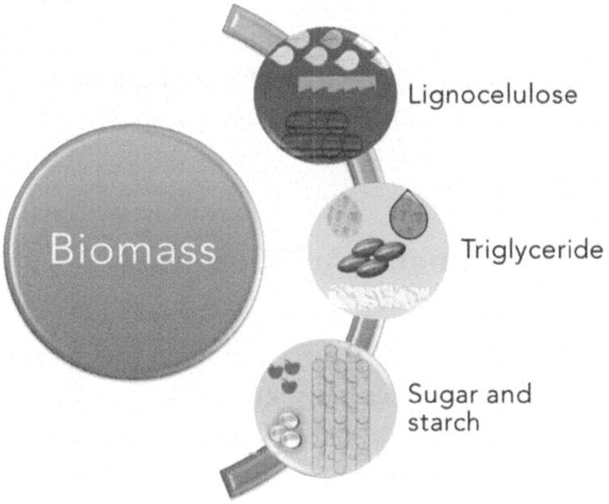

Fig. 15.2 Biomass classification considering the chemical nature. (Own elaboration)

origin, physical state, edibility, previous usage, generation, and also chemical composition. In particular, the classification based on chemical composition allows to integrate easily the conversion platforms in spite of their origin, edibility, previous usage, or generation. It is important to mention that the first biomasses used for biofuel production were edibles, and this causes ethical concerns regarding food safety. There are also biomasses that are nonedible; these kinds of biomasses compete in an indirect way with edible biomasses due to both types require land and water to be obtained. Because of the aforementioned, the use of residual biomasses as well as those which not demand arable lands (like microalgae, insects, and bacteria, among others) has been preferred for the biofuel production.

Considering the chemical nature of the biomass, this can be classified as triglyceride, lignocellulosic as well as sugar and starchy [30], Fig. 15.2. It is important to mention that in most cases these fractions are part of the complete biomass, but one of them is predominant. The triglyceride feedstock is integrated by fatty acids and triglycerides, which are presented in vegetable oils, animal fats, and also wastes. Examples of triglyceride feedstocks include oils from *Jatropha curcas*, microalgae, waste cooking, as well as fats from chicken, beef, and pork. On the other hand, lignin, cellulose, and hemicellulose are the main components of lignocellulosic materials, which are present in wastes from forestry, agriculture, and pruning, among others. Examples of lignocellulosic biomasses are rice husks, wastes from pruning citrus trees, and agave bagasse, among others. Finally, sugar and starchy

feedstock is composed of glucose and fructose, which is presented in most edible biomasses such as sugarcane, beet, and corn.

Therefore, considering this biomass classification, the main conversion pathways to produce sustainable aviation fuel are shown in Fig. 15.3. From Fig. 15.3, it can be observed that the most direct conversion pathway is the hydroprocessing of triglyceride feedstock. For the case of lignocellulose, sugar, and starchy feedstocks, additional processing is performed in order to modify the biomass composition, and then its hydroprocessing can be possible.

The triglyceride feedstock usually is pretreated in order to increase its pressure and temperature before entering the reactive zone of the process. This kind of feedstock contains high amounts of oxygen, which must be removed, at high temperatures and pressures with hydrogen, to produce long linear hydrocarbons. The oxygen is removed as water, carbon dioxide, and carbon monoxide. The long linear hydrocarbons are then isomerized and cracked, at high temperatures and pressures, in order to generate hydrocarbons in the range of C8–C16, which correspond to SAF. After the reactive zone, the renewable hydrocarbon streams must be distilled in order to generate light gases, naphtha, green diesel, and SAF. In the reactive zones, the temperature range is 300–350 °C, while the pressure is greater than 20 bar [12]. The ratio H_2 NPT ml/oil ml varies according to the composition of the triglyceride feedstock, but a value is 1500 is recommended to avoid coke formation [12]. The catalysts used are those conventionally used for the hydroprocessing in the oil industry, so catalysts based on Pt, Pd, or Mo have a good performance in the hydroprocessing of triglyceride feedstock [12].

For lignocellulose feedstock, a reduction of particle size is performed, which is sometimes accompanied by a drying step. Then, this treated biomass can be

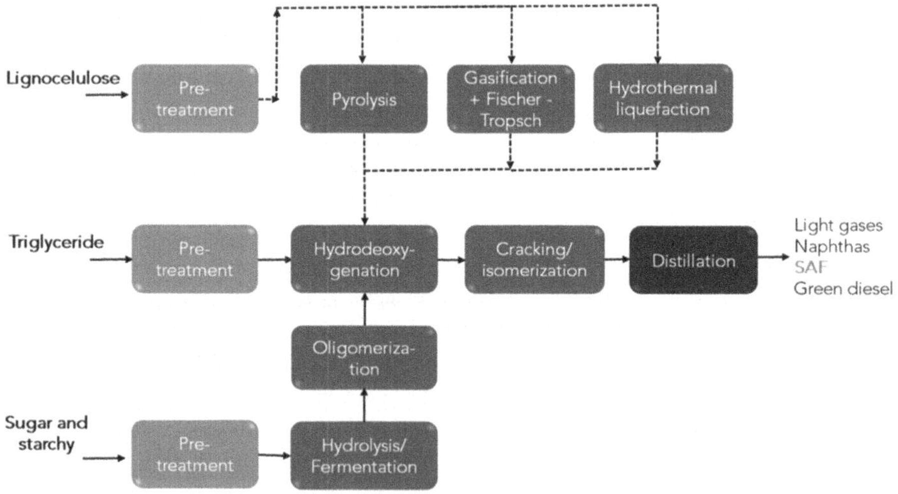

Fig. 15.3 Conversion pathways for the production of sustainable aviation fuel considering lignocellulose, triglyceride, sugar, and starchy feedstocks. (Own elaboration)

pyrolyzed in order to obtain bio-oil, whose chemical composition is very complex as it contains thousands of different, mostly oxygen-containing, compounds with a wide distribution of physical and chemical properties, and concentrations [42]; nevertheless, the key oxygen-containing functional groups present in the bio-oils, including aldehydes, ketones, carboxylic acids, phenols, carbohydrates, among others [42]. Then, bio-oil can be hydroprocessed. On the other hand, the pretreated lignocellulosic biomass can be gasified to obtain synthesis gas (also called syngas), which contains varying amounts of carbon monoxide and hydrogen generated by the gasification of a carbon-containing fuel to a gaseous product with a heating value [41]. Then, the syngas can be converted to hydrocarbons through a Fischer-Tropsch synthesis. It is important to mention that depending on the biomass composition, the effluent of Fischer-Tropsch synthesis could require to be hydroprocessed. Finally, the third alternative to process wet lignocellulosic material is hydrothermal liquefaction, which is a thermochemical process for obtaining liquid product, often called bio-oil from biomass in the presence of water as a solvent at a medium-temperature, and high-pressure [24]. The liquid product is also known as bio-crude, and then it can be hydroprocessed.

For the starchy feedstock, the pretreatment relies on hydrolysis in order to release the sugar compounds content in it. In the case of sugar feedstock, sometimes pressing or solid liquid extractions are required to separate the sugar fraction. Once the sugars are released, they are fermented with the aid of yeast to produce alcohol and carbon dioxide [35]. The produced alcohol is then oligomerized in order to produce renewable hydrocarbons, which could require additional hydroprocessing stages in order to obtain SAF.

Despite the chemical nature of biomass, its conversion through the aforementioned conversion pathways allows to obtain SAF. As it was mentioned before, the main difference between the SAF obtained with these different processes is the content of aromatic compounds. This content establishes the maximum percentage in volume that can be used in mixtures with fossil jet fuels. These percentages are established in the technical standards approved by the American Society of Testing and Materials (ASTM), which must be satisfied for the use of SAF in commercial aircraft. Table 15.2 shows the certified conversion pathways to produce SAF, the year of certification, and the maximum percentage in volume that can be used in mixtures with fossil jet fuel.

It is important to mention that in 2018, it was approved the coprocessing of renewable lipids with crude oil-derived middle distillates in petroleum refineries [6, 12]. Moreover, these production processes were assessed based on the ASTM D4054 standard [2], which evaluation processes consist of three phases. The certification process takes between 3 and 5 years, and almost 900,000 L of sustainable aviation fuel are required for all the tests involved in the evaluation process [12]. Nevertheless, it is not enough to satisfy the technical standards established by ASTM for the use of sustainable aviation fuel in commercial flights. In addition, sustainable aviation fuel must fulfill a sustainability standard, which is described next.

In 2016, the ICAO agreed on the adoption of a global market-based scheme to limit international aviation CO_2 emissions, which is known as the Carbon Offsetting

Table 15.2 Certified processes for the production of sustainable aviation fuel

Production process	Maximum percentage in volume	Certification year
Fischer-Tropsch synthetic paraffinic kerosene (FT-SPK)	50	2009
Hydroprocessed esters and fatty acids synthetic paraffinic kerosene (HEFA-SPK)	50	2011
Hydroprocessed fermented sugars to synthetic isoparaffins (HFS-SIP)	10	2014
Fischer-Tropsch synthetic paraffinic kerosene with aromatics (FT-SPK/A)	50	2015
Alcohol to jet synthetic paraffinic kerosene (ATJ-SPK)	50	2016
Catalytic hydrothermolysis synthesized kerosene (CH-SK, or CHJ)	50	2020
Hydroprocessed hydrocarbons, esters, and fatty acids synthetic paraffinic kerosene (HHC-SPK or HC-HEFA-SPK)	10	2020
Alcohol to jet synthetic kerosene with aromatics (ATJ-SKA)	50	2023

Own elaboration with data from CAAFI [6]

and Reduction Scheme for International Aviation (CORSIA) [17, 36]. CORSIA is the first global market-based measure for any sector and represents a cooperative approach that moves away from a "patchwork" of national or regional regulatory initiatives through the implementation of a global scheme that has been developed through global consensus among governments, industry, and international organizations [18]. The main objective of CORSIA is the reduction of emissions of commercial aviation around the globe, excluding airplane operators with less than 10,000 tonnes of annual CO_2 emissions, airplanes with less than 5700 kg take-off weight, and humanitarian, medical, and firefighting operations [18]. So, CORSIA allows the use of SAF that reduces carbon dioxide emissions by at least 10%compared to the use of fossil jet fuel [14]. In order to evaluate this value, CORSIA employs a life cycle assessment, which considers specific sustainability items [36]:

1. Use of life-cycle accounting for GHG emissions
2. Inclusion of induced land use change
3. Safeguards to prevent deforestation
4. Crediting of practices that mitigate the risk of land use change

The inclusion of these particular aspects allows to guarantee that effectively the use of SAF is contributing to the decrease in carbon dioxide emissions in the aviation sector. Also, this assessment is a pioneer at the international level, since there is no other sector with a similar proposal. Due to this, it is possible that the most challenging sector to be decarbonized will be the first one. At the moment, two Sustainability Certification Schemes are approved by the ICAO Council as meeting the requirements included in the ICAO document "CORSIA Eligibility Framework and Requirements for Sustainability Certification Schemes"; so, the authorized

companies are International Sustainability and Carbon Certification (ISCC) and Roundtable on Sustainable Biomaterials (RSB) [19], both of which were approved on June 2023 [19].

Regarding the sustainability certification, two important facts are highlighted. The first one is the "additionality" of the raw material. According to IATA, a key requirement is that the CO_2 reduction or removal used as an offset be "additional" to business-as-usual activity of the raw material used for SAF production [14]. This implies that for SAF production materials that release carbon dioxide emissions due to their natural decomposition process are preferred. The second one is that the sustainability certification considers the complete supply chain, not just the production process. Due to this, aspects related to induced land use change and deforestation are considered in the scheme. In this regard, some studies have been developed for the SAF supply chain in México [8, 9, 31], Brazil [23], and United States [28]. The reported studies indicated that the supply chains must be local, in order to reduce the environmental impact associated with the transportation of raw materials and/or intermediate products. Moreover, the use of residues is preferable since the environmental impact regarding its procurement is less than the cultivated crops, which require water, fertilizers, and pesticides; in this matter, those residues that present additionality are preferred. Nevertheless, the recollection of waste also has an environmental impact that needs to be considered. The analysis of these supply chains brings valuable information to identify the main contributors and propose alternatives to improve sustainability.

3 Advances in the Production of Sustainable Aviation Fuel

In order to present the advances in the production of sustainable aviation fuel, first the scientific advances will be discussed. As it was mentioned before, in 2009, 187 scientific documents were reported in a search conducted at ScienceDirect [10] with the keywords *sustainable aviation fuel*; in the same search, 2203 documents were reported in 2024. The complete graphic is presented in Fig. 15.4. From Fig. 15.4, we can observe how the number of documents increases with a given ratio until 2020, when the rate is augmented. We can correlate this tendency with the phases in which CORSIA will be implemented. According to IATA [14], from 2027 all international flights will be subject to offsetting requirements. This implies that SAF will need to be produced around the world.

Thus, it is required to evaluate different feedstocks for the production of sustainable aviation fuel, which can be grown or collected in differing regions around the world, depending on the natural environment, wherever the aviation industry needs it [15, 32]. According to ICAO, this diversification of feedstocks facilitates the production of sustainable aviation fuel, allowing the establishment of industries [16, 32]. Nevertheless, it cannot be expected that a single raw material can be enough to supply the world's demand for renewable aviation fuel [32]. Due to this, the analysis

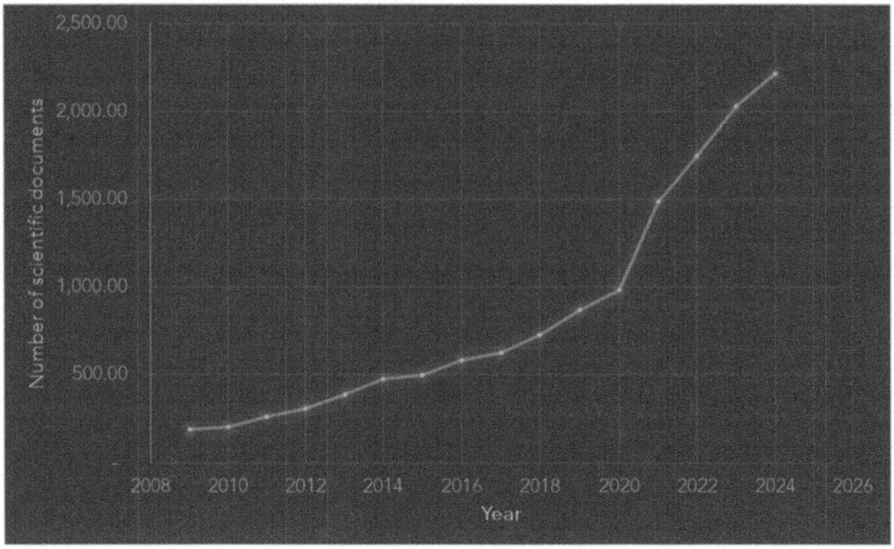

Fig. 15.4 Evolution of the number of scientific documents from 2009 to 2024. (Own elaboration with data from ScienceDirect [10])

of different feedstocks for SAF production was the main objective of the scientific community, as well as the improvement of the processing pathways.

Regarding the conversion of triglycerides, two main lines were the improvement of catalyst and the evaluation of nonedible and waste triglyceride feedstocks. In the line of the catalysts, the main advance was the development of multifunctional catalysts for the one-step hydroprocessing. Once this achievement was made, the improvements in catalyst development allowed to decrease in the operating conditions. One of the first hydroprocessing studies of *Jatropha curcas* oil indicated that the operating conditions were around 80 bar and 320 °C (hydrodeoxygenation) and 80 bar and 480 °C (hydroisomerization and hydrocracking), as in the studies reported by Sharma et al. [37] and Shayegh et al. [38], respectively; the combination of these two steps, allows to achieve selectivity of 19% to SAF. On the other hand, in 2020 Tao et al. reported a multifunctional catalyst for the one-step hydroprocessing of *Jatropha curcas* oil at 330 °C and 45 bar, which allows to obtain a selectivity of 74.21% to SAF. It is important to note how the yield to SAF increased significantly, and the new operation pressure was 56% of the previous one (80 bar). This data is reported for *Jatropha curcas* oil, nevertheless, this is not the only raw material studied. Moreover, the hydroprocessing of microalga *Chlorella vulgaris* oil through a one-step process highlights due to the operating conditions (410 °C and 50 bar) and the obtained selectivity to SAF of 78.5% [45].

On the other hand, the use of waste triglyceride feedstock has also been widely explored for SAF production; this is due to the low cost of this type of raw materials, usually associated only to the supply chain required to its recollection. It can be listed as raw materials such as waste cooking oils, chicken fat, beef tallow, and pork

lard [12]. Similar to the other raw materials, it highlights the development of one-step hydroprocessing. Hanafi et al. [13] reported the one-step hydroprocessing of oil derived from chicken fat. The operating conditions were 6 MPa and 400 °C, with a H_2/oil molar ratio of 450 v/v. The resulting conversion was 94.2%. Among the analyzed triglyceride feedstock, it can be listed those presented in Table 15.3.

Regarding the conversion of sugar and starchy feedstock, the most explored route is the fermentation to produce alcohols, and then upgraded to SAF through the process of alcohol to jet; it is important to mention that alcohols generated from sugars and starchy are called first-generation alcohols (FGA). In this conversion pathway, the first articles related to the production of SAF from sugars were reported in 2017 [48]. Tao et al. [44] reported a techno-economic analysis for the production of SAF through the fermentation of corn grains with a plant scale of 2000 dry metric tons per day. The alcohol produced was ethanol, and the minimum selling price was 4.20 USD per liter [44]. In the same year, Yao et al. [51] reported a stochastic techno-economic analysis of the SAF production via fermentation considering sugarcane and corn grain as raw materials. They found that the SAF prices were 3.65 and 3.84 USD/gal for sugarcane and corn grain, respectively. Later, Tanzil et al. [43] presented the evaluation of a dry corn ethanol biorefinery concept for the production of SAF, finding reductions to 63% in the minimum selling price. In the SAF production through the conversion pathway alcohol to jet, the main bottleneck is the limited yield in FGA production. The maximum reported yields for FGA production were 192 L/ton of sugarcane molasses (sugar feedstock) and 475 L/ton of sweet sorghum grains (starchy feedstock) [22]; these data were reported in 2016 and 2017, respectively, and they are still the maximum values reported. Therefore, recent advancements in bioethanol production concentrate primarily on technical development to achieve increased ethanol yield per unit of biomass, decreased processing costs and process time, and, more significantly, minimized process steps [26]. Thus, SAF production via this pathway remains economically unfeasible due to the use of high-value alcohol as feedstock [27].

Regarding the conversion of lignocellulosic feedstock, there are two main branches. The first one is related to the conversion of lignocellulosic feedstock through its hydrolysis and fermentation to generate second-generation alcohols (SGA) and then applied the alcohol to the jet process to obtain SAF. In this production branch, similar to sugar and starchy feedstock, the main bottleneck is alcohol

Table 15.3 Feedstocks analyzed for the production of sustainable aviation fuel

Feedstock	Type	Feedstock
Triglyceride	Vegetable oils	Palm oil, sunflower oil, coconut oil, soybean oil, jatropha curcas oil, castor oil, rapeseed oil, camelina oil, corn oil, carinata oil
	Residual oils	Waste cooking oil, waste oil
	Residual fats	Chicken fat, beef tallow, pork lard
	Microorganisms oils	Microalgae such as *Chlorella vulgaris* and *Botryococcus braunii*

Own elaboration with data from Song et al. [40] and Lau et al. [27]

generation. The maximum reported yield for SGA production was 330 L/ton of corn stover, reported in 2015 and 2018 [22]; nevertheless, in general, the SGA yields are lower than those reported for FGA. Moreover, these yields fall dramatically if pre-treatment is not applied. For instance, sugarcane bagasse with pretreatment reached SGA yields of 38.02 and 82.38 L per ton, while values of 1.46 and 2.53 L per ton were obtained for untreated sugarcane bagasse [22]. In spite of the pretreatments increased the yields, it is important to consider the additional costs associated with every additional treatment in the process. Thus, SAF production via this pathway remains economically unfeasible due to the use of high-value alcohol as feedstock and the extremely low yields of SGA [27].

The second branch to produce SAF from lignocellulosic feedstock considers thermochemical processes, such as pyrolysis and gasification, previous to its conversion to SAF. In the case of pyrolysis, the lignocellulosic material is heated without oxygen, in sometimes other atmospheres are used, to produce biochar, bio-oil, and bio-gases. In this process, bio-oil is of interest because it can be hydroprocessed to obtain SAF. For the hydroprocessing of bio-oil, the conditions for the process are 320 °C, 40 bar, and a hydrogen flow rate of 3 h^{-1} [5]. Nevertheless, in 2012 Xiu and Shahbazi indicated that taking into account capital and operating costs, the preparation of bio-oil by hydrotreatment was not profitable. Thus, Xiu and Shahbazi [49] recommended the development of novel integrated refinery processes to systematically upgrade bio-oils into transportation fuels that have desirable qualities. Then, Zhang et al. [53] reported the catalytic pyrolysis of biomass into low-carbon hydrocarbons coupled with alkylation of aromatics, obtaining a selectivity of 92.4%. In the study of Zhang et al. [53], the raw material was sawdust, and the SAF met the technical requirements. Recently, it was reported the conversion of palm fatty materials through pyrolysis followed by simultaneous deoxygenation, cracking, and isomerization occurred over $NbOPO_4$ [7]. In this last study, operating conditions were 350 °C and 10 bar, and a conversion of 95% was reached. It is important to mention that with this processing route, the conversion of plastic waste to produce SAF has also been explored [46].

In the case of gasification, the lignocellulosic material is heated with a limited amount of oxygen; in other words, it is a partial oxidation [41]. As a result of gasification processes syngas is obtained, which is a gas mixture consisting primarily of hydrogen (H_2), carbon monoxide (CO), and very often some carbon dioxide (CO_2) [50]. Syngas is then conditioned, followed by acid treatment, Fischer-Tropsch production, and syn-crude refining [52]. The most challenging aspect of the FT process is syn-crude refining in the integrated system of biomass gasification and requires additional development to ensure a good-quality FT feed while achieving significant cost reduction [33]. In this conversion pathway, several works have been reported. The main raw materials include wheat and barley straw [29], rice husk [47], wood [3], agricultural and forest residues [4], sugarcane bagasse [25], among others. In general, the yields to sustainable aviation fuels range from 5.45% to 14% for the aforementioned biomasses; indeed, the sustainable aviation fuel fraction is still small, and the production cost is higher than that of biodiesel yield from the same feedstocks [11].

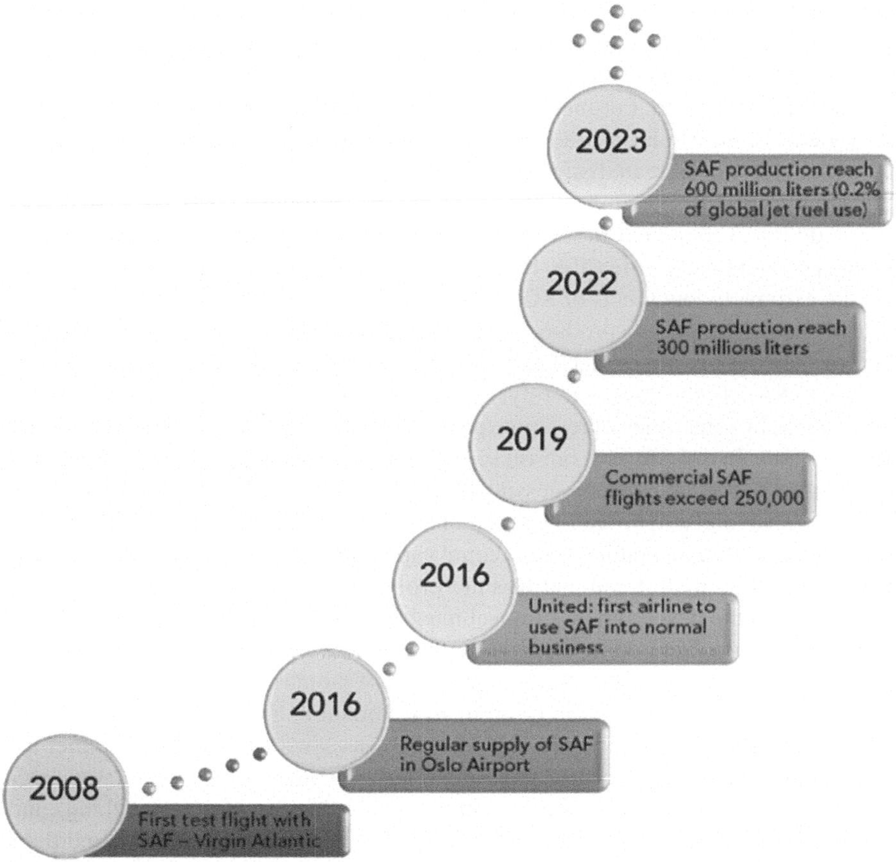

Fig. 15.5 Historic milestones related to sustainable aviation fuel. (Own elaboration with data from IATA [15])

On the other hand, there is also an evolution of the milestones of the aviation sector regarding this matter. This evolution can be observed in Fig. 15.5.

From Fig. 15.5, it can be observed how the first test flight was reported in 2008, but the regular supply in an airport was reported until 2016. In the same year, United was the first airline that introduce SAF in its normal operations. In spite of that, the SAF production has increased, in 2023 only represented 0.2% of the global jet fuel. On the other hand, until 2023 there were eight conversion pathways to produce sustainable aviation fuel, which are certified by ASTM [2]. During this period of time, there were also different resolutions approved in IATA meetings; nevertheless, 2021 highlighted the commitment to achieve net zero carbon emissions by 2050 in the aviation sector. This commitment is expected to detonate the production of sustainable aviation fuel around the world.

4 Challenges in the Production of Sustainable Aviation Fuel

At the moment, the aviation industry is quite unique in the transport sector due to its interest in guaranteeing its sustainable development. At the moment, the aviation sector has established goals, strategies, and directives, as well as a regulatory framework to certify that reported reductions are really contributing to sustainability. In a parallel way, aviation-related companies have focused their efforts on the development of more efficient engines, lighter aircraft, and also the synthesis of sustainable aviation fuel. Indeed, the main producers of aircraft engines are already on the path to develop motors that can operate with 100% of SAF.

In the case of sustainable aviation fuel, there are several certified conversion pathways for its production considering different raw materials. The first production process was certified in 2009; nevertheless, it is important to highlight that the certification process takes between 3 and 5 years. In spite of this, the development of new production processes continues, being the last process certified in 2023. This is an important fact, considering that sustainable aviation fuel must be produced near the place where it is going to be used; therefore, several raw materials and conversion pathways must be explored and optimized. This implies that the supply chain must be as short and sustainable as possible, in order to guarantee the fulfillment of CORSIA requirements.

The most explored production process, and also the most mature technology, is hydroprocessing. This is the most direct conversion pathway, while the other ones include processing units, which finally "converge" to the hydroprocessing. The main challenge in this production process is the cost, and also the availability of raw materials. In this line, the use of waste materials has been explored; nevertheless, the collection of these wastes can be expensive, and they can be competent in price due to preexisting uses of them. Thus, the use of microorganisms, as microbial, and insects, as black soldier fly larvae, represent a new alternative to transform organic wastes into triglyceride feedstock. Regarding this strategy, the standardization of the cultivation process as well as the industrial scale production remains the main challenge. Moreover, the pyrolysis of lignocellulosic biomass is an alternative to employ this abundant and low-cost raw material. Pyrolysis is a technically feasible alternative, where the main focus must be on the development of catalysts and lower operating conditions; this will allow to reduce the environmental impact of the generation of bio-oil. Recently, hydrothermal liquefaction has been explored for the conversion of materials with high moisture content into bio-crude. This strategy allows to generate other triglyceride feedstock suitable for its hydroprocessing. Nevertheless, high operating conditions are required; thus, the main challenge is the development of catalysts that allow reducing the operating conditions. In the case of sugar and starchy feedstock, the limiting step is alcohol production, which is the direction where the efforts must be focused. In order to overcome this limitation, microorganisms modified genetically, as well as intensified processes, where the fermentation and alcohol removal are performed in the same vessel, seem the most promissory strategy.

In general, for all production pathways, the application of process intensification and energy integration tools will allow to reduce the space requirements, increase the yield, and decrease energy consumption as well as the use of external utility services, respectively. In addition, the incorporation of renewable energies will decrease the dependence of the process on fossil fuels, increasing the resilience of the production process.

Moreover, the complete use of biomasses as well as mixtures of them is another interesting alternative, for which biorefinery processing schemes are required. A biorefinery is a processing scheme to convert complete biomasses into bioenergy, biofuels, and value-added products; for this, different conversion platforms are required, which may include biochemical, chemical, biological, mechanical, and thermochemical operations. As a result, the financial viability of biomass conversion depends on several products. It is important to highlight that in some cases, the biorefineries can satisfy their own energy requirements, and generate energy surplus in other cases. In spite of this, it is important to improve the processing through the application of process intensification, energy integration, and optimization strategies.

Finally, it is important to highlight that the establishment of a supply chain for the production of sustainable aviation fuel must consider the most abundant and available biomasses, especially if they present the "additionality" characteristic. It is also important to note that the production cost to obtain SAF depends on the processing scale, and this data depends on the supply chain size. Therefore, an interesting competing interest is observed in biofuel production that must be analyzed, since it is not present in the case of petrochemical refineries.

5 Concluding Remarks

In this chapter, a recapitulation of the advances in the production of sustainable aviation fuel was presented, along with a discussion of where future efforts should be focused. From the scientific point of view, the first works were reported in 2009; however, in 2020 the publication rate was augmented. On the other hand, the milestones in the aviation sector indicate how in the beginning the flight tests were developed by many airlines, with different feedstocks; in all cases, the hydroprocessing technology is the predominant one.

From the processes point of view, the improvement of pyrolysis and hydrothermal liquefaction is the main opportunity area, mainly to decrease its environmental impact in the use of dry and wet biomasses, respectively. Moreover, in the case of biological processes, such as microbial and black soldier fly larvae cultivation, standardization and industrial-scale production are the big challenges. Finally, the use of biorefineries, which generate at least one bioenergy type, one biofuel, and one value-added product, is a good strategy to use the complete biomasses; this will help in two main directions. The first one is that the products are diversified, which improves the economic feasibility of the process. The second one is that this

processing scheme is aligned with Circular Economy Strategies, reducing waste generation.

In all cases, the production of sustainable aviation fuel must fulfill technical, economic, and sustainability criteria, in order to guarantee the energy transition in the aviation sector without any change in the aircraft fleet.

Acknowledgments The authors thank the Consejo Nacional de Humanidades, Ciencia y Tecnología (CONAHCYT) and the Universidad Autónoma de Querétaro (UAQ) for the financial support provided through Laboratorio Nacional CONAHCYT de Biocombustibles Sólidos (BIOENER)—Apoyo LNC-2023-40 for the realization of this work. This research was also supported by the Bioproducts Sustainable Production Processes Research Group from the Universidad Autónoma de Querétaro (UAQ), Facultad de Ingeniería.

References

1. ASTM. (2022). *ASTM 1655-22. Standard specification for aviation turbine fuels*. American Society of Testing and Materials. Available from: https://www.astm.org/d1655-22.html. Accessed on 11.08.2024.
2. ASTM. (2022). *ASTM D7566-22. Standard specification for aviation turbine fuel containing synthesized hydrocarbons*. American Society of Testing and Materials. Available from: https://www.astm.org/d7566-22.html. Accessed on 11.08.2024.
3. Atsonios, K., Kougioumtzis, M. A., Panopoulos, K. D., & Kakaras, E. (2015). Alternative thermochemical routes for aviation biofuels via alcohols synthesis: Process modeling, techno-economic assessment and comparison. *Applied Energy, 138*, 346–366.
4. Brandt, K. L., Martinez-Valencia, L., & Wolcott, M. P. (2022). Cumulative impact of federal and state policy on minimum selling price of sustainable aviation fuel. *Frontiers in Energy Research, 10*, 828789.
5. Burov, N. O., Savelenko, V. D., Ershov, M. A., Vikhritskaya, A. O., Tikhomirova, E. O., Klimov, N. A., Kapustin, V. M., Chernysheva, E. A., Sereda, A. V., Abdellatief, T. M. M., Ramadan, M., & Abdelkareem, M. A. (2023). Knowledge contribution from science to technology in the conceptualization model to produce sustainable aviation fuels from lignocellulosic biomass. *Renewable Energy, 215*, 118898.
6. CAAFI. (2024). *Fuel qualification*. Commercial Aviation Alternative Fuels Initiative. Available from: https://www.caafi.org/focus_areas/fuel_qualification.html. Accessed on 11.08.2024.
7. de Sousa, F. P., dos Reis, G. P., & Pasa, V. M. D. (2024). Catalytic pyrolysis of vegetable oils over NbOPO4 for SAF and green diesel production. *Journal of Analytical and Applied Pyrolysis, 177*, 106314.
8. Domínguez-García, S., Gutiérrez-Antonio, C., De Lira-Flores, J. A., & Ponce-Ortega, J. M. (2017). Optimal planning for the supply chain of biofuels for aviation in Mexico. *Clean Technologies and Environmental Policy, 19*, 1387–1402.
9. Domínguez-García, S., Gutiérrez-Antonio, C., De Lira-Flores, J. A., Ponce-Ortega, J. M., & El-Halwagi, M. M. (2017). Strategic planning for the supply chain of aviation biofuel with consideration of hydrogen production. *Industrial & Engineering Chemistry Research, 56*(46), 13812–13830.
10. Elsevier. (2024). *Search in ScienceDirect database with keyword sustainable aviation fuel*. Available from: https://www.sciencedirect.com/search?qs=sustainable%20aviation%20fuel. Accessed on 11.08.2024.

11. Emmanouilidou, E., Mitkidou, S., Agapiou, A., & Kokkinos, N. C. (2023). Solid waste biomass as a potential feedstock for producing sustainable aviation fuel: A systematic review. *Renewable Energy, 206*, 897–907.
12. Gutiérrez-Antonio, C., Romero-Izquierdo, A. G., Gómez-Castro, F. I., & Hernández, S. (2021). *Production processes of renewable aviation fuel – Present technologies and future trends*. Elsevier. ISBN: 9780128197196.
13. Hanafi, S. A., Elmelawy, M. S., Shalaby, N. H., El-Syed, H. A., Eshaq, G., & Mostafa, M. S. (2016). Hydrocracking of waste chicken fat as a cost effective feedstock for renewable fuel production: A kinetic study. *Egyptian Journal of Petroleum, 25*(4), 531–537.
14. IATA. (2020). *Carbon offsetting for international aviation*. International Air Transport Association. Available from: https://www.iata.org/contentassets/fb745460050c48089597a3ef1b9fe7a8/paper-offsetting-for-aviation.pdf. Accessed on 11.08.2024.
15. IATA. (2024). *Developing sustainable aviation fuel (SAF)*. International Air Transport Association. Available from: https://www.iata.org/en/programs/sustainability/sustainable-aviation-fuels/. Accessed on 12.08.2024.
16. ICAO. (2018). *Sustainable aviation fuels guide – Transforming global aviation collection*. Secretariat, International Civil Aviation Organization. Available from: https://www.icao.int/environmental-protection/Documents/Sustainable%20Aviation%20Fuels%20Guide_100519.pdf. Accessed on 12.08.2024.
17. ICAO. (2019). *Introduction to the ICAO basket of measures to mitigate climate change. ICAO environmental report 2019* (pp. 111–115). Secretariat, International Civil Aviation Organization. Available from: https://www.icao.int/environmental-protection/Documents/EnvironmentalReports/2019/ENVReport2019_pg111-115.pdf. Accessed on 11.08.2024.
18. ICAO. (2019). *Climate change mitigation: CORSIA. ICAO environmental report 2019* (pp. 206–250). Secretariat, International Civil Aviation Organization. Available from: https://www.icao.int/environmental-protection/CORSIA/Documents/ICAO%20Environmental%20Report%202019_Chapter%206.pdf. Accessed on 11.08.2024.
19. ICAO. (2023). *CORSIA approved sustainability certification schemes*. Secretariat, International Civil Aviation Organization. Available from: https://www.icao.int/environmental-protection/CORSIA/Documents/CORSIA_Eligible_Fuels/ICAO%20document%2004%20-%20Approved%20SCSs%20-%20June%202023.pdf. Accessed on 11.08.2024.
20. IEA. (2024). *CO_2 emissions in 2023*. International Energy Agency. Available from: https://iea.blob.core.windows.net/assets/33e2badc-b839-4c18-84ce-f6387b3c008f/CO2Emissionsin2023.pdf. Accessed on 10.08.24.
21. IEA. (2024). *Transport*. International Energy Agency. Available from: https://www.iea.org/energy-system/transport. Accessed on 10.08.24.
22. Jain, S., & Kumar, S. (2024). A comprehensive review of bioethanol production from diverse feedstocks: Current advancements and economic perspectives. *Energy, 296*, 131130.
23. Kamali, F. P., Rossi Borges, J. A., Osseweijer, P., & Posada, J. A. (2018). Towards social sustainability: Screening potential social and governance issues for biojet fuel supply chains in Brazil. *Renewable and Sustainable Energy Reviews, 92*, 50–61.
24. Kang, S., Li, X., Fan, J., & Chang, J. (2013). Hydrothermal conversion of lignin: A review. *Renewable and Sustainable Energy Reviews, 27*, 546–558.
25. Klein, B. C., Chagas, M. F., Junqueira, T. L., Ferreira Rezende, M. C. A., de Fátima Cardoso, T., Cavalett, O., & Bonomi, A. (2018). Techno-economic and environmental assessment of renewable jet fuel production in integrated Brazilian sugarcane biorefineries. *Applied Energy, 209*, 290–305.
26. Lamichhane, G., Acharya, A., Poudel, D. K., Aryal, B., Gyawali, N., Niraula, P., Phuyal, S. R., Budhathoki, P., & Parajuli, N. (2021). Recent advances in bioethanol production from Lignocellulosic biomass. *International Journal of Green Energy, 18*(7), 731–744.
27. Lau, J. I. C., Wang, Y. S., Ang, T., Seo, J. C. F., Khadaroo, S. N. B. A., Chew, J. J., Lup, A. N. K., & Sunarso, J. J. (2024). Emerging technologies, policies and challenges toward implementing sustainable aviation fuel (SAF). *Biomass and Bioenergy, 186*, 107277.

28. Leila, M., Whalen, J., & Bergthorson, J. (2018). Strategic spatial and temporal design of renewable diesel and biojet fuel supply chains: Case study of California, USA. *Energy, 156*, 181–195.
29. López-Fernández, A., Bolonio, D., Amez, I., Castells, B., Ortega, M. F., & García-Martínez, M. J. (2021). Design and pinch analysis of a GFT process for production of biojet fuel from biomass and plastics. *Energies, 14*(19), 6035.
30. Maity, S. K. (2015). Opportunities, recent trends and challenges of integrated biorefinery: Part I. *Renewable and Sustainable Energy Reviews, 43*, 1427–1445.
31. Matos-Ríos, D. M., Martínez-Guido, S. I., Ponce Ortega, J. M., Gómez-Castro, F. I., García-Trejo, J. F., & Gutiérrez-Antonio, C. (2022). Optimal supply chain design for renewable aviation fuel production in México considering the full use of nonedible biomass. *ACS Sustainable Chemistry and Engineering, 10*(30), 9770–9786.
32. Moreno-Gómez, A. L., Gutiérrez-Antonio, C., Gómez-Castro, F. I., & Hernández, S. (2021). Modelling, simulation and intensification of the hydroprocessing of chicken fat to produce renewable aviation fuel. *Chemical Engineering and Processing – Process Intensification, 159*, 108250.
33. Nilanjana, B. (2023). Biomass to energy – An analysis of current technologies, prospects, and challenges. *Bioenergy Research, 16*(2), 683–716.
34. Pixabay. (2024). *Aviación*. Available from: https://pixabay.com/es/vectors/avi%C3%B3n-de-pasajeros-avi%C3%B3n-volador-8886817/. Accessed on 11.08.2024.
35. Prabhjot Kaur, P., Ghoshal, G., & Banerjee, U. C. (2019). 3 – Traditional bio-preservation in beverages: fermented beverages. In A. M. Grumezescu & A. M. Holban (Eds.), *Preservatives and preservation approaches in beverages* (pp. 69–113). Academic Press.
36. Prussi, M., Lee, U., Wang, M., Malina, R., Valin, H., Taheripour, F., Velarde, C., Staples, M. D., Lonza, L., & Hileman, J. I. (2021). CORSIA: The first internationally adopted approach to calculate life-cycle GHG emissions for aviation fuels. *Renewable and Sustainable Energy Reviews, 150*, 111398.
37. Sharma, R. K., Ananda, M., Rana, B. S., Kumar, R., Farooqui, S. A., Sibi, M. G., & Sinha, A. K. (2012). Jatropha-oil conversion to liquid hydrocarbon fuels using mesoporous titanosilicate supported sulfide catalysts. *Catalysis Today, 198*(1), 314–320.
38. Shayegh, F., Farshi, A., & Dehgan, A. (2012). A kinetics lumped model for VGO catalytic cracking in a fluidized bed reactor. *Petroleum Science and Technology, 30*(9), 945–957.
39. Singh, J. (1996). Economic and environmental aspects of bio-energy productivity. In P. Chartier, G. L. Ferrero, U. M. Henius, S. Hultberg, J. Sachau, & M. Wiinblad (Eds.), *Biomass for energy and the environment* (pp. 1865–1868). Pergamon.
40. Song, M., Zhang, X., Chen, Y., Zhang, Q., Chen, L., Liu, J., & Ma, L. (2023). Hydroprocessing of lipids: An effective production process for sustainable aviation fuel. *Energy, 283*, 129107.
41. Speight, J. G. (2020). 14 – A biorefinery. In J. G. Speight (Ed.), *The refinery of the future* (2nd ed., pp. 515–548). Gulf Professional Publishing.
42. Staš, M., Auersvald, M., Kejla, L., Vrtiška, D., Kroufek, J., & Kubička, D. (2020). Quantitative analysis of pyrolysis bio-oils: A review. *TrAC Trends in Analytical Chemistry, 126*, 115857.
43. Tanzil, A. H., Zhang, X., Wolcott, M., Brandt, K., Stöckle, M., Murthy, G., & Garcia-Perez, M. (2021). Evaluation of dry corn ethanol bio-refinery concepts for the production of sustainable aviation fuel. *Biomass and Bioenergy, 146*, 105937.
44. Tao, L., Markham, J. N., Haqb, Z., & Biddya, M. J. (2017). Techno-economic analysis for upgrading the biomass-derived ethanol-to-jet blendstocks. *Green Chemistry, 19*, 1082–1101.
45. Verma, D., Kumar, R., Ranaa, B. S., & Sinha, A. K. (2011). Aviation fuel production from lipids by a single-step route using hierarchical mesoporous zeolites. *Energy & Environmental Science, 5*, 667–1671.
46. Wang, Z., Burra, K. G., Lei, T., & Gupta, A. K. (2021). Co-pyrolysis of waste plastic and solid biomass for synergistic production of biofuels and chemicals-A review. *Progress in Energy and Combustion Science, 84*, 100899.

47. Wang, W.-C., Liu, Y.-C., & Nugroho, R. A. A. (2022). Techno-economic analysis of renewable jet fuel production: The comparison between Fischer-Tropsch synthesis and pyrolysis. *Energy, 239*(Part A), 121970.
48. Watson, M. J., Machado, P. G., da Silva, A. V., Saltar, Y., Ribeiro, C. O., Nascimento, C. A. O., & Dowling, A. W. (2024). Sustainable aviation fuel technologies, costs, emissions, policies, and markets: A critical review. *Journal of Cleaner Production, 449*, 141472.
49. Xiu, S., & Shahbazi, A. (2012). Bio-oil production and upgrading research: A review. *Renewable and Sustainable Energy Reviews, 16*(7), 4406–4414.
50. Yang, L., & Ge, X. (2016). Chapter three – Biogas and syngas upgrading. In Y. Li & X. Ge (Eds.), *Advances in bioenergy* (Vol. 1, pp. 125–188). Elsevier.
51. Yao, G., Staples, M. D., Malina, R., & Tyner, W. E. (2017). Stochastic techno-economic analysis of alcohol-to-jet fuel production. *Biotechnoly for Biofuels, 10*, 18.
52. Zahid, I., Nazir, M. H., Chiang, K., Christo, F., & Ameen, M. (2024). Current outlook on sustainable feedstocks and processes for sustainable aviation fuel production. *Current Opinion in Green and Sustainable Chemistry, 49*, 100959.
53. Zhang, Y., Bi, P., Wang, J., Jiang, P., Wu, X., Xue, H., Liu, J., Zhou, X., & Li, Q. (2015). Production of jet and diesel biofuels from renewable lignocellulosic biomass. *Applied Energy, 150*, 128–137.

Index

M. Aslam et al. (eds.), *Sustainable Aviation Fuels*, Sustainable Aviation,
https://doi.org/10.1007/978-3-031-83721-0

T

The manufacturer's authorised representative in the EU is Springer Nature Customer Service Centre GmbH, Europaplatz 3, 69115 Heidelberg, Germany. If you have any concerns regarding our products, please contact ProductSafety@springernature.com

Printed and bound by CPI Group (UK) Ltd, Croydon, CR0 4YY
01/10/2026
02225909-0013